Oracle9*i*: SQL with an Introduction to PL/SQL

Lannes L. Morris-Murphy, Ph.D.

THOMSON
COURSE TECHNOLOGY

Australia • Canada • Mexico • Singapore • Spain • United Kingdom • United States

Oracle9*i*: SQL
with an Introduction to PL/SQL

by Lannes L. Morris-Murphy, Ph.D.

Senior Vice President, Publisher:
Kristen Duerr

Executive Editor:
Jennifer Locke

Product Manager:
Barrie Tysko

Developmental Editor:
DeVona Dors

Associate Production Manager:
Jennifer Goguen

Associate Product Manager:
Janet Aras

Marketing Manager:
Jason Sakos

Editorial Assistant:
Christy Urban

Composition House:
GEX Publishing Services

Cover Designer:
Betsy Young

Cover Art:
Rakefet Kenaan

Printed in Canada

1 2 3 4 5 6 7 8 9 WC 07 06 05 04 03

For more information, contact Course Technology, 25 Thomson Place, Boston, Massachusetts, 02210.

Or find us on the World Wide Web at: www.course.com

ISBN 0-619-06475-7

BRIEF

Contents

TABLE OF Contents

Preface

The last few decades have seen a proliferation of organizations that rely heavily on information technology. These organizations store their data in databases, and many choose Oracle® database management systems to access their data. The current Oracle database version, Oracle9*i*, is an object-oriented database management system that allows users to create, manipulate, and retrieve data. In addition, Oracle9*i* has increased database security and includes new features that ease the administrative duties associated with databases.

To allow user interaction, most database management systems support some aspect of the industry standard ANSI-SQL. Oracle9*i* introduces the JOIN keyword to support the ANSI approach to linking tables—a major change from previous versions of the Oracle database. The purpose of this textbook is to introduce the student to basic SQL commands used to interact with an Oracle9*i* database in a business environment. In addition, concepts relating specifically to the objectives of the Oracle9*i* SQL certification exams have been incorporated in the text for those individuals wishing to pursue certification.

The Intended Audience

This textbook has been designed for students in technical two-year or four-year programs who need to learn how to interact with Oracle9*i* databases. Although it is preferred that students have some understanding of database design, an introductory chapter has been included to review the basic concepts of E-R Modeling and the normalization process.

Oracle Certification Program (OCP)

This textbook covers the objectives of *Exam 1Z0-001, Introduction to Oracle: SQL and PL/SQL* and *Exam 1Z0-007, Introduction to Oracle9i: SQL*. Successful completion of Exam 1Z0-001 can also be applied toward certification as an Oracle8*i* or Oracle9*i* Database Administrator. Those pursuing certification as an Oracle9*i* Database Administrator can elect to take Exam 1Z0-007 rather than Exam 1Z0-001. Exam 1Z0-007 is administered on-line. Information about registering for these exams, along with other reference material, can be found at **www.oracle.com/education/certification**.

The Approach

The concepts introduced in this textbook are presented in the context of a hypothetical "real world" business—an online book retailer named JustLee Books. First, the business operation and the database structure are introduced and analyzed. Then, as commands are introduced throughout the text, examples using the JustLee Books' database model the commands. This allows students to not only learn the syntax of a command, but also how it can be used in a real-world environment. In addition, a script file that generates the database is available to allow students hands-on practice in re-creating the examples and practicing variations of SQL commands to enhance their understanding.

Because the majority of real-world database operations involve data retrieval, the initial focus of this textbook is on the SELECT command. In Chapters 2 through 7, students learn how to retrieve data from the database, using SELECT statements, functions, and subqueries. Then, beginning with Chapter 8, students create tables and learn how to perform data manipulation operations.

To reinforce the material presented, each chapter includes a chapter summary and, when appropriate, a syntax guide of the commands covered in the chapter. In addition, at the end of each chapter, groups of activities are presented that test students' knowledge and challenge them to apply that knowledge to solving business problems.

Overview of This Book

The examples, assignments, and cases in this book will help students to achieve the following objectives:

- Issue SQL commands that will retrieve data based on criteria specified by the user.
- Use SQL commands to join tables and retrieve data from the joined tables.
- Perform calculations based on data contained within the database.
- Use subqueries to retrieve data based on unknown conditions.
- Create, modify, and drop database tables.
- Manipulate data stored in database tables.
- Enforce business rules through the use of table constraints.
- Create users and assign the privileges necessary for a user to complete various tasks.
- Create printable reports through various SQL*Plus commands.
- Create PL/SQL blocks with execution and iterative controls.

The contents of the chapters build in complexity while reinforcing previous ideas. **Chapter 1** introduces basic database management concepts, including database design. **Chapter 2** shows how to retrieve data from a table. **Chapter 3** demonstrates how to restrict rows retrieved from a table, based on a given condition. **Chapter 4** presents how to link tables with common columns. **Chapter 5** details the various single-row functions supported by Oracle9*i*. **Chapter 6** covers multiple-row functions used to derive a single value for a group of rows—and how to restrict groups of rows. **Chapter 7** covers the use of subqueries to retrieve rows based on an unknown condition already contained within the database. **Chapter 8** presents how to create new database tables. **Chapter 9** addresses the use of constraints to enforce business rules and to ensure the integrity of table data. **Chapter 10** explicates adding data to a table, modify existing data, and deleting data. **Chapter 11** covers the use of views to restrict access to data and reduce the complexity of certain types of queries. **Chapter 12** shows how to use a sequence to generate numbers, create indexes to speed up data retrieval, and create synonyms to provide aliases for tables and views. **Chapter 13** steps the reader through creating user accounts and roles and demonstrates how to grant (and revoke) privileges to those accounts and roles. **Chapter 14** reveals how to create printable reports, including grouping and subtotaling within the report. **Chapter 15** provides an introduction to PL/SQL, including the structure of a PL/SQL block. **Chapter 16** shows how to create explicit cursors to retrieve multiple rows through a PL/SQL block.

The Appendices provide support and reinforcement for the chapter materials. **Appendix A** provides a printed version of the tables and data in the JustLee Books' database. This database serves as a sustained example from chapter to chapter. **Appendix B** provides a syntax guide of the commands presented within each chapter. **Appendix C** contains five practice exams that can be used to prepare for certification exams or to assess comprehension of groups of chapters. **Appendix D** provides a list of Oracle resources.

Features

To enhance students' learning experience, each chapter in this book includes the following elements:

- **Chapter Objectives:** Each chapter begins with a list of the concepts to be mastered by the chapter's conclusion. This list provides a quick overview of chapter contents as well as a useful study aid.
- **Running Case:** A sustained example, the business operation of JustLee Books, serves as the basis for introducing new commands and for practicing the material introduced in each chapter.
- **Methodology:** As new commands are presented in each chapter, the syntax of the command is presented and then an example, using the JustLee Books' database, illustrates the command in the context of a business operation. This methodology shows the student not only *how* the command is used but also *when* and *why* it is used. The script file used to create the database is available so students can work through the examples in this textbook, engendering a hands-on environment in which students can reinforce their knowledge of chapter material.

- **Help?:** Additional help, designated by the *Help?* icon, is provided at points when students might encounter problems getting the desired output. The potential problem(s) is targeted and possible solutions are presented.
- **Tip:** This feature, designated by the *Tip* icon, provides students with practical advice. In some instances, Tips explain how a concept applies in the workplace.
- **Note:** These explanations, designated by the *Note* icon, provide further information about loading files and operations with the JustLee Books database.
- **Caution:** This warning, designated by the *Caution* icon, appears in the second half of this textbook. These warnings indicate database operations that, if misused, could have devastating results.
- **Chapter Summaries:** Each chapter's text is followed by a summary of chapter concepts. These summaries are a helpful recap of chapter contents.
- **Syntax Summaries:** Beginning with Chapter 2, a Syntax Guide table is given after each Chapter Summary. It recaps the command syntax presented in the chapter.
- **Review Questions:** End-of-chapter assessment begins with a set of 10 review questions that reinforce the main ideas introduced in each chapter. These questions ensure that students have mastered the concepts and understand the information presented.
- **Multiple Choice Questions:** Each chapter contains 20 multiple choice questions that cover the material presented within the chapter. Oracle certification-type questions are included to prepare students for the type of questions that can be expected on a certification exam, as well as to measure the students' level of understanding.
- **Hands-on Assignments:** Along with conceptual explanations and examples, each chapter provides 10 hands-on assignments related to the chapter's contents. The purpose of these assignments is to provide students with practical experience. In most cases, the assignments are based on the JustLee Books' database and serve as a continuation of the examples given within the chapter.
- **A Case for Oracle9*i*:** One major case is presented at the end of each chapter. These cases are designed to help students apply what they have learned to real-world situations. The cases give students the opportunity to independently synthesize and evaluate information, examine potential solutions, and make recommendations, much as students will do in an actual business situation.
- The Course Technology Kit for Oracle9*i* Software, available when purchased as a bundle with this book, provides the Oracle database software on CDs, so users can install on their own computers all the software needed to complete the in-chapter examples, Hands-on Assignments, and Case. The software included in the kit can be used with Microsoft Windows NT, 2000, or XP operating systems. However, only certain editions of Oracle 9*i* Release 2 will be available for installation. The edition available will depend on the operating system installed on your computer. The installation instructions for Oracle9*i* and the log in procedures are available at **www.course.com/cdkit** on the Web page for this book's title.

Teaching Tools

The following supplemental materials are available when this book is used in a classroom setting. All teaching tools available with this book are provided to the instructor on a single CD-ROM.

- **Electronic Instructor's Manual:** The Instructor's Manual that accompanies this textbook includes the following elements:

  — Additional instructional material to assist in class preparation, including suggestions for lecture topics.

  — Commands and solutions for all end-of-chapter assignments.

  — When appropriate, information about potential problems that can occur in networked environments are identified.

- **ExamView®:** This objective-based test generator lets the instructor create paper, LAN, or Web-based tests from testbanks designed specifically for this Course Technology text. Instructors can use the QuickTest Wizard to create tests in fewer than five minutes by taking advantage of Course Technology's question banks—or create customized exams.
- **PowerPoint Presentations:** Microsoft PowerPoint slides are included for each chapter. Instructors might use the slides in three ways: As teaching aids during classroom presentations, as printed handouts for classroom distribution, or as network-accessible resources for chapter review. Instructors can add their own slides for additional topics introduced to the class.
- **Data Files:** The script file necessary to create the JustLee Books' database is provided through the Course Technology Web site at **www.course.com**, and is also available on the Teaching Tools CD-ROM. Additional script files needed for chapters 9–16 are also available through the Web site and the Instructor's Resource Kit.
- **Solution Files:** Solutions to the chapter examples, end-of-chapter review questions and multiple-choice questions, Hands-On Assignments, and the Case are provided on the Teaching Tools CD-ROM. Solutions may also be found on the Course Technology Web site at **www.course.com**. The solutions are password protected.

ACKNOWLEDGMENTS

Considering how many times I changed my undergraduate major to get out of writing term papers, I cannot take sole credit for the publication of this textbook. It is the result of an amazing effort by a large number of people, many of whom I never had to opportunity to thank personally. In particular, Bill Larkin, Acquisition Editor; Jennifer Locke, Executive Editor; Jennifer Goguen, Associate Production Manager; Nicole Ashton, Manuscript Quality Assurance Lead.

I would like to express my appreciation to Barrie Tysko, the Product Manager who spent an unknown number of hours keeping my "spurts of creativity" in check (even though she would not give me her home phone number), and DeVona Dors, Developmental Editor, who had the unfortunate task of trying to civilize my southern rendition of SQL. Also, Serge Palladino and Chris Scriver deserve credit for having to work through the examples and material presented in each chapter during the two Quality Assurance stages of the process.

In addition to several of my students, the following reviewers also provided helpful suggestions and insight into the development of this textbook: Munther Alraban, Montgomery College, Germantown Campus; Denise M. Copeland, Old Dominion University; Randy Langston, Northern Virginia Community College; William McClure, DeVry University, Irving, Texas; Angela Mattia, J. Sargeant Reynolds Community College; Christopher G. Olson, Wake Forest University; Michele Smeeton, Tidewater Community College; and Richard M. Smith, Austin Community College; David Welch, Nashville State Technical Community College. A special thanks to Angela Mattia who pulled double-duty by also reviewing the final drafts of the chapters and provided valuable suggestions (and a lot of encouragement) throughout the process.

Read This Before You Begin

To the User

Data Files

To work through the examples and complete the projects in this book, you will need to load the data files created for this book. Your instructor will provide you with those data files, or you can obtain them electronically from the Course Technology Web site by accessing **www.course.com** and then searching for this book's title. The data files are designed to provide you with the same data shown in the chapter examples, so you can have hands-on practice re-creating the example queries and their output. The tables in the database can be reset if you encounter problems, such as accidentally deleting data. It is highly recommended that you work through all the examples to re-enforce your learning.

The database used throughout this book is created with the **Bookscript.sql** file. The file is located in the JustLee Database folder on your Data Disk. When you begin Chapter 2, you should run this file in Oracle9*i*. (A printed version of the tables and data generated by that file are provided in Appendix A.) You will continue using that file for Chapters 3–8. Each chapter in Chapters 9–16 has its own script file that uses the same underlying database created in the **Bookscript.sql** file. The script files for Chapters 9–16 are found in their respective chapter folders (Chapter09, Chapter10, etc.) on your Data Disk and have the file names **prech#.sql**, where the number sign (#) is replaced by the chapter number e.g., **prech10.sql**. A reminder at the beginning of each of these chapters indicates the script file that should be executed. If the computer in your school lab—or your own computer—has Oracle9*i* database software installed, you can work through the chapter examples and complete the Hands-on Assignments and Case projects. At a minimum, you will need the Oracle9*i* Release 2 Personal Edition of the software to complete the examples and assignments in this textbook.

Using Your Own Computer

To use your own computer to work through the chapter examples and to complete the Hands-on Assignments and Case projects, you will need the following:

- Hardware: A computer capable of using the Microsoft Windows NT, 2000 Professional, or XP Professional operating system. You should have at least 256MB of RAM and between 2.75GB and 4.75GB of hard disk space available before installing the software.

- Software: Oracle9*i* Release 2 Personal Edition. The Course Technology Kit for Oracle9*i* Software contains the database software necessary to perform all the tasks shown in this textbook. Detailed installation, configuration, and logon information for the software in this kit are provided at **www.course.com/cdkit** on the Web page for this title.

  When you install the Oracle9*i* software, you will be prompted to change the password for certain default administrative user accounts. Make certain that you record the names and passwords of the accounts because you may need to log in to the database with one of these administrative accounts in later chapters. After you install Oracle9*i*, you will be required to enter a user name and password to access the software. One default user name created during the installation process is "scott". The default password for the user name is "tiger". If you have installed the Personal Edition of Oracle9*i*, you will not need to enter a Connect String during the log in process.

- Data files: You will not be able to use your own computer to work through the chapter examples and complete the projects in this book unless you have the data files. You can get the data files from your instructor, or you can obtain the data files electronically by accessing the Course Technology Web site at **www.course.com** and then searching for this book's title.

  When you download the data files, they should be stored in a directory separate from any other files on your hard drive or diskette. You will need to remember the path or folder containing the files, because the file name of each script must be prefixed with its location when it is executed in SQL*Plus. (SQL*Plus is the interface tool you will use to interact with the database.)

Visit Our World Wide Web Site

Additional materials designed especially for you might be available on the World Wide Web. Go to **www.course.com** periodically and search this site for more details.

To the Instructor

To complete the chapters in this book, your users must have access to a set of data files. These files are included in the Instructor's Resource Kit. They may also be obtained electronically through the Course Technology Web site at **www.course.com**.

The set of data files consists of the **Bookscript.sql** file that creates the database tables necessary to work through the examples and to complete the activities in this textbook. Each user should execute the **Bookscript.sql** to have a copy of the tables stored in his or her schema. Chapters 9–16 also have script files that must be executed at the beginning of each chapter to ensure the data are consistent with the data presented in the textbook. Users should be granted with the DBA role to complete various tasks. If you prefer not to grant the DBA role to users, the Instructor's Manual provides a list of the minimum privileges necessary to complete each chapter. In addition, for Chapter 13, the Instructor's Manual presents factors that should be considered when using the Enterprise Edition in a networked, multiple-user environment. After the files are copied, you should instruct your users in how to copy the files to their own computers or workstations.

The chapters and projects in this book were tested using the Microsoft Windows 2000 Professional operating system with Oracle9*i* Release 2 (9.2.0.1.0.) Enterprise Edition and the Microsoft Windows XP Professional operating system with Oracle9*i* Release 2 (9.2.0.1.0) Personal Edition.

Course Technology Data Files

You are granted a license to copy the data files to any computer or computer network used by individuals who have purchased this book.

CHAPTER

1

OVERVIEW OF DATABASE CONCEPTS

Objectives

After completing this chapter, you should be able to do the following:

- Identify the purpose of a database management system (DBMS)
- Distinguish a field from a record and a column from a row
- Identify the basic components of an Entity-Relationship Model
- Define the three types of relationships that can exist between entities
- Identify the problem associated with many-to-many relationships and the appropriate solutions
- Explain the purpose of normalization
- Describe the role of a primary key
- Identify partial dependency and transitive dependency in the normalization process
- Explain the purpose of a foreign key
- Determine how to link data in different tables through the use of a common field
- Explain the purpose of a structured query language (SQL)

Imagine that it is 20 years ago, and you work in the Billing Department of a local telephone company. One day, your supervisor requests a list of all customers. The list must include each customer's name, telephone number, and current account balance. For most of today's businesses, this would be a simple request. However, this telephone company's investment in technology has been spent on faster and clearer telephone connection services for its customers, not on information management for its Billing Department. Thus, customer information—names, addresses, balance due, payments, etc.—for all of the company's 55,000 customers is stored on index cards. To create the list for your supervisor, you must take the names and telephone numbers from the index cards and type the list on a typewriter.

Two weeks later, after the list is completed, your supervisor returns and requests a list of all telephone numbers. The list must contain the names of all customers and their telephone numbers; however, the data must be sorted and presented in order of the telephone numbers. Because all the index cards are organized alphabetically by customers' last names, you can expect at least three more months of job security. Of course, by the time your list is completed, it will be out of date because there may be new customers not on the list, individuals who are no longer customers, or customers who have requested new telephone numbers. This scenario describes an ideal setting for a database. A **database** is a storage structure that allows data to be entered, manipulated, and retrieved in a variety of formats—all without having to retype the data each time they are needed.

The database used throughout this textbook is based on the activities of a hypothetical business, an online bookseller named JustLee Books. The company sells books via the Internet to customers throughout the United States. When a new customer places an order, a customer service representative collects data regarding the customer's name, billing and shipping addresses, and the items ordered. The company also keeps data about the books in inventory.

To access the data required for the operation of JustLee Books, management relies upon a **database management system (DBMS)**. A DBMS is used to create and maintain the structure of a database, and then to enter, manipulate, and retrieve the data it stores. Creating an efficient database design is the key to effectively using a database to support an organization's business operations.

This chapter will introduce basic database terminology and discuss the process of designing JustLee Books' database.

DATABASE TERMINOLOGY

Whenever a customer opens an account with a company, certain data must be collected. In many cases, the customer will complete an online form that asks for the customer's name, address, etc., as shown in Figure 1–1.

While completing the form, the customer or a service representative will fill in each blank with a series of characters. A **character** is the basic unit of data, and it can be a letter, number, or special symbol. A group of related characters (e.g., the characters that make up a customer's name) is called a **field**. A field represents one attribute or characteristic (e.g., the name) of the customer. A collection of fields (e.g., name, address, city, state, and zip code) about one customer is called a **record**. A group of records about the same type of entity (e.g., customers, inventory items) is stored in a **file**. A collection of interrelated files, such as those relating to customers, their purchases, and their payments, is stored in a **database**. Although these terms relate to the logical database design, in many cases, they are used interchangeably with the terminology for the physical database design. When creating the physical database, a field is commonly referred to as a **column**; a record is called a **row**. A file is known as a **table**.

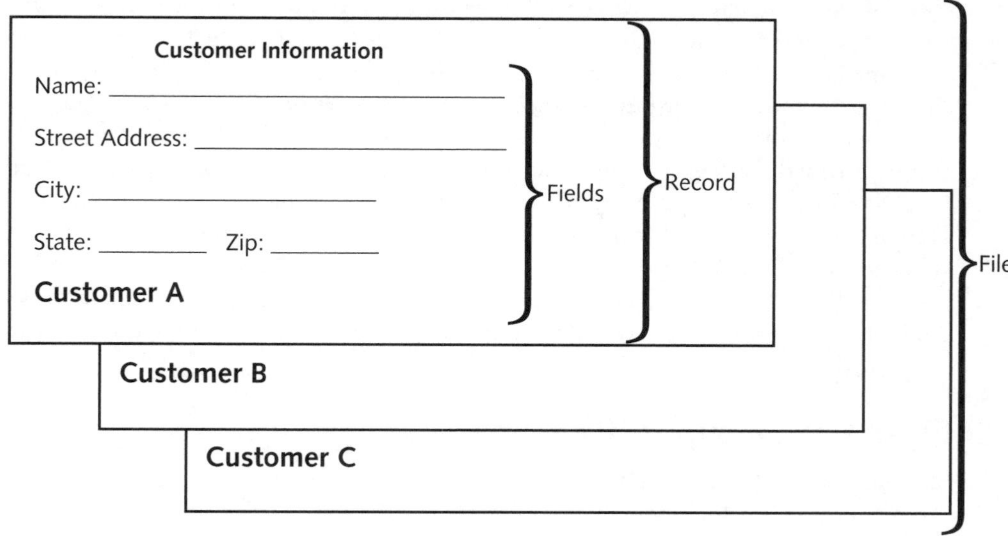

Figure 1-1 A customer's information creates data fields and a record, and then becomes part of a file.

Review of Database Design

To determine the most appropriate structure of fields, records, and files in a database, developers go through a design process. The design and development of a system is accomplished through a series of steps. The process is formally called the **Systems Development Life Cycle (SDLC)** and consists of the following steps:

1. *Systems investigation*—understanding the problem
2. *Systems analysis*—understanding the solution to the previously identified problem
3. *Systems design*—creating the logical and physical components
4. *Systems implementation*—placing the completed system into operation
5. *Systems maintenance and review*—evaluating the implemented system

Although the SDLC is a methodology designed for any type of system needed by an organization, this chapter specifically addresses the development of a DBMS. For the purposes of this discussion, assume that the problem identified was the need to collect and maintain data about customers and their orders. The identified solution was to use a database to store all needed data. The discussion that follows will present the steps necessary to design the database.

To design a database, the requirements—inputs, processes, and outputs—of the database must first be identified. Usually the first question asked is, "What information, or output, must come from this database?" or "What questions should this database be able to answer?" By understanding the necessary output, the designer can then determine what information should be stored in the database. For example, if the organization wants to send out birthday cards to its customers, then the database must store the birth date of each customer. Once the requirements of a database have been identified, an **Entity-Relationship (E-R) Model** is usually drafted to obtain a better understanding of the data to be stored in the database. Specifically, an E-R Model is a diagram that identifies the entities (customers, books, orders, etc.) in the database, and it shows how the data link the various entities together. It serves as the logical representation of the physical system to be built.

Entity-Relationship (E-R) Model

In an E-R Model, an **entity** is any person, place, or thing with characteristics or attributes that will be included in the system. In an E-R Model, an entity is usually represented as a square or rectangle.

A line depicts how an entity's data relates to another entity, as shown in Figure 1–2. If the line connecting two entities is solid, the relationship between the entities is mandatory. However, if the relationship between two entities is optional, then a dashed line is used.

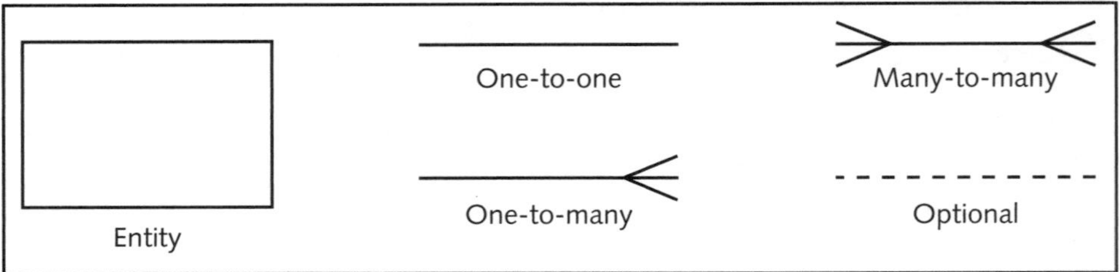

Figure 1-2 E-R Model notations

As shown in Figure 1–2, the following types of relationships can exist between two entities:

1. *One-to-one:* In a one-to-one relationship, each occurrence of data in one entity is represented by only one occurrence of data in the other entity. For example, each individual has just one Social Security Number (SSN), and each SSN is assigned to just one person. This type of relationship is depicted in an E-R Model as a simple straight line.
2. *One-to-many:* In a one-to-many relationship, each occurrence of data in one entity can be represented by many occurrences of the data in the other entity.

For example, a class has only one instructor, but an instructor may teach many classes. A one-to-many relationship is represented by a straight line with a "crowfoot" at the "many" end to indicate "many."

3. *Many-to-many:* In a many-to-many relationship, data can have multiple occurrences in both entities. For example, a class can consist of more than one student, and a student can take more than one class. A straight line with a "crowfoot" at each end indicates a many-to-many relationship.

Figure 1–3 shows a simplified E-R Model for the JustLee Books database used throughout this textbook. A more thorough E-R Model would include a list of attributes for each entity.

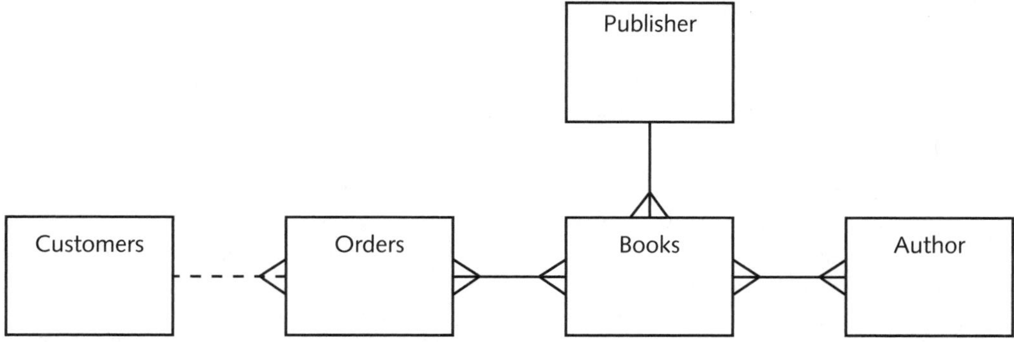

Figure 1-3 An E-R Model for JustLee Books

The following relationships are defined in this E-R Model:

1. Customers can place multiple orders, but each order can only be placed by one customer (one-to-many). The dashed line between Customers and Orders means a customer may exist in the database without having a current order stored in the ORDERS table.
2. An order can consist of more than one book, and a book can appear on more than one order (many-to-many).
3. Books can have more than one author, and an author can write more than one book (many-to-many).
4. A book can have only one publisher, but a publisher can publish more than one book (one-to-many).

Although some E-R modeling approaches are much more complex, the simplified notations used in this chapter do point out the important relationships among the entities, and using them helps the designer identify potential problems in table layouts. Upon examination of the E-R Model in Figure 1–3, you should have noticed the two many-to-many relationships. Before creating the physical database, all many-to-many relationships must be reduced to a set of one-to-many relationships, as you will learn.

The identification of entities and relationships in the database design process is important because the entities in the database are usually represented as a table, and the relationships can reveal whether additional tables are needed in the database. If the problem arising from the many-to-many relationship in the E-R Model is not apparent to the designer at this point, it will become clear during the normalization process.

Database Normalization

The purpose of database **normalization** is to reduce or control data redundancy (i.e., the needless duplication of data) and to avoid data anomalies. In general, normalization helps the database designer to determine which attributes, or fields, belong to each entity. In turn, this helps to determine which fields belong in each table(s). Normalization is a multistep process that allows the designer to take the raw data to be collected about an entity and evolve the data into a structured, normalized form that will reduce the risks associated with data redundancy. Data redundancy presents a special problem in databases because storing the same data in different places can cause problems when updates or changes to the data are required.

It is difficult for most novices to understand the impact of trying to store unnormalized data in a database. Here's an example. Suppose you work for a large company and submit a change-of-address form to the Human Resources (HR) Department. If all the data stored by HR is normalized, then a data-entry clerk would only need to update the EMPLOYEES master table with your new address. However, if the data used by HR is *not* stored in a normalized format, it is likely the data-entry clerk will need to enter the change in each table that contains your address—your EMPLOYEE RECORD table, HEALTH INSURANCE table, CAFETERIA PLANS table, SICK LEAVE table, ANNUAL TAX INFORMATION table, etc.—even though all this data is stored in the same database. Thus, if your mailing address is stored in different tables (or even duplicated in the same table) and the data-entry clerk fails to make the change in one table, you might get a paycheck that shows one address and, at the end of the year, have your W-2 form mailed to a different address! Storing the data in a normalized format should require only one update to reflect the new address, and it should always be the one that appears whenever your mailing address is needed.

A portion of the database for JustLee Books will be used to step through the normalization process—specifically, the books that are sold to customers. For each book, you need to store the book's International Standard Book Number (ISBN), title, publication date, wholesale cost, retail price, category (literature, self-help, etc.), publisher name, the person at the publisher to contact for reordering the book (and telephone number), and the name of the book's author or authors.

Figure 1–4 shows a sample of the data that must be maintained. For ease of illustration, the publishers' telephone numbers are eliminated, and the authors' first initial are used.

ISBN	TITLE	PUBLICATION DATE	COST	RETAIL	CATEGORY	PUBLISHER	CONTACT	AUTHOR
8843172113	Database Implementation	04-JUN-99	31.40	55.95	Computer	American Publishing	Davidson	T. Peterson, J. Austin, J. Adams
1915762492	Handcranked Computers	21-JAN-01	21.80	25.00	Computer	American Publishing	Davidson	W. White, L. White

Figure 1-4 JustLee Books' BOOKS table

The first step in determining the data that should be contained in each table is to identify a **primary key**. A primary key is a field that serves to uniquely identify each record. You might select the ISBN to identify each book because no two books will ever have the same ISBN.

However, note that in Figure 1–4, if a book has more than one author, the Author field will contain more than one data value. When a record contains repeating groups (i.e., multiple entries for a single column), it is considered **unnormalized**. To convert the record to **first-normal form (1NF)**, simply remove the repeating group by making each author entry a separate record, as shown in Figure 1–5.

ISBN	TITLE	PUBLICATION DATE	COST	RETAIL	CATEGORY	PUBLISHER	CONTACT	AUTHOR
8843172113	Database Implementation	04-JUN-99	31.40	55.95	Computer	American Publishing	Davidson	T. Peterson
8843172113	Database Implementation	04-JUN-99	31.40	55.95	Computer	American Publishing	Davidson	J. Austin
8843172113	Database Implementation	04-JUN-99	31.40	55.95	Computer	American Publishing	Davidson	J. Adams
1915762492	Handcranked Computers	21-JAN-01	21.80	25.00	Computer	American Publishing	Davidson	W. White
1915762492	Handcranked Computers	21-JAN-01	21.80	25.00	Computer	American Publishing	Davidson	L. White

Figure 1-5 BOOKS table from Figure 1–4 after conversion to 1NF. Fields containing multiple values are eliminated.

In Figure 1–5, the repeating group of authors' names is eliminated—each record now contains no more than one data value for the Author field. Notice that you are no longer able to use the book's ISBN as the primary key, because more than one record will have the same value in the ISBN field. *The only combination of fields that will uniquely identify each record is the ISBN and Author fields together.* When more than one field is used as the primary key for a table, the combination of fields is usually referred to as a **composite primary key**. Now that the repeating groups have been eliminated and the records can be uniquely identified, the data is in 1NF, but a few design problems remain. When the primary key consists of more than one field, another problem may occur—partial dependency may exist.

Partial dependency means that the fields contained within a record (row) are only dependent upon one portion of the primary key. For example, a book's title, publication date, publisher name, etc., are all dependent upon the book itself, not upon who wrote the book (the author). *The simplest way to resolve a partial dependency is to break the composite primary key into two parts—each representing a separate table.* In this case, you can create a table for books and a table for authors. By removing the partial dependency, you have converted the BOOKS table to **second-normal form (2NF)**, as shown in Figure 1–6.

ISBN	TITLE	PUBLICATION DATE	COST	RETAIL	CATEGORY	PUBLISHER	CONTACT
8843172113	Database Implementation	04-JUN-99	31.40	55.95	Computer	American Publishing	Davidson
1915762492	Handcranked Computers	21-JAN-01	21.80	25.00	Computer	American Publishing	Davidson

Figure 1-6 The BOOKS table, after elimination of partial dependency, is in 2NF.

Now that the BOOKS records are in 2NF, you must look for any transitive dependencies. A **transitive dependency** means that at least one of the values in the record is not dependent upon the primary key, but upon another field in the record. In this case, the contact person from the publisher's office is actually dependent upon the publisher of the book, not the book itself. To remove the transitive dependency from the BOOKS table, remove the contact information and place it in a separate table. Because the table was in 2NF and has had all transitive dependencies removed, the BOOKS table is now in **third-normal form (3NF)**, as shown in Figure 1–7.

ISBN	TITLE	PUBLICATION DATE	COST	RETAIL	CATEGORY	PUBLISHER
8843172113	Database Implementation	04-JUN-99	31.40	55.95	Computer	American Publishing
1915762492	Handcranked Computers	21-JAN-01	21.80	25.00	Computer	American Publishing

Figure 1-7 The BOOKS table in 3NF—transitive dependency of the publisher's contact person is removed.

In actuality, there are several levels of normalization. However, with a few exceptions, tables in the "real world" are only normalized to 3NF. A summary of the normalization steps presented in this section is as follows:

1. *1NF:* Eliminate all repeating groups, and identify a primary key or primary composite key.

2. *2NF:* Make certain the table is in 1NF, and eliminate any partial dependencies.
3. *3NF:* Make certain the table is in 2NF, and remove any transitive dependencies.

Linking Tables Within the Database

After the BOOKS table is in 3NF, you can then normalize each of the remaining tables of the database. Once each table has been normalized, make certain all links among the tables have been established. For example, you will need a way of determining the author(s) for each book in the BOOKS table. Because the authors' names are stored in a separate table, there must be some way to join data together. In most cases, a connection between two tables is established through a **common field**. A common field is a field that exists in both tables. In many cases, the common field will be a primary key for one of the tables. In the second table, it is referred to as a **foreign key**. The purpose of a foreign key is to establish a relationship or link with another table or tables. The foreign key will appear in the "many" side of a one-to-many relationship.

Also, an accepted industry standard is to use an ID code (numbers and/or letters) to represent an entity to reduce the chances of data-entry errors. For example, rather than entering the entire name of each publisher into the BOOKS table, you can assign each publisher an ID code in the PUBLISHER table, and then list that ID code in the BOOKS table as a foreign key to retrieve the publisher's name for each book. In this case, the publisher ID code could be the primary key in the PUBLISHER table and a foreign key in the BOOKS table.

During the normalization of JustLee Books' database, the many-to-many relationships prevented data from being normalized to 3NF. The unnormalized version of the data had repeating groups for authors in the BOOKS table and for books in the ORDERS table. As part of the conversion of the data into 3NF, two additional tables were created that did not appear on the original E-R Model: ORDERITEMS and BOOKAUTHOR.

A many-to-many relationship cannot exist in a relational database. The most common approach used to eliminate a many-to-many relationship is to create two one-to-many relationships through the addition of a **bridging table**. This type of table is placed between the original entities and serves as a "filter" for the data. The ORDERITEMS table, a bridging table, created one-to-many relationships with the ORDERS and BOOKS tables. The BOOKAUTHOR table, another bridging table, created one-to-many relationships with the BOOKS and AUTHOR tables.

After normalization, the final table structures are as shown in Figure 1–8. Notice the following about the table structures:

- The name of each table is shown in all capital letters.
- The underlined field(s) in each table indicates the primary key for that particular table. As previously mentioned, the primary key is the field that uniquely identifies each record in the table.

- For the bridging tables that were added, note that composite primary keys uniquely identify each record. The composite primary key for the BOOKAUTHOR table was created using the primary key from each table it joins together (BOOKS and AUTHOR). The arrowed lines connecting the various tables indicate the fields used to join the tables together.

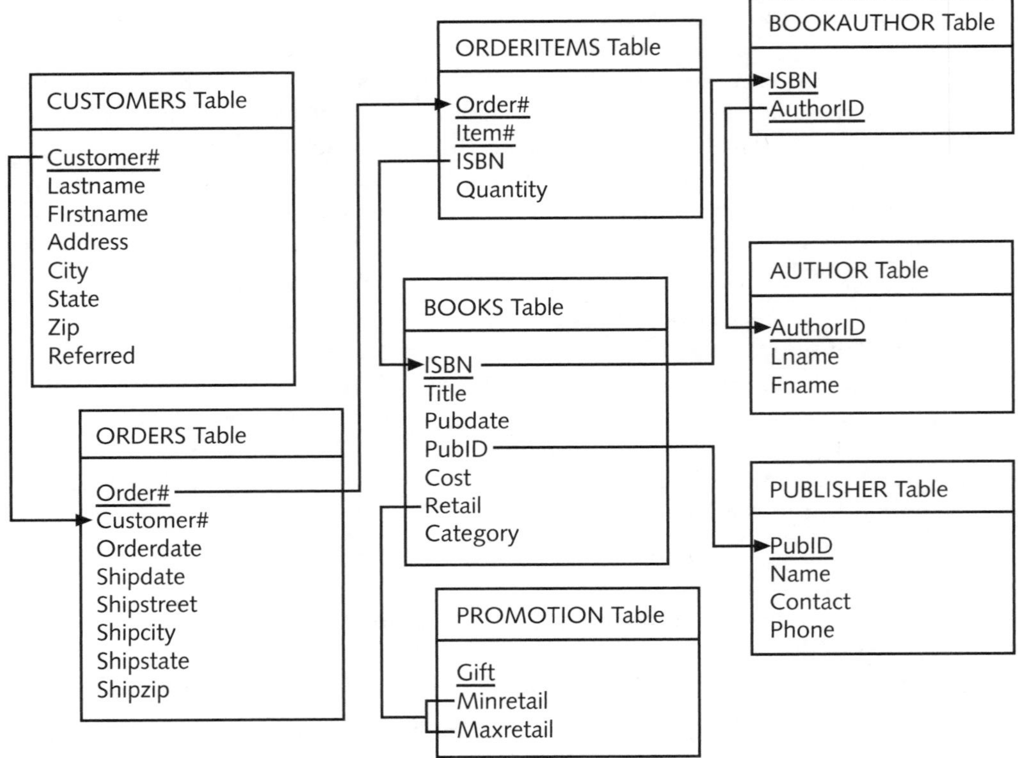

Figure 1-8 JustLeeBook's table structures after normalization

Figure 1–9 shows a portion of the BOOKS table and the fields it contains after normalization. As mentioned previously, each field represents a characteristic, or attribute, that is being collected for an entity. The group of attributes for a specific occurrence (e.g., a customer or a book) is called a record. In Oracle9*i*, a list of a table's contents will use *columns to represent fields and rows to represent records*. These terms will be used interchangeably throughout this textbook.

ISBN	TITLE	PUBDATE	...	CATEGORY
1059831198	BODYBUILD IN 10 MINUTES A DAY	21-JAN-01	...	FITNESS
0401140733	REVENGE OF MICKEY	14-DEC-01	...	FAMILY LIFE
4981341710	BUILDING A CAR WITH TOOTHPICKS	18-MAR-02	...	CHILDREN
8843172113	DATABASE IMPLEMENTATION	04-JUN-99	...	COMPUTER ← Record
...	...	...	...	...
0299282519	THE WOK WAY TO COOK	11-SEP-00	...	COOKING
8117949391	BIG BEAR AND LITTLE DOVE	08-NOV-01	...	CHILDREN
0132149871	HOW TO GET FASTER PIZZA	11-NOV-02	...	SELF HELP
9247381001	HOW TO MANAGE THE MANAGER	09-MAY-99	...	BUSINESS
2147428890	SHORTEST POEMS	01-MAY-01	...	LITERATURE

Field (arrow pointing to PUBDATE column)

Figure 1-9 A portion of the BOOKS table after normalization

JustLee Books' Database Description

The initial organization of the database structure for JustLee Books—the case study for this textbook—is shown in Figure 1–8. The database is used first to record customers' orders. Customers and JustLee's employees can identify a book by its ISBN, title, or author's name(s). Employees can also determine when a particular order was placed and when, or if, the order was shipped. The database also stores the publisher contact information so the bookseller can reorder a book.

Basic Assumptions

Three assumptions made when designing the database are as follows:

1. An order is not shipped until all items are available (i.e., there are no back orders or partial order shipments).
2. All addresses are within the United States; otherwise, the Address/Zip Code fields would need to be altered.
3. Only orders for the current month or orders from previous months that have not yet shipped are stored in the ORDERS table. At the end of each month, all completed orders are transferred to an annual SALES table. This allows for faster processing of data within the ORDERS table; when necessary, users can still access information pertaining to previous orders through the annual SALES table.

In addition to recording data, management also wants to have the ability to track the type of books that customers purchase. Although databases were originally developed to record, maintain, and report data for their collected purpose, organizations have realized the importance of having data to support other business functions. These secondary data are used for a purpose other than originally collected. Organizations that deal with thousands (even millions) of sales transaction each month usually store copies of transactions in a separate database for various types of research. Analyzing historical sales data and other information stored in an organization's database is generally referred to as **data mining**. Thus, the bookseller's database also includes data to be used by the Marketing Department to determine which categories of books customers frequently purchase. By knowing a buyer's purchasing habits, new items in inventory can be promoted to customers who frequently purchase that type of book. For example, if a customer has placed several orders for children's books, then that customer might purchase similar books in the future. The Marketing Department can thus target promotions for other children's books to that customer, knowing there's an increased likelihood of a purchase.

Tables Within JustLee Books' Database

Next, let's consider each of the tables within JustLee Books' database. Refer to the table structures in Figure 1–8.

CUSTOMERS table: Notice that the CUSTOMERS table is the first table in Figure 1–8. It serves as a master table for storing the basic data relating to any customer who has placed an order with JustLee Books. It stores the customer's name and mailing address, plus the name of the person who referred that customer to the company. As a promotion to obtain new customers, the bookstore sends a 10 percent discount coupon to any customer who refers a friend who makes a purchase.

You should also notice in Figure 1–8 that there is a Customer# field in the CUSTOMERS table. Why? Because you might have two customers with the same name, and by assigning each customer a number, you can uniquely identify each person. Using account numbers, codes, etc., can also decrease the likelihood of data-entry errors due to incorrect spelling or abbreviations.

BOOKS table: The BOOKS table stores the ISBN, title, publication date, ID of the publisher, wholesale cost, and retail price of each book. The table also stores a category name for each book (e.g., Fitness, Children, Cooking) to track customers' purchasing patterns, as previously mentioned. At the present time, the actual name of the category is entered into the database. Because someone might spell the name of one of the categories incorrectly, a CATEGORY table will be created in Chapter 8, and only the code for each category will be stored in the BOOKS table.

AUTHOR/BOOKAUTHOR tables: As shown in Figure 1–8, the AUTHOR table maintains a list of the authors' names. Since a many-to-many relationship originally existed between the books entity and the author entity, the BOOKAUTHOR table was created as a bridging table between the two entities. The BOOKAUTHOR table stores the ISBN and the author's ID for each book. If you need to know who wrote a particular book, you would first have the DBMS look up the ISBN of the book in the BOOKS table, then look up each entry of the ISBN in the BOOKAUTHOR table, and finally trace the name of the author(s) back to the AUTHORS table through the AuthorID field.

ORDERS/ORDERITEMS tables: Data about a customer's order are divided into two tables: ORDERS and ORDERITEMS. The ORDERS table identifies which customer placed each order, the date on which the order was placed, and the date on which it was shipped. Because the shipping address might be different from a customer's billing address, the shipping address is also stored within the ORDERS table. If a customer orders two or more books in one order, the ORDERS table could contain a repeating group. Therefore, the individual items purchased on each order are stored in the ORDERITEMS table.

The ORDERITEMS table records the order number, the ISBN of the book being purchased, and the quantity for each book. To uniquely identify each item purchased in an order for multiple products, the table includes an Item# field that corresponds to the item's position in the sequence of products ordered. For example, if a customer places an order for three different books, the first book listed in the order would be assigned an Item# of 1, the second book listed would be Item# 2, etc. A variation of this table could use the combination of the Order# and the book's ISBN to identify each product for a particular order. However, the concept of item# or line# is widely used in industry to identify various line items on an invoice or in a transaction, and thus it has been included in this table to familiarize you with the concept.

PUBLISHER table: The PUBLISHER table contains the publisher's ID code, the name of the publisher, the publisher's contact person, and the publisher's telephone number. The PUBLISHER table can be joined to the BOOKS table through the PubID field, which is contained in both tables. The linked data from the PUBLISHER and BOOKS table would allow you to know which publisher to contact when books need to be reordered by identifying which books are obtained from each publisher.

PROMOTION table: The last table in Figure 1–8 is the PROMOTION table. JustLee Books has an annual promotion that includes a gift with each book purchased. The gift is based upon the retail price of the book. Books that cost less than $12 will receive a certain gift, whereas books costing between $12.01 and $25 will receive a different gift. The PROMOTION table identifies the gift, the minimum retail value of the range, and the maximum retail value. There is no exact value that matches the Retail field in the BOOKS table; therefore, to determine the appropriate gift, you will need to determine whether a retail price falls within a particular range, as you will see in Chapter 4.

An actual online bookseller's database would contain thousands of customers and books. It would, naturally, be much more complex than the database shown in this textbook. For example, this database does not track data such as the quantity on-hand for each book, discounted prices, and sales tax. Furthermore, to simplify the display of the data on the screen and in reports, each table only contains a few records.

A complete list of the tables that will be referenced throughout this book can be found in Appendix A at the end of this textbook.

Now that you've had a review of database basics, let's look at how users can interact with a database management system.

STRUCTURED QUERY LANGUAGE (SQL)

The industry standard for interacting with a relational database is through a **structured query language (SQL)**—pronounced "sequel." SQL is not considered a programming language such as COBOL or Java. It is a data sublanguage, and unlike a programming language, it processes sets of data as groups, and it can navigate the data stored within various tables.

Through the use of SQL statements, users can instruct the DBMS to create and modify tables, enter and maintain data, and retrieve data for a variety of situations. You will be issuing the SQL commands used in this textbook through Oracle9*i*'s **SQL*Plus**® tool, which is the interface that allows users to create, maintain, and search stored data. As will be discussed in Chapter 3, SQL*Plus has additional commands available beyond simple SQL commands. These commands let users perform interactive searches and set environmental variables.

Two industry-accepted committees set the industry standards for SQL: the **American National Standards Institute (ANSI)** and the **International Standards Organization (ISO)**. The use of industry-established standards allows the user to transfer skills among various relational database management systems and enables various applications to communicate with different databases without major redevelopment efforts. The benefit this creates for users (and students) is that the SQL statements learned with Oracle9*i* can be transferred to another DBMS program, such as Informix. To work correctly in another environment, you may need to substitute a square bracket for a parenthesis in a SQL statement, but the basic structure and keywords should remain the same.

The remaining chapters of this textbook introduce the SQL statements and concepts you will need to know for the Oracle9*i* SQL exam and to use Oracle9*i* in the workplace. In addition, each chapter also includes information and tips for using databases in the "real world." As each SQL command is introduced, it will be demonstrated using the JustLee Books database presented in this chapter and the Appendix. A copy of the script to be used to create the tables and the data is also available for download from the publisher's Web site (*www.course.com*). The script works with any edition of Oracle9*i*.

Working through the examples presented in each chapter and completing the assignments will enhance your learning process.

Chapter Summary

- A DBMS is used to create and maintain a database.
- A database is composed of a group of interrelated files.
- A file is a group of related records. A file is also called a table in the physical database.
- A record is a group of related fields regarding one specific entity. A record is commonly called a row.
- Before building a database, designers must look at the input, processing, and output requirements of the system. Tables to be included in the database can be identified through the E-R Model. An entity in the E-R Model will usually represent a table in the physical system.
- Through the normalization process, designers can determine whether additional tables are needed, and which attributes or fields belong in each table.
- A record is considered unnormalized if it contains repeating groups.
- A record is in first-normal form (1NF) if no repeating groups exist and it has a primary key.
- Second-normal form (2NF) is achieved if the record is in 1NF and has no partial dependencies.
- Once a record is in 2NF and all transitive dependencies have been removed, then it will be in third-normal form (3NF), which is generally sufficient for most databases.
- A primary key is used to uniquely identify each record.
- A common field is used to join data contained in different tables.
- A foreign key is a common field that exists between two tables but is also a primary key in one of the tables.
- A structured query language (SQL) is a data sublanguage that navigates the data stored within a database's tables. Through the use of SQL statements, users can instruct the DBMS to create and modify tables, enter and maintain data, and retrieve data for a variety of situations.

Review Questions

1. What is the purpose of an E-R Model?
2. What is an entity?
3. Give an example of three entities that might exist in a database for a medical office and some of the attributes that would be stored in a table for each entity.
4. Define a one-to-many relationship.

5. Discuss the problems that can be caused by data redundancy.
6. Explain the role of a primary key.
7. Identify a foreign key.
8. List the steps of the normalization process.
9. What type of relationship cannot be stored in a relational database? Why?
10. Identify at least three reasons an organization might analyze historical sales data stored in its database.

Multiple Choice

1. Which of the following represents a row in a table?
   a. an attribute
   b. a characteristic
   c. a field
   d. a record
2. Which of the following defines a relationship in which each occurrence of data in one entity is represented by multiple occurrences of the data in the other entity?
   a. one-to-one
   b. one-to-many
   c. many-to-many
   d. none of the above
3. An entity is represented in an E-R Model as a(n):
   a. arrow
   b. crowfoot
   c. dashed line
   d. none of the above
4. Which of the following is not a valid E-R Model relationship?
   a. sometimes-to-always
   b. one-to-one
   c. one-to-many
   d. many-to-many

5. Which of the following symbols represents a many-to-many relationship in an E-R Model?
   a. a straight line
   b. a dashed line
   c. a straight line with a crowfoot at both ends
   d. a straight line with a crowfoot at one end
6. Which of the following may contain repeating groups?
   a. unnormalized data
   b. 1NF
   c. 2NF
   d. 3NF
7. Which of the following defines a relationship in which each occurrence of data in one entity is represented by only one occurrence of data in the other entity?
   a. one-to-one
   b. one-to-many
   c. many-to-many
   d. none of the above
8. Which of the following has no partial or transitive dependencies?
   a. unnormalized data
   b. 1NF
   c. 2NF
   d. 3NF
9. Which of the following symbols represents a one-to-many relationship in an E-R Model?
   a. a straight line
   b. a dashed line
   c. a straight line with a crowfoot at both ends
   d. a straight line with a crowfoot at one end
10. Which of the following has no partial dependencies but may contain transitive dependencies?
    a. unnormalized data
    b. 1NF
    c. 2NF
    d. 3NF

11. Which of the following has no repeating groups but may contain partial or transitive dependencies?
    a. unnormalized data
    b. 1NF
    c. 2NF
    d. 3NF
12. The unique identifier for a record is called the:
    a. foreign key
    b. primary key
    c. turn key
    d. common field
13. Which of the following is normally a primary key in another table when you are joining two tables together?
    a. foreign key
    b. primary key
    c. turn key
    d. repeating group
14. A unique identifier that consists of more than one field is commonly called a:
    a. primary plus key
    b. composite key
    c. foreign key
    d. none of the above
15. Which of the following symbols represents an optional relationship in an E-R Model?
    a. a straight line
    b. a dashed line
    c. a straight line with a crowfoot at both ends
    d. a straight line with a crowfoot at one end
16. Which of the following will indicate the need for an additional table if it exists in an E-R Model?
    a. sometimes-to-always relationship
    b. one-to-one relationship
    c. one-to-many relationship
    d. many-to-many relationship

17. Which of the following represents a field in a table?
    a. a record
    b. a row
    c. a column
    d. an entity
18. Which of the following defines a relationship in which data can have multiple occurrences in each entity?
    a. one-to-one
    b. one-to-many
    c. many-to-many
    d. none of the above
19. When part of the data in a table depends upon a field within the table which is not the table's primary key, this is known as:
    a. transitive dependency
    b. partial dependency
    c. psychological dependency
    d. a foreign key
20. Which of the following is used to join data contained in two or more tables?
    a. primary key
    b. unique identifier
    c. common field
    d. foreign key

HANDS-ON ASSIGNMENTS

To perform these activities, refer to the table structures in Figure 1–8 or the tables in Appendix A.

1. Which tables and fields would you access to determine which books have been purchased by a customer in the current month's orders?
2. How would you determine which orders have not yet been shipped to the customer?
3. If management needed to determine which book category generated the most sales for the first week in April, which tables and fields would be consulted to derive this information?

4. How would you determine how much profit was generated from orders placed in the current month?
5. If a customer inquired about a book written in 1999 by an author named Thompson, which access path (tables and fields) would you need to follow to find the list of books meeting the customer's request?

A CASE FOR ORACLE9*i*

To perform this activity, refer to the table structures in Figure 1–8 or the tables in Appendix A.

In this chapter, the normalization process was shown just for the BOOKS table. The other tables in JustLee Books' database are shown after normalization. Because the database needs to contain data for each customer's order, perform the necessary steps to normalize the following data elements to 3NF:

- customer's name and billing address
- quantity and retail price of each item ordered
- shipping address for each order
- the date each order was placed and the date it was shipped

Assume that the unnormalized data in the list are all stored in one table. Provide your instructor with a list of the table(s) that have been identified at each step of the normalization process (i.e., 1NF, 2NF, 3NF) and the attributes, or fields, contained in each table. Remember that each customer may place more than one order, each order may contain more than one item, and an item may appear on more than one order.

CHAPTER

2

BASIC SQL SELECT STATEMENTS

Objectives

After completing this chapter, you should be able to do the following:

- Distinguish between an RDBMS and an ORDBMS
- Identify keywords, mandatory clauses, and optional clauses in a SELECT statement
- Select and view all columns of a table
- Select and view one column of a table
- Display multiple columns of a table
- Use a column alias to clarify the contents of a particular column
- Perform basic arithmetic operations in the SELECT clause
- Remove duplicate lists, using either the DISTINCT or UNIQUE keyword
- Combine fields, literals, and other data
- Format output

In Chapter 1, you reviewed database structures and were introduced to the concept of using SQL statements to enter, manipulate, and retrieve data through a DBMS. DBMS is a generic term that applies to software that allows users to interact with a database. When you are working with relational databases, however, the DBMS software is considered to be a **relational database management system (RDBMS)**. The RDBMS is the software program used to create the database and allows you to enter, manipulate, and retrieve data. Most RDBMSs include capabilities to create forms for user input screens and reports to display output. When trying to retrieve data, most RDBMSs provide the user with an option to interact with the database through a graphical user interface (GUI) or through SQL statements. When a GUI is used, the RDBMS actually converts entries made by the user into SQL statements that are subsequently executed to perform the desired operation.

In this textbook, Oracle9*i* Database is used to interact with the database for JustLee Books. Oracle9*i* is an **object relational database management system (ORDBMS)** because it can be used to reference not only individual data elements but also objects (e.g., object fields and maps), which can be composed of individual data elements. However, since the data stored in the database for JustLee Books are composed of simple alphanumeric characters, the examples and concepts presented throughout this textbook also apply to traditional RDBMSs. The use of objects is usually addressed in advanced application development courses.

Oracle9*i* Database comes in three editions: Enterprise Edition, Standard Edition, and Personal Edition. All the concepts and examples presented in this textbook can be used with any of these editions. However, they were specifically written for the Personal Edition because it is the only edition that can be used on many home computers. This will allow you to practice the SQL commands at home if you have a computer available, reducing the amount of time you must spend in a computer lab.

In this chapter, you will begin learning about Oracle9*i* by using SELECT statements. You might wonder why this chapter doesn't begin with creating tables—a logical beginning. This is the reason: When students begin with creating tables, they usually are not able to understand what is actually happening in the database—or the impact of their actions on naming conventions, data types, missing constraints, data inserts, etc. By contrast, beginning with the SELECT statement (which accounts for about 80 percent of most database operations) allows students to learn many database concepts *before* actually creating a table. Thus, by using the SELECT statement first, the student becomes familiar with problems that can be encountered when creating tables.

Chapter 2 will present the SQL commands shown in Figure 2–1.

Command Description	Basic Syntax Structure	Example
Command to view all columns of a table	`SELECT *` `FROM tablename;`	`SELECT *` `FROM books;`
Command to view one column of a table	`SELECT columnname` `FROM tablename;`	`SELECT title` `FROM books;`
Command to view multiple columns of a table	`SELECT columnname, columnname,…` `FROM tablename;`	`SELECT title, pubdate` `FROM books;`

Command to assign an alias to a column during display	`SELECT columnname [AS] alias`
	`FROM tablename;`
		`SELECT title AS titles`
		`FROM books;`
		or
		`SELECT title titles`
		`FROM books;`

Command to perform arithmetic operations during retrieval	`SELECT arithmetic expression` `FROM tablename;`	`SELECT retail – cost` `FROM books;`

Figure 2-1 Overview of chapter contents

Command Description	Basic Syntax Structure	Example

Command to eliminate duplication in output	`SELECT DISTINCT` *`columnname`*
	`FROM` *`tablename`*`;`
	or
	`SELECT UNIQUE` *`columnname`*
	`FROM` *`tablename`*`;`
		`SELECT DISTINCT state`
		`FROM customers;`
		or
		`SELECT UNIQUE state`
		`FROM customers;`

Command to perform concatenation of column contents during display	`SELECT` *`columnname`* `||`
	`columname`
	`FROM` *`tablename`*`;`
		`SELECT firstname ||`
		`lastname`
		`FROM customers;`

Command to view the structure of a table	`DESCRIBE` *`tablename`*	`DESCRIBE books`

Figure 2-1 Overview of chapter contents (continued)

Before beginning this chapter, you will need to open and run the data file named **Bookscript.sql**. This script file will create the tables to be used in the examples and Hands-on Assignments.

1. Go to the JustLee Database folder in your data files.
2. Select the **Bookscript.sql** script file.
3. To execute the script file, enter **start d:\Bookscript.sql** at the SQL> prompt inside Oracle9*i*. The ***d:*** should be substituted with the appropriate drive letter and path name, if applicable. Then press **Enter**.
4. After running the script, you can verify and view the structure of each table by entering **`describe`** ***`tablename`*** (where *`tablename`* is substituted with the actual name of the table presented in this chapter) at the SQL> prompt.

The script assumes you have the correct privileges for your user name to perform the commands included in the script. If you would like to see the contents of the script, it can be opened with any word-processing program.

If you open the **Bookscript.sql** file with a word-processing program and then accidentally click **Save**, your word-processing program may save the file with a different extension than "sql," and Oracle9*i* will be unable to use the script. The script file must have the extension "sql" to execute properly.

SELECT STATEMENT SYNTAX

As mentioned in this chapter's introduction, the majority of the SQL operations performed on a database in the average organization are SELECT statements. Basic **SELECT statements** allow the user to retrieve data from tables. The user can view all the fields and records within a table or specify only certain fields and records to be displayed. In essence, the SELECT statement asks the database a question, also known as a **query**.

After querying a database, the results that are displayed can be based on certain conditions specified in the SELECT statement. In other words, what is displayed is basically the answer to the question asked by the user. For example, in this chapter, you will learn the basic structure of a SELECT statement and how to display only certain fields from a table. In Chapter 3, you will learn how to modify the SELECT statement to display only certain rows.

The **syntax** for a SQL statement gives the basic structure, or rules, required to execute the statement. The syntax for the SELECT statement is shown in Figure 2-2.

```
SELECT  [DISTINCT | UNIQUE] (*, columnname [ AS alias], …)
        FROM        tablename
        [WHERE      condition]
        [GROUP BY group_by_expression]
        [HAVING     group_condition]
        [ORDER BY columnname];
```

Figure 2-2 Syntax for the SELECT statement

The capitalized words (SELECT, FROM, WHERE, etc.) in Figure 2-2 are **keywords** (i.e., words that have a predefined meaning in Oracle9*i*). Each section of the example that begins with a keyword is referred to as a **clause** (SELECT clause, FROM clause, WHERE clause, etc.). Note these important points about SELECT statements:

- The only clauses required for the SELECT statement are SELECT and FROM. (These are the only clauses in Figure 2-2 to be discussed in this chapter.)
- Square brackets are used to indicate portions of the statement that are optional. (Optional clauses will be discussed in subsequent chapters.)
- SQL statements can be entered over several lines (as shown in Figure 2-2) or on one line. Most SQL statements are entered with each clause on a separate line to improve readability and make editing easier. As various SELECT commands are demonstrated in this chapter, you will see variations on spacing, number of lines used, and capitalization. These variations will be pointed out as they are encountered within the text.
- To execute a SQL statement after it is entered, you have two options. Usually the SQL statement is executed by entering a semicolon (;) at the end of the last line of the statement (as given in the syntax example). If you forget to enter the semicolon and press the **Enter** key, you can still execute the statement by entering a slash (/) at the SQL> prompt.

When a SQL statement is entered at the SQL> prompt, it is stored in the **SQL buffer** for execution. The SQL buffer is a portion of the computer's memory that holds the SQL statement being executed. The statement will remain in the buffer until another SQL statement is entered (i.e., the buffer only holds one SQL statement at a time, and it is temporary).

To view what is currently being held in the SQL buffer, enter either a semicolon (;) or the letter **L** at the SQL> prompt, and then press the **Enter** key. If you would like to execute the SQL statement currently stored in the buffer, simply type **run**, an **r**, or a slash (/) at the SQL> prompt, and then press the **Enter** key.

Selecting All Data in a Table

To have the SELECT statement return *all* data from a specific table, type an asterisk (*) after SELECT. Enter the query shown in Figure 2–3 at the SQL> prompt.

```
SELECT *
FROM customers;
```

Figure 2-3 Command to select all data within a table

To move from the first line to the second, press **Enter** after typing the asterisk. The asterisk is a symbol that instructs Oracle9*i* to include all columns in the table. The symbol can only be used in the SELECT clause of a SELECT statement. If you need to view or display all columns in a table, it is much simpler to type an asterisk than to type the name of each column.

By entering the semicolon at the end of the second line, the statement will be executed after pressing **Enter**. If you press **Enter** before entering the semicolon, simply enter a slash (/) on the third line that appears, and then press **Enter**.

The results of the SELECT statement should look like those shown in Figure 2–4.

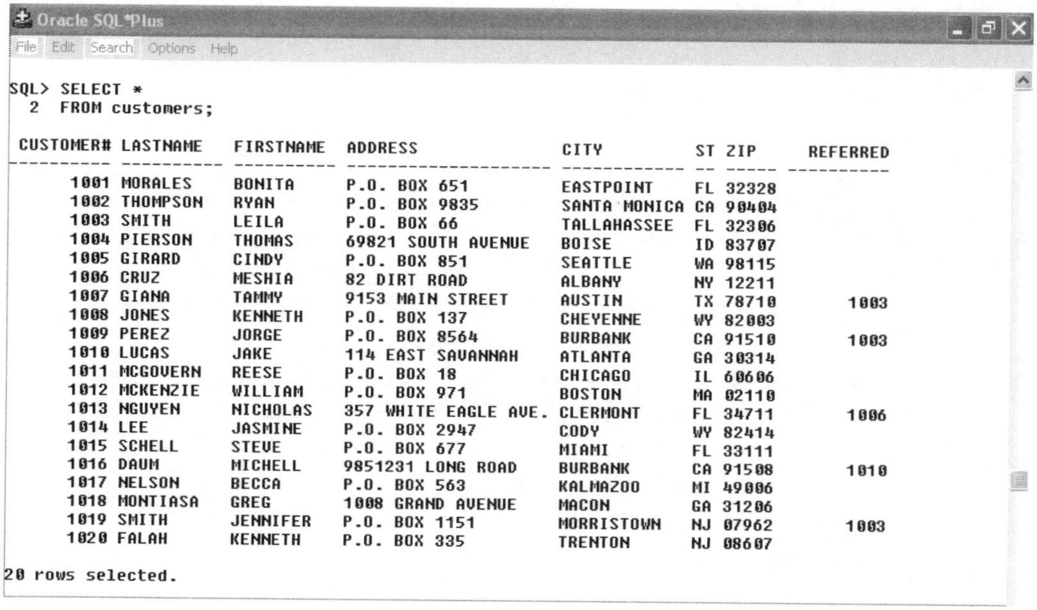

```
Oracle SQL*Plus
File  Edit  Search  Options  Help

SQL> SELECT *
  2  FROM customers;

 CUSTOMER# LASTNAME   FIRSTNAME  ADDRESS              CITY         ST ZIP     REFERRED
---------- ---------- ---------- -------------------- ------------ -- ----- ----------
      1001 MORALES    BONITA     P.O. BOX 651         EASTPOINT    FL 32328
      1002 THOMPSON   RYAN       P.O. BOX 9835        SANTA MONICA CA 90404
      1003 SMITH      LEILA      P.O. BOX 66          TALLAHASSEE  FL 32306
      1004 PIERSON    THOMAS     69821 SOUTH AVENUE   BOISE        ID 83707
      1005 GIRARD     CINDY      P.O. BOX 851         SEATTLE      WA 98115
      1006 CRUZ       MESHIA     82 DIRT ROAD         ALBANY       NY 12211
      1007 GIANA      TAMMY      9153 MAIN STREET     AUSTIN       TX 78710       1003
      1008 JONES      KENNETH    P.O. BOX 137         CHEYENNE     WY 82003
      1009 PEREZ      JORGE      P.O. BOX 8564        BURBANK      CA 91510       1003
      1010 LUCAS      JAKE       114 EAST SAVANNAH    ATLANTA      GA 30314
      1011 MCGOVERN   REESE      P.O. BOX 18          CHICAGO      IL 60606
      1012 MCKENZIE   WILLIAM    P.O. BOX 971         BOSTON       MA 02110
      1013 NGUYEN     NICHOLAS   357 WHITE EAGLE AVE. CLERMONT     FL 34711       1006
      1014 LEE        JASMINE    P.O. BOX 2947        CODY         WY 82414
      1015 SCHELL     STEVE      P.O. BOX 677         MIAMI        FL 33111
      1016 DAUM       MICHELL    9851231 LONG ROAD    BURBANK      CA 91508       1010
      1017 NELSON     BECCA      P.O. BOX 563         KALMAZOO     MI 49006
      1018 MONTIASA   GREG       1008 GRAND AVENUE    MACON        GA 31206
      1019 SMITH      JENNIFER   P.O. BOX 1151        MORRISTOWN   NJ 07962       1003
      1020 FALAH      KENNETH    P.O. BOX 335         TRENTON      NJ 08607

20 rows selected.
```

Figure 2-4 List of all customers in the CUSTOMERS table

If the results of your SELECT statement appear to wrap to two lines, enter **`SET LINESIZE 100`** at the SQL> prompt, and then press **Enter**. This will reset the number of characters that can be displayed on one line to 100 characters. If you notice a second column heading displayed in the middle of the results, enter **`SET PAGESIZE 100`** at the SQL> prompt, and press **Enter**. This will extend the number of lines to be displayed per page of output.

After making any necessary changes, retype the SELECT statement, and your results should look like those shown in Figure 2-4. These changes will only last as long as your current session. To make the settings permanent, the database administrator will need to make the changes to your login script.

When looking at the results of the SELECT statement, pay attention to the column headings. The column heading for the State field has been truncated and only ST is displayed. The column was created as a **character field** and, therefore, the heading will be no longer than the width of the data stored in the field. If a field is defined as a **numeric column**, the entire column heading will be displayed, regardless of the width of the field (as shown by the Customer# field). Because the State field was defined to store only two characters, only the first two characters of the column name will be displayed in the column heading. However, when you refer to the State field in any SQL statement, you will still need to specify the entire column name, not just ST.

The exact name of each column can be viewed by entering **`DESCRIBE`** ***`tablename`*** at the SQL> prompt. To view the exact name of each column in the CUSTOMERS table, simply enter **`DESCRIBE customers`** at the SQL> prompt, and then press **Enter**.

Selecting One Column from a Table

In the previous example, an asterisk was used to indicate that all columns in the table should be displayed. When displaying a table containing a large number of fields, the results may look cluttered. In certain cases, there may be sensitive data you do not want other users to see. In these situations, you can instruct Oracle9*i* to return only specific columns in the results. Choosing specific columns in a SELECT statement is called **projection**. You can select one column—or as many as all the columns—contained within the table.

For example, suppose that you would like to view the titles of all books in inventory. The data regarding books are stored in the BOOKS table. The name of the column you need is Title. As shown in Figure 2-5, you can list the name of the desired column after the SELECT keyword. Type the statement shown in Figure 2-5.

```
SELECT title
FROM books;
```

Figure 2-5 Command to select a single column

If you receive an error message rather than the results of the query, there may have been a typing error. The error message should display the line in which the error occurred. An asterisk beneath the line serves as an indicator of the error; however, it is not always the exact cause. If the error message indicates that the error is in the second line of the statement, then BOOKS may have been entered as BOOK. Simply retype the statement with the correction, and it should return the desired results.

Results returned from the query should look like those shown in Figure 2–6.

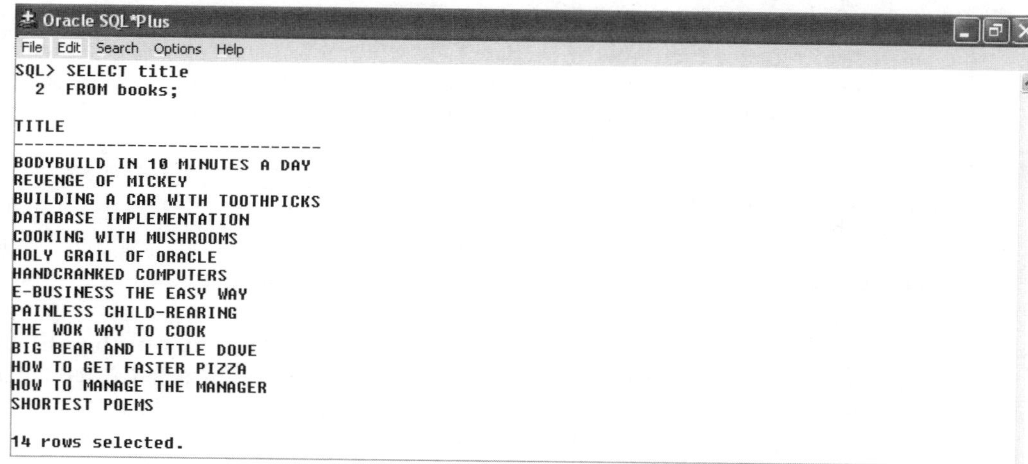

Figure 2-6 List of all book titles in the BOOKS table

The results only display the field specified, which was Title. You may want to practice some variations of the same SELECT statement. Try entering the examples shown in Figure 2–7 at the SQL> prompt, and notice that the results are the same.

```
SQL> SELECT TITLE FROM BOOKS;

SQL> select title from books;

SQL> SELECT title FROM books;

SQL> SELECT TITLE
     FROM BOOKS
     /
```

Figure 2-7 The SELECT statement can be entered on one or more lines.

As shown in these examples, the statement can be entered on one or more lines. *Keywords, table names, and column names are not case sensitive.* To distinguish between keywords and other parts of the SELECT statement, the keywords are capitalized. Just remember, this is *not* a requirement of Oracle9*i*; it is simply a convention used to improve readability.

Selecting Multiple Columns from a Table

In most cases, displaying only one column from a table is not sufficient output on which to base decisions. If you want to know the date on which each book was published, you could retrieve all the fields from the BOOKS table and manually extract the needed fields. As an alternative, you could issue one SELECT statement to retrieve the title field, another to retrieve the publication date field, and then match up the two results. Because your time is valuable, it would be much more practical to simply issue a query requesting both the title and the publication date for each book, as shown in Figure 2–8.

```
SELECT title, pubdate
FROM books;
```

Figure 2-8 Command to select multiple columns from a table

When specifying more than one column in the SELECT clause of the SELECT statement, commas should separate the columns listed. Although a space has been entered after the comma, it is not required. The space serves to improve the readability of the statement and is not part of the required syntax of the SELECT statement. The example shown in Figure 2–9 would return the same results.

```
SELECT title,pubdate FROM books;
```

Figure 2-9 Multiple clauses of the SELECT statement on one line

The data returned from the query should look like that shown in Figure 2–10.

When looking at the results of this query, notice the order in which the columns are listed in the output. In this case, Title is listed first, followed by Pubdate. Oracle9*i* sequences the columns in the display in the same order that the user sequences them in the SELECT clause of the SELECT statement. To change the order and display the Pubdate column first, simply reverse the order of the columns listed in the SELECT statement, as shown in Figure 2–11.

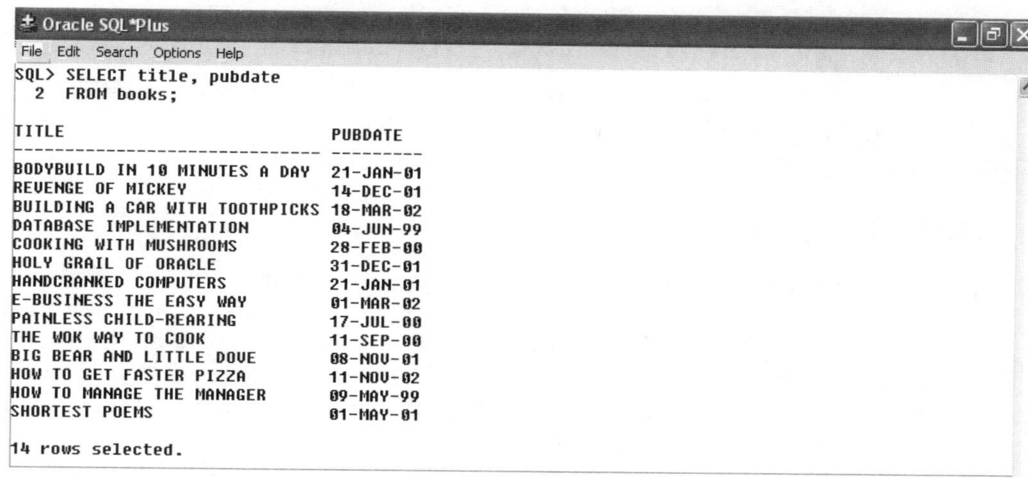

Figure 2-10 Display of Title and publication date of books in the BOOKS table

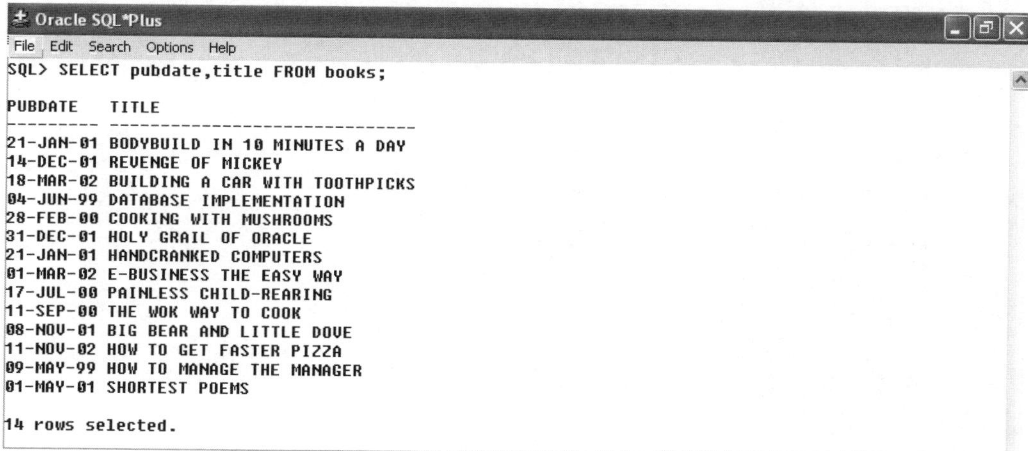

Figure 2-11 Reversed column sequence in a SELECT clause

If your screen begins to look cluttered, you can clear the screen by pressing **SHIFT+DEL** and then selecting **Ok** from the dialog window that appears.

OPERATIONS WITHIN THE SELECT STATEMENT

Now that you've selected columns from tables, let's look at some other operations. In this section, you'll learn how to use column aliases, employ arithmetic operations, eliminate duplicate output, and display rows on multiple lines.

Using Column Aliases

In some cases, a column name can be a vague indicator of the data displayed in a particular column. To better describe the data listed, you can substitute a **column alias** for the column name in the results of a query. For example, if you are presenting a list of all books stored in the database, you may want the column heading to read Title of Books. To instruct the software to use a column alias, simply list the column alias next to the column name in the SELECT clause. Figure 2–12 shows the title and category for each book in the BOOKS table, but it adds a column alias for the title. The **optional keyword** of **AS** has been included in this example to distinguish between the column name and the column alias.

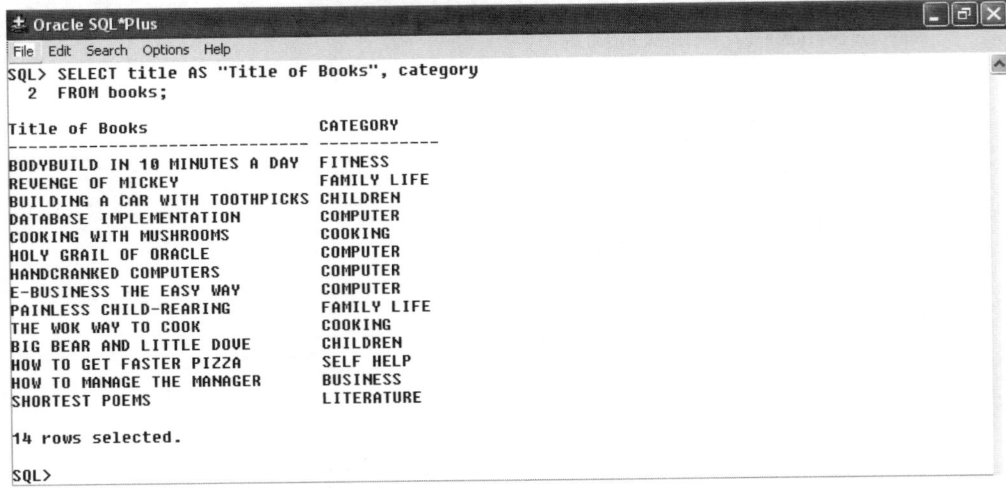

Figure 2-12 Using a column alias in a SELECT clause

There are guidelines you will need to keep in mind when using a column alias. If the column alias contains spaces, special symbols, or if you do not want it to appear in all capital letters, *it must be enclosed in double quotation marks* (" "). By default, the column headings shown in the results of queries are capitalized. The use of the double quotation marks will override the default for the column heading. However, notice that the case of the data displayed *within* the column was not altered.

As shown in the SELECT statement, *you must separate the list of field names with commas*. If you forget a comma, Oracle9*i* will interpret the subsequent field name as a column alias, and you will not receive the intended results.

If the column alias consists of only one word without special symbols, it does not need to be enclosed in double quotation marks. In Figure 2–13, the Retail field has been assigned the column alias of Price. Also note that the optional keyword AS used in Figure 2–12 has not been included. Because a comma does not separate the words *retail* and *price*, Oracle9*i* assumes that Price is the column alias for the Retail column.

```
Oracle SQL*Plus
File Edit Search Options Help
SQL> SELECT title, retail price
  2  FROM books;

TITLE                               PRICE
------------------------------ ----------
BODYBUILD IN 10 MINUTES A DAY       30.95
REVENGE OF MICKEY                      22
BUILDING A CAR WITH TOOTHPICKS      59.95
DATABASE IMPLEMENTATION             55.95
COOKING WITH MUSHROOMS              19.95
HOLY GRAIL OF ORACLE                75.95
HANDCRANKED COMPUTERS                  25
E-BUSINESS THE EASY WAY              54.5
PAINLESS CHILD-REARING              89.95
THE WOK WAY TO COOK                 28.75
BIG BEAR AND LITTLE DOVE             8.95
HOW TO GET FASTER PIZZA             29.95
HOW TO MANAGE THE MANAGER           31.95
SHORTEST POEMS                      39.95

14 rows selected.
```

Figure 2-13 Using a column alias without the AS keyword

As you look at the results in Figure 2–13, notice the alignment of the column headings:

- By default, the data for text, or character, fields are left-aligned. Likewise, the column heading for text, or character, fields is left-aligned.
- Data from a numeric field are right-aligned, and so is the column heading. If you look at the column heading "PRICE" in the results of the query, it aligns to the right rather than at the beginning of the line separating the column heading from the data.
- Notice how Oracle9*i* does not display insignificant zeros (i.e., zeros that do not affect the value of the number being displayed). The retail price of the book *Handcranked Computers* is $25.00. Because the zeros in the two decimal positions are insignificant, Oracle9*i* does not display them. To force Oracle9*i* to display a specific number of decimal positions, formatting codes are required. These are presented in Chapter 5.

Using Arithmetic Operations

If needed, simple arithmetic operations such as multiplication (*), division (/), addition (+), and subtraction (-) can be used in the SELECT clause of a query. Keep in mind that Oracle9*i* adheres to the standard order of operations:

1. Moving from left to right within the arithmetic equation, any required multiplication and division operations are solved first.
2. Any addition and subtraction operations are solved after multiplication and division, again moving from left to right within the equation.

To override the order of operations, you can use parentheses to enclose any portion of the equation that should be completed first.

Oracle9*i* does not explicitly support exponents in the SELECT statement. For example, in some application programs, the user can enter `number^3` to raise a number to the power of three. A number raised to the power of three simply means to multiply a number by itself three times (e.g., 5^3 equals 5 * 5 * 5). With Oracle9*i*, if you need to use an exponential operation in the SELECT statement, simply break it down into its multiplication equivalent.

Next, you would like to determine the profit generated by the sale of each book. The BOOKS table contains two fields you can use to derive the profit: Cost and Retail. A book's profit is the difference (subtraction) between the amount paid for the book by the bookstore (cost) and the selling price of the book (retail). To clarify the contents of the column, assign the column alias "profit" to the calculated field, as shown in Figure 2–14.

```
Oracle SQL*Plus
File  Edit  Search  Options  Help
SQL> SELECT title, retail-cost profit
  2  FROM books;

TITLE                              PROFIT
------------------------------ ----------
BODYBUILD IN 10 MINUTES A DAY        12.2
REVENGE OF MICKEY                     7.8
BUILDING A CAR WITH TOOTHPICKS      22.15
DATABASE IMPLEMENTATION             24.55
COOKING WITH MUSHROOMS               7.45
HOLY GRAIL OF ORACLE                 28.7
HANDCRANKED COMPUTERS                 3.2
E-BUSINESS THE EASY WAY              16.6
PAINLESS CHILD-REARING              41.95
THE WOK WAY TO COOK                  9.75
BIG BEAR AND LITTLE DOVE             3.63
HOW TO GET FASTER PIZZA              12.1
HOW TO MANAGE THE MANAGER           16.55
SHORTEST POEMS                       18.1

14 rows selected.
```

Figure 2-14 Using a column alias for an arithmetic expression in the SELECT clause

Using DISTINCT and UNIQUE

Suppose that you want to know the states in which your customers reside so you can focus a marketing campaign on a particular region of the country. You want a list to identify only the states, not customer names, addresses, etc. One option would be to select the State column from the CUSTOMERS table. What you would quickly notice is that some states are listed more than once if more than one customer lives in that particular state. If you are working with only 20 records, it would not be a problem simply to cross out any duplicate states on a printout. However, if you are dealing with thousands of records, it would be a cumbersome task.

To eliminate duplicate listings, you can use the DISTINCT option in your SELECT statement. For example, suppose that you have five customers living in Texas (TX). Without the DISTINCT option, TX would appear in your results five times. If you include the DISTINCT option, TX would only appear once. To use the DISTINCT option, simply use the keyword DISTINCT between the SELECT keyword and the first column of the column list, as shown in Figure 2–15.

Figure 2-15 List of distinct states stored in the CUSTOMERS table

In Figure 2–15, the database was queried to determine the states in which customers live. Although there are 20 customers in the CUSTOMERS table, they live in only 12 states. You could use this information to determine where you are most likely to attract more customers, or identify geographical areas that are not responding to a current, nationwide marketing effort.

When using the DISTINCT keyword, it is applied to all columns listed in the SELECT clause, even though it is only stated directly after the SELECT keyword. In the previous example, if you had also included CITY in the SELECT clause, each different combination of city and state would have been listed once in the output. In other words, if no two customers in the database live in the same city and state, you still would have had 20 rows of output—one row for each customer.

You can also use the UNIQUE option to eliminate duplicates. It works the same way as the DISTINCT keyword. The following would return the same results as the example in Figure 2–15:

```
SELECT UNIQUE state
FROM customers;
```

Creating Concatenation

In previous examples, if an output list contained more than one field, each field was placed in a separate column. In some situations, however, you might want the contents of each field to be displayed next to each other, without much blank space. For example, for a list of customer names, you might prefer to have them combined to appear as a single column, rather than as separate first name and last name columns.

Combining the contents of two or more columns is known as **concatenation**. To instruct Oracle9*i* to concatenate the output of a query, use two vertical bars or pipes (| |). (On a keyboard, this symbol is located above the backslash (\).) In the following example, the goal is to have the last name of the customer listed immediately after the first name, rather than in a separate column.

As you look at the results in Figure 2–16, the first thing you should notice is that the first name and last name of each customer run together, and it is difficult to tell where one name ends and the other begins. To make the results more readable, you need to include a blank space between the Firstname and Lastname fields.

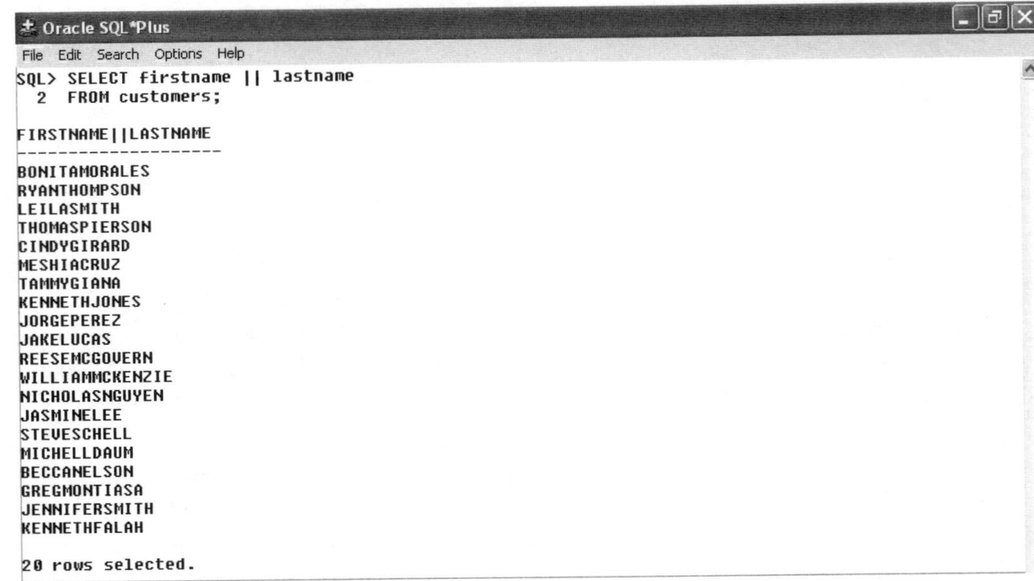

Figure 2-16 Concatenation of two columns in the SELECT clause

To have Oracle9*i* insert a blank space, you must concatenate, or combine, the Firstname and Lastname fields with a **string literal**. A string literal instructs Oracle9*i* to interpret what you have entered "literally" and not to consider it as a keyword or command. *A string literal must be enclosed within single quotation marks (' ').* Whenever you use a string literal, the character or set of characters that you have typed should appear in the output exactly as you have typed them. In this instance, the string literal is a blank space. Figure 2–17 shows how the customers list looks after including a blank space in the output.

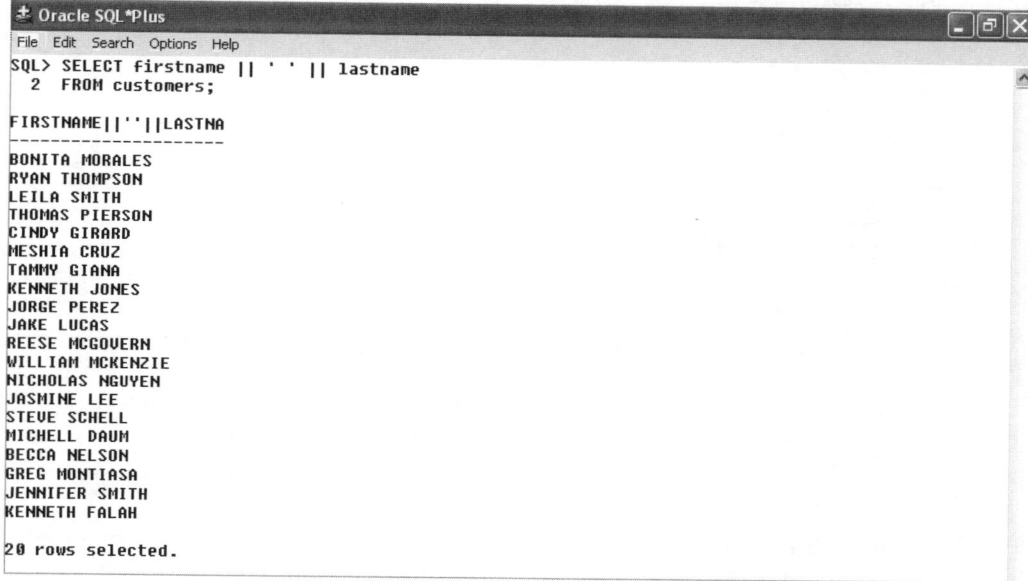

Figure 2-17 Insertion of a string literal into the concatenation of two columns

Although you now have a readable list of all customer names, the display has an unusual column heading. The column heading shown in the results is exactly what you entered for the field list—including the concatenation symbols and the literal. If this list is for management or some individual who is not familiar with Oracle9*i*, you might want to give your output a more professional appearance. Rather than have this unusual and unappealing column heading, you can have Oracle9*i* substitute a column alias, as you did previously. The query in Figure 2–18 substitutes Customer Name as the column heading in the results.

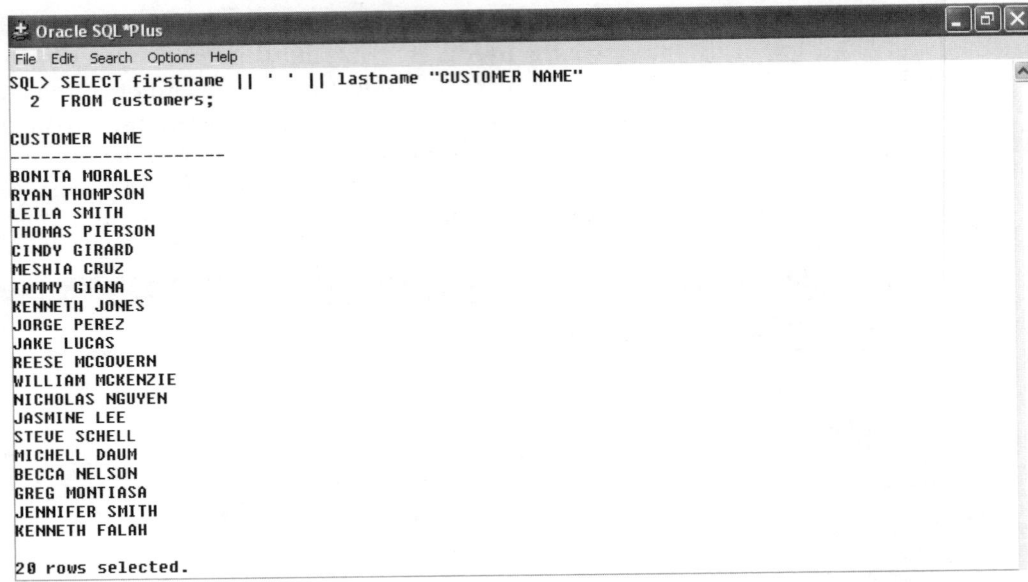

Figure 2-18 Using a column alias to describe the contents of the concatenated columns

If you get an error message, make sure the blank space is in single quotation marks and the column alias is in double quotation marks.

Suppose that management says they need a list of all customers, showing each customer's number and name on separate lines. Also, management requests that the customer names appear as "last name, first name." To display customer names as specified by management, simply modify the query used in Figure 2–18 to switch the order of the columns and include a comma in the string literal. To have the customer number and name appear on separate lines, include **CHR(10)** after the customer number to instruct Oracle9*i* to insert a line break. The CHR(10) code indicates to the computer that a line break should occur at that location. Anything listed after the CHR(10) code is displayed on the next line.

You can use the CHR(10) code whenever you would like output results to be displayed over several lines. As shown in Figure 2–19, the vertical bars are used to include the line break in the column list. Since CHR(10) is not considered a column, Oracle9*i* will return an error message, indicating it expected to find the FROM keyword, unless the function is concatenated with the columns from the CUSTOMERS table.

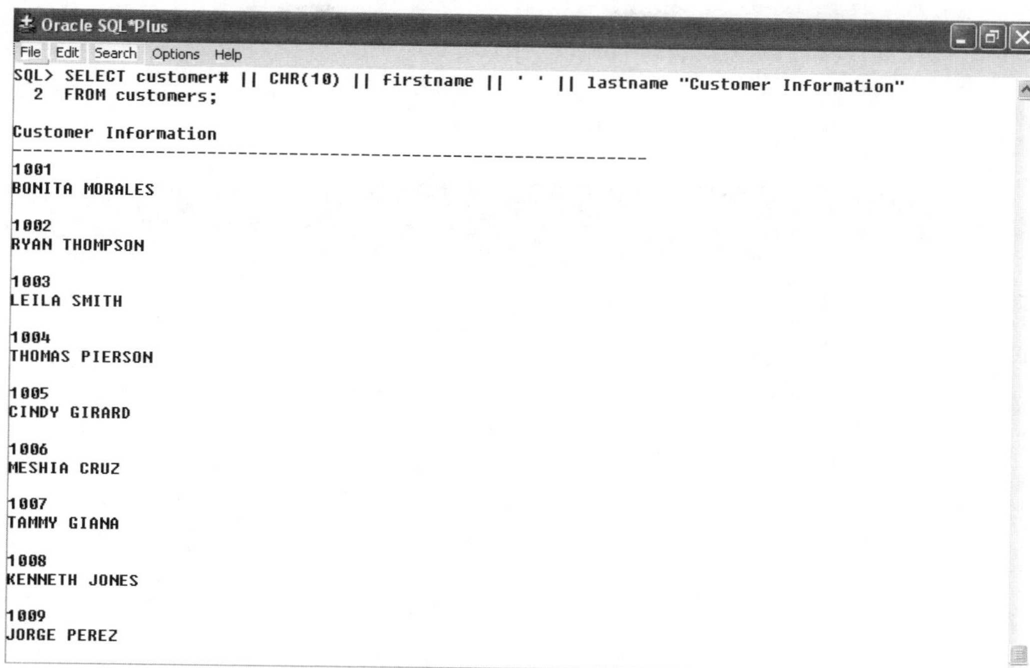

Figure 2-19 First screen of the display of customer information using line breaks

CHAPTER SUMMARY

- Oracle9*i* is an ORDMBS.
- A basic query in Oracle9*i* SQL includes the SELECT and FROM clauses. These are the only mandatory clauses in a SELECT statement.
- To view all columns in the table, specify an asterisk (*) or list all the column names individually in the SELECT clause.
- When listing column names in the SELECT clause, a comma must separate column names.
- A column alias can be used to clarify the contents of a particular column. If the alias contains spaces or special symbols, or if you wish to display the column with any lower-case letters, you must enclose the column alias in double quotation marks (" ").
- Basic arithmetic operations can be performed in the SELECT clause.
- To remove duplicate listings, include either the DISTINCT or UNIQUE keyword.
- To specify which table contains the desired columns, the name of the table must be listed after the keyword FROM.
- Use vertical bars (| |) to combine, or concatenate, fields, literals, and other data.

Chapter 2 Syntax Summary

The following table presents a summary of the syntax that you have learned in this chapter. You can use the table as a study guide and reference.

<table>
<tr><th colspan="3">Syntax Guide</th></tr>
<tr><th>Element</th><th>Description</th><th>Example</th></tr>
<tr><td>SELECT clause</td><td>Identify the column(s) for retrieval in a SELECT command</td><td><code>SELECT title</code></td></tr>
<tr><td>FROM clause</td><td>Identify the table containing selected columns</td><td><code>FROM books</code></td></tr>
<tr><td>SELECT statement</td><td>View column(s) in a table</td><td><code>SELECT title</code>
<code>FROM books;</code></td></tr>
<tr><td>,</td><td>Separate column names in a list when retrieving multiple columns from a table</td><td><code>SELECT title, pubdate</code>
<code>FROM books;</code></td></tr>
<tr><td>*</td><td>Return all data in a table when used in a SELECT clause</td><td><code>SELECT *</code>
<code>FROM books;</code></td></tr>
<tr><td>AS</td><td>Indicate a column alias to change the heading of a column in output</td><td><code>SELECT title AS titles,</code>
<code>pubdate</code>
<code>FROM books;</code></td></tr>
<tr><td></td><td>Create a column alias to change the heading of a column in output without using AS</td><td><code>SELECT title titles,</code>
<code>pubdate</code>
<code>FROM books;</code></td></tr>
<tr><td>" "</td><td>Preserve spaces, symbols, or case in an output column heading alias</td><td><code>SELECT title AS "Book Name"</code>
<code>FROM books;</code></td></tr>
<tr><td>* multiplication
/ division
+ addition
- subtraction</td><td>Solve arithmetic operations (Oracle9i first solves * and /, then solves + and -)</td><td><code>SELECT title, retail-cost</code>
<code>profit</code>
<code>FROM books;</code></td></tr>
<tr><td>DISTINCT</td><td>Eliminate duplicate lists</td><td><code>SELECT DISTINCT state</code>
<code>FROM customers;</code></td></tr>
<tr><td>UNIQUE</td><td>Eliminate duplicate lists</td><td><code>SELECT UNIQUE state</code>
<code>FROM customers;</code></td></tr>
<tr><td>||
(concatenation)</td><td>Combine display of content from multiple columns into a single column</td><td><code>SELECT city || state</code>
<code>FROM customers;</code></td></tr>
<tr><td>' '
(string literal)</td><td>Indicate the exact set of characters, including spaces, to be displayed</td><td><code>SELECT city || ' ' || state</code>
<code>FROM customers;</code></td></tr>
<tr><td>CHR(10)</td><td>Insert a line break</td><td><code>SELECT customer# ||CHR(10)</code>
<code>||city || ' ' || state</code>
<code>FROM customers;</code></td></tr>
</table>

SQL* Plus Syntax		
Element	**Description**	**Example**
DESCRIBE	Display structure of a table	`DESCRIBE books`

; *or* /	Execute a SQL statement	`SELECT Zip`
		`FROM customers;`
		or
		`FROM customers/`

;, **L,** *or* **LIST**	View contents in a buffer	`SQL>;` *or* `SQL> list`
		or
		`SQL>L`

type **RUN, r**, or / and press **Enter**	Execute a SQL statement stored in the buffer	`SQL>run (Enter)`
		or
		`SQL>r (Enter)`
		or
		`SQL>`
		`/ (Enter)`

Review Questions

1. What is an RDBMS? An ORDBMS?
2. What are the two required clauses for a SELECT statement?
3. What is the purpose of the SELECT statement?
4. What does the use of an asterisk (*) in the SELECT clause of a SELECT statement represent?
5. What is the purpose of a column alias?
6. How do you indicate that a column alias should be used?
7. When is it appropriate to use a column alias?
8. What are the guidelines to keep in mind when using a column alias?
9. How can you concatenate columns in a query?
10. How do you indicate that a line break should occur in the output of a query?

MULTIPLE CHOICE

To determine the exact name of the fields used in the tables for these questions, either refer to Appendix A or use the **DESCRIBE *tablename*** *command to view the structure of the appropriate table.*

1. Which of the following SELECT statements will provide a list of customer names from the CUSTOMERS table?
   a. `SELECT customer names FROM customers;`
   b. `SELECT "Names" FROM customers;`
   c. `SELECT firstname, lastname FROM customers;`
   d. `SELECT firstname, lastname, FROM customers;`
   e. `SELECT firstname, lastname, "Customer Names" FROM customers;`
2. Which clause is required in a SELECT statement?
   a. WHERE
   b. ORDER BY
   c. GROUP BY
   d. all of the above
   e. none of the above
3. Which of the following is *not* a valid SELECT statement?
   a. `SELECT lastname, firstname FROM customers;`
   b. `SELECT * FROM orders;`
   c. `Select FirstName NAME from CUSTOMERS;`
   d. `SELECT lastname Last Name FROM customers;`
4. Which of the following symbols is used to represent concatenation?
   a. *
   b. ||
   c. [ ]
   d. ‘ ’
5. Which of the following SELECT statements will return all the fields in the ORDERS table?
   a. `SELECT customer#, order#, orderdate, shipped, address FROM orders;`
   b. `SELECT * FROM orders;`
   c. `SELECT ? FROM orders;`
   d. none of the above

6. Which of the following symbols is used for a column alias that contains spaces?
   a. ' '
   b. ||
   c. " "
   d. / /
7. Which of the following is a valid SELECT statement?
   a. `SELECT TITLES *TITLE! FROM BOOKS;`
   b. `SELECT "customer#" FROM books;`
   c. `SELECT title AS "Book Title" from books;`
   d. all of the above
8. Which of the following instructs Oracle9*i* to execute the current SQL statement?
   a. \
   b. /
   c. go
   d. start
9. Which of the following is *not* a valid SELECT statement?
   a. `SELECT Cost-Retail FROM books;`
   b. `SELECT Retail+Cost FROM books;`
   c. `SELECT retail*retail*retail FROM books;`
   d. `SELECT retail^3 from books;`
10. When must a comma be used in the SELECT clause of a query?
    a. when a field name is followed by a column alias
    b. to separate the SELECT clause and the FROM clause when only one field is selected
    c. It is never used in the SELECT clause.
    d. when listing more than one field name and the fields are not concatenated
    e. when an arithmetic expression is being included in the SELECT clause
11. Which of the following commands will display a listing of the category for each book in the BOOKS table?
    a. `SELECT title books, category`
       `FROM books;`
    b. `SELECT title, books, category FROM books;`
    c. `SELECT title, cat FROM books;`
    d. `SELECT books, ||category "Categories" FROM books;`

12. Which clause is *not* required in a SELECT statement?
    a. SELECT
    b. FROM
    c. WHERE
    d. All of the above clauses are required.
13. Which of the following lines of the SELECT statement contains an error?

```
1. SELECT title, isbn,
2. Pubdate "Date of Publication"
3. FROM books;
```

    a. line 1
    b. line 2
    c. line 3
    d. There are no errors—the statement returns the intended results.
14. Which of the following lines of the SELECT statement contains an error?

```
1. SELECT ISBN,
2. retail-cost
3. FROM books;
```

    a. line 1
    b. line 2
    c. line 3
    d. There are no errors—the statement returns the intended results.
15. Which of the following lines of the SELECT statement contains an error?

```
1. SELECT title, cost,
2. cost*2
3. 'With 200% MarkUp'
4. FROM books;
```

    a. line 1
    b. line 2
    c. line 3
    d. line 4
    e. There are no errors—the statement returns the intended results.

16. Which of the following lines of the SELECT statement contains an error?

```
1. SELECT name, contact,
2. "Person to Call", phone
3. FROM publisher;
```

   a. line 1
   b. line 2
   c. line 3
   d. There are no errors—the statement returns the intended results.

17. Which of the following lines of the SELECT statement contains an error?

```
1. SELECT ISBN, || ' is the ISBN for the book named ' ||
2. title
3. FROM books
4. /
```

   a. line 1
   b. line 2
   c. line 3
   d. There are no errors—the statement returns the intended results.

18. Which of the following lines of the SELECT statement contains an error?

```
1. SELECT title, category
2. FORM books
3. /
```

   a. line 1
   b. line 2
   c. line 3
   d. There are no errors—the statement returns the intended results.

19. Which of the following lines of the SELECT statement contains an error?

```
1. SELECT name, contact, CHR(10),
2. "Person to Call", phone
3. FROM publisher;
```

   a. line 1
   b. line 2
   c. line 3
   d. There are no errors-the statement returns the intended results.

20. Which of the following lines of the SELECT statement contains an error?

    1. `SELECT *`
    2. `FROM publishers;`

    a. line 1

    b. line 2

    c. There are no errors-the statement returns the intended results.

HANDS-ON ASSIGNMENTS

To determine the exact name of the fields used in the tables for these exercises, either refer to Appendix A or use the **DESCRIBE** ***tablename*** *command to view the structure of the appropriate table.*

1. Display a list of all data contained within the BOOKS table.
2. From the BOOKS table, list only the title of all the books available in inventory.
3. From the BOOKS table, list the title and publication date for each book in the table. Use the column heading of Publication Date for the Pubdate field.
4. List the customer number for each customer in the CUSTOMERS table, and the city and state in which they reside.
5. Create a list containing the name of each publisher, the person usually contacted, and the telephone number of the publisher. Rename the column containing the contact person, and call it Contact Person in the displayed results. (*Hint*: Use the PUBLISHER table.)
6. Determine which categories are represented by the current inventory. List each category only once. (*Hint*: Use the DISTINCT or UNIQUE keyword.)
7. From the ORDERS table, list the customer number for each customer who has placed an order with the bookstore. List each customer number only once.
8. From the BOOKS table, create a list of each book stored in the table and the category in which each book belongs. However, reverse the sequence of the columns so the category of each book is listed first.
9. From the AUTHOR table, list the first and last name of each author.
10. Create a list of authors that displays the last name followed by the first name for each author. The last names and first names should be separated by a comma and a blank space.

A CASE FOR ORACLE9*i*

The management of JustLee Books has submitted two requests. The first is for a mailing list of all customers stored in the CUSTOMERS table. The second is for a list of the percentage of profit generated by each book in the BOOKS table. The requests are as follows:

1. Create a mailing list from the CUSTOMERS table. The mailing list should display the name, address, city, state, and zip code for each customer. The name of each customer should be listed in order of first name followed by last name. The name of the customer should appear on the first line, the address on the second line, and the city, state, and zip code on the third line.
2. To determine the percentage of profit for a particular item, simply subtract the cost for the item from the retail price to obtain the dollar amount of profit, and then divide the profit by the cost of the item. The solution is then multiplied by 100 to determine the profit percentage for each book. Using a SELECT statement, display the title of each book and its percentage of profit. For the column displaying the percentage markup, use Profit % as the column heading.

Required: Determine the SQL statements needed to perform the two required tasks. Each statement should be tested to ensure its validity. Submit the appropriate documentation of the commands and their results using the format specified by your instructor.

CHAPTER

3

RESTRICTING ROWS AND SORTING DATA

Objectives

After completing this chapter, you should be able to do the following:

- Use a WHERE clause to restrict the rows returned by a query
- Create a search condition using mathematical comparison operators
- Use the BETWEEN...AND comparison operator to identify records within a range of values
- Specify a list of values for a search condition using the IN comparison operator
- Search for patterns using the LIKE comparison operator
- Identify the purpose of the % and _ wildcard characters
- Join multiple search conditions using the appropriate logical operator
- Perform searches for NULL values
- Specify the order for the presentation of query results, using ORDER BY, DESC, ASC, and the SELECT clause
- Use SQL*Plus editing commands to edit the contents of the SQL*Plus buffer

In the previous chapter, you learned how to retrieve specific fields from a table. However, unless you used the DISTINCT or UNIQUE keyword, your results included every record. In some instances, you will only want to see the records that meet a certain condition or conditions—a process referred to as **selection**. Because selection reduces the number of records retrieved by a query, it might be easier to locate a particular record in the output. Similarly, it might be easier to identify trends if the data are presented in a sorted order. This chapter will explain how to perform queries using search conditions and methods for sorting results. In particular, you'll see how the WHERE clause of the SELECT statement can be used as a search condition; the ORDER BY clause can be used to present results in a specific sequence. Figure 3–1 provides an overview of this chapter's topics.

Element	Description
WHERE clause	Used to specify condition(s) that must be true for a record to be included in the query results
ORDER BY clause	Used to specify the sorted order for presenting the results of a query
Mathematical comparison operators (=, <, >, <=, >=, <>, !=, ^=)	Used to indicate how a record should relate to a specific search value
Other comparison operators (BETWEEN...AND, IN, LIKE, IS NULL)	Used in conditions with search values that include patterns, ranges, or null values
Logical operators (AND, OR, NOT)	Used to join multiple search conditions (AND, OR) or to reverse the meaning of a given search condition (NOT)
SQL*Plus editing commands	Used to edit or manipulate SQL commands currently active in the SQL*Plus buffer

Figure 3-1 Overview of chapter contents

Go to the JustLee Database folder in your Data Files. If you run the **Bookscript.sql** file, you will be able to work through all the queries shown in this chapter. Your output should match the output shown.

WHERE CLAUSE SYNTAX

To have Oracle9*i* retrieve records based upon a given condition, the WHERE clause is added to the SELECT statement. As indicated by square brackets ([ ]), the WHERE clause is optional. When used, it should be listed beneath the FROM clause in the SELECT statement, as shown in Figure 3–2.

```
SELECT     [DISTINCT|UNIQUE] (*, column [ AS alias], …)
FROM       table
[WHERE     condition] ◄────────
[GROUP BY group_by_expression]
[HAVING    group_condition]
[ORDER BY column];
```

Figure 3-2 Syntax of the SELECT statement

A **condition** identifies what must exist or a requirement that must be met. Oracle9*i* searches through each record to determine whether the condition is TRUE. If a record meets the given condition, it will be returned in the results of the query. For a simple search of a table, the condition portion of the WHERE clause uses the following format:

```
<column name> <comparison operator> <another named column
or a value>
```

For example, suppose that you need the last name of every customer living in the state of Florida. You would use the SQL statement shown in Figure 3-3.

```
SELECT lastname, state
FROM customers
WHERE state = 'FL';
```

Figure 3-3 Query to perform a simple search based upon a given condition

As shown in Figure 3-3, the query specifies the Lastname and State columns stored in the CUSTOMERS table as the data to list in the output. However, you only want to see the records of the customers who have the letters FL stored in the State field. Thus, in the clause, WHERE is the keyword, State is the name of the column to be searched, the comparison operator "equal to" (=) means it must contain the exact value specified, and the specified value is FL. Notice the single quotation marks around FL. They designate FL as a string literal. Figure 3-4 shows the output of Figure 3-3.

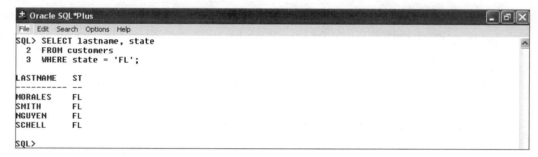

```
Oracle SQL*Plus
File Edit Search Options Help
SQL> SELECT lastname, state
  2  FROM customers
  3  WHERE state = 'FL';

LASTNAME   ST
---------- --
MORALES    FL
SMITH      FL
NGUYEN     FL
SCHELL     FL

SQL>
```

Figure 3-4 List of all customers living in Florida

If you receive an error message, double-check that the FL value was entered with single and not double quotation marks. (Double quotation marks were used in Chapter 2 for column aliases.) If no rows are returned in the results, make certain the letters FL are capitalized. The data for JustLee Books were originally entered in upper-case characters and are, therefore, stored in the database tables in upper-case characters. Any value entered in a string literal (i.e., within single quotation marks) is evaluated *exactly* as entered—both in spacing and case of the characters. Therefore, if a string literal is entered for a search condition, it must be in the same case as the data being searched, or no rows will be returned in the results. Although Oracle9*i* is not case sensitive when evaluating keywords, table names, and column names, the *evaluation of data contained within a record is case sensitive.*

The results returned from the query list the last name and state for each customer living in Florida. Notice that only four rows were returned, even though our table contained

20 customers. The WHERE clause restricts the number of records returned in the results to only those meeting the given condition of **state = 'FL'**.

Rules for Character Strings

As shown in the SQL command in Figure 3–3, the value of FL is shown within single quotation marks. Whenever you use a string literal as part of a search condition, the value must be enclosed within single quotation marks and, as a result, will be interpreted exactly as listed within the single quotation marks. By contrast, if the target field had consisted only of *numbers*, single quotation marks would not have been required. If you try issuing the command given in Figure 3–3 without the single quotation marks, you will receive an error message rather than the expected query results.

Next, suppose that you want to see all the data stored in the CUSTOMERS table for customer 1010. You could issue the SQL statement and receive the output shown in Figure 3–5.

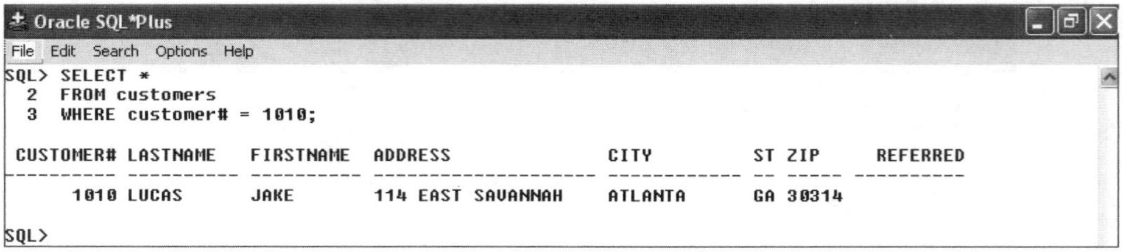

```
SQL> SELECT *
  2  FROM customers
  3  WHERE customer# = 1010;

 CUSTOMER# LASTNAME   FIRSTNAME  ADDRESS              CITY         ST ZIP   REFERRED
---------- ---------- ---------- -------------------- ------------ -- ----- ----------
      1010 LUCAS      JAKE       114 EAST SAVANNAH    ATLANTA      GA 30314

SQL>
```

Figure 3-5 Search results for Customer 1010

In this example, the value of 1010 for the Customer# column is not enclosed in single quotation marks because the Customer# column has been defined to store only numbers. Thus, single quotation marks are not necessary.

Next, let's use the WHERE statement to search for a book with the ISBN of 1915762492. Figure 3–6 shows the input and the output of that query.

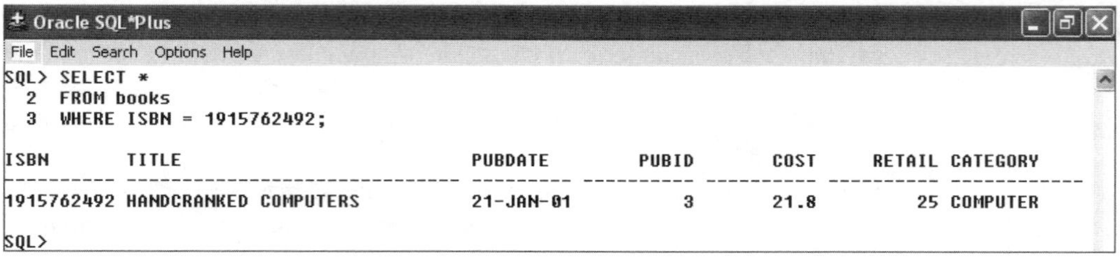

```
SQL> SELECT *
  2  FROM books
  3  WHERE ISBN = 1915762492;

ISBN       TITLE                          PUBDATE        PUBID       COST     RETAIL CATEGORY
---------- ------------------------------ --------- ---------- ---------- ---------- ------------
1915762492 HANDCRANKED COMPUTERS          21-JAN-01          3       21.8         25 COMPUTER

SQL>
```

Figure 3-6 Search results for ISBN 1915762492

The ISBN column of the BOOKS table is defined as a character field, or text field, rather than a numeric field because some ISBNs could contain letters. In this particular instance, however, none of the values stored within the ISBN column contain any letters. Therefore, you were able to search the field using a search condition specified as a numeric value without any quotation marks. However, if the table had contained even *one* record that had a letter in the ISBN column, Oracle9*i* would have returned an error message. In other words, it may have worked in this one case, but it might not always work. Using single quotation marks ultimately depends upon whether the field is defined to hold text or only numeric data. Therefore, *always use single quotation marks if the column is defined as anything other than a numeric field.*

If you don't know if a column is defined to hold only numeric values, issue the `DESCRIBE` *`tablename`* command to see how table columns have been defined. Columns that can only store numeric data will be defined as NUMBER columns.

Rules for Dates

Sometimes, you might need to use a date as a search condition. Oracle9*i* displays dates in the default format of DD-MON-YY, with MON being the standard three-letter abbreviation for the month. Because the Pubdate field contains letters and hyphens, it is not considered a numeric value when Oracle9*i* performs searches. Therefore, the date value must be enclosed in single quotation marks. Figure 3–7 shows a query for books published on January 21, 2001.

```
SQL> SELECT *
  2  FROM books
  3  WHERE pubdate = '21-JAN-01';

ISBN       TITLE                          PUBDATE        PUBID       COST     RETAIL CATEGORY
---------- ------------------------------ --------- ---------- ---------- ---------- ------------
1059831198 BODYBUILD IN 10 MINUTES A DAY  21-JAN-01          4      18.75      30.95 FITNESS
1915762492 HANDCRANKED COMPUTERS          21-JAN-01          3       21.8         25 COMPUTER

SQL>
```

Figure 3-7 Simple query based on a date condition

Comparison Operators

Thus far in this chapter, you have used an equal sign, or **equality operator**, to evaluate search conditions; basically, you instructed Oracle9*i* to return only results containing the *exact* value you provided. However, there are many situations that are not based on an "equal to" condition. For example, suppose management needs a list of books for a proposed marketing campaign. The Marketing Department wants to include a gift with the purchase of any book that has a retail price of more than $55.00. Management wants to know which books can be specifically mentioned in an advertisement of this marketing campaign. The equality operator would not be appropriate in this situation; you would need a different comparison operator. A **comparison operator** indicates how the data should relate to the given search value (equal to, greater than, less than, etc.). In this case, you would need to use a comparison operator that means "greater than" (>) to determine which books meet the "more than $55.00" requirement. Figure 3–8 shows that SQL statement.

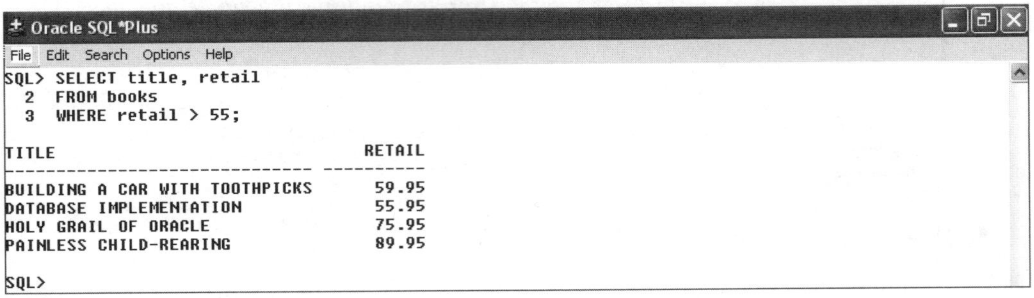

```
Oracle SQL*Plus
File Edit Search Options Help
SQL> SELECT title, retail
  2  FROM books
  3  WHERE retail > 55;

TITLE                              RETAIL
------------------------------ ----------
BUILDING A CAR WITH TOOTHPICKS      59.95
DATABASE IMPLEMENTATION             55.95
HOLY GRAIL OF ORACLE                75.95
PAINLESS CHILD-REARING              89.95

SQL>
```

Figure 3-8 Search for books with a retail price greater than $55

Based on these results, you know that four books meet the condition for this sales promotion. Notice that you entered **55** as the value for the Retail price condition. This value could have also been entered as **55.00**. Oracle9*i* will accept a period to indicate decimal positions without considering the entry to be a character value rather than a numeric value. However, if you had entered the dollar sign ($) or a comma (to indicate a thousands position), you would have received an error message indicating that the Retail field is numeric and the value entered was an "invalid character" ($55.00 is not equivalent to 55.00). Unlike some other database management systems, Oracle9*i* does not have a currency datatype, and it will regard the comma and the dollar sign as characters.

The "greater than" (**>**) comparison operator can also be used with text fields. Suppose you are about to take a physical inventory of all books in stock. The procedure JustLee Books uses for taking a physical inventory is to give each employee a list of books, and then have the person record the quantity on hand. When creating the list of books, each person is responsible for a portion of the alphabet. For example, one person may be responsible for all books with titles falling in the A through D range.

Figure 3–9 shows how to create the list of books for the person who has been assigned to take an inventory of all books with a title that alphabetically occurs after the letters ST.

The results of this SQL statement returned one book because only one book had a title that would alphabetically occur after the letters ST.

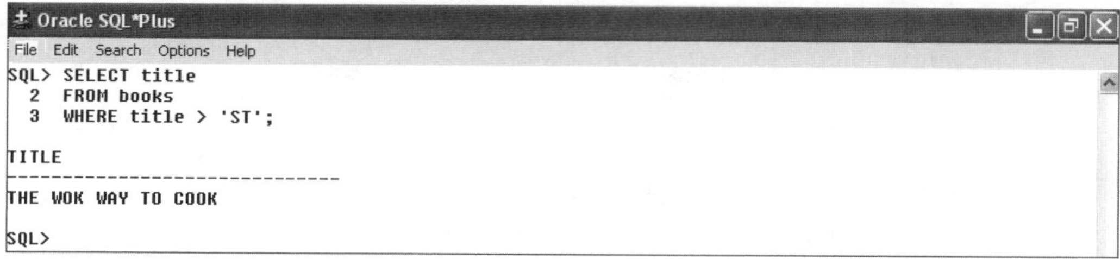

Figure 3-9 Search for books with a title greater than the letters ST

Now that you've examined "equal to" and "greater than" comparison operators, let's examine some others. Figure 3–10 shows comparison operators commonly used in Oracle9*i*.

Comparison Operators	
	Mathematical Comparison Operators
=	equality or "equal to"—e.g., `cost = 55.95`
>	greater than—e.g., `cost > 20`
<	less than—e.g., `cost < 20`
<>, !=, or ^=	not equal to—e.g., `cost <> 55.95` or `cost != 55.95` or `cost^= 55.95`
<=	less than or equal to—e.g., `cost <= 20`
>=	greater than or equal to—e.g., cost >= `20`
	Other Comparison Operators
[NOT] BETWEEN x AND y	Allows user to express a range—e.g., searching for numbers BETWEEN 5 and 10. The optional NOT would be used when searching for numbers that are `NOT BETWEEN 5 AND 10`.
[NOT] IN(x,y,...)	Similar to the OR logical operator. Can search for records which meet at least one condition contained within the parentheses—e.g., `Pubid IN (1, 4, 5)`, only books with a publisher id of 1, 4, or 5 will be returned. The optional NOT keyword instructs Oracle to return books not published by Publisher 1, 4, or 5.
[NOT] LIKE	Can be used when searching for patterns if you are not certain how something is spelled—e.g. title `LIKE 'TH%'`. Using the optional NOT indicates that records that do contain the specified pattern should not be included in the results.
IS [NOT] NULL	Allows user to search for records which do not have an entry in the specified field—e.g., `Shipdate IS NULL.` If you include the optional NOT, it would find records that do have an entry in the field—e.g., `Shipdate IS NOT NULL.`

Figure 3-10 Comparison operators

In contrast to the "greater than" (>) operator that only returns rows with a value higher than the value in the stated condition, the "less than" (<) operator only returns values that are less than the stated condition. For example, the management of JustLee Books would like a list of all books that have a profit of less than 20 percent of the book's cost. Because the profit for a book would be determined by subtracting the cost from the retail price of the book, this calculated value can be compared against the cost of the book multiplied by .20, or the decimal version of 20 percent. As shown in Figure 3–11, there is only one book that generates less than a 20 percent profit margin.

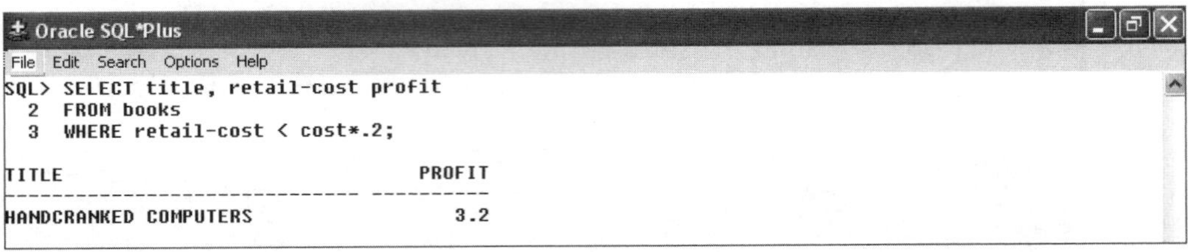

Figure 3-11 Search using the "less than" operator

The "greater than" and "less than" operators will not include values that exactly match the given condition. For example, if you had wanted the results in Figure 3–11 to also include books that return a 20 percent profit, then the comparison operator would have to be changed to the "less than or equal to" operator (<=). Using the <= operator, any book that returned exactly 20 percent profit would also have been shown in the results.

For example, suppose the Marketing Department is sorting paper files and requests a list of all customers who live in Georgia or in a state that alphabetically appears before the state of Georgia (i.e., A through GA). The simplest way to identify those customers would be to search for all customers using the condition **state <= 'GA'**, as shown in Figure 3–12.

Oracle SQL*Plus

File Edit Search Options Help

```
SQL> SELECT firstname, lastname, state
  2  FROM customers
  3  WHERE state <= 'GA';

FIRSTNAME  LASTNAME   ST
---------- ---------- --
BONITA     MORALES    FL
RYAN       THOMPSON   CA
LEILA      SMITH      FL
JORGE      PEREZ      CA
JAKE       LUCAS      GA
NICHOLAS   NGUYEN     FL
STEVE      SCHELL     FL
MICHELL    DAUM       CA
GREG       MONTIASA   GA

9 rows selected.

SQL>
```

Figure 3-12 Search using the "less than or equal to" operator

Later, the Marketing Department requests that you identify all customers who live in Georgia or in a state that has a state abbreviation "greater than" Georgia's abbreviation of GA. Although you may think this is a little unusual because the previous list already included customers living in Georgia, you nevertheless create the list using the condition **`state >= 'GA'`** and obtain the names of 13 customers who meet that condition, as shown in Figure 3–13.


```
Oracle SQL*Plus
File Edit Search Options Help
SQL> SELECT firstname, lastname, state
  2  FROM customers
  3  WHERE state >= 'GA';

FIRSTNAME  LASTNAME   ST
---------- ---------- --
THOMAS     PIERSON    ID
CINDY      GIRARD     WA
MESHIA     CRUZ       NY
TAMMY      GIANA      TX
KENNETH    JONES      WY
JAKE       LUCAS      GA
REESE      MCGOVERN   IL
WILLIAM    MCKENZIE   MA
JASMINE    LEE        WY
BECCA      NELSON     MI
GREG       MONTIASA   GA
JENNIFER   SMITH      NJ
KENNETH    FALAH      NJ

13 rows selected.

SQL>
```

Figure 3-13 Search using the "greater than or equal to" operator

As you anticipated, the Marketing Department later calls and requests a list of all customers who *do not* live in the state of Georgia, so they can sort out the overlap between the two previous listings. Although it would seem more reasonable to use a list of customers who *do* live in the state of Georgia to see which customers are appearing on both lists, you perform the task requested and use the condition **`state <> 'GA'`** to generate the requested list, as shown in Figure 3–14.

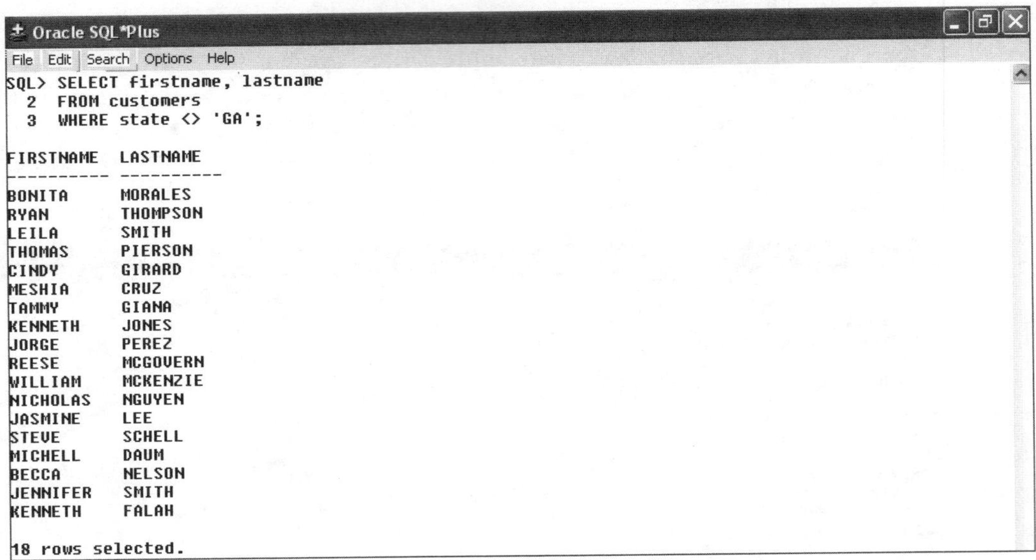

```
SQL> SELECT firstname, lastname
  2  FROM customers
  3  WHERE state <> 'GA';

FIRSTNAME  LASTNAME
---------- ----------
BONITA     MORALES
RYAN       THOMPSON
LEILA      SMITH
THOMAS     PIERSON
CINDY      GIRARD
MESHIA     CRUZ
TAMMY      GIANA
KENNETH    JONES
JORGE      PEREZ
REESE      MCGOVERN
WILLIAM    MCKENZIE
NICHOLAS   NGUYEN
JASMINE    LEE
STEVE      SCHELL
MICHELL    DAUM
BECCA      NELSON
JENNIFER   SMITH
KENNETH    FALAH

18 rows selected.
```

Figure 3-14 Search using the "not equal to" operator

Using != or ^= in the query shown in Figure 3–14 would have returned the same results as using <> for the "not equal to" operator.

BETWEEN...AND Operator

Next, let's look at how the BETWEEN...AND comparison operator works. The **BETWEEN...AND** comparison operator is used when searching a field for values that fall within a specific range. Figure 3–15 on the next page shows a query to find any book whose publisher has an assigned ID between 1 and 3. Notice in the results that the range is inclusive and includes any publisher with the ID of 1, 2, or 3.

IN Operator

The **IN** operator returns records that match one of the values given in the listed values. Oracle9*i* *syntax requires the items listed to be separated by commas, and the entire list must be enclosed in parentheses.* The output of the query in Figure 3–16 shows that there are seven books currently in inventory that were published by Publisher 1, 2, or 5.

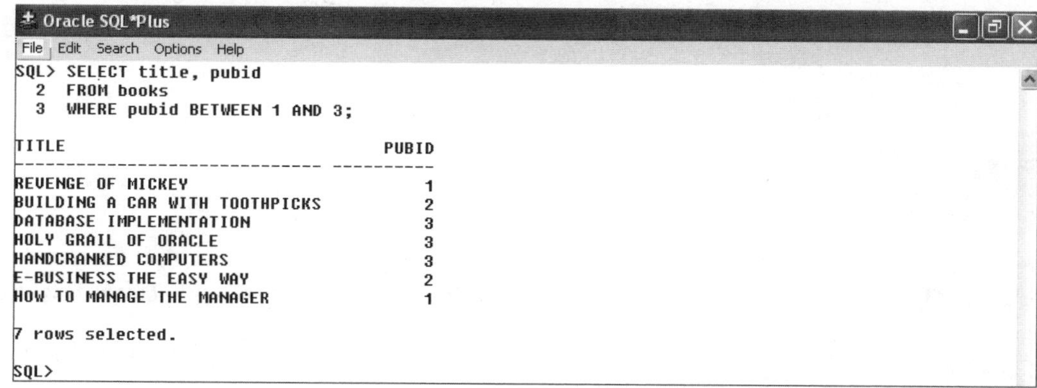

```
Oracle SQL*Plus
File Edit Search Options Help
SQL> SELECT title, pubid
  2  FROM books
  3  WHERE pubid BETWEEN 1 AND 3;

TITLE                               PUBID
------------------------------ ----------
REVENGE OF MICKEY                       1
BUILDING A CAR WITH TOOTHPICKS          2
DATABASE IMPLEMENTATION                 3
HOLY GRAIL OF ORACLE                    3
HANDCRANKED COMPUTERS                   3
E-BUSINESS THE EASY WAY                 2
HOW TO MANAGE THE MANAGER               1

7 rows selected.

SQL>
```

Figure 3-15 Search using the BETWEEN...AND operator

```
Oracle SQL*Plus
File Edit Search Options Help
SQL> SELECT title, pubid
  2  FROM books
  3  WHERE pubid IN (1, 2, 5);

TITLE                               PUBID
------------------------------ ----------
REVENGE OF MICKEY                       1
BUILDING A CAR WITH TOOTHPICKS          2
E-BUSINESS THE EASY WAY                 2
PAINLESS CHILD-REARING                  5
BIG BEAR AND LITTLE DOVE                5
HOW TO MANAGE THE MANAGER               1
SHORTEST POEMS                          5

7 rows selected.

SQL>
```

Figure 3-16 Search using the IN operator

LIKE Operator

By contrast, the **LIKE** operator is used with **wildcard characters** to search for patterns. Wildcard characters can be used to represent one or more alphanumeric characters. The wildcard characters available for pattern searches in Oracle9*i* are the percent sign (%) and the underscore symbol (_). The percent sign is used to represent *any number of characters*; the underscore symbol represents exactly *one character*. For example, if you were trying to find any customer whose last name started with P and did not care about the remaining letters of the last name, you could enter the SQL statement shown in Figure 3–17.

```
SQL> SELECT lastname
  2  FROM customers
  3  WHERE lastname LIKE 'P%';

LASTNAME
----------
PIERSON
PEREZ

SQL>
```

Figure 3-17 Search using the LIKE operator with the percent sign wildcard character

The results include two customers whose last names begin with a P. If, however, you had been searching for customers whose last names contain a P in any position, you would have changed the search pattern to '%P%'. The software would interpret the pattern to mean, "I don't really care what is before the letter P or what comes after, just that a P is somewhere in the Lastname column."

Suppose that you are having difficulty reading the printout of an order for a customer because someone spilled coffee on the printed order form. You can tell that the first two digits of the Customer# are a "1" and a "0," and the last digit is "9." However, you can't read the third number. In this case, you could use an underscore character to represent the missing digit, as shown in Figure 3–18.

```
SQL> SELECT customer#, lastname, firstname
  2  FROM customers
  3  WHERE customer# LIKE '10_9';

 CUSTOMER# LASTNAME   FIRSTNAME
---------- ---------- ----------
      1009 PEREZ      JORGE
      1019 SMITH      JENNIFER

SQL>
```

Figure 3-18 Search using the LIKE operator with the underscore sign wildcard character

Oracle9*i* interprets the search condition in Figure 3–18 as, "Look for any customer number that begins with a '1' and a '0', is followed by any character, and then ends with a '9'." The results return two customers, 1009 and 1019.

The percent sign and underscore symbol can also be combined in the same search condition to create more complex search patterns. For illustrative purposes, suppose that you need to identify every book ISBN that has the numeral 4 as its second numeral and ends with a 0. The actual pattern you are trying to identify can be stated as '_4%0' because you know something comes before the number 4 and that there will be additional numbers after the 4; you don't care what the numbers are or how many there are—as long as the last digit is 0. As shown in Figure 3–19, this search pattern will identify two books from the BOOKS table.

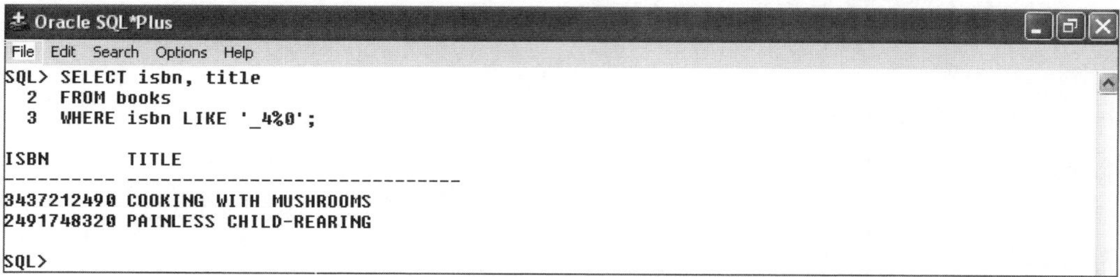

Figure 3-19 Search using the LIKE operator with both the percent sign and underscore sign wildcard characters

Although NULL and IS NOT NULL are comparison operators, they specifically address searches based on a column having or not having a NULL value, respectively. Therefore, the discussion of these two operators will be presented in a later section of this chapter.

LOGICAL OPERATORS

There may be times when you need to search for records based on two or more conditions. Then, you can use **logical operators** to combine search conditions. The logical operators **AND** and **OR** are commonly used to combine search conditions. (The **NOT** operator, mentioned in Figure 3–10, is also a logical operator available in Oracle9*i*, but it is used to reverse the meaning of search conditions, rather than to combine them.) When queried with a WHERE clause, each record in the table is compared to the stated condition. If the condition is TRUE when compared to a record, the record is included in the results. When the AND operator is used in the WHERE clause, both conditions combined by the AND operator must be evaluated as being TRUE, or the record is not included in the results.

Figure 3–20 shows a query for titles of books published by Publisher 3 *and* that are in the Category column and are in *Computer* Category. Because the search is for books meeting both conditions, the conditions are combined with the AND operator.

Remember that all data for JustLee Books are stored in the tables in upper-case letters because the data were entered in upper-case letters when the tables were originally created. If no rows are returned, make certain COMPUTER was typed in all-capital letters.

On the other hand, if you wanted a list of books that either were published by Publisher 3 *or* are in the Computer Category, you could use the OR operator, as shown in Figure 3–21.

```
Oracle SQL*Plus
File Edit Search Options Help
SQL> SELECT title, pubid, category
  2  FROM books
  3  WHERE pubid = 3
  4  AND category = 'COMPUTER';

TITLE                               PUBID CATEGORY
------------------------------ ---------- ------------
DATABASE IMPLEMENTATION                 3 COMPUTER
HOLY GRAIL OF ORACLE                    3 COMPUTER
HANDCRANKED COMPUTERS                   3 COMPUTER

SQL>
```

Figure 3-20 Search using the AND logical operator

```
Oracle SQL*Plus
File Edit Search Options Help
SQL> SELECT title, pubid, category
  2  FROM books
  3  WHERE pubid = 3
  4  OR category = 'COMPUTER';

TITLE                               PUBID CATEGORY
------------------------------ ---------- ------------
DATABASE IMPLEMENTATION                 3 COMPUTER
HOLY GRAIL OF ORACLE                    3 COMPUTER
HANDCRANKED COMPUTERS                   3 COMPUTER
E-BUSINESS THE EASY WAY                 2 COMPUTER

SQL>
```

Figure 3-21 Search using the OR logical operator

With the OR operator, only one of the conditions must be evaluated as TRUE to have the record included in the results. In Figure 3–21, the first three records met both conditions. However, the last record only met the **`category='COMPUTER'`** condition; the **`pubid=3`** condition was evaluated as FALSE for the last record. Because the OR operator was used, only one condition had to be TRUE, so even though that book was not published by Publisher 3, it was included in the output.

Using a series of OR logical operators to join conditions that are based on the same column is identical to using the IN comparison operator. For example, `state = 'GA' OR state = 'CA'` is the same as `IN ('GA', 'CA')`.

Let's also consider the order of logical operators. Because the WHERE clause can contain multiple types of operators, you need to understand the order in which they are resolved.

- Arithmetic operations are solved first.
- Comparison operators (<, >, =, LIKE, etc.) are solved next.
- Logical operators have a lower precedence and are evaluated last—in the order of NOT, AND, and finally OR.

If you need to change the order of evaluation, simply use parentheses to indicate the operators to be resolved first.

Look at the results of the query shown in Figure 3–22. The list includes books that are published by Publisher 4 *and* that cost more than $15.00. The list also includes any book from the Family Life Category. Although the OR operator was actually listed first in the WHERE clause, Oracle9*i* first evaluated the Pubid and Cost conditions that are combined with the AND logical operator. Once that was solved, the Category condition that preceded the OR logical operator was then considered.


```
Oracle SQL*Plus
File Edit Search Options Help
SQL> SELECT * FROM books
  2  WHERE category = 'FAMILY LIFE'
  3  OR pubid = 4
  4  AND cost>15;

ISBN       TITLE                          PUBDATE        PUBID       COST     RETAIL CATEGORY
---------- ------------------------------ --------- ---------- ---------- ---------- ------------
1059831198 BODYBUILD IN 10 MINUTES A DAY  21-JAN-01          4      18.75      30.95 FITNESS
0401140733 REVENGE OF MICKEY              14-DEC-01          1       14.2         22 FAMILY LIFE
2491748320 PAINLESS CHILD-REARING         17-JUL-00          5         48      89.95 FAMILY LIFE
0299282519 THE WOK WAY TO COOK            11-SEP-00          4         19      28.75 COOKING
0132149871 HOW TO GET FASTER PIZZA        11-NOV-02          4      17.85      29.95 SELF HELP

SQL>
```

Figure 3-22 Search using the AND and OR logical operators

Suppose that after examining the results of the previous query, you realize the order in which the logical operators were evaluated did not yield the output you wanted—to find any book that cost more than $15.00 and is either published by Publisher 4 or is in the Family Life Category. To have Oracle9*i* evaluate the conditions in the desired order, you must use parentheses first to identify any book that is published by Publisher 4 or is categorized as Family Life. After the books meeting either the Category or Publisher condition are found, the Cost condition is then evaluated, and only those records that cost more than $15.00 are displayed. Notice how the query in Figure 3–23 returns results different from those previously shown in Figure 3–22.

```
Oracle SQL*Plus
File Edit Search Options Help
SQL>  SELECT * FROM books
  2   WHERE (category = 'FAMILY LIFE'
  3   OR pubid = 4)
  4   AND cost>15;

ISBN       TITLE                          PUBDATE        PUBID       COST     RETAIL CATEGORY
---------- ------------------------------ --------- ---------- ---------- ---------- ------------
1059831198 BODYBUILD IN 10 MINUTES A DAY  21-JAN-01          4      18.75      30.95 FITNESS
2491748320 PAINLESS CHILD-REARING         17-JUL-00          5         48      89.95 FAMILY LIFE
0299282519 THE WOK WAY TO COOK            11-SEP-00          4         19      28.75 COOKING
0132149871 HOW TO GET FASTER PIZZA        11-NOV-02          4      17.85      29.95 SELF HELP

SQL>
```

Figure 3-23 Search using parentheses to override the order of evaluation

TREATMENT OF NULL VALUES

When performing arithmetic operations or search conditions, NULL values can return unexpected results. A **NULL value** simply means that no value has been stored in that particular field. Do not confuse a NULL value with a blank space: *A NULL is the absence of data; a field containing a blank space does contain a value—a blank space.* When searching for NULL values, you cannot use the equal sign (=) because there is not a value available to use for comparison in the search condition. If you need to identify records containing a NULL value, you must use the **IS NULL** comparison operator.

For example, when an order is shipped to a customer, the date on which the order is shipped is entered into the ORDERS table. If a date does not appear in the Shipdate field, then the order has not yet shipped. To find any order that has not yet been sent to the customer, use the query shown in Figure 3–24.

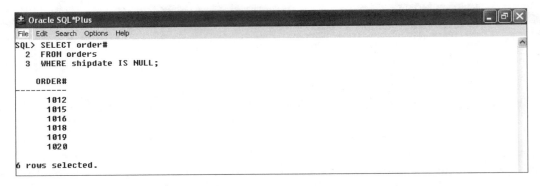

```
SQL> SELECT order#
  2  FROM orders
  3  WHERE shipdate IS NULL;

    ORDER#
----------
      1012
      1015
      1016
      1018
      1019
      1020

6 rows selected.
```

Figure 3-24 Search using IS NULL operator

As shown in Figure 3–24, there are currently six orders outstanding. Notice that when searching for a NULL value, you simply state the field to be searched followed by the words IS NULL in the WHERE clause. If you want a list of all orders that have shipped (i.e., the Shipdate column contains an entry), simply add the logical operator NOT. When searching for a field that is not NULL, you instruct Oracle9*i* to return any records with data available in the named field, as shown in Figure 3–25.

```
Oracle SQL*Plus
File Edit Search Options Help
SQL> SELECT order#
  2  FROM orders
  3  WHERE shipdate IS NOT NULL;

    ORDER#
----------
      1000
      1001
      1002
      1003
      1004
      1005
      1006
      1007
      1008
      1009
      1010
      1011
      1013
      1014
      1017

15 rows selected.

SQL>
```

Figure 3-25 Search using IS NOT NULL operator

ORDER BY CLAUSE SYNTAX

The **ORDER BY** clause is used for displaying the results of a query in a sorted order. The ORDER BY clause is listed at the end of the SELECT statement, as shown in Figure 3–26.

```
SELECT     [DISTINCT|UNIQUE] (*, column [ AS alias], …)
FROM       table
[WHERE     condition]
[GROUP BY  group_by_expression]
[HAVING    group_condition]
[ORDER BY  column]; ←
```

Figure 3-26 Syntax of the SELECT statement

To see a list of all publishers sorted by Name, enter the SQL statement shown in Figure 3–27.

In the results of the query, the second column (NAME) is listed in ascending order. Note these important points:

- When sorting in ascending order, values will be listed in this order:
  1. numeric values
  2. character values
  3. NULL values
- Unless you specify "desc" for descending, the ORDER BY clause sorts in ascending order by default.

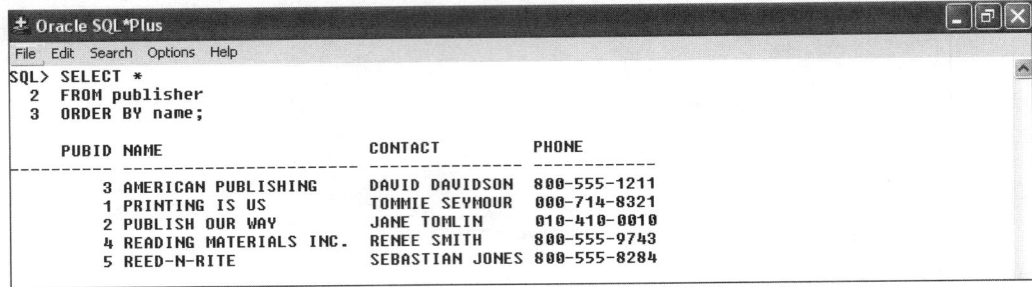

Figure 3-27 Results presented in ascending order by contents of Name column

To view the publishers in descending order by name, simply enter **desc** after the name of the column. After changing the sort order to descending, you will get the results shown in Figure 3–28.

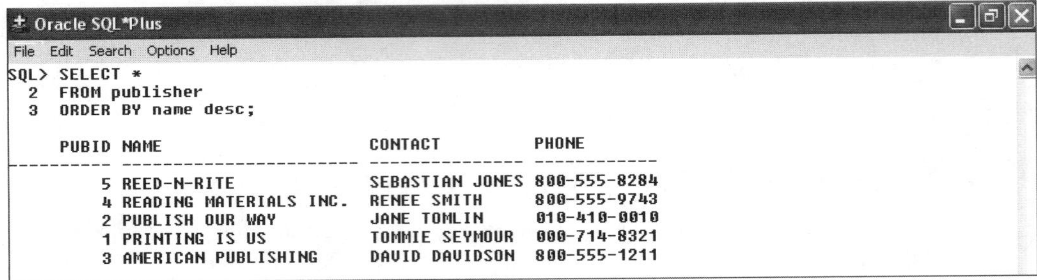

Figure 3-28 Results presented in descending order by contents of Name column

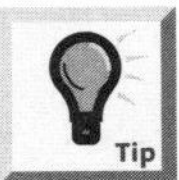

It is not uncommon in the "real world" to store commonly used queries in a file that can later be accessed by other employees to perform similar searches. If someone is reviewing your SQL commands and you wish to make clear that a column is to be sorted in ascending order, you can specify **asc** after the column name.

If a column alias is given to a field in the SELECT clause and that field is also used in the ORDER BY clause, you can use the same column alias in the ORDER BY clause—although this is not required. Note the example in Figure 3–29.

Remember, if the column alias contains a space, it must be enclosed in double quotation marks. This remains true even if the column alias is being used in the ORDER BY clause.

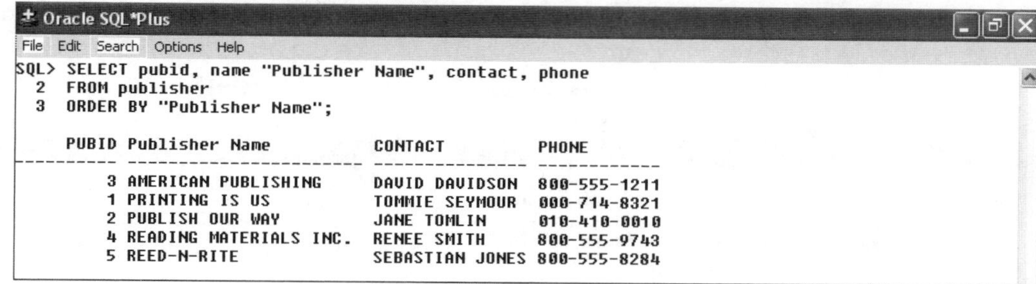

```
SQL> SELECT pubid, name "Publisher Name", contact, phone
  2  FROM publisher
  3  ORDER BY "Publisher Name";

     PUBID Publisher Name          CONTACT         PHONE
---------- ----------------------- --------------- ------------
         3 AMERICAN PUBLISHING     DAVID DAVIDSON  800-555-1211
         1 PRINTING IS US          TOMMIE SEYMOUR  000-714-8321
         2 PUBLISH OUR WAY         JANE TOMLIN     010-410-0010
         4 READING MATERIALS INC.  RENEE SMITH     800-555-9743
         5 REED-N-RITE             SEBASTIAN JONES 800-555-8284
```

Figure 3-29 Reference of a column alias in the ORDER BY clause

The ORDER BY clause can also be used with optional **NULLS FIRST** or **NULLS LAST** keywords to change the order for the listing of NULL values. By default, nulls are listed *last* when the results are sorted in ascending order, and *first* when they are sorted in descending order. The query in Figure 3–30 lists the last and first name of each customer and the customer number of the person who referred that customer to JustLee Books. The results are sorted in ascending order by the Referred column.

```
SQL> SELECT lastname, firstname, referred
  2  FROM customers
  3  ORDER BY referred;

LASTNAME   FIRSTNAME    REFERRED
---------- ---------- ----------
GIANA      TAMMY            1003
PEREZ      JORGE            1003
SMITH      JENNIFER         1003
NGUYEN     NICHOLAS         1006
DAUM       MICHELL          1010
MORALES    BONITA
THOMPSON   RYAN
SMITH      LEILA
LUCAS      JAKE
LEE        JASMINE
FALAH      KENNETH
MONTIASA   GREG
NELSON     BECCA
SCHELL     STEVE
MCKENZIE   WILLIAM
MCGOVERN   REESE
JONES      KENNETH
PIERSON    THOMAS
GIRARD     CINDY
CRUZ       MESHIA

20 rows selected.
```

Figure 3-30 Results presented in ascending order by Referred column

Suppose, however, that you would like the results sorted in ascending order, but you need to have the NULL values listed first. To override the placement of the NULL values, you can add NULLS FIRST in the ORDER BY clause; this instructs Oracle9*i* to place the NULL values at the beginning of the list, and sort the remaining records in ascending order, as shown in Figure 3–31. (If you had used the descending sequence, you would use NULLS LAST to override the order for the NULL values.)

```
Oracle SQL*Plus
File  Edit  Search  Options  Help
SQL> SELECT lastname, firstname, referred
  2  FROM customers
  3  ORDER BY referred NULLS FIRST;

LASTNAME   FIRSTNAME    REFERRED
---------- ---------- ----------
MORALES    BONITA
THOMPSON   RYAN
SMITH      LEILA
PIERSON    THOMAS
JONES      KENNETH
LUCAS      JAKE
MCKENZIE   WILLIAM
LEE        JASMINE
SCHELL     STEVE
FALAH      KENNETH
MONTIASA   GREG
NELSON     BECCA
MCGOVERN   REESE
GIRARD     CINDY
CRUZ       MESHIA
GIANA      TAMMY            1003
PEREZ      JORGE            1003
SMITH      JENNIFER         1003
NGUYEN     NICHOLAS         1006
DAUM       MICHELL          1010

20 rows selected.
```

Figure 3-31 Results presented in order of NULLS FIRST

Secondary Sort

When you only specify one column in the ORDER BY clause, this is referred to as a **primary sort**. In some cases, you may also want to include a secondary sort. A **secondary sort** provides an alternative field to use if an exact match occurs between two or more rows in the primary sort. For example, telephone books list residential customers alphabetically by last name. However, when two or more customers have the same last name, the customers are listed in alphabetical order by their first name. In other words, a primary sort is performed on last name and then, when necessary, a secondary sort is performed on first name.

The limit on the number of columns that can be used in the ORDER BY clause is 255.

For illustrative purposes, the query in Figure 3–32 shows that customers are to be listed in descending order by state. When there is more than one customer living in a particular state, customers are to be sorted by city—in ascending order.

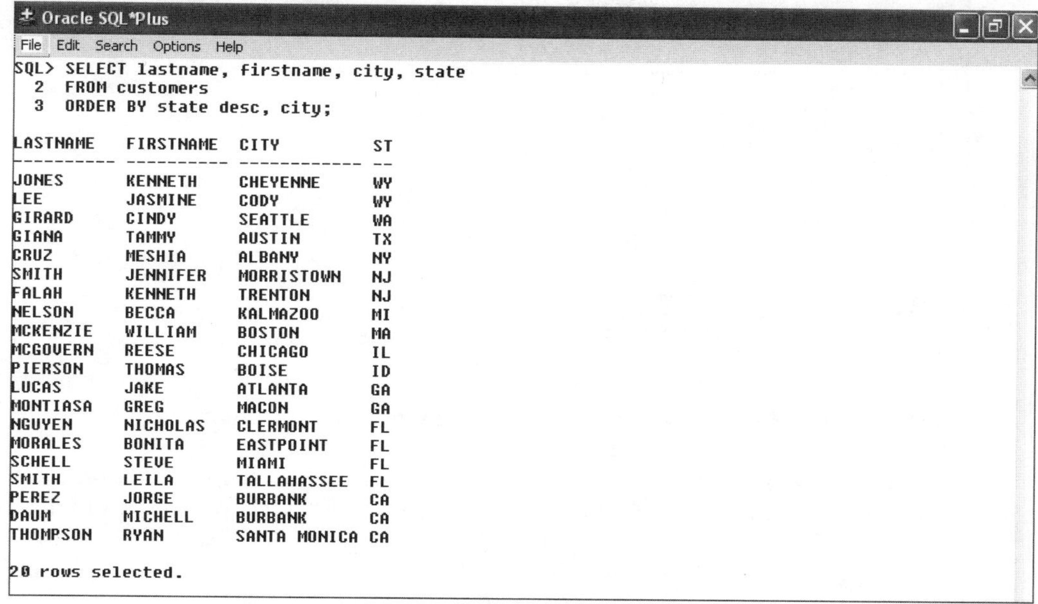

Oracle SQL*Plus
File Edit Search Options Help
SQL> SELECT lastname, firstname, city, state
2 FROM customers
3 ORDER BY state desc, city;
LASTNAME FIRSTNAME CITY ST
JONES KENNETH CHEYENNE WY
LEE JASMINE CODY WY
GIRARD CINDY SEATTLE WA
GIANA TAMMY AUSTIN TX
CRUZ MESHIA ALBANY NY
SMITH JENNIFER MORRISTOWN NJ
FALAH KENNETH TRENTON NJ
NELSON BECCA KALMAZOO MI
MCKENZIE WILLIAM BOSTON MA
MCGOVERN REESE CHICAGO IL
PIERSON THOMAS BOISE ID
LUCAS JAKE ATLANTA GA
MONTIASA GREG MACON GA
NGUYEN NICHOLAS CLERMONT FL
MORALES BONITA EASTPOINT FL
SCHELL STEVE MIAMI FL
SMITH LEILA TALLAHASSEE FL
PEREZ JORGE BURBANK CA
DAUM MICHELL BURBANK CA
THOMPSON RYAN SANTA MONICA CA
20 rows selected.

Figure 3-32 Results of using City column as a secondary sort

When looking at the results of the query, you can see that several states have multiple residents. Within those states, notice the customers are sorted in ascending order, according to the city in which they live, as instructed by the SQL statement. The sort order of descending applied only to the column after which it was listed. Since City did not reference a sort order, the default value of ascending was assumed.

In the examples shown, the column used for sorting was also listed in the SELECT clause. This is not a requirement. When necessary, you can reference a field in the ORDER BY clause that was not previously used in the SELECT clause. However, if the DISTINCT or UNIQUE keyword was used in the SELECT clause, you may **only** use those columns specifically listed in the SELECT clause for sorting.

Sorting by SELECT Order

The query statement in Figure 3–33 requests a list of customers who live in states CA and FL. The statement also specifies that the output should be listed with a primary sort on State and a secondary sort on City.

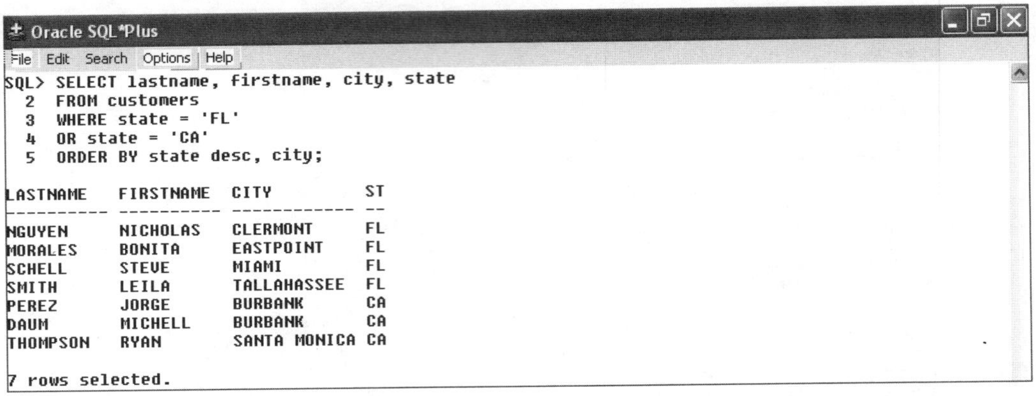

```
Oracle SQL*Plus
File Edit Search Options Help
SQL> SELECT lastname, firstname, city, state
  2  FROM customers
  3  WHERE state = 'FL'
  4  OR state = 'CA'
  5  ORDER BY state desc, city;

LASTNAME   FIRSTNAME  CITY         ST
---------- ---------- ------------ --
NGUYEN     NICHOLAS   CLERMONT     FL
MORALES    BONITA     EASTPOINT    FL
SCHELL     STEVE      MIAMI        FL
SMITH      LEILA      TALLAHASSEE  FL
PEREZ      JORGE      BURBANK      CA
DAUM       MICHELL    BURBANK      CA
THOMPSON   RYAN       SANTA MONICA CA

7 rows selected.
```

Figure 3-33 Results presented in order of State and City columns

Oracle9*i* also provides an abbreviated method for referencing the column to use for sorting if the field name is previously used in the SELECT clause. In the previous example, State and City were used in both the SELECT and ORDER BY clauses. Rather than relisting the field name in the ORDER BY clause, you can reference a field by its position in the column list of the SELECT clause. Since State is listed fourth in the SELECT clause and City is third, you can modify the SQL statement, as shown in Figure 3–34, and receive the same results.

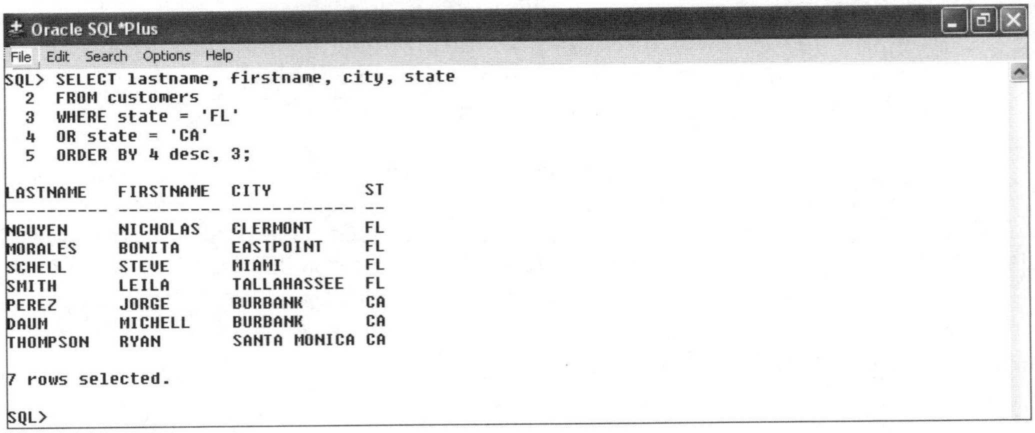

```
Oracle SQL*Plus
File Edit Search Options Help
SQL> SELECT lastname, firstname, city, state
  2  FROM customers
  3  WHERE state = 'FL'
  4  OR state = 'CA'
  5  ORDER BY 4 desc, 3;

LASTNAME   FIRSTNAME  CITY         ST
---------- ---------- ------------ --
NGUYEN     NICHOLAS   CLERMONT     FL
MORALES    BONITA     EASTPOINT    FL
SCHELL     STEVE      MIAMI        FL
SMITH      LEILA      TALLAHASSEE  FL
PEREZ      JORGE      BURBANK      CA
DAUM       MICHELL    BURBANK      CA
THOMPSON   RYAN       SANTA MONICA CA

7 rows selected.

SQL>
```

Figure 3-34 ORDER BY clause references the column positions in the SELECT clause

EDITING IN SQL*PLUS

By this time, you should have noticed that the SQL statements you are entering are getting longer. As the statements get longer and more complex, the chance of making a typing error increases. If you make a mistake, you can simply retype the statement and then execute the corrected version. However, other options also exist. One approach is to enter your statements into an editor. An **editor** is usually a word-processing program. The default editor for most Windows operating systems is Notepad. If you wish to use another program (or if no editor is defined), simply enter **DEFINE_EDITOR = NOTEPAD** (or another program name) at the SQL> prompt, and press the **Enter** key.

If you attempt to change the default editor and receive a permissions error message from the operating system, you may not be allowed to make changes to the system setting, and you should notify your instructor.

To enter your commands in the editor, use the following steps:

1. To access the editor, type **edit** at the SQL> prompt. Because you did not provide a file name, the contents of the buffer (the SQL statement you last executed) will appear in the editor using the file name **afiedt.buf**. This will allow you to use any features—such as cut and paste, insert, and delete—that are available in the editor to make changes to your SQL statement.
2. After you close the editor, the SQL statement that was in the editor when you exited will be displayed on the SQL*Plus screen.
3. To execute the statement, simply type a slash (/) at the SQL> prompt, and press the **Enter** key. The statement will then be executed.

If you prefer not to use an editor, you can use the SQL*Plus editing commands. **SQL*Plus** is a tool enabling you to interact with the database. Through SQL*Plus, you are able not only to enter SQL commands, but also to set or alter environmental variables (e.g., line size of the displays, length of a page of display), display the structure of your tables (DESCRIBE command), and execute interactive scripts.

The script used to build the tables for JustLee Books consists of a series of SQL statements. The script could have been modified to prompt the user to give a name to each table as it was created—hence making it interactive. Appendix B at the end of the text provides a command syntax guide. As you look at Appendix B, notice that the commands are categorized as either SQL or SQL*Plus. The commands that will be used for editing in this section are SQL*Plus commands; whereas the SELECT statement previously discussed is a SQL command.

To view what is currently in the SQL*Plus buffer, do the following:

1. Enter the letter **L**, the word **LIST**, or a semicolon (**;**) at the SQL> prompt.
2. To simply display the last line stored in the buffer, type **LIST LAST** and press the **Enter** key.
3. If you need to delete a line from the buffer, type **DEL** followed by the line number and press **Enter**.

4. You can add lines to the stored SQL statement by typing **INPUT** or the letter **I** and pressing **Enter**. You will then be able to add the remaining lines of text. To add text to the end of the current line in the buffer, enter **APPEND** or the letter **A** and the text to be added, and then press the **Enter** key.

You will experiment with the editing commands by entering the SQL statement shown in Figure 3–35.

```
SELECT title, cost
FROM books;
```

Figure 3-35 Simple SELECT statement

Suppose that after pressing the **Enter** key, you realize that you only wanted the books from Publisher 4. To revise the query you just entered, do the following:

1. Type **LIST** to redisplay the SQL statement you just entered. Notice the asterisk (*) next to line 2. This indicates that the current line is line 2.
2. Since you want to add another line to the SQL statement, type **INPUT** and press **Enter**. After pressing the **Enter** key, you will be able to enter additional lines to the statement immediately after the current line. Line 2 was the current line when you entered INPUT, so a **3** appeared after pressing **Enter**.
3. At this point, type **WHERE pubid=4;**. If you included the semicolon as indicated, the results shown in Figure 3–36 should be displayed.

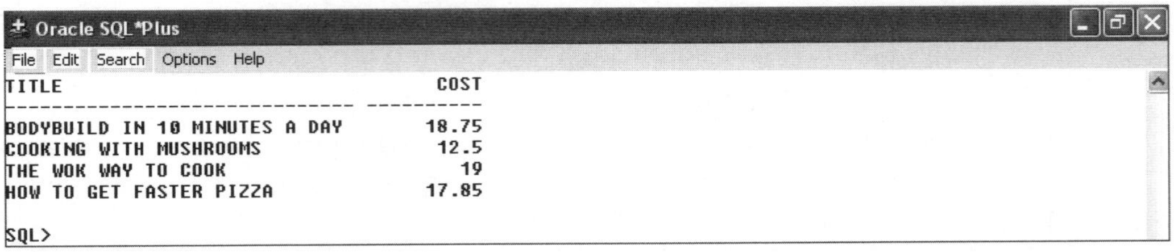

```
Oracle SQL*Plus
File Edit Search Options Help
TITLE                              COST
------------------------------ ----------
BODYBUILD IN 10 MINUTES A DAY     18.75
COOKING WITH MUSHROOMS             12.5
THE WOK WAY TO COOK                  19
HOW TO GET FASTER PIZZA           17.85

SQL>
```

Figure 3-36 Results of adding a WHERE clause to a SELECT statement

After looking over the results, you realize you forgot to include the retail price for each book. To include Retail in the display, you would need to alter the first line of the SQL statement you entered. Here's how to make that change:

1. If you type **LIST** to review the current statement in the buffer, you should notice that the asterisk is now indicating that line 3 is the current line. If you append the additional column name at this time, it would be appended to line 3.

2. To set line 1 as the current line, type a **1** at the SQL> prompt and press **Enter**. Line 1 will be redisplayed with an asterisk, indicating this is now the current line.
3. At the SQL> prompt, type **`A , retail`** and press **Enter**. This will add a comma and the word retail to the end of line 1. After pressing Enter, line 1 will be redisplayed with the additional text displayed.
4. If you type **`L`** or **`LIST`** and press **Enter**, the revised SQL statement shown in Figure 3–37 will be displayed.
5. To execute the revised SQL statement, enter a slash (/) or type **`RUN`**, and then press **Enter**.

```
Oracle SQL*Plus
File  Edit  Search  Options  Help
SQL> LIST
  1  SELECT title, cost, retail
  2  FROM books
  3* WHERE pubid = 4
SQL>
```

Figure 3-37 Display of revised SQL statement

Suppose that after examining the new output, you realize that the ISBN of each book, not its title, is needed. You can use the **CHANGE** command to perform a simple search-and-replace operation to change Title to ISBN in the first line of the SQL statement. The format for the **CHANGE** command is **`C\old\new\`** (the last backslash is optional). Here's how to make that change:

1. To make the correction to the first line, you need to make line 1 the current line by entering a **1** at the SQL> prompt and pressing **Enter**.
2. Once the first line is displayed, enter **`C\title\ISBN\`** and press **Enter**.
3. The revision for line 1 is displayed. If it is correct, then run the statement and examine the new output.

If you are trying to distinguish between SQL and SQL*Plus commands, keep two things in mind: (1) SQL keywords cannot be abbreviated like SQL*Plus commands, and (2) a semicolon is not needed to execute a SQL*Plus command.

CHAPTER SUMMARY

- The WHERE clause can be included in a SELECT statement to restrict the rows returned by a query to only those meeting a specified condition.
- When searching a non-numeric field, the search values should be enclosed in single quotation marks.

- Comparison operators are used to indicate how the record should relate to the search value.
- Mathematical comparison operators include =, >, <, >=, and <=; and the "not equal to" conditions <>, !=, and ^=.
- The BETWEEN...AND comparison operator is used to search for records that fall within a certain range of values.
- The IN comparison operator is used to identify a list of values that should be used for the search condition. A record must contain one of the values in the list to be included in the query results.
- The LIKE comparison operator is used with the percent and underscore symbols to establish search patterns.
- Logical operators such as AND and OR can be used to combine several search conditions.
- Logical operators are always evaluated in the order of NOT, AND, and finally OR. Parentheses can be used to override the order of evaluation.
- When using the AND operator, all conditions must be TRUE for a record to be returned in the results. However, with the OR operator, only one condition must be TRUE.
- A NULL value is the absence of data.
- The IS NULL comparison operator is used to match NULL values. The IS NOT NULL comparison operator is used to find records that do not contain NULL values in the indicated column.
- The results of queries can be sorted through the use of an ORDER BY clause. When used, the ORDER BY clause should be listed last in the SELECT statement.
- By default, records are sorted in ascending order. Entering DESC directly after the column name will sort the records in descending order.
- Multiple columns can be used for sorting by listing each column name to be used in a single ORDER BY clause separated by commas. The first column listed will be used as the primary sort. If an exact match occurs between two or more records, the next column listed will be used to determine the correct order, and so on. Columns can be specified by column name or by their position in the SELECT clause.
- A column does not have to be listed in the SELECT clause to serve as a basis for sorting.
- SQL*Plus is a tool that can be used to interact with a database. It has its own set of commands, including various commands that can be used to edit SQL statements.

Chapter 3 Syntax Summary

The following table presents a summary of the syntax that you have learned in this chapter. You can use the table as a study guide and reference.

Syntax Guide		
Element	**Description**	**Example**

WHERE clause	Specifies a search condition	`SELECT *`
		`FROM customers`
		`WHERE state = 'GA';`

ORDER BY clause	Specifies presentation order of the results	`SELECT *`
		`FROM publisher`
		`ORDER BY name;`

Mathematical Comparison Operators		
=	Equality operator—requires an exact match of the record data and the search value	`WHERE cost = 55.95`
>	"Greater than" operator—requires record to be greater than the search value	`WHERE cost > 55.95`
<	"Less than" operator—requires a record to be less than the search value	`WHERE cost < 55.95`
<>, !=, ^=	"Not equal to" operator—requires a record not to match the search value	`WHERE cost <> 55.95` or `WHERE cost != 55.95` or `WHERE cost ^= 55.95`
<=	"Less than or equal to" operator—requires a record to be less than or an exact match with the search value	`WHERE cost <= 55.95`
>=	"Greater than or equal to" operator—requires a record to be greater than or an exact match with the search value	`WHERE cost >= 55.95`
Other Comparison Operators		
[NOT] BETWEEN x AND y	Searches for records in a specified range of values	`WHERE cost BETWEEN 40 AND 65`
[NOT] IN(x,y,...)	Searches for records that match one of the items in the list	`WHERE cost IN(22, 55.95, 13.50)`

Syntax Guide		
Element	**Description**	**Example**
[NOT] LIKE	Searches for records that match a search pattern—used with wildcard characters	`WHERE lastname LIKE '_A%'`
IS [NOT] NULL	Searches for records with NULL value in the indicated column	`WHERE referred IS NULL`
Wildcard Characters		
%	Percent sign represents any number of characters	`WHERE lastname LIKE '%R%'`
_	Underscore represents exactly one character in the indicated position	`WHERE lastname LIKE '_A';`
Logical Operators		
AND	Combines two conditions together—record must match both conditions	`WHERE cost > 20 AND retail < 50`
OR	Requires a record to only match one of the search conditions	`WHERE cost > 20 OR retail < 50`
SQL*Plus Editing Commands		
A or APPEND	Adds the entered text to the end of the active line	`A WHERE retail < 42.89`
C *old**new*\\ or CHANGE *old**new*\\	Finds an indicated string of characters and replaces it with a new string of characters	`C \42.89\52.35\` or `CHANGE \42.89\52.35\`
DEL *n* or DELETE *n*	Deletes the indicated line number from the SQL statement	`DEL 3` or `DELETE 3`
I or INPUT	Allows new rows to be entered after the current active line	`I OR cost > 15` or `INPUT OR cost > 15`
L or LIST	Displays the current command in the SQL*Plus buffer	`L` or `LIST`
n	Sets the current active line in the SQL*Plus buffer	`2`
RUN	Executes the SQL statement currently stored in the SQL*Plus buffer	`RUN`
/	Executes the SQL statement currently stored in the	`/`

Review Questions

1. Which clause is used to restrict the number of rows returned from a query?
2. Which clause displays the results of a query in a specific sequence?
3. Which operator would be used to find any books with a retail price of at least $24.00?
4. Which operator is used to find NULL values?
5. The IN comparison operator is similar to which logical operator?
6. When should single quotation marks be used in a WHERE clause?
7. What is the effect of using the NOT operator in a WHERE clause?
8. When should a percent sign (%) be used with the LIKE operator?
9. When should an underscore symbol (_) be used with the LIKE operator?

Multiple Choice

To answer the following questions, consult Appendix A if necessary.

1. Which of the following SQL statements is not valid?

   a.
   ```
   SELECT address || city || state || zip 'Address'
   FROM customers
   WHERE lastname = 'SMITH';
   ```
   b.
   ```
   SELECT * FROM publisher ORDER BY contact;
   ```
   c.
   ```
   SELECT address, city, state, zip
   FROM customers
   WHERE lastname = "SMITH";
   ```
   d. All of the above are valid and return the expected results.

2. Which clause is used to restrict rows?

   a. SELECT

   b. FROM

   c. WHERE

   d. ORDER BY

3. Which of the following SQL statements is valid?

   a. `SELECT order# FROM orders WHERE shipdate = NULL;`

   b. `SELECT order# FROM orders WHERE shipdate = 'NULL';`

   c. `SELECT order# FROM orders WHERE shipdate = "NULL";`

   d. None of the statements are valid.

4. Which of the following will return a list of all customers' names, sorted in descending order by city within state?
   a. `SELECT name FROM customers ORDER BY desc state, city;`
   b. `SELECT firstname, lastname FROM customers`
      `SORT BY desc state, city;`
   c. `SELECT firstname, lastname FROM customers`
      `ORDER BY state desc, city;`
   d. `SELECT firstname, lastname FROM customers`
      `ORDER BY state desc, city desc;`
   e. `SELECT firstname, lastname FROM customers`
      `ORDER BY 5 desc, 6 desc;`
5. Which of the following will NOT return a customer with the last name of THOMPSON in its results?
   a. `SELECT lastname FROM customers`
      `WHERE lastname = "THOMPSON";`
   b. `SELECT * FROM customers;`
   c. `SELECT lastname FROM customers`
      `WHERE lastname > 'R';`
   d. `SELECT * FROM customers`
      `WHERE lastname <'V';`
6. Which of the following will display all books published by Publisher 1 and having a retail price of at least $25.00?
   a. `SELECT * FROM books`
      `WHERE pubid = 1 AND retail >= 25;`
   b. `SELECT * FROM books`
      `WHERE pubid = 1 AND retail > 25;`
   c. `SELECT * FROM books`
      `WHERE pubid = 1 AND WHERE retail > 25;`
   d. `SELECT * FROM books`
      `WHERE pubid=1, retail >=25;`
   e. `SELECT * FROM books`
      `WHERE pubid = 1, retail >= $25.00;`
7. What is the default sort sequence for the ORDER BY clause?
   a. ascending
   b. descending
   c. There is no default sort sequence.
   d. the order the records are stored in the table

8. Which of the following will NOT display books published by Publisher 2 and having a retail price of at least $35.00?

   a.
   ```
   SELECT * FROM books
   WHERE pubid = 2, retail >= $35.00;
   ```
   b.
   ```
   SELECT * FROM books
   WHERE pubid = 2 AND NOT retail< 35;
   ```
   c.
   ```
   SELECT * FROM books
   WHERE pubid IN (1, 2, 5) AND retail NOT BETWEEN 1 AND
   29.99;
   ```
   d. All of the above will display the specified books.

9. Which of the following will include a customer with the first name of BONITA in the results?

   a. `SELECT * FROM customers WHERE firstname = 'B%';`

   b. `SELECT * FROM customers WHERE firstname LIKE '%N%';`

   c. `SELECT * FROM customers WHERE firstname = '%N%';`

   d. `SELECT * FROM customers WHERE firstname LIKE '_B%';`

10. Which of the following characters or symbols is used to represent exactly one character during a pattern search?

    a. C

    b. ?

    c. _

    d. %

    e. none of the above

11. Which of the following will return the book titled *HANDCRANKED COMPUTERS* in the results?

    a. `SELECT * FROM books WHERE title = 'H_N_%';`

    b. `SELECT * FROM books WHERE title LIKE "H_N_C%";`

    c. `SELECT * FROM books WHERE title LIKE 'H_N_C%';`

    d. `SELECT * FROM books WHERE title LIKE '_H%';`

12. Which of the following clauses is used to present the results of a query in a sorted order?

    a. WHERE

    b. SELECT

    c. SORT

    d. ORDER

    e. none of the above

13. Which of the following SQL statements will return all books published after March 20, 1998?
    a. `SELECT * FROM books WHERE pubdate> 03–20–1998;`
    b. `SELECT * FROM books WHERE pubdate> '03–20–1998';`
    c. `SELECT * FROM books WHERE pubdate> '20–MAR–98';`
    d. `SELECT * FROM books WHERE pubdate> 'MAR–20–98';`
14. Which of the following will list all books published before June 2, 1999, *and* all books either published by Publisher 4 or in the Fitness category?
    a. ```
       SELECT * FROM books
       WHERE category = 'FITNESS' OR pubid = 4
       AND pubdate < '06–02–1999';
       ```
    b. ```
       SELECT * FROM books
       WHERE category = 'FITNESS' AND pubid = 4
       OR pubdate < '06–02–1999';
       ```
    c. ```
       SELECT * FROM books
       WHERE category = 'FITNESS' OR
       (pubid = 4 AND pubdate < '06–02–1999');
       ```
    d. ```
       SELECT * FROM books
       WHERE category = 'FITNESS'
       OR pubid = 4, pubdate < '06–02–99';
       ```
    e. none of the above
15. Which of the following will find all orders placed before April 5, 2003 but have not yet shipped?
    a. ```
       SELECT * FROM orders WHERE orderdate < '04–05–03'
       AND shipdate = NULL;
       ```
    b. ```
       SELECT * FROM orders WHERE orderdate < '05–04–03'
       AND shipdate IS NULL;
       ```
    c. ```
       SELECT * FROM orders WHERE orderdate < 'Apr–05–03'
       AND shipdate IS NULL;
       ```
    d. ```
       SELECT * FROM orders WHERE orderdate < '05–Apr–03'
       AND shipdate IS NULL;
       ```
    e. none of the above
16. Which of the following is used during pattern searches to represent any number of characters?
    a. *
    b. ?
    c. %
    d. _

17. Which of the following will list books generating at least $12.00 in profit?
    a. `SELECT * FROM books WHERE retail—cost > 12;`
    b. `SELECT * FROM books WHERE retail—cost < 12;`
    c. `SELECT * FROM books WHERE profit => 12;`
    d. `SELECT * FROM books WHERE retail-cost => 12.00;`
    e. none of the above

18. Which of the following will list each book having a profit of at least $10.00 in descending order by profit?
    a.
    ```
    SELECT * FROM books
    WHERE profit => 10.00
    ORDER BY "Profit" desc;
    ```
    b.
    ```
    SELECT title, retail-cost "Profit"
    FROM books
    WHERE profit => 10.00
    ORDER BY "Profit" desc;
    ```
    c.
    ```
    SELECT title, retail-cost "Profit"
    FROM books
    WHERE "Profit" => 10.00
    ORDER BY "Profit" desc;
    ```
    d.
    ```
    SELECT title, retail-cost profit
    FROM books
    WHERE retail-cost >= 10.00
    ORDER BY "PROFIT" desc;
    ```
    e.
    ```
    SELECT title, retail-cost "Profit"
    FROM books
    WHERE profit => 10.00
    ORDER BY 3 desc;
    ```

19. Which of the following will include the book HOW TO GET FASTER PIZZA in its results?
    a. `SELECT * FROM books WHERE title LIKE '%AS_E%';`
    b. `SELECT * FROM books WHERE title LIKE 'AS_E%';`
    c. `SELECT * FROM books WHERE title = '%AS_E%'`
    d. `SELECT * FROM books WHERE title = 'AS_E%';`

20. Which of the following will return all books published after March 20, 1998?
    a. `SELECT * FROM books WHERE pubdate> 03—20—1998;`
    b. `SELECT * FROM books WHERE pubdate> '03—20—1998';`
    c. `SELECT * FROM books WHERE pubdate NOT< '20—MAR—98';`
    d. `SELECT * FROM books WHERE pubdate NOT < 'MAR—20—98';`
    e. none of the above

Hands-on Assignments

To perform these activities, refer to the tables in Appendix A.

Give the SQL statements that determine the following:

1. Which customers live in New Jersey?
2. Which orders were shipped after April 1, 2003?
3. Which books are not in the Fitness Category?
4. Which customers live in either Georgia or New Jersey? Put the results in ascending order by last name.
5. Which orders were placed before April 2, 2003?
6. List all authors whose last name contains the letter pattern "IN". Put the results in order of last name, then first name.
7. List all customers who were referred to the bookstore by another customer.
8. Use a search pattern to list all books in the Children and Cooking Categories. Do not use any logical operators in the WHERE clause.
9. Use a search pattern to find any book where the title has an "A" for the second letter in the title, and an "N" for the fourth letter.
10. List the title of any computer book that was published in 2001.

A Case for Oracle9*i*

To perform these activities, refer to the tables in Appendix A.

During the course of an afternoon at work, you receive various requests for data stored in the database. As each request is completed, you decide to document the SQL statements used to obtain the data so each person will know how the results were obtained. The following are two of the requests that were made:

1. One of the managers at JustLee Books requests a list of the titles of all books that generate a profit of at least $10.00. It is preferred that the results be listed in descending order, based upon the profit returned by each book.
2. One of the customer service representatives is trying to identify all books that are either in the Computer or Family Life Category *and* were published by Publisher 1 or Publisher 3. However, the results should not include any book that sells for less than $45.00.

For each request, create a memo that identifies the SQL statement used to generate the needed data and provide the results of the query.

CHAPTER

4

JOINING MULTIPLE TABLES

Objectives

After completing this chapter, you should be able to do the following:

- Create a Cartesian join
- Create an equality join using the WHERE clause
- Create an equality join using the JOIN keyword
- Create a non-equality join using the WHERE clause
- Create a non-equality join using the JOIN...ON approach
- Create a self-join
- Distinguish an inner join from an outer join
- Create an outer join using the WHERE clause
- Create an outer join using the OUTER keyword
- Use set operators to combine the results of multiple queries
- Join three or more tables

The main advantage of using a relational database is that you can virtually eliminate data redundancy by structuring data in multiple tables. This chapter focuses on creating the access paths needed to combine or join data that are stored in more than one table.

Traditionally, people have used join conditions in the WHERE clause to specify how data are related. Beginning with Oracle9*i*, support for the JOIN keyword has been introduced for use in the FROM clause to enable you to explicitly create specific types of joins. The **JOIN** keyword instructs the software to join two or more tables. The specific type of join to be performed is indicated by combining the JOIN keyword with other keywords that will be discussed throughout this chapter. Inclusion of the JOIN keyword in the FROM clause permits the user to reserve the WHERE clause strictly for the purpose of restricting the rows being returned from the tables.

In this chapter, you'll examine several kinds of joins. For each join, you'll examine the syntax for creating the join, first using the traditional WHERE clause approach, then using the JOIN keyword. You will need to understand both approaches to creating joins to support legacy Oracle systems and to pass the Oracle9*i* SQL exam. Figure 4–1 provides an overview of this chapter's topics.

Element	Description
Cartesian Join Also known as a Cartesian product or cross join	Replicates each row from the first table with every row from the second table. Creates a join between tables by displaying every possible record combination. Can be created by two methods: (1) not including a joining condition in a WHERE clause (2) using the JOIN method with the CROSS JOIN keywords
Equality Join Also known as equijoin, inner join, or simple join	Creates a join through a commonly named and defined column. Can be created by two methods: (1) using the WHERE clause (2) using the JOIN method with the NATURAL JOIN or JOIN...ON or JOIN...USING keywords
Non-Equality Join	Joins tables when there are no equivalent rows in the tables to be joined, e.g., to match values in one column of a table with a range of values in another table. Can be created by two methods: (1) using the WHERE clause (2) using the JOIN method with the JOIN...ON keywords
Self-Join	Joins a table to itself. Can be created by two methods: (1) using the WHERE clause (2) using the JOIN method with the JOIN...ON keywords
Outer Join	Includes records of a table in output when there is no matching record in the other table. Can be created by two methods: (1) using the WHERE clause (2) using the JOIN method with the OUTER JOIN keywords, and also the keywords LEFT, RIGHT, or FULL
Outer Join Operator (+)	Used to indicate the table containing the deficient rows. The operator is placed next to the table that should have null rows added to create a match.
Set Operators	Used to combine results of multiple SELECT statements. Includes the keywords UNION, UNION ALL, INTERSECT, and MINUS.

Figure 4-1 Overview of chapter contents

Go to the JustLee Database folder in your Data Files. If you run the **Bookscript.sql** file, you will be able to work through all the queries shown in this chapter. Your output should match the output shown.

CARTESIAN JOINS

In a **Cartesian join**, also called a **Cartesian product** or **cross join**, each record in the first table is matched with each record in the second table. This type of join is useful when performing certain statistical procedures for data analysis. Thus, if you have three records in the first table and four in the second table, the first record from the first table would be matched with each of the four records in the second table. Then, the second record of the first table would be matched with each of the four records from the second table, and so on. You can always identify a Cartesian join because the results will display $m * n$ rows.

As shown in Figure 4–2, a Cartesian join of Table 1 (m) and Table 2 (n) would result in 12 records being displayed.

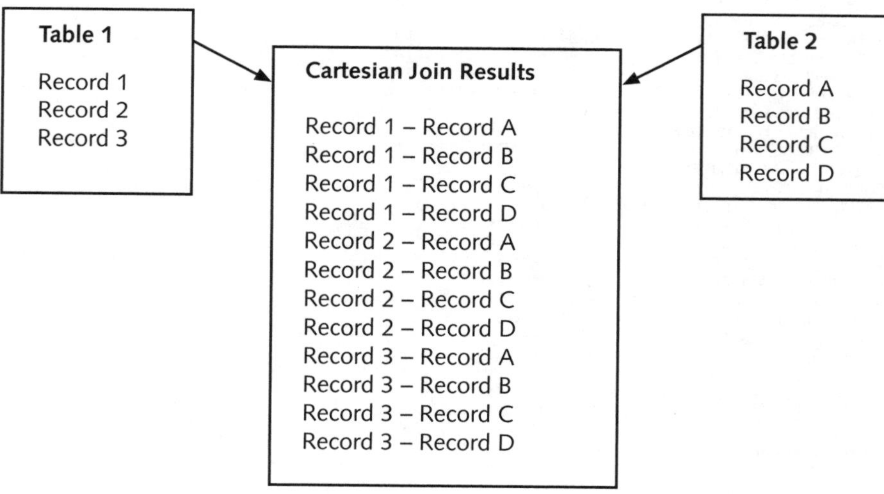

Figure 4-2 Results of a Cartesian join

Cartesian Join—Traditional Method

Suppose that you need to find the publisher's name for each book in inventory. The following SELECT statement, Figure 4–3, instructs Oracle9*i* to list the Title column, which is stored in the BOOKS table, and the Name column, which is stored in the PUBLISHER table.

```
SELECT title, name
FROM books, publisher;
```

Figure 4-3 Selecting columns from two tables

The results of the query in Figure 4–3 are shown in Figure 4–4. (Only the last screen of output is shown to conserve space.) Although there are only 14 book titles in the database, 70 records were returned! This should lead you to be suspicious of the nature of the output.

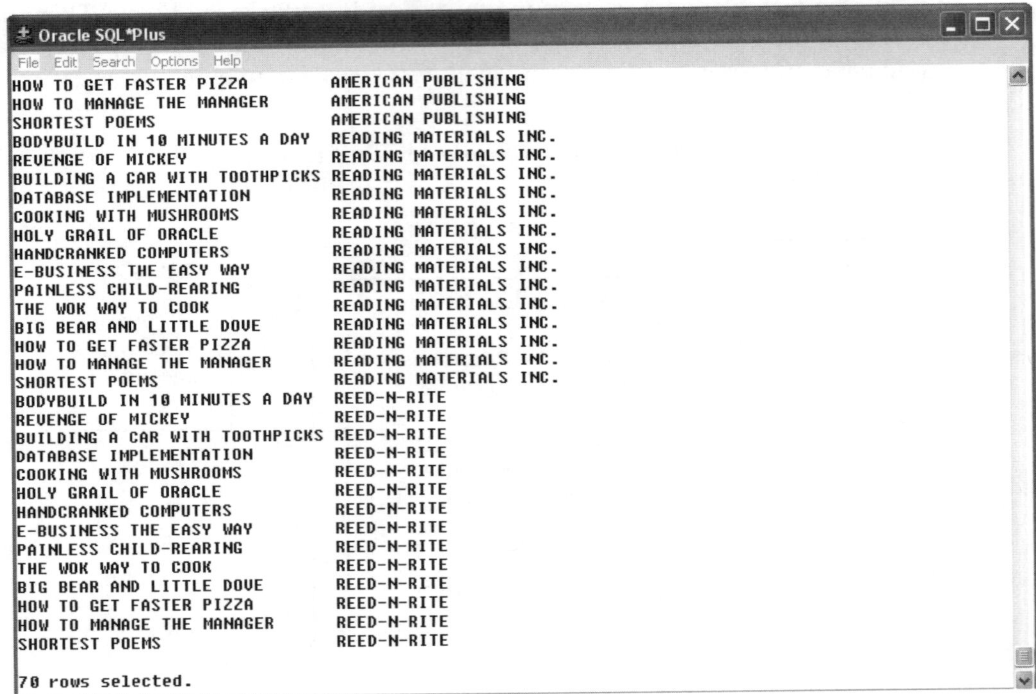

```
HOW TO GET FASTER PIZZA         AMERICAN PUBLISHING
HOW TO MANAGE THE MANAGER       AMERICAN PUBLISHING
SHORTEST POEMS                  AMERICAN PUBLISHING
BODYBUILD IN 10 MINUTES A DAY   READING MATERIALS INC.
REVENGE OF MICKEY               READING MATERIALS INC.
BUILDING A CAR WITH TOOTHPICKS  READING MATERIALS INC.
DATABASE IMPLEMENTATION         READING MATERIALS INC.
COOKING WITH MUSHROOMS          READING MATERIALS INC.
HOLY GRAIL OF ORACLE            READING MATERIALS INC.
HANDCRANKED COMPUTERS           READING MATERIALS INC.
E-BUSINESS THE EASY WAY         READING MATERIALS INC.
PAINLESS CHILD-REARING          READING MATERIALS INC.
THE WOK WAY TO COOK             READING MATERIALS INC.
BIG BEAR AND LITTLE DOVE        READING MATERIALS INC.
HOW TO GET FASTER PIZZA         READING MATERIALS INC.
HOW TO MANAGE THE MANAGER       READING MATERIALS INC.
SHORTEST POEMS                  READING MATERIALS INC.
BODYBUILD IN 10 MINUTES A DAY   REED-N-RITE
REVENGE OF MICKEY               REED-N-RITE
BUILDING A CAR WITH TOOTHPICKS  REED-N-RITE
DATABASE IMPLEMENTATION         REED-N-RITE
COOKING WITH MUSHROOMS          REED-N-RITE
HOLY GRAIL OF ORACLE            REED-N-RITE
HANDCRANKED COMPUTERS           REED-N-RITE
E-BUSINESS THE EASY WAY         REED-N-RITE
PAINLESS CHILD-REARING          REED-N-RITE
THE WOK WAY TO COOK             REED-N-RITE
BIG BEAR AND LITTLE DOVE        REED-N-RITE
HOW TO GET FASTER PIZZA         REED-N-RITE
HOW TO MANAGE THE MANAGER       REED-N-RITE
SHORTEST POEMS                  REED-N-RITE

70 rows selected.
```

Figure 4-4 Partial display of data from the joined tables

The problem with the SQL statement in Figure 4–3 is that you have specified the columns to be retrieved from the two tables, but not that the tables have a common field, which should be used to join the tables. The 70 rows that were returned are the result of a Cartesian join. With a Cartesian join, every row in the BOOKS table is returned for every individual row in the PUBLISHER table. Because the BOOKS table contains 14 records and the PUBLISHER table contains 5 records, you receive 14 * 5, or 70, records. If you ever *do* need to perform a Cartesian join, simply list the columns to be included in the results in the SELECT clause, and the tables containing the columns in the FROM clause. Then, because the software does not "know" how the tables are related, it will automatically replicate every possible combination of records.

Cartesian Join—JOIN Method

Beginning with Oracle9*i*, the **CROSS** keyword, combined with the JOIN keyword, can be used in the FROM clause to explicitly instruct Oracle9*i* to create a Cartesian, or cross, join. The CROSS keyword instructs Oracle9*i* to create cross products, using all the records of the tables listed. The first screen of that output is shown in Figure 4–5.

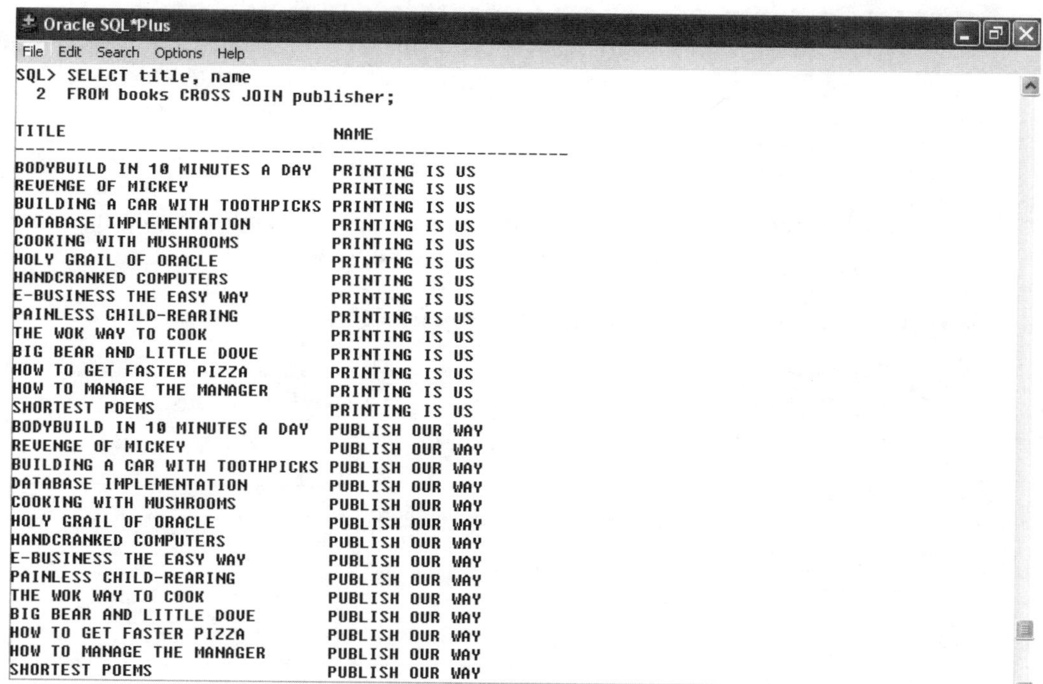

Figure 4-5 SELECT statement using the CROSS JOIN keywords (partial output shown)

Notice the syntax of the SQL statement given in Figure 4–5. In the FROM clause, the names of the tables to be used in the Cartesian join are separated by the **CROSS JOIN** keywords. Do not use commas to separate any parts of the FROM clause, as you would using the traditional method.

If you receive an error message, make certain there is *not* a comma entered after the BOOKS table name in the FROM clause.

EQUALITY JOINS

The query in Figure 4–3 returned a Cartesian join because the software did not know what data the two tables had in common. The most common type of join that you will use in the workplace is based upon two (or more) tables having equivalent data stored in a common column. Such joins are called **equality joins**. They may also be referred to as **equijoins**, **inner joins**, or **simple joins**.

A **common column** is a column with equivalent data that exists in two or more tables. For example, the BOOKS and PUBLISHER tables both have a common column, called Pubid, that contains an identification code assigned to each publisher. Thus, when you wanted a list of publishers for each book in the BOOKS table, you wanted to match the publisher ID stored in the record of each book in the BOOKS table with the corresponding publisher ID in the PUBLISHER table. The results *should* have only included the name of the publisher whenever there was a match between the Pubid columns stored in each table.

The following section will demonstrate how to create joins based upon a common column containing equivalent data stored in multiple tables.

Equality Joins—Traditional Method

The traditional method of avoiding the problem of a Cartesian join is to use the WHERE clause, introduced in Chapter 3. The WHERE clause is used to define the access path Oracle9*i* needs to correctly join tables.

```
SELECT       [DISTINCT|UNIQUE] (*, column [AS alias], …)
FROM         table
[WHERE       condition] ←
[GROUP BY    group_by_expression]
[HAVING      group_condition]
[ORDER BY    column];
```

Figure 4-6 Syntax for the SELECT statement

Using the same scenario given in the section on Cartesian joins, let's include the WHERE clause to retrieve the correct results, as shown in Figure 4–7.

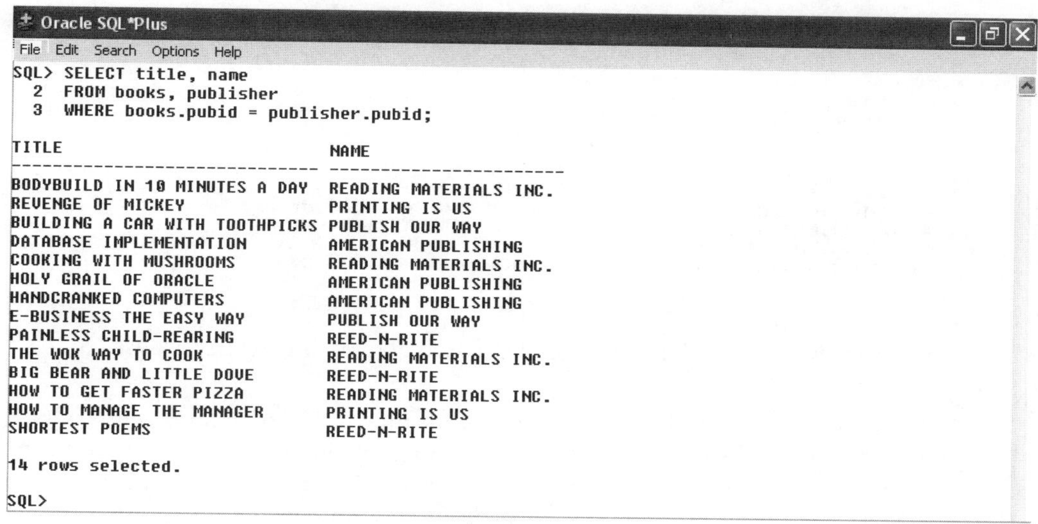

```
Oracle SQL*Plus
File Edit Search Options Help
SQL> SELECT title, name
  2  FROM books, publisher
  3  WHERE books.pubid = publisher.pubid;

TITLE                          NAME
------------------------------ -----------------------
BODYBUILD IN 10 MINUTES A DAY  READING MATERIALS INC.
REVENGE OF MICKEY              PRINTING IS US
BUILDING A CAR WITH TOOTHPICKS PUBLISH OUR WAY
DATABASE IMPLEMENTATION        AMERICAN PUBLISHING
COOKING WITH MUSHROOMS         READING MATERIALS INC.
HOLY GRAIL OF ORACLE           AMERICAN PUBLISHING
HANDCRANKED COMPUTERS          AMERICAN PUBLISHING
E-BUSINESS THE EASY WAY        PUBLISH OUR WAY
PAINLESS CHILD-REARING         REED-N-RITE
THE WOK WAY TO COOK            READING MATERIALS INC.
BIG BEAR AND LITTLE DOVE       REED-N-RITE
HOW TO GET FASTER PIZZA        READING MATERIALS INC.
HOW TO MANAGE THE MANAGER      PRINTING IS US
SHORTEST POEMS                 REED-N-RITE

14 rows selected.

SQL>
```

Figure 4-7 An equality join

The WHERE clause lets Oracle9*i* know that the BOOKS table and the PUBLISHER table are related by the Pubid column. The equal sign is used to specify that the contents of the Pubid column in each table must be exactly equal for the rows to be joined and returned in the results. Also notice that in the WHERE clause in Figure 4–7, the Pubid column names were prefixed with their corresponding table names. Any time Oracle9*i* references multiple tables having the same column name, the column name *must* be prefixed with the table name. You will receive an error message if your query is ambiguous and does not specify exactly which column is the common column. By entering **`publisher.pubid`**, you are specifying the Pubid column within the PUBLISHER table. This is known as "qualifying" the column name. A **column qualifier** indicates the table containing the column being referenced.

Suppose that you wanted some additional information—the publisher ID—to be included in the output. If the publisher ID had also been listed in the SELECT clause to be included in the query output, the table name prefix would be needed to avoid an ambiguity error message, as shown in Figure 4–8.

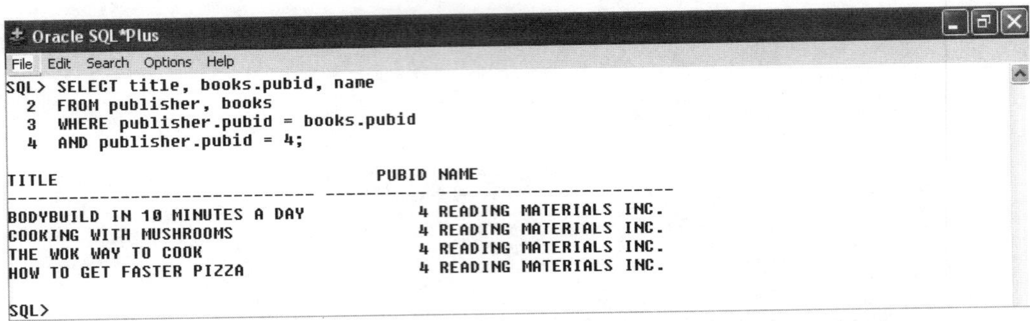


```
SQL> SELECT title, books.pubid, name
  2  FROM publisher, books
  3  WHERE publisher.pubid = books.pubid
  4  AND publisher.pubid = 4;

TITLE                              PUBID NAME
------------------------------ ---------- -----------------------
BODYBUILD IN 10 MINUTES A DAY          4 READING MATERIALS INC.
COOKING WITH MUSHROOMS                 4 READING MATERIALS INC.
THE WOK WAY TO COOK                    4 READING MATERIALS INC.
HOW TO GET FASTER PIZZA                4 READING MATERIALS INC.

SQL>
```

Figure 4-8 Using a table alias to avoid ambiguity in the SELECT clause

Figure 4–8 includes the publisher ID from the BOOKS table in the results. Notice the AND logical operator in the WHERE clause. The inclusion of this operator limits query results to only those from Publisher 4. Any of the search conditions used in Chapter 3 can still be issued in the WHERE clause when you are joining a table—without creating an error message. However, always check the logic of your search condition, or you may get unexpected results.

In Figure 4–9, the SELECT statement requests the title, publisher's ID number, and publisher's name for any book costing less than $15.00 or any book from Publisher 1.

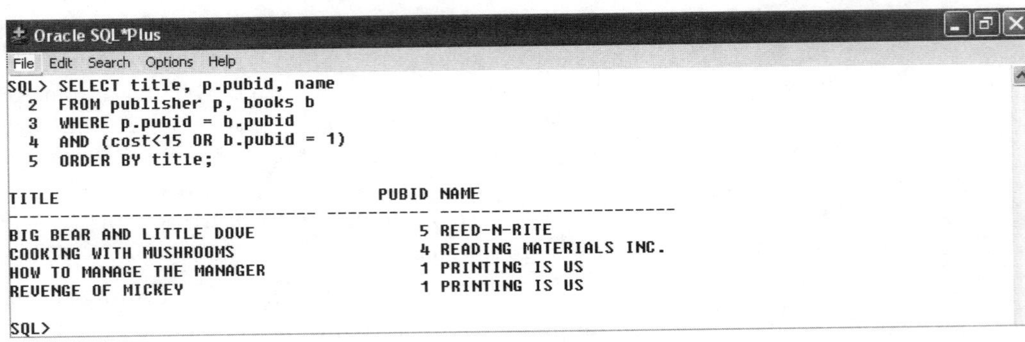


```
SQL> SELECT title, p.pubid, name
  2  FROM publisher p, books b
  3  WHERE p.pubid = b.pubid
  4  AND (cost<15 OR b.pubid = 1)
  5  ORDER BY title;

TITLE                              PUBID NAME
------------------------------ ---------- -----------------------
BIG BEAR AND LITTLE DOVE               5 REED-N-RITE
COOKING WITH MUSHROOMS                 4 READING MATERIALS INC.
HOW TO MANAGE THE MANAGER              1 PRINTING IS US
REVENGE OF MICKEY                      1 PRINTING IS US

SQL>
```

Figure 4-9 Equality join using a WHERE clause

Let's take a look at some of the elements in Figure 4–9.

- The SELECT clause not only lists the columns to be displayed, it also includes a **table alias** for the publisher (p). A period is used to separate a table alias from a column name. This alias was assigned in the FROM clause (and is discussed next).
- The table alias in the FROM clause works like a column alias by temporarily giving a different name to a table. You can use a table alias for various reasons, the most important being the reduction in memory requirements. You might

also use it to reduce the number of keystrokes needed when specifying a table throughout the SQL statement (although a table alias can have as many as 30 characters). There is one important rule you must remember when using a table alias: *If a table alias is assigned in the FROM clause, it must be used any time the table is referenced in that SQL statement.*

- The WHERE clause includes not only the access path to join the BOOKS and PUBLISHER tables, but also other search conditions using the AND and OR logical operators.
- The statement concludes with an ORDER BY clause to display the results in a sorted order.

Make certain to use the letter "p" or the letter "b" before the column name, or you will receive an error message.

Equality Joins—JOIN Method

There are three approaches you can use to create an equality join that uses the JOIN keyword: NATURAL JOIN, JOIN...USING, and JOIN...ON.

1. The **NATURAL JOIN** keywords will automatically create a join between two tables that have a commonly named and defined field.
2. Joins based on a column that has the same name and definition in both tables can be created with the **USING** clause.
3. When the tables to be joined in a USING clause do not have a commonly named and defined field, you must either add a WHERE clause (as with the traditional method) or add the **ON** clause to the JOIN keyword to specify how the tables are related.

The query in Figure 4–10 uses the NATURAL JOIN keywords to instruct Oracle9*i* to list the book title of each book in the BOOKS table—and the corresponding publisher ID number and publisher name.

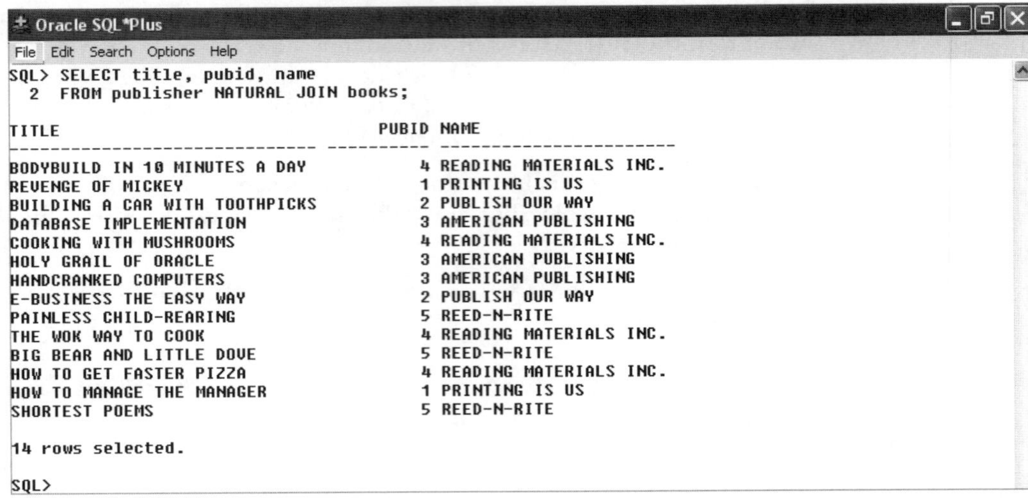

Figure 4-10 Creating a join using the NATURAL JOIN keywords

Because both the BOOKS and the PUBLISHER tables contained the Pubid column, this column is considered to be a common column and should be used to relate the two tables. When using the NATURAL JOIN keywords, you are not required to specify which column(s) the two tables have in common. The NATURAL keyword implies that the two specified tables have at least one column in common with the same name and contain the same datatype. Oracle9*i* will compare the two tables and use the common column to join the table.

Unlike the traditional method, you are not allowed to use a column qualifier for the column used to create the join. In essence, because the data value in a column is equivalent in both tables when the records match, it does not make sense to identify the column from only one of the tables. Thus, Oracle9*i* will return an error message if a qualifier is used anywhere in a SELECT statement that includes the NATURAL JOIN keywords, as shown in Figure 4–11.

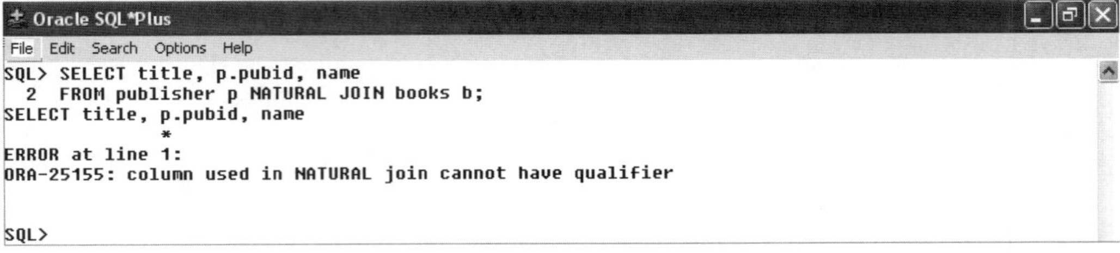

Figure 4-11 Error returned when column qualifier is used with the NATURAL JOIN keywords

There is another option available when joining two tables through a common field having the same column name in each table: You can include a USING clause immediately after the FROM clause. As with the NATURAL JOIN keywords, a column referenced by a USING clause cannot contain a column qualifier anywhere in the SELECT statement. In addition, the column referenced in the USING clause must be enclosed in parentheses.

Figure 4–12 reissues the query from Figure 4–10, using the JOIN...USING keywords. Notice that the query returns the same results as the NATURAL JOIN keywords.

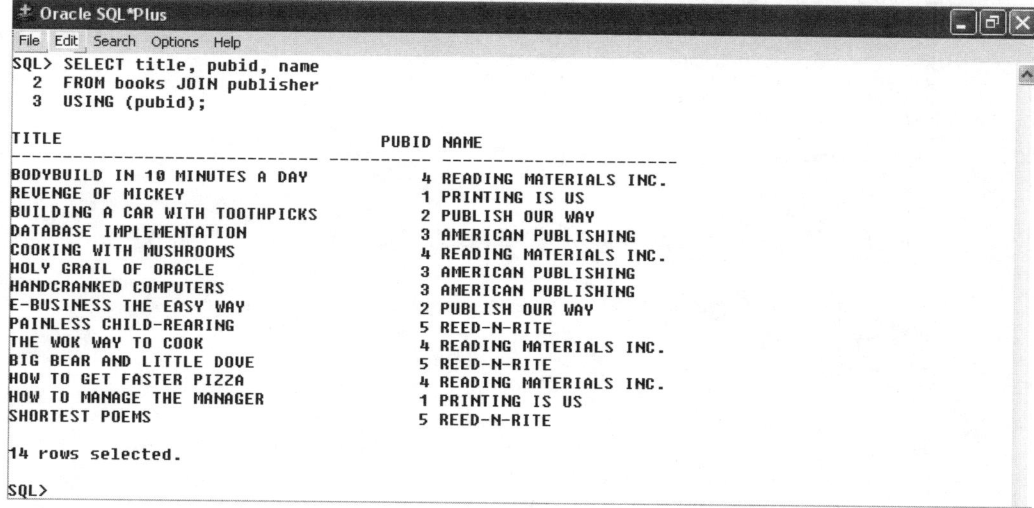

Figure 4-12 Joining the PUBLISHER and BOOKS tables with the JOIN...USING keywords

There may be instances in which you have created tables that have common columns, but you did not give columns the same columns names. When related columns have different names, the software will not "know" how the tables are related, and you will receive an error message. When there are no commonly named columns to use in a join, use the JOIN keyword in the FROM clause, and simply add an ON clause immediately after the FROM clause to specify which fields are related. Because the tables do not contain related columns with different names, Figure 4–13 uses the same scenario from previous examples—listing the Title and publisher Name columns from the BOOKS and PUBLISHER tables.

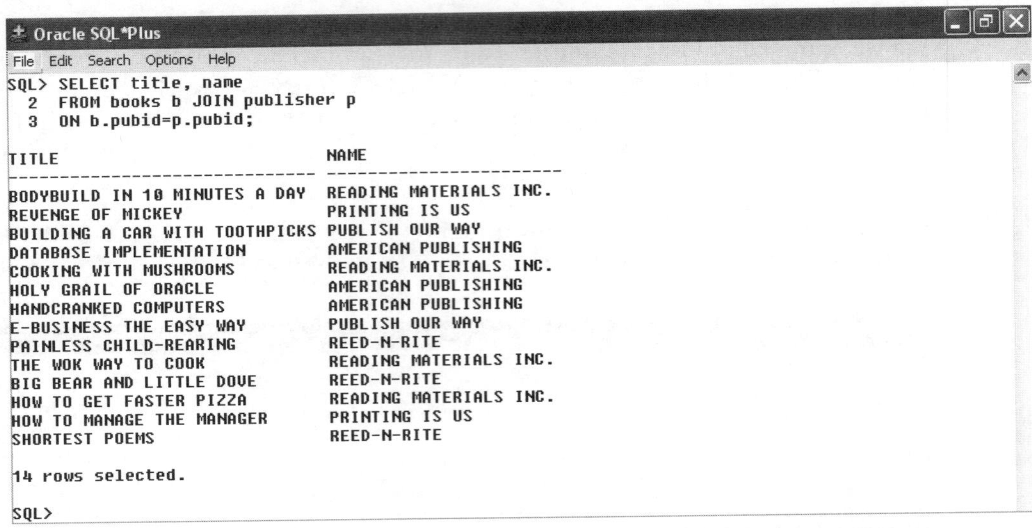

Figure 4-13 Joining the BOOKS and PUBLISHER tables using the JOIN...ON keywords

Don't forget to add the table alias before the column names in the ON clause, or you will receive an ambiguity error.

Because it is possible that ambiguity can exist when referencing columns with the JOIN...ON keywords, Oracle9*i* permits the use of column qualifiers to avoid ambiguity. Using the ON clause in a SELECT statement provides you with the freedom of using the WHERE clause exclusively for restricting the rows to be included in the results. This can improve the readability of complex SELECT statements for you and for individuals who are unfamiliar with the traditional method of joining tables. Figure 4–14 shows how to use the JOIN...ON approach to return the title, publisher ID (Pubid), and publisher name for all books published by Publisher 4.

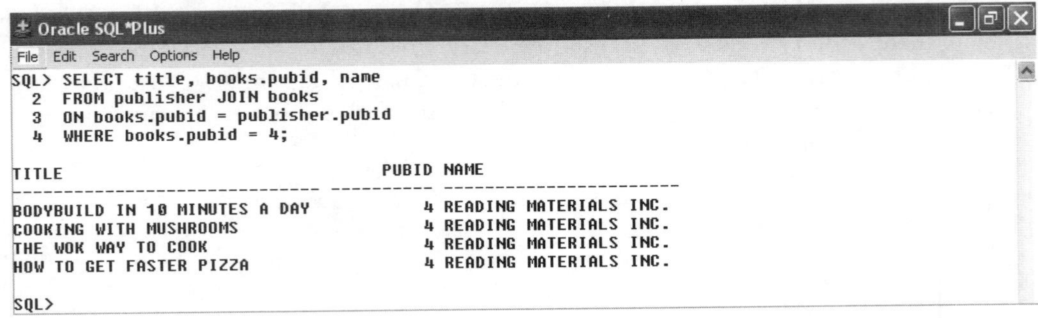

Figure 4-14 Using the WHERE clause to restrict rows returned from joined tables

There are two main differences between using the USING and ON clauses with the JOIN keyword.

1. The USING clause can *only* be used if the tables being joined have a common column with the same name. This is not a requirement for the ON clause.
2. A condition is specified in the ON clause; this is not allowed in the USING clause. The USING clause can only contain the name of the common column.

NON-EQUALITY JOINS

With an equality join, the data value of a record stored in the common column for the first table must match the data value in the second table. However, there are many cases in which there will be no exact match. A **non-equality join** is used when the related columns cannot be joined through the use of an equal sign—i.e., there are no equivalent rows in the tables to be joined. For example, the shipping fee charged by many freight companies is based on the weight of the item being shipped. To use a database table to determine shipping fees, you could store every possible weight and its corresponding fee in a table. Then, whenever an item is shipped, you could use an equality join to match the weight of that particular item to the equivalent weight stored in the table and find the correct fee. However, most shipping fees are based on a scale, or range, of weights. For example, an item weighing between three and five pounds might have one fee, whereas an item weighing between five and eight pounds might have another fee.

A non-equality join enables you to store the minimum value for a range in one column of a record, and the maximum value for the range in another column. Thus, instead of finding a column-to-column match, you can use a non-equality join to determine whether the item being shipped falls between minimum and maximum ranges in the columns. If the join does find a matching range for the item, the corresponding shipping fee can be returned in the results.

As with the traditional method of equality joins, a non-equality join can be performed in a WHERE clause. In addition, the JOIN keyword can be used with the ON clause to specify the relevant columns for the join. Let's first look at creating a non-equality join with the WHERE clause, and then with the JOIN...ON approach.

Non-Equality Joins—Traditional Method

Once a year, JustLee Books offers a week-long promotion in which customers receive a gift based on the value of each book purchased. If a customer purchases a book with a retail price of $12 or less, the customer receives a bookmark. If the retail price is more than $12 but less than or equal to $25, the customer receives a box of book-owner labels. For books retailing for more than $25 and less than or equal to $56, the customer is entitled to a free book cover. For books retailing for more than $56, the customer gets free shipping. Figure 4–15 shows the PROMOTION table.

```
SQL> SELECT *
  2  FROM promotion;

GIFT             MINRETAIL  MAXRETAIL
--------------- ---------- ----------
BOOKMARKER               0         12
BOOK LABELS          12.01         25
BOOK COVER           25.01         56
FREE SHIPPING        56.01     999.99

SQL>
```

Figure 4-15 Contents of the PROMOTION table

Because the rows between the BOOKS and PROMOTION tables do not contain equivalent values, you are required to use a non-equality join to determine which gift a customer will receive during the promotion, as shown in Figure 4–16.

```
SQL> SELECT title, gift
  2  FROM books, promotion
  3  WHERE retail
  4  BETWEEN minretail AND maxretail;

TITLE                          GIFT
------------------------------ ---------------
BIG BEAR AND LITTLE DOVE       BOOKMARKER
REVENGE OF MICKEY              BOOK LABELS
COOKING WITH MUSHROOMS         BOOK LABELS
HANDCRANKED COMPUTERS          BOOK LABELS
BODYBUILD IN 10 MINUTES A DAY  BOOK COVER
DATABASE IMPLEMENTATION        BOOK COVER
E-BUSINESS THE EASY WAY        BOOK COVER
THE WOK WAY TO COOK            BOOK COVER
HOW TO GET FASTER PIZZA        BOOK COVER
HOW TO MANAGE THE MANAGER      BOOK COVER
SHORTEST POEMS                 BOOK COVER
BUILDING A CAR WITH TOOTHPICKS FREE SHIPPING
HOLY GRAIL OF ORACLE           FREE SHIPPING
PAINLESS CHILD-REARING         FREE SHIPPING

14 rows selected.

SQL>
```

Figure 4-16 A traditional non-equality join with the BOOKS and PROMOTION tables

As shown in Figure 4–16, the BETWEEN operator is used in the WHERE clause to determine the range in which the retail price of the book falls. Then, based on the range set between the Minretail and Maxretail columns, the query determines which gift is appropriate for each purchase.

Note that when you use a non-equality join to determine where a value falls within a range, *you must make certain none of the values overlap*. If you select all the records from the PROMOTION table to see the values stored in each field, you will notice that the Minretail value in one row of the PROMOTION table does not equal the Maxretail value in another row. If any of the values did overlap, a customer could be returned twice in the results (and receive two gifts rather than one). Any results from a non-equality join should always be double-checked to make certain that rows are not unintentionally appearing more than once in the results.

Although the non-equality join presented in Figure 4-16 uses the BETWEEN operator, you can, theoretically, use any of the comparison operators for a non-equality join, with the exception of the equal sign (=). Use of the equal sign would signal an equality join, which is not logical.

Non-Equality Joins—JOIN Method

A non-equality join using the JOIN keyword has the same syntax as an equality join with the JOIN keyword. The only difference is that an equal sign is not used to establish the relationship in the ON clause. Figure 4-17 displays the gift for each book using the JOIN keyword. Notice that the joining condition used in the ON clause is the same as the one used in the WHERE clause in Figure 4-16.

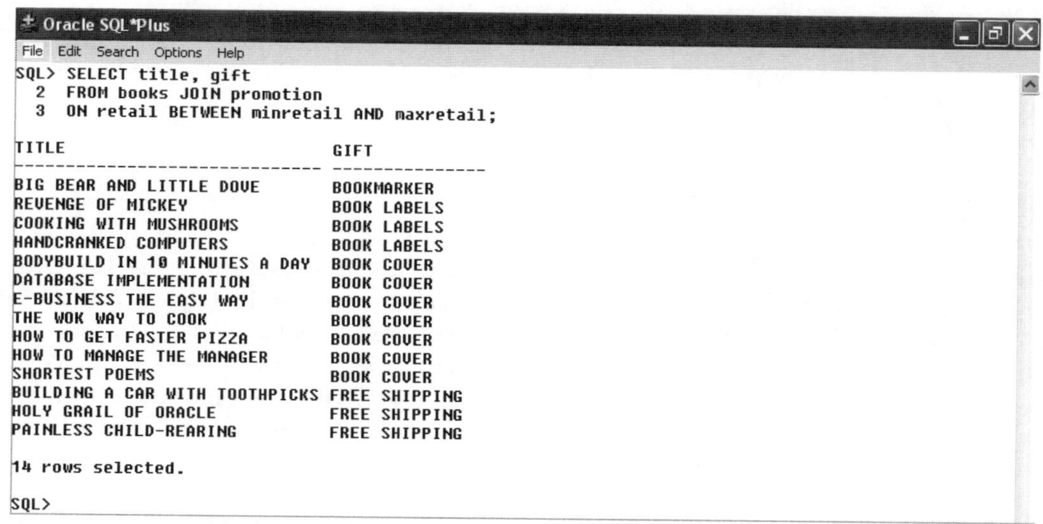

```
SQL> SELECT title, gift
  2  FROM books JOIN promotion
  3  ON retail BETWEEN minretail AND maxretail;

TITLE                          GIFT
------------------------------ ---------------
BIG BEAR AND LITTLE DOVE       BOOKMARKER
REVENGE OF MICKEY              BOOK LABELS
COOKING WITH MUSHROOMS         BOOK LABELS
HANDCRANKED COMPUTERS          BOOK LABELS
BODYBUILD IN 10 MINUTES A DAY  BOOK COVER
DATABASE IMPLEMENTATION        BOOK COVER
E-BUSINESS THE EASY WAY        BOOK COVER
THE WOK WAY TO COOK            BOOK COVER
HOW TO GET FASTER PIZZA        BOOK COVER
HOW TO MANAGE THE MANAGER      BOOK COVER
SHORTEST POEMS                 BOOK COVER
BUILDING A CAR WITH TOOTHPICKS FREE SHIPPING
HOLY GRAIL OF ORACLE           FREE SHIPPING
PAINLESS CHILD-REARING         FREE SHIPPING

14 rows selected.

SQL>
```

Figure 4-17 Using the JOIN...ON keywords to create a non-equality join

The NATURAL JOIN keywords can also be used to determine the gift a customer will receive with each purchase. To re-create the query using the NATURAL JOIN approach, you would simply add the keyword NATURAL to the JOIN keyword in the FROM clause, and change the ON clause to a WHERE clause. Although this query will work, using NATURAL JOIN is not a standard approach for formulating a non-equality join. The concept behind using the NATURAL JOIN keywords is that there is a common column between the tables being joined. Because you are still required to include a WHERE clause to identify the relationship between the two tables, the preferred strategy is to use either the traditional approach or the JOIN...ON method in the query.

SELF-JOINS

Sometimes, data in a table reference other data stored within the same table. For example, customers who refer a new customer to JustLee Books receive a discount certificate for a future purchase. The Referred column of the CUSTOMERS table stores the customer number of the individual who referred the new customer.

If you need to determine the name of the customer who referred another customer, you face a problem: The CUSTOMERS table serves as the master table for all customer information. Thus, the Referred column in the CUSTOMERS table actually relates to other rows within the same table. To retrieve all the information you need, you must join a table to itself. This is known as a **self-join**. You can create a self-join either with a WHERE clause or by using the JOIN keyword with the ON clause.

Keep this in mind: When you execute a query that contains a self-join, the software must search the same table twice, which can take a long time if the table is very large.

Self-Joins—Traditional Method

To perform a self-join, list the CUSTOMERS table twice in the FROM statement. However, to create the join, you must make it appear as if the query is referencing two different tables. Therefore, you must assign each listing of the CUSTOMERS table a different table alias. In Figure 4–18, the query uses the table alias "c" to identify the table containing the information for the new customer, and the table alias "r" to identify the table storing the individual who referred the new customer. Because the table aliases are different, Oracle9*i* is able to examine different records within the same table while executing the query.

```
Oracle SQL*Plus
File Edit Search Options Help
SQL> SELECT r.firstname, r.lastname, c.lastname referred
  2  FROM customers c, customers r
  3  WHERE c.referred = r.customer#;

FIRSTNAME  LASTNAME   REFERRED
---------- ---------- ----------
LEILA      SMITH      GIANA
LEILA      SMITH      PEREZ
MESHIA     CRUZ       NGUYEN
JAKE       LUCAS      DAUM
LEILA      SMITH      SMITH

SQL>
```

Figure 4-18 A self-join constructed in the WHERE clause

If an error message is returned, make certain the CUSTOMERS table is listed twice in the FROM clause, each with a different table alias. Also remember to precede each column with the table alias so there are no ambiguity errors.

Self-Joins—JOIN Method

Regardless of the method used, the concept behind a self-join is the same—to make it appear that two different tables are being joined through the use of table aliases. To demonstrate a self-join using the JOIN keyword, let's use the same circumstances as discussed in the previous section—determining which customers referred new customers to JustLee Books. The self-join query using the JOIN keyword is shown in Figure 4–19.

```
SQL> SELECT r.firstname, r.lastname, c.lastname referred
  2  FROM customers c JOIN customers r
  3  ON c.referred = r.customer#;

FIRSTNAME  LASTNAME   REFERRED
---------- ---------- ----------
LEILA      SMITH      GIANA
LEILA      SMITH      PEREZ
MESHIA     CRUZ       NGUYEN
JAKE       LUCAS      DAUM
LEILA      SMITH      SMITH

SQL>
```

Figure 4-19 A self-join using the JOIN...ON keywords

As previously discussed, the table is listed twice in the FROM clause, but each listing is given a different table alias to mimic using two different tables. The columns used to relate the two occurrences of the table are identified in the ON clause. Using the JOIN...ON approach to create a self-join will still allow you to place row restrictions in a WHERE clause, so anyone viewing the statement will know which portion of the statement is being used to join the tables, and which portion is actually limiting the number of rows being returned.

Outer Joins

When performing equality, non-equality, and self-joins, a row was only returned if there was a corresponding record in each table queried. These types of joins can be categorized as **inner joins** because records are only listed in the results if a match is found in each table. In fact, the default **INNER** keyword can be included with the JOIN keyword to specify that only records having a matching row in the corresponding table should be returned in the results. However, suppose that you wanted a list of *all* customers *and* the order number(s) of any order the customer has recently placed? (Recall that the CUSTOMERS table lists all customers who have ever placed an order, but the

ORDERS table lists just the current month's orders—or unfilled orders from previous months.) An inner join might not give you the exact results you desire, because some customers might not have placed a recent order.

The query in Figure 4–20 shows an equality join that will return all order numbers stored in the ORDERS table and the name of the customer placing the order.

```
SQL> SELECT lastname, firstname, order#
  2  FROM customers c, orders o
  3  WHERE c.customer# = o.customer#
  4  ORDER BY c.customer#;

LASTNAME   FIRSTNAME      ORDER#
---------- ---------- ----------
MORALES    BONITA           1003
MORALES    BONITA           1018
SMITH      LEILA            1006
SMITH      LEILA            1016
PIERSON    THOMAS           1008
GIRARD     CINDY            1000
GIRARD     CINDY            1009
GIANA      TAMMY            1007
GIANA      TAMMY            1014
JONES      KENNETH          1020
LUCAS      JAKE             1001
LUCAS      JAKE             1011
MCGOVERN   REESE            1002
LEE        JASMINE          1013
SCHELL     STEVE            1017
NELSON     BECCA            1012
MONTIASA   GREG             1005
MONTIASA   GREG             1019
SMITH      JENNIFER         1010
FALAH      KENNETH          1004
FALAH      KENNETH          1015

21 rows selected.

SQL>
```

Figure 4-20 Matching records from the CUSTOMERS and ORDERS tables

Although this query will identify any customer who has placed an order that is stored in the ORDERS table, it does not list customers who have not recently placed an order. Your instructions specified a list of *all* customers, so you will need to make a change to the query you just issued. When you need to include records in the results of a joining query that exist in one table but do not have a corresponding row in the other table, you will need to use an outer join. The keywords **OUTER JOIN** instruct Oracle9*i* to include records of a table in the output, even if there is no matching record in the other table. In essence, Oracle9*i* will join the "dangling" record to a NULL record in the other table. An outer join can be created in either the WHERE clause with an **outer join operator** (+) or by using the OUTER JOIN keywords.

Outer Joins—Traditional Method

To tell Oracle9*i* to create NULL rows for records that do not have a matching row, use an outer join operator, which is a plus sign within parentheses (+). It is placed in the joining condition of the WHERE clause, immediately after the column name of the table that is missing the corresponding row (which basically tells the software to create a NULL row in that table to join with the row in the other table).

Figure 4–21 corrects the problem with the previous customer query. It shows a list of all customers, and for those that have placed orders, it shows the corresponding order number(s).

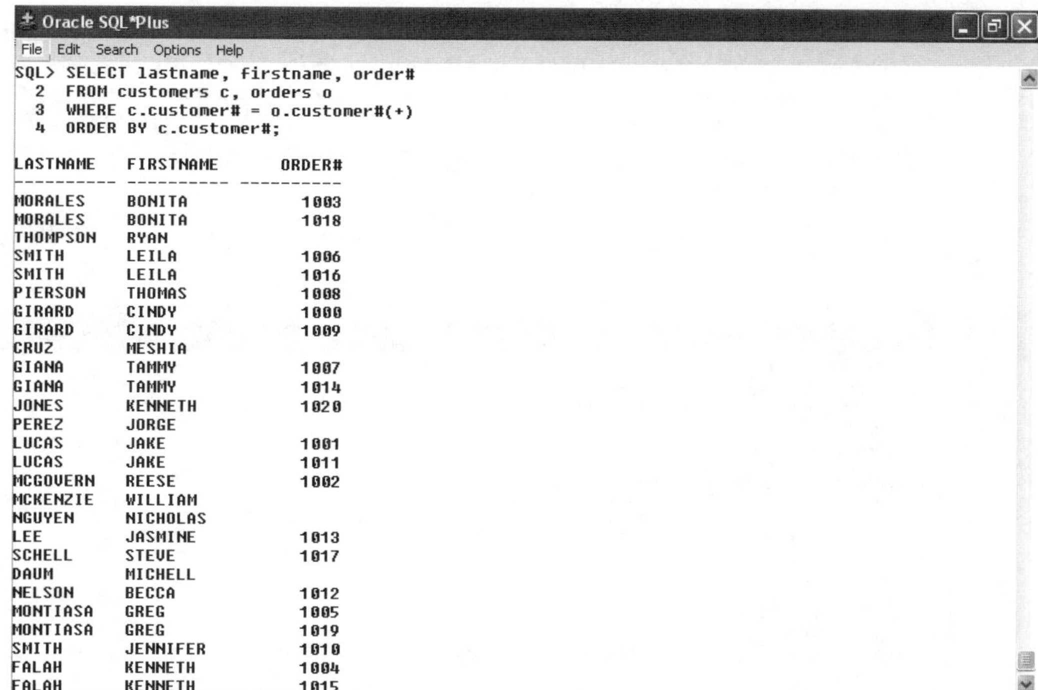

Figure 4-21 An outer join using the outer join operator (+)

If a customer is in the CUSTOMERS table but has not placed a recent order, the ORDERS table will lack a corresponding row—or will be the deficient table (i.e., the table with the missing data). Therefore, the outer join operator (+) is placed immediately after the portion of the joining condition in the WHERE clause that references the deficient ORDERS table.

There are two rules you will need to remember when working with the traditional approach to outer joins:

1. The outer join operator can only be used for *one* table in the joining condition. In other words, using the traditional approach, you cannot create NULL rows in both tables at the same time.
2. A condition that includes the outer join operator cannot use the IN or the OR operator because this would imply that a row should be shown in the results if it matches a row in the other table, or if it matches some other given condition.

Outer Joins—JOIN Method

When creating an outer join with the outer join operator, the outer join can only be applied to one table—not both. However, with the JOIN keyword, you can create a left, right, or full join. *By default, use of the JOIN keyword creates an inner join.* To use the JOIN keyword to create an outer join, you include the keyword **LEFT**, **RIGHT**, or **FULL** with the JOIN keyword to identify the join type. You can also include the OUTER keyword to request an outer join; however, it is optional.

Figure 4–22 re-creates the query used in Figure 4–21 by using the LEFT OUTER JOIN keywords.

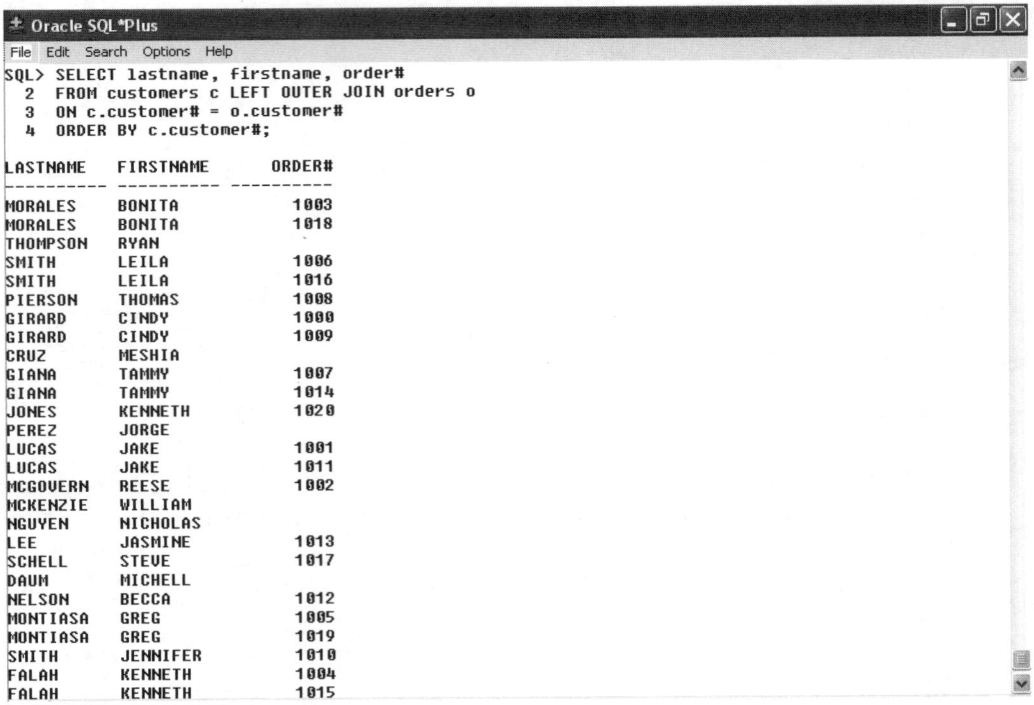

```
Oracle SQL*Plus
File Edit Search Options Help
SQL> SELECT lastname, firstname, order#
  2  FROM customers c LEFT OUTER JOIN orders o
  3  ON c.customer# = o.customer#
  4  ORDER BY c.customer#;

LASTNAME   FIRSTNAME      ORDER#
---------- ---------- ----------
MORALES    BONITA           1003
MORALES    BONITA           1018
THOMPSON   RYAN
SMITH      LEILA            1006
SMITH      LEILA            1016
PIERSON    THOMAS           1008
GIRARD     CINDY            1000
GIRARD     CINDY            1009
CRUZ       MESHIA
GIANA      TAMMY            1007
GIANA      TAMMY            1014
JONES      KENNETH          1020
PEREZ      JORGE
LUCAS      JAKE             1001
LUCAS      JAKE             1011
MCGOVERN   REESE            1002
MCKENZIE   WILLIAM
NGUYEN     NICHOLAS
LEE        JASMINE          1013
SCHELL     STEVE            1017
DAUM       MICHELL
NELSON     BECCA            1012
MONTIASA   GREG             1005
MONTIASA   GREG             1019
SMITH      JENNIFER         1010
FALAH      KENNETH          1004
FALAH      KENNETH          1015
```

Figure 4-22 Query using the LEFT OUTER JOIN keywords

The use of the LEFT JOIN keywords means that if the table listed on the left side of the joining condition given in the ON clause has an unmatched record, it should be matched with a NULL record and displayed in the results. As shown in Figure 4–23, had you used the RIGHT JOIN keywords, Oracle9*i* would have interpreted the query to mean the results should include any order that did not have a corresponding match in the CUSTOMERS table (i.e., the customer record had been deleted from the CUSTOMERS table, but the order placed still existed in the ORDERS table). Since a customer exists for every order that has been placed recently, no NULL rows were created. However, notice that the customers who have not placed an order currently in the ORDERS table are no longer displayed (Thompson, Cruz, McKenzie, etc.). These customers are no longer displayed since the CUSTOMERS table is referenced in the left side of the joining condition, and this statement is executing a RIGHT OUTER JOIN query.

4

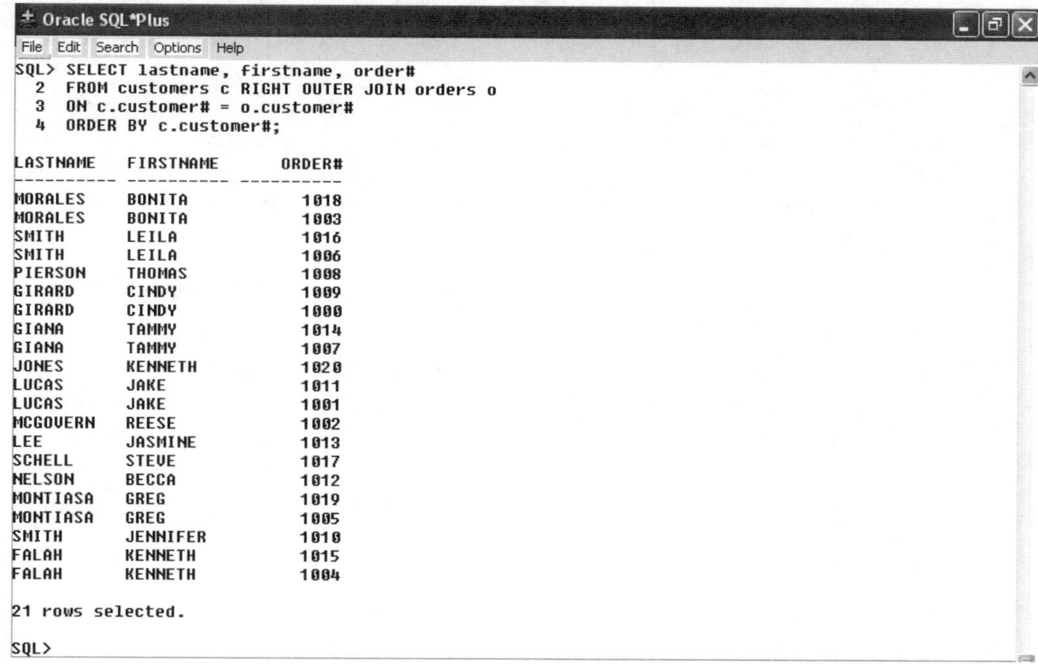
```
SQL> SELECT lastname, firstname, order#
  2  FROM customers c RIGHT OUTER JOIN orders o
  3  ON c.customer# = o.customer#
  4  ORDER BY c.customer#;

LASTNAME   FIRSTNAME      ORDER#
---------- ---------- ----------
MORALES    BONITA           1018
MORALES    BONITA           1003
SMITH      LEILA            1016
SMITH      LEILA            1006
PIERSON    THOMAS           1008
GIRARD     CINDY            1009
GIRARD     CINDY            1000
GIANA      TAMMY            1014
GIANA      TAMMY            1007
JONES      KENNETH          1020
LUCAS      JAKE             1011
LUCAS      JAKE             1001
MCGOVERN   REESE            1002
LEE        JASMINE          1013
SCHELL     STEVE            1017
NELSON     BECCA            1012
MONTIASA   GREG             1019
MONTIASA   GREG             1005
SMITH      JENNIFER         1010
FALAH      KENNETH          1015
FALAH      KENNETH          1004

21 rows selected.

SQL>
```

Figure 4-23 Query using the RIGHT OUTER JOIN keywords

Substituting the FULL JOIN keywords would instruct Oracle9*i* to return records from either table that do not have a matching record in the other table. A FULL JOIN is not an option that is available when creating an outer join with the operator in the WHERE clause for the traditional approach; it is only available with the JOIN keyword.

SET OPERATORS

Set operators are used to combine the results of two (or more) SELECT statements. Valid set operators in Oracle9*i* are UNION, UNION ALL, INTERSECT, and MINUS. When used with two SELECT statements, the **UNION** set operator will return the results of both queries. However, if there are any duplicates, they will be removed, and the duplicated record will only be listed once. To include duplicates in the results, use the **UNION ALL** set operator. **INTERSECT** will only list records that were returned by both queries; the **MINUS** set operator will remove the results of the second query from the output. Figure 4–24 provides a summary of the set operators.

Set Operator	Description
UNION	Returns the results of the combined SELECT statements. Suppresses duplicates.
UNION ALL	Returns the results of the combined SELECT statements. Does not suppress duplicates.
INTERSECT	Returns only the rows included in the results of both SELECT statements.
MINUS	Removes the results of the second query that are also found in the first query and only displays the rows that were uniquely returned by only the first query.

Figure 4-24 Summary of set operators

Suppose that you want a list of all customers in the CUSTOMERS table who have recently placed an order.

As previously mentioned, the UNION set operator will display all the rows returned by both queries. In Figure 4–25, the customer number is retrieved from the CUSTOMERS table in the first SELECT statement and then retrieved from the ORDERS table in the second SELECT statement. Because the two SELECT statements are combined using the UNION set operator, each customer number is listed only once—even if a number appears several times in the ORDERS table.

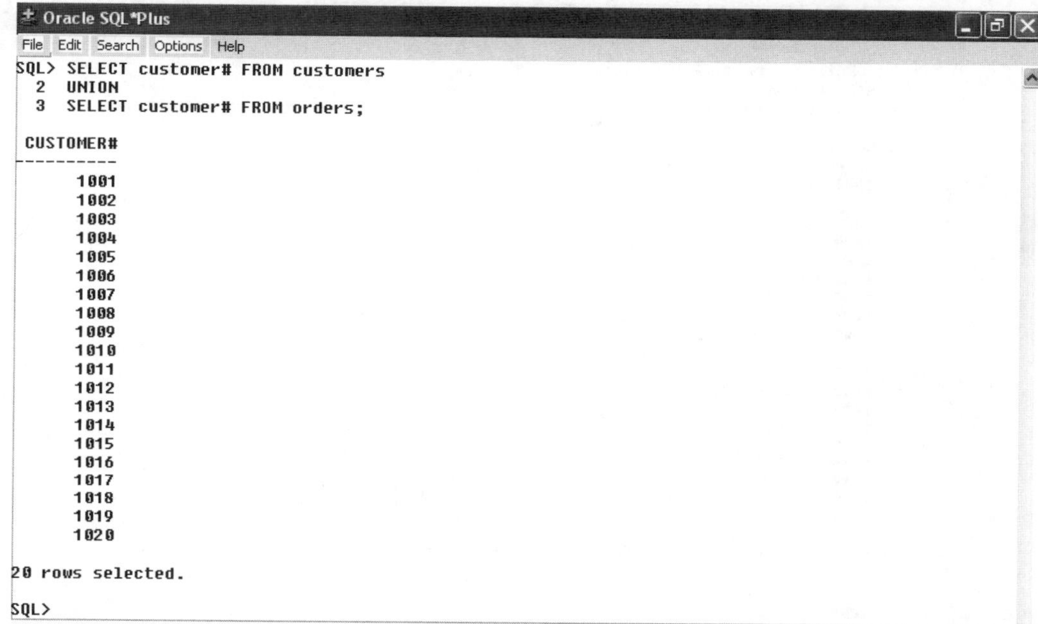

Figure 4-25 Combine SELECT statements using the UNION set operator

Unlike the UNION set operator, the UNION ALL set operator displays every row returned by the combined SELECT statements. In Figure 4–26, the query from Figure 4–25 is re-executed using the UNION ALL set operator. However, the results now include 41 rows in the results, compared to the 20 rows previously returned (partial output shown). Why? The UNION ALL set operator does not suppress duplicate rows, so customer numbers may be displayed more than once if that individual has recently placed more than one order.

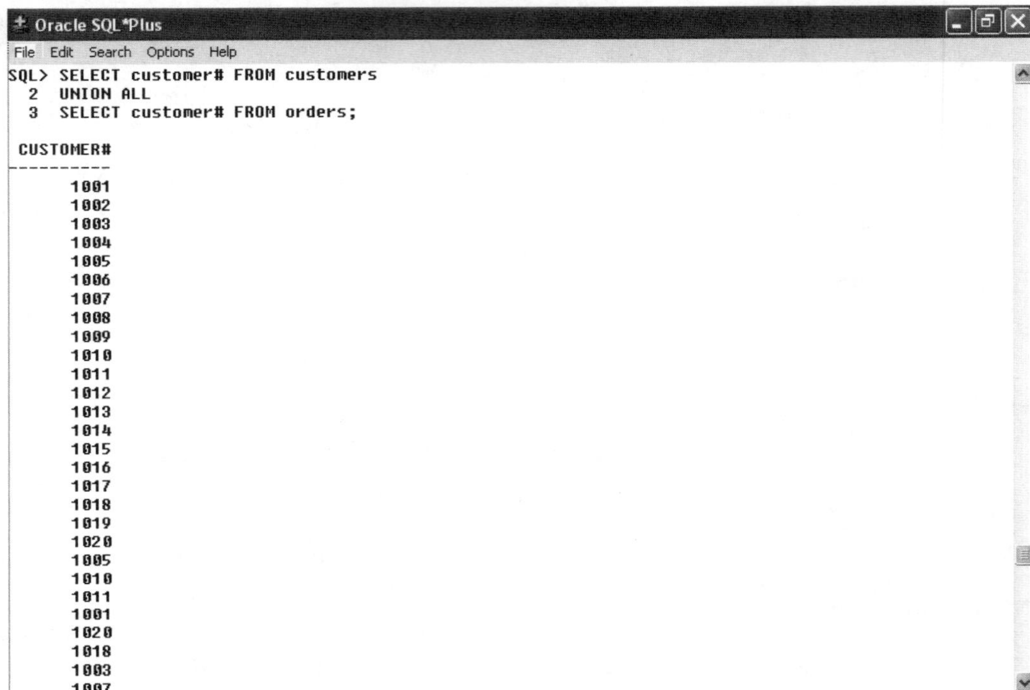

Figure 4-26 Combined SELECT statements using the UNION ALL set operator (partial output shown)

The query in Figure 4–27 contains two SELECT statements. The first SELECT statement asks for all the customer numbers in the CUSTOMERS table—basically a list of all customers. The second SELECT statement lists all the customer numbers for customers who have recently placed an order with the bookstore. By using the INTERSECT set operator to combine the two SELECT statements, you are instructing Oracle9*i* to provide a list of all customers who have recently placed an order and who exist in the CUSTOMERS table (i.e., records returned by both queries).

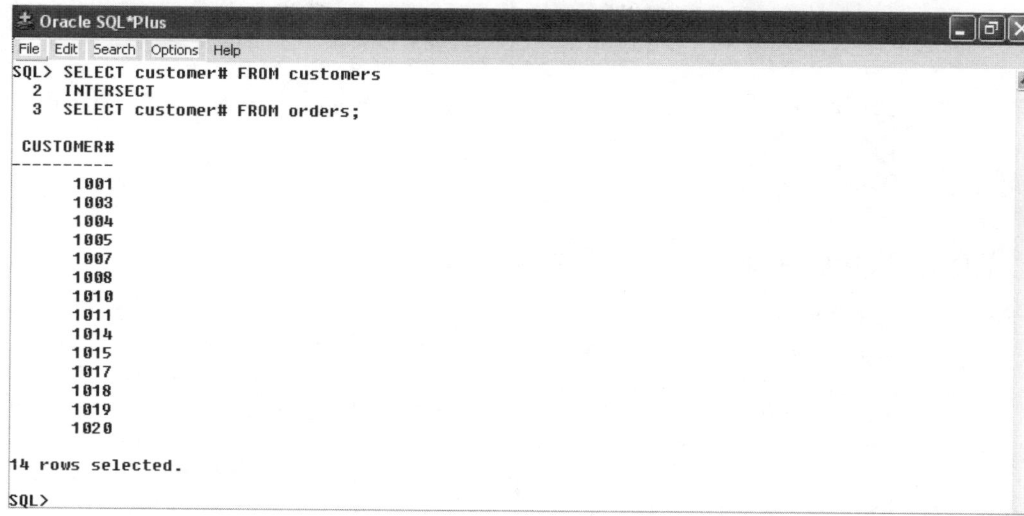

Figure 4-27 Combined SELECT statements using the INTERSECT set operator

The query shown in Figure 4–28 requests a list of customer numbers for those customers who are stored in the CUSTOMERS table but who have not recently placed an order. Because the customers who have not recently placed an order will not be listed in the ORDERS table, you can use the MINUS set operator to remove any customer numbers that are returned by the second SELECT statement from the results of the first SELECT statement.

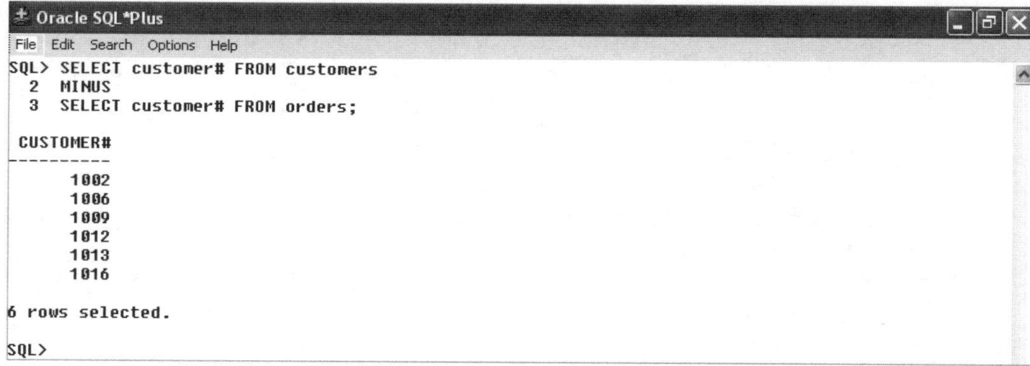

Figure 4-28 Combined SELECT statements using the MINUS set operator

Twenty customer numbers were returned in response to the first SELECT statement; however, only 14 of those customers had placed orders. Because you used the MINUS set operator, the 14 customers who had placed orders were deleted from the results. As you can see in the output, there are six remaining customers who are listed in the CUSTOMERS tables but who are not also stored in the ORDERS table.

Another set operator, EXISTS is also available and will be discussed in Chapter 7.

JOINING THREE OR MORE TABLES

You will probably encounter many situations in which the data you need to retrieve are stored in more than two tables. For example, what if you want to know the name of each book that has been ordered by each customer? First, you would need to know which customer has placed which order (CUSTOMERS and ORDERS tables), then which items were on each order (ORDERS and ORDERITEMS tables), and finally the name of the book represented by each item ordered (ORDERITEMS and BOOKS tables). This will require you to query four tables to determine the name of each customer and the title of each book the customer ordered.

To get the desired results using the various approaches presented in this chapter, you will first join multiple tables using the WHERE clause, then the JOIN...ON approach, and finally the NATURAL JOIN keywords to access data in several tables.

Joining Three or More Tables—Traditional Method

With the traditional method of accessing multiple tables, you use the WHERE clause to identify the access path Oracle9*i* should take to relate the various tables. In Figure 4–29, notice that you are required to use three conditions to establish the relationship among the four tables. Whenever you relate multiple tables in a WHERE clause, you will always have *one less joining condition than the number of tables being joined* (i.e., # of tables - 1). Figure 4–29 shows the first screen of output from such a join.

In Figure 4–29, the first portion of the WHERE clause establishes a relationship between the CUSTOMERS and ORDERS tables, based on the Customer# column. The second portion of the WHERE clause, given on line 4 of the example, joins the ORDERS and ORDERITEMS tables through the Order# column. The last portion of the WHERE clause uses the ISBN column to join the ORDERITEMS and BOOKS tables. Once you have established the relationships among the four tables, you will be able to join the customer's name with the book title ordered.

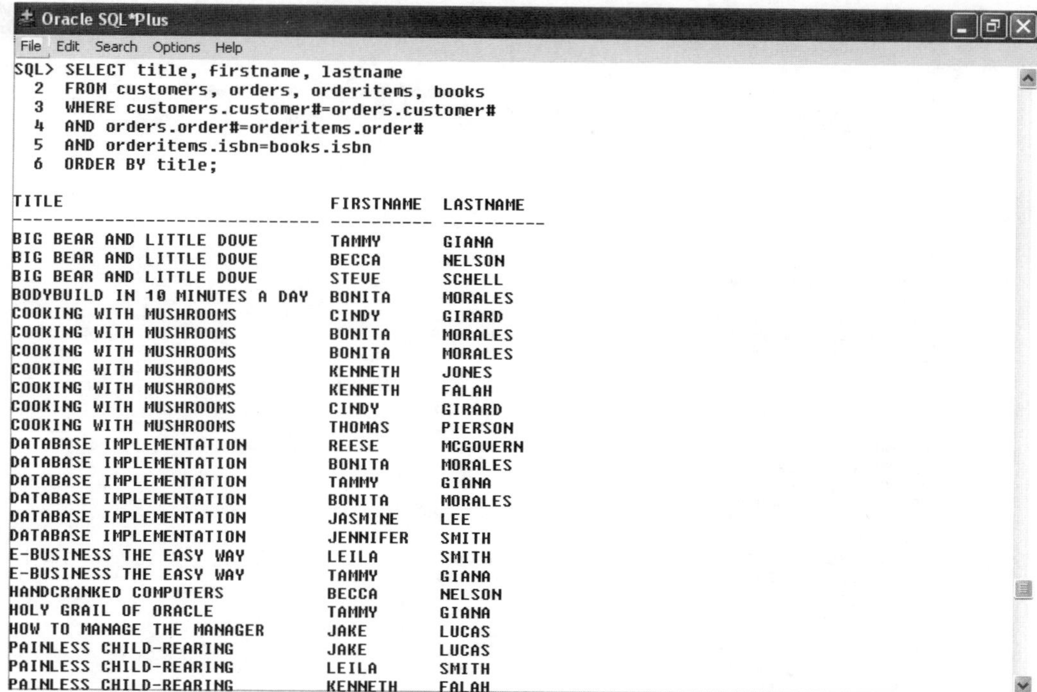

Figure 4-29 Retrieving data from tables, using the WHERE clause (partial output shown)

Joining Three or More Tables—JOIN Method

To use the JOIN...ON method to join the four tables in Figure 4–29, you will need to create three joins in the FROM clause to specify how the tables are related. The key to using the JOIN method for multiple joins is that an ON clause must be provided *each time* the JOIN keyword is used. Notice how the join is structured for the second and third joins, versus the first join in Figure 4–30. With the first join, the JOIN keyword is preceded by the name of the first table, CUSTOMERS, and is then followed by the name of the second table, ORDERS. However, neither the second nor the third JOIN keyword is preceded by a specific table name. This is simply because the ORDERITEMS table specified after the second join is actually combined with the results of the first join, and the BOOKS table is combined with the results of the previous joins. Figure 4–30 shows the first screen of output.

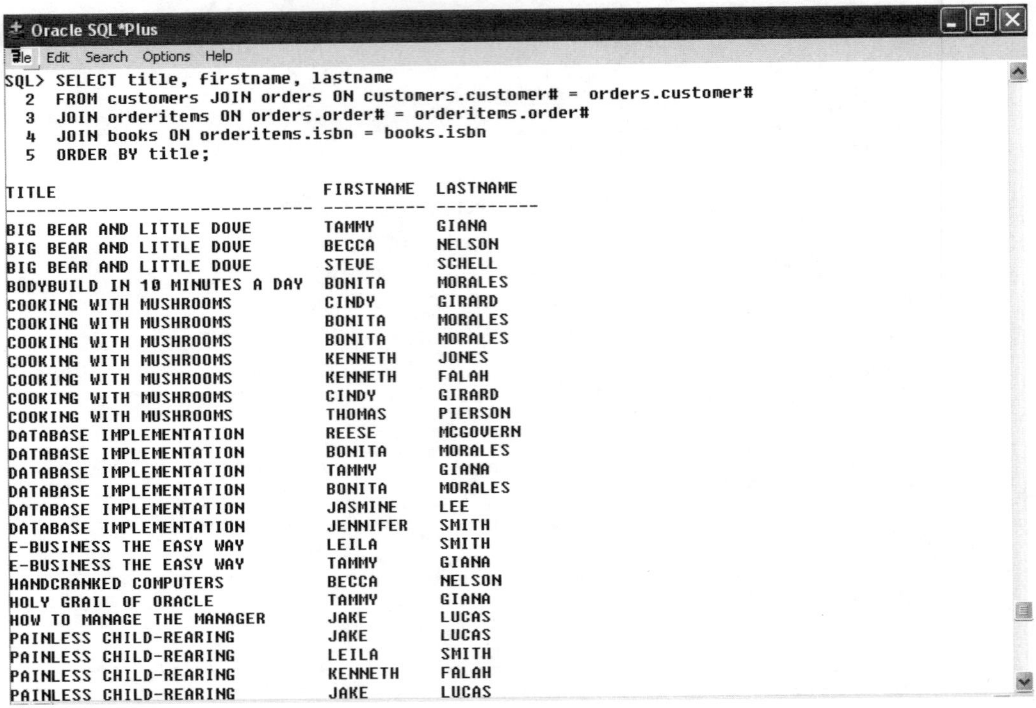

```
Oracle SQL*Plus
File Edit Search Options Help
SQL> SELECT title, firstname, lastname
  2  FROM customers JOIN orders ON customers.customer# = orders.customer#
  3  JOIN orderitems ON orders.order# = orderitems.order#
  4  JOIN books ON orderitems.isbn = books.isbn
  5  ORDER BY title;

TITLE                          FIRSTNAME  LASTNAME
------------------------------ ---------- ----------
BIG BEAR AND LITTLE DOVE       TAMMY      GIANA
BIG BEAR AND LITTLE DOVE       BECCA      NELSON
BIG BEAR AND LITTLE DOVE       STEVE      SCHELL
BODYBUILD IN 10 MINUTES A DAY  BONITA     MORALES
COOKING WITH MUSHROOMS         CINDY      GIRARD
COOKING WITH MUSHROOMS         BONITA     MORALES
COOKING WITH MUSHROOMS         BONITA     MORALES
COOKING WITH MUSHROOMS         KENNETH    JONES
COOKING WITH MUSHROOMS         KENNETH    FALAH
COOKING WITH MUSHROOMS         CINDY      GIRARD
COOKING WITH MUSHROOMS         THOMAS     PIERSON
DATABASE IMPLEMENTATION        REESE      MCGOVERN
DATABASE IMPLEMENTATION        BONITA     MORALES
DATABASE IMPLEMENTATION        TAMMY      GIANA
DATABASE IMPLEMENTATION        BONITA     MORALES
DATABASE IMPLEMENTATION        JASMINE    LEE
DATABASE IMPLEMENTATION        JENNIFER   SMITH
E-BUSINESS THE EASY WAY        LEILA      SMITH
E-BUSINESS THE EASY WAY        TAMMY      GIANA
HANDCRANKED COMPUTERS          BECCA      NELSON
HOLY GRAIL OF ORACLE           TAMMY      GIANA
HOW TO MANAGE THE MANAGER      JAKE       LUCAS
PAINLESS CHILD-REARING         JAKE       LUCAS
PAINLESS CHILD-REARING         LEILA      SMITH
PAINLESS CHILD-REARING         KENNETH    FALAH
PAINLESS CHILD-REARING         JAKE       LUCAS
```

Figure 4-30 Retrieving data from four tables, using the JOIN...ON keywords (partial output shown)

As previously discussed, the NATURAL JOIN keywords can be used to create joins among tables that have commonly named and defined fields. Because the CUSTOMERS, ORDERS, ORDERITEMS, and BOOKS tables are all related through columns having the same name and datatype, using the NATURAL JOIN keywords is perhaps the simplest way to determine the title of each book purchased by each customer.

As shown in Figure 4–31, the NATURAL JOIN keywords in the FROM clause establish the relationships among the tables. You will need to remember to repeat the keywords in the FROM clause for each set of tables. Note that only the first screen of output is shown.

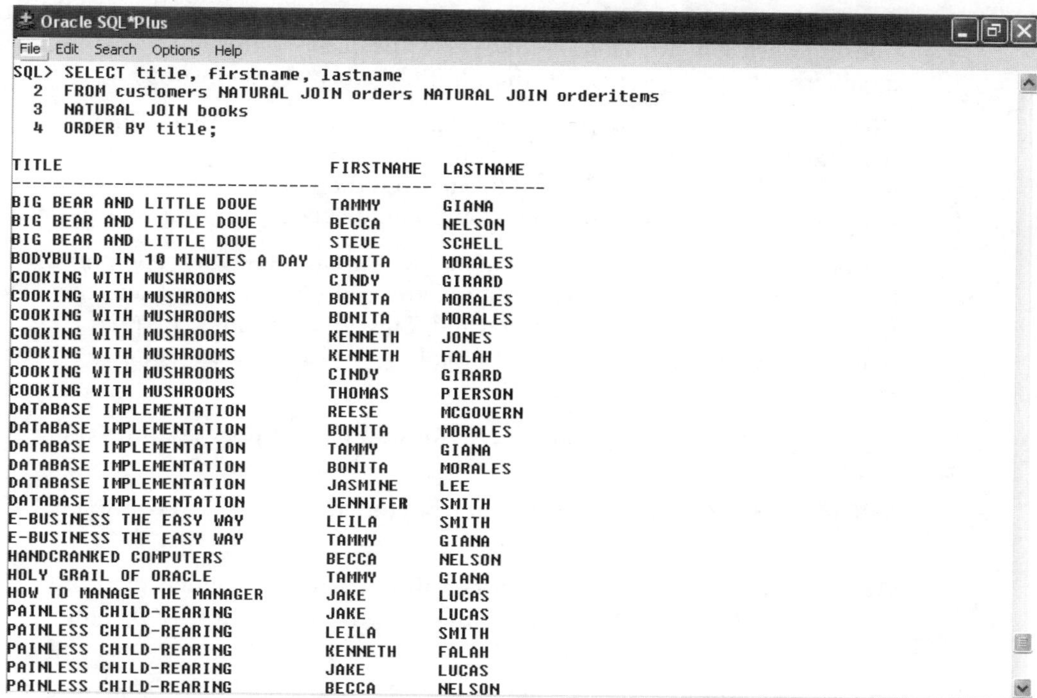

```
Oracle SQL*Plus
File Edit Search Options Help
SQL> SELECT title, firstname, lastname
  2  FROM customers NATURAL JOIN orders NATURAL JOIN orderitems
  3  NATURAL JOIN books
  4  ORDER BY title;

TITLE                          FIRSTNAME  LASTNAME
------------------------------ ---------- ----------
BIG BEAR AND LITTLE DOVE       TAMMY      GIANA
BIG BEAR AND LITTLE DOVE       BECCA      NELSON
BIG BEAR AND LITTLE DOVE       STEVE      SCHELL
BODYBUILD IN 10 MINUTES A DAY  BONITA     MORALES
COOKING WITH MUSHROOMS         CINDY      GIRARD
COOKING WITH MUSHROOMS         BONITA     MORALES
COOKING WITH MUSHROOMS         BONITA     MORALES
COOKING WITH MUSHROOMS         KENNETH    JONES
COOKING WITH MUSHROOMS         KENNETH    FALAH
COOKING WITH MUSHROOMS         CINDY      GIRARD
COOKING WITH MUSHROOMS         THOMAS     PIERSON
DATABASE IMPLEMENTATION        REESE      MCGOVERN
DATABASE IMPLEMENTATION        BONITA     MORALES
DATABASE IMPLEMENTATION        TAMMY      GIANA
DATABASE IMPLEMENTATION        BONITA     MORALES
DATABASE IMPLEMENTATION        JASMINE    LEE
DATABASE IMPLEMENTATION        JENNIFER   SMITH
E-BUSINESS THE EASY WAY        LEILA      SMITH
E-BUSINESS THE EASY WAY        TAMMY      GIANA
HANDCRANKED COMPUTERS          BECCA      NELSON
HOLY GRAIL OF ORACLE           TAMMY      GIANA
HOW TO MANAGE THE MANAGER      JAKE       LUCAS
PAINLESS CHILD-REARING         JAKE       LUCAS
PAINLESS CHILD-REARING         LEILA      SMITH
PAINLESS CHILD-REARING         KENNETH    FALAH
PAINLESS CHILD-REARING         JAKE       LUCAS
PAINLESS CHILD-REARING         BECCA      NELSON
```

Figure 4-31 Retrieving data from four tables, using the NATURAL JOIN keywords (partial output shown)

The same rule used in the traditional approach applies when using the JOIN approach—the number of joins required is one less than the number of tables to be joined.

Chapter Summary

- Data stored in multiple tables regarding a single entity can be reconstructed through the use of joins.
- A Cartesian join between two tables will return every possible combination of rows from the tables. The resulting number of rows will always be $m * n$.
- An equality join is created when the data joining the records from two different tables are an exact match (i.e., an equality condition creates the relationship). The traditional approach uses an equal sign as the comparison operator in the WHERE clause. The JOIN approach can use the NATURAL JOIN, JOIN...USING, or JOIN...ON keywords.

- The NATURAL JOIN keywords do not require a condition to establish the relationship between two tables. However, a common column must exist. Column qualifiers cannot be used with the NATURAL JOIN keywords.
- The JOIN...USING approach is similar to the NATURAL JOIN approach, except that the common column is specified in the USING clause. A condition cannot be given in the USING clause to indicate how the tables are related. Column qualifiers cannot be used for the common column specified in the USING clause.
- The JOIN...ON approach joins tables based upon a specified condition. The JOIN keyword in the FROM clause indicates the tables to be joined, and the ON clause indicates how the two tables are related. This approach must be used if the tables being joined do not have a common column with the same column name in each table.
- A non-equality join establishes a relationship based upon anything other than an equal condition.
- Self-joins are used when a table must be joined to itself to retrieve any needed data.
- Broadly speaking, a join can be either an inner join, in which the only records returned in the results have a matching record in all tables, or an outer join, where a matching record is not required.
- Inner joins are categorized as being equality, non-equality, or self-joins.
- An outer join is created when records need to be included in the results but do not have corresponding records in the join tables. Basically the record is matched with a NULL record so it will be included in the output.
- When using the WHERE clause to create an outer join, only records from one table can be matched to a null record. The outer join operator (+) is placed next to the table that does not contain rows that match existing rows in the other table.
- With the OUTER JOIN keyword, you have the FULL option available that will include records from either table that do not have a corresponding record in the other table.
- Set operators such as UNION, UNION ALL, INTERSECT, and MINUS can be used to combine the results of multiple queries.

CHAPTER 4 SYNTAX SUMMARY

The following tables present a summary of the syntax and information that you have learned in this chapter. You can use the tables as a study guide and reference.

Syntax Guide		
Element	**Description**	**Example**
WHERE clause	In the traditional approach, the WHERE clause can be used to indicate which column(s) should be used to join tables.	SELECT columnname [,...] FROM tablename1, tablename2 WHERE tablename1.columnname <comparison operator> tablename2.columnname;
NATURAL JOIN keywords	These keywords are used in the FROM clause to join tables containing a common column with the same name and definition.	SELECT columnname[,...] FROM tablename1 NATURAL JOIN tablename2;
JOIN...USING	The JOIN keyword is used in the FROM clause and, combined with the USING clause, it identifies the common column to be used to join the tables. It is normally used if the tables have more than one commonly named column, and only one is being used for the join.	SELECT columnname [,...] FROM tablename1 JOIN tablename2 USING (columnname);
JOIN...ON	The JOIN keyword is used in the FROM clause. The ON clause identifies the column to be used to join the tables.	SELECT columnname [,...] FROM tablename1 JOIN tablename2 ON tablename1.columnname <comparison operator> tablename2.columnname;
OUTER JOIN Can be a RIGHT, LEFT, or FULL OUTER JOIN	This indicates that at least one of the tables does not have a matching row in the other table.	SELECT columnname [,...] FROM tablename1 [RIGHT|LEFT|FULL] OUTER JOIN tablename2 ON tablename1.columnname = tablename2.columnname;

Kinds of Joins		
Join	**Traditional Method**	**JOIN Method**

Cartesian Join
• Also known as a **Cartesian product** or **cross join**
• Matches each record in one table with each record in another table
	Example:
	`SELECT title, name`
	`FROM books, publisher;`
		Uses keywords **CROSS JOIN**
		Example:
		`SELECT title, name`
		`FROM books CROSS JOIN`
		`publisher;`

Equality Join
• Also known as an **equijoin**, **inner join**, or **simple join**
• Joins data in tables having equivalent data in a **common column**
• Might need to create **table aliases**
	Uses the keyword **WHERE**
	Example:
	`SELECT title, books.pubid,`
	`name`
	`FROM publisher, books`
	`WHERE publisher.pubid =`
	`books.pubid`
	`AND publisher.pubid = 4;`
		Uses keywords NATURAL JOIN or JOIN...USING to create a join between tables having a commonly defined field
		Uses keywords JOIN...ON when tables don't have a commonly defined field. The column qualifier ON tells Oracle9*i* how tables are related
		Examples:
		`SELECT title, pubid, name`
		`FROM publisher NATURAL`
		`JOIN books;`
		
		`SELECT title, pubid, name`
		`FROM publisher JOIN books`
		`USING (pubid);`
		
		`SELECT title, name`
		`FROM books b JOIN`
		`publisher p ON b.pubid =`
		`p.pubid;`

Non-Equality Join
• Joins tables when there are no equivalent rows in the tables to be joined (i.e., to match values in one column with a range of values in another column)
	Uses the keyword **WHERE**
	Example:
	`SELECT title, gift`
	`FROM books, promotion`
	`WHERE retail`
	`BETWEEN minretail AND`
	`maxretail;`
		Uses keywords **JOIN...ON**—and the same syntax as JOIN method equality join
		Example:
		`SELECT title, gift`
		`FROM books JOIN promotion`
		`ON retail BETWEEN`
		`minretail AND maxretail;`

Kinds of Joins (Continued)		
Join	**Traditional Method**	**JOIN Method**
• Can use any comparison operator *except* the equals sign (=)		

Self-Join
• Joins a table to itself so columns within the table can be joined
• Must create table alias
	Uses the keyword WHERE
	Example:
	`SELECT r.firstname, r.lastname, c.lastname referred`
	`FROM customers c, customers r`
	`WHERE c.referred = r.customer#;`
		Uses the keywords **JOIN...ON**
		Example:
		`SELECT r.firstname, r.lastname, c.lastname referred`
		`FROM customers c`
		`JOIN customers r ON c.referred = r.customer#;`

Outer Join
• Includes records of a table in output when there is nomatching record in the other table
	Uses the keyword **WHERE**
	Uses the outer join operator (+) to create NULL rows in the deficient table for records that do not have a matching row
	Example:
	`SELECT lastname, firstname, order#`
	`FROM customers c, orders o`
	`WHERE c.customer#=o.customer#(+)`
	`ORDER BY c.customer#;`
		Includes the keyword **LEFT**, **RIGHT**, or **FULL** with the **OUTER JOIN** keywords
		Example:
		`SELECT lastname, firstname, order#`
		`FROM customers c LEFT OUTER JOIN orders o`
		`ON c.customer# = o.customer#`
		`ORDER BY c.customer#;`

Operators		
Operator	**Description**	**Example**

Set Operators
Includes UNION, UNION ALL, INTERSECT, MINUS
	Used to combine results of multiple SELECT statements.	`SELECT customer# FROM customers`
		`UNION`
		`SELECT customer# FROM orders;`

Outer Join Operator (+)	Used to indicate the table containing the deficient rows. The operator is placed next to the table that should have NULL rows added to create a match.	`SELECT lastname, firstname, order#`
		`FROM customers c, orders o`
		`WHERE c.customer# = o.customer#(+);`

Review Questions

To answer these questions, refer to the tables in Appendix A.

1. Explain the difference between an inner join and an outer join.
2. How many rows will be returned in a Cartesian join between one table having 5 records and a second table having 10 records?
3. How would you restructure the query issued in Figure 4–16 to use the NATURAL JOIN keywords and produce the same results?
4. Why are the NATURAL JOIN keywords not an option for producing a self-join? (*Hint:* Think about what happens if you use a table alias with the NATURAL JOIN keywords.)
5. What is the purpose of a column qualifier? When are you required to use a column qualifier?
6. In an OUTER JOIN query, the outer join operator (+) is placed next to which table?
7. Which operators cannot be used in a condition that includes the outer join operator?
8. How many join conditions would be needed for a query that joins five tables?
9. What is the difference between an equality and a non-equality join?
10. What are the differences between using the JOIN...USING and JOIN...ON approaches for joining tables?

Multiple Choice

To answer the following questions, refer to the tables in Appendix A.

1. Which of the following queries will create a Cartesian join?
   a. `SELECT title, authorid FROM books, bookauthor;`
   b. `SELECT title, name FROM books CROSS JOIN publisher;`
   c. `SELECT title, gift FROM books NATURAL JOIN promotion;`
   d. all of the above
2. Which of the following operators is not allowed in an outer join?
   a. AND
   b. =
   c. OR
   d. >

3. Which of the following is an example of an equality join?
   a. `SELECT title, authorid FROM books, bookauthor`
      `WHERE books.isbn = bookauthor.isbn AND retail > 20;`
   b. `SELECT title, name FROM books CROSS JOIN publisher;`
   c. `SELECT title, gift FROM books, promotion`
      `WHERE retail>=minretail AND retail<=maxretail;`
   d. none of the above
4. Which of the following is an example of a non-equality join?
   a. `SELECT title, authorid FROM books, bookauthor`
      `WHERE books.isbn = bookauthor.isbn AND retail > 20;`
   b. `SELECT title, name FROM books JOIN publisher USING`
      `(pubid);`
   c. `SELECT title, gift FROM books, promotion`
      `WHERE retail>=minretail AND retail<=maxretail;`
   d. none of the above
5. The following SQL statement represents which type of query?

```
SELECT title, order#, quantity
FROM books FULL JOIN orderitems
ON books.isbn = orderitems.isbn;
```

   a. equality
   b. self-join
   c. non-equality
   d. outer join
6. Which of the following queries is valid?
   a.
   ```
   SELECT b.title, b.retail, o.quantity
   FROM books b NATURAL JOIN orders od
   NATURAL JOIN orderitems o
   WHERE od.order#=1005;
   ```
   b.
   ```
   SELECT b.title, b.retail, o.quantity
   FROM books b, orders od, orderitems o
   WHERE orders.order#=orderitems.order#
   AND orderitems.isbn=book.isbn
   AND od.order#=1005;
   ```
   c.
   ```
   SELECT b.title, b.retail, o.quantity
   FROM books b, orderitems o
   WHERE o.isbn = b.isbn AND o.order#=1005;
   ```
   d. none of the above

7. Given the following query:

```
SELECT zip, order# FROM customers NATURAL JOIN orders;
```

Which of the following queries is equivalent?

a.
```
SELECT zip, order#
FROM customers JOIN orders
WHERE customers.customer#=orders.customer#;
```

b.
```
SELECT zip, order#
FROM customers, orders
WHERE customers.customer#=orders.customer#;
```

c.
```
SELECT zip, order#
FROM customers, orders
WHERE customers.customer#=orders.customer#(+);
```

d. none of the above

8. Which line in the following SQL statement contains an error?

```
1 SELECT name, title
2 FROM books NATURAL JOIN publisher
3 WHERE category = 'FITNESS'
4 OR
5 books.pubid=4
6 /
```

a. line 1

b. line 2

c. line 3

d. line 4

e. line 5

9. Given the following query:

```
SELECT lastname, firstname, order#
FROM customers c LEFT OUTER JOIN orders o
ON c.customer# = o.customers#
ORDER BY c.customers#;
```

Which of the following queries will return the same results?

a.
```
SELECT lastname, firstname, order#
FROM customers c OUTER JOIN orders o
ON c.customer# = o.customers#
ORDER BY c.customers#;
```

b.
```
SELECT lastname, firstname, order#
FROM orders o RIGHT OUTER JOIN customers c
ON c.customer# = o.customers#
ORDER BY c.customers#;
```

c.
```
SELECT lastname, firstname, order#
FROM customers c, orders o
ON c.customer# = o.customers#(+)
ORDER BY c.customers#;
```

d. none of the above

10. Given the following query:

```
SELECT DISTINCT zip, category
FROM customers NATURAL JOIN orders NATURAL JOIN orderitems
NATURAL JOIN books;
```

Which of the following queries is equivalent?

a.
```
SELECT zip FROM customers
UNION
SELECT category FROM books;
```

b.
```
SELECT DISTINCT zip, category
FROM customers c, orders o, orderitems oi, books b
WHERE c.customer#=o.customer# AND o.order#=oi.order#
AND oi.isbn=b.isbn;
```

c.
```
SELECT DISTINCT zip, category
FROM customers c JOIN orders o JOIN orderitems oi
JOIN books b
ON c.customer#=o.customer# AND o.order#=oi.order#
AND oi.isbn=b.isbn;
```

d. all of the above

e. none of the above

11. Which line in the following SQL statement contains an error?

```
1 SELECT name, title
2 FROM books JOIN publisher
3 WHERE books.pubid = publisher.pubid
4 AND
5 cost <45.95
6 /
```

   a. line 1
   b. line 2
   c. line 3
   d. line 4
   e. line 5

12. Given the following query:

```
SELECT title, gift FROM books CROSS JOIN promotion;
```

Which of the following queries is equivalent?

   a. `SELECT title, gift FROM books NATURAL JOIN promotion;`
   b. `SELECT title FROM books INTERSECT SELECT gift FROM promotion;`
   c. `SELECT title FROM books UNION ALL SELECT gift FROM promotion;`
   d. all of the above

13. If the PRODUCTS table contains seven records and the INVENTORY table has eight records, how many records would the following query produce?

```
SELECT * FROM products CROSS JOIN inventory;
```

   a. 0
   b. 8
   c. 7
   d. 15
   e. 56

14. Which of the following SQL statements is not valid?
    a. `SELECT b.isbn, p.name FROM books b NATUAL JOIN publisher p;`
    b.
    ```
    SELECT isbn, name
    FROM books b, publisher p
    WHERE b.pubid = p.pubid;
    ```
    c.
    ```
    SELECT isbn, name
    FROM books b JOIN publisher p
    ON b.pubid=p.pubid;
    ```
    d.
    ```
    SELECT isbn, name
    FROM books JOIN publisher
    USING (pubid);
    ```
    e. None—all of the above are valid SQL statements.
15. Which of the following will produce a list of all books published by Printing Is Us?
    a.
    ```
    SELECT title FROM books NATURAL JOIN publisher
    WHERE name LIKE 'PRIN%';
    ```
    b.
    ```
    SELECT title FROM books, publisher
    WHERE pubname = 1;
    ```
    c.
    ```
    SELECT * FROM books b, publisher p
    JOIN tables ON b.pubid = p.pubid;
    ```
    d. none of the above
16. Which of the following SQL statements is not valid?
    a.
    ```
    SELECT isbn FROM books
    MINUS
    SELECT isbn FROM orderitems;
    ```
    b.
    ```
    SELCET isbn, name FROM books, publisher
    WHERE books.pubid (+) = publisher.pubid (+);
    ```
    c. `SELECT title, name FROM books NATURAL JOIN publisher`
    d. None—all of the above SQL statements are valid.
17. Which of the following statements regarding an outer join between two tables is true?
    a. If the relationship between the tables is established through a WHERE clause, both tables can include the outer join operator.
    b. To include unmatched records in the results, the record is paired with a NULL record in the deficient table.
    c. The RIGHT, LEFT, and FULL keywords are equivalent keywords.
    d. all of the above
    e. none of the above

18. Which line in the following SQL statement contains an error?

```
1 SELECT name, title
2 FROM books b, publisher p
3 WHERE books.pubid = publisher.pubid
4 AND
5 retail > 25 OR retail-cost > 18.95;
```

   a. line 1
   b. line 3
   c. line 4
   d. line 5

19. What is the maximum number of characters allowed in a table alias?
   a. 10
   b. 30
   c. 255
   d. 256

20. Which of the following SQL statements is valid?
   a.
   ```
   SELECT books.title, orderitems.quantity
   FROM books b, orderitems o
   WHERE b.isbn= o.ibsn;
   ```
   b.
   ```
   SELECT title, quantity
   FROM books b JOIN orderitems o;
   ```
   c.
   ```
   SELECT books.title, orderitems.quantity
   FROM books JOIN orderitems
   ON books.isbn = orderitems.isbn;
   ```
   d. none of the above

Hands-on Assignments

To perform these activities, refer to the tables in Appendix A.

For each of the following tasks, determine (a) the SQL statement needed to perform the stated task using the traditional approach and (b) the SQL statement needed to perform the stated task using the JOIN keyword.

1. Create a list that displays the title of each book and the name and phone number of the person at the publisher's office whom you would need to contact to reorder each book.
2. Determine which orders have not yet shipped and the name of the customer that placed each order. Sort the results by the date on which the order was placed.

3. List the customer number and names of all individuals who have purchased books in the Fitness Category.
4. Determine which books Jake Lucas has purchased.
5. Determine the profit of each book sold to Jake Lucas. Sort the results by the date of the order. If more than one book was ordered, have the results sorted by the profit amount in descending order.
6. Which book was written by an author with the last name of Adams?
7. What gift will a customer who orders the book *Shortest Poems* receive?
8. Identify the author(s) of the books ordered by Becca Nelson.
9. Display a list of all books in the BOOKS table. If a book has been ordered by a customer, also list the corresponding order number(s) and the state in which the customer resides.
10. Produce a list of all customers who live in the state of Florida and have ordered books about computers.

A Case for Oracle9*i*

To perform this activity, refer to the tables in Appendix A.

The Marketing Department of JustLee Books is preparing for its annual sales promotion. Each customer who places an order during the promotion will receive a free gift with each book purchased. The actual gift received will be based upon the retail price of the book.

JustLee Books also participates in co-op advertising programs with certain publishers. If the publisher's name is included in advertisements, JustLee Books is reimbursed a certain percentage of the advertisement costs. To determine the projected costs of this year's sales promotion, the Marketing Department needs the publisher's name, the profit, and the free gift for each book held in inventory by JustLee Books. Create a memo that includes the necessary SQL statement and the output requested by the Marketing Department.

CHAPTER

5

SELECTED SINGLE-ROW FUNCTIONS

Objectives

After completing this chapter, you should be able to do the following:

- Use the UPPER, LOWER, and INITCAP functions to change the case of field values and character strings
- Extract a substring using the SUBSTR function
- Determine the length of a character string, using the LENGTH function
- Use the LPAD and RPAD functions to pad a string to a certain width
- Use the LTRIM and RTRIM functions to remove specific character strings
- Round and truncate numeric data, using the ROUND and TRUNC functions
- Calculate the number of months between two dates, using the MONTHS_BETWEEN function
- Identify and correct problems associated with calculations involving NULL values, using the NVL function
- Display dates and numbers in a specific format, using the TO_CHAR function
- Determine the current date setting, using the SYSDATE keyword
- Nest functions inside other functions
- Identify when to use the DUAL table

In this chapter, you will learn about single-row SQL functions. A **function** is a predefined block of code that accepts one or more **arguments**—values listed within parentheses—and then returns a single value as output. The nature of an argument depends on the **syntax**, or structure, of the function being executed. **Single-row functions** return one row of results for each record processed. By contrast, **multiple-row functions** return only one result per group or category of rows processed—e.g., counting the number of books published by each publisher. (Multiple-row functions are presented in Chapter 6.)

Single-row functions range from performing character case conversions to calculating the difference between two dates. The functions presented in this chapter have been grouped into character functions (case conversion functions and character manipulation functions), number functions, date functions, and other functions. Where appropriate, examples will include using a dummy table as well as data from the JustLee Books database. Figure 5–1 presents an overview of the functions discussed in this chapter.

Type of Functions	Functions
Case Conversion Functions	UPPER, LOWER, INITCAP
Character Manipulation Functions	SUBSTR, LENGTH, LPAD/RPAD, RTRIM/LTRIM, REPLACE
Numeric Functions	ROUND, TRUNC
Date Functions	MONTHS_BETWEEN, ADD_MONTHS, NEXT_DAY, TO_DATE
Other Functions	NVL, NVL2, TO_CHAR, DECODE, SOUNDEX

Figure 5-1 Overview of chapter contents

Go to the JustLee Database folder in your Data Files. If you run the **Bookscript.sql** file, you will be able to work through all the queries shown in this chapter. Your output should match the output shown.

Case Conversion Functions

You can use **character functions** to change the case of characters (e.g., convert upper-case letters to lower-case letters) or to manipulate characters (e.g., substitute characters). Although most database administrators rarely need to use character functions, application developers frequently include them to create user-friendly database interfaces. Let's first examine case conversion functions.

Case conversion functions alter the case of data stored in a field or character string. The case conversion is only temporary—it does not affect how data are stored, only how data are viewed by Oracle9*i* during the execution of a specific query. The character conversion functions supported by Oracle9*i* are LOWER, UPPER, and INITCAP.

LOWER Function

The data in JustLee Books' database are in upper-case letters. However, when users perform data searches, they may forget and enter character strings for search conditions in lower-case letters. If this happens, no rows will be returned because table data are in upper-case letters. There are several ways to solve this problem. Perhaps the easiest way is to use the LOWER function. The **LOWER** function converts character strings to lower-case letters.

Because some users might enter data for searches in lower-case characters, you can temporarily convert a table's data to lower-case characters during a query's execution, as shown in Figure 5–2.

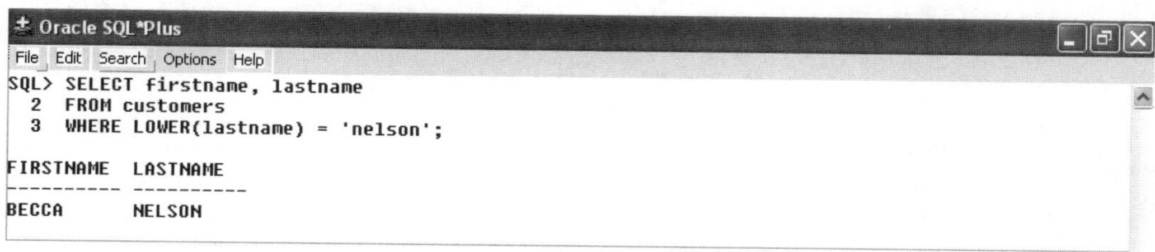

Figure 5-2 LOWER function in a WHERE clause

In Figure 5–2, you are searching for a customer with the last name Nelson. The syntax for the LOWER function is `LOWER(c)`, where *c* is the field or character string to be converted. In this case, you need the data in the Lastname field converted to lower-case characters during the search. Therefore, the Lastname field is inserted between the parentheses.

Notice that the results of the query in Figure 5–2 are still displayed in upper-case letters. Because the LOWER function was not used in the SELECT clause of the SELECT statement, the first and last name are displayed in the same case in which they are stored. If you would like data to be displayed in lower-case characters, simply include the LOWER function for each field to be converted in the SELECT clause. Figure 5–3 shows the LOWER function in a SELECT and a WHERE clause.

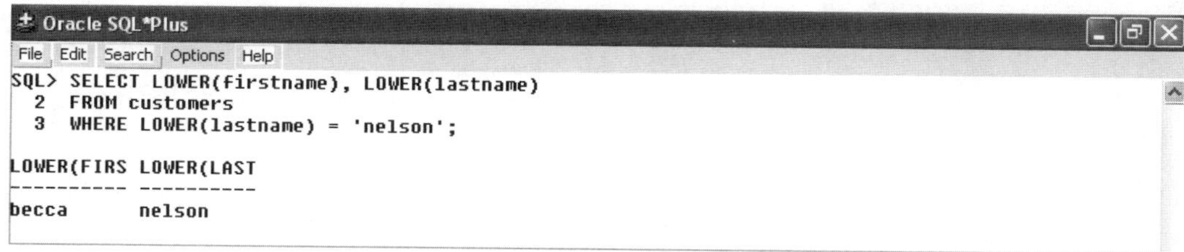

Figure 5-3 LOWER function in SELECT and WHERE clauses

In Figure 5–3, the LOWER function is used in the SELECT clause to display the query results in lower-case characters. However, you still must include the LOWER function in the WHERE clause because the character string `'nelson'` is, for comparison purposes, still entered in its lower-case form. In other words, *when a function is used in a SELECT clause, it only affects how the data are displayed in the results*. By contrast, when a function is used in a WHERE clause, it is only used during the specified comparison operation.

UPPER Function

Another technique commonly employed to make searches more "user friendly" is to include a character string within the UPPER function. In previous examples, you used the LOWER function to convert field data during the execution of the SELECT statement. However, what if you want to enter a search string in mixed case—that is, in both upper case and lower case?

To assist users with this problem, most application developers prefer to include the search string in an UPPER function. The **UPPER** function simply converts the characters indicated into upper-case characters. The syntax for the UPPER function is `UPPER(c)`, where *c* is the character string or field to be converted into upper-case characters. Figure 5–4 shows the UPPER function.

Figure 5-4 UPPER function in a WHERE clause

In Figure 5–4, the UPPER function is included in the condition portion of the WHERE clause to instruct Oracle9*i* to convert the user-entered character string of *nelson* to upper-case characters while executing the SELECT query. It is much more efficient to convert the single-search condition to the same case as the data stored in the table for the following reasons:

1. You don't need to be concerned with whether the user knows the exact case to use when entering search strings.
2. Oracle9*i* does not need to convert the case of all the data contained within the field during the execution of the query, thus reducing the processing burden placed on the Oracle server.

INITCAP Function

Although having the table data and the search criteria in the same case enables you to find the record(s) you seek, the output may not be presented in an appealing manner. It is generally easier for most people to read data displayed in mixed-case letters, rather than in all upper-case or lower-case letters. Oracle9*i* includes the **INITCAP** function to convert character strings to mixed case, with each word beginning with a capital letter—for example, *Great Mushroom Recipes*. The syntax of the INITCAP function is `INITCAP(c)`, where *c* represents the field or character string to be converted. Thus, the INITCAP function converts the first (INITial) letter of each word in the character string to upper case (CAPital letter) and the remaining characters into lower case, as shown in Figure 5–5.

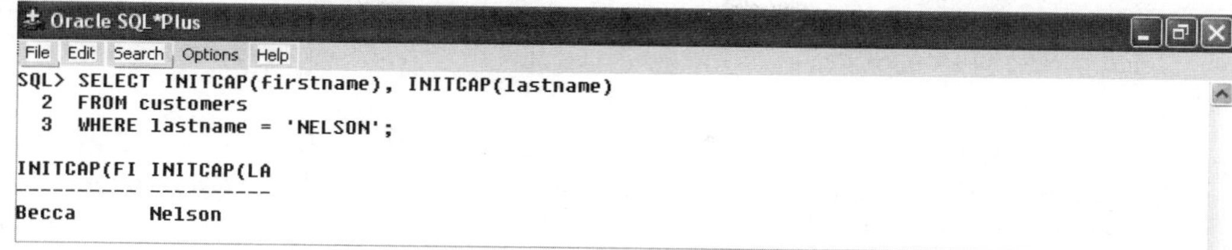

Figure 5-5 INITCAP function in a SELECT clause

In Figure 5–5, the INITCAP function is used in the SELECT clause to convert data in the Firstname and Lastname columns to mixed case, with the first letter of the word in each field being in upper case and the remaining letters in lower case. Although the INITCAP function can also be used in a WHERE clause, this is rare because all users may not be consistent in the case they use for entering data. Also notice headings for the Firstname and Lastname columns in the output. Because the columns contain the results of the INITCAP function, the column headings are displayed as the actual function used in the SELECT clause. Oracle9*i* includes the function in the column heading to indicate that the data were manipulated or altered before they were listed in the output. If you do not consider such a column heading appealing, simply include a column alias after the function in the SELECT clause, and it will be displayed in the results.

CHARACTER MANIPULATION FUNCTIONS

Although most data needed by the management of JustLee Books are already stored in the appropriate form in the database, sometimes data might need to be manipulated to yield the desired query output. For example, there may be times when you will need to determine the length of a string, extract portions of a string, or reposition a string, using **manipulation functions**.

The following sections will explain some of the more commonly used manipulation functions; however, you should realize that Oracle9*i* supports more than 150 different functions. Documentation of the functions supported by Oracle9*i* can be viewed on the Oracle Web site at *http://otn.oracle.com/docs/content.html.*

SUBSTR Function

The **SUBSTR** function is used to return a **substring**, or portion of a string. Many organizations code their inventory, general ledger accounts, etc., according to some type of coding scheme. For example, the area code of a customer's telephone number indicates a region within a state.

One way to determine where a customer resides is to look at the first three numbers of a customer's zip code. The United States Postal Service assigns the same first three digits of the zip code for a geographical distribution area within each state. The Marketing Department can use these data to determine where to concentrate certain promotional campaigns. Thus, they can use the SUBSTR function to extract the first three digits of the zip code stored for each customer.

The syntax for this function is `SUBSTR(c, p, l)`, where *c* represents the character string, *p* represents the beginning character position for the extraction, and *l* represents the length of the string to be returned in the results of the query, as shown in Figure 5–6.

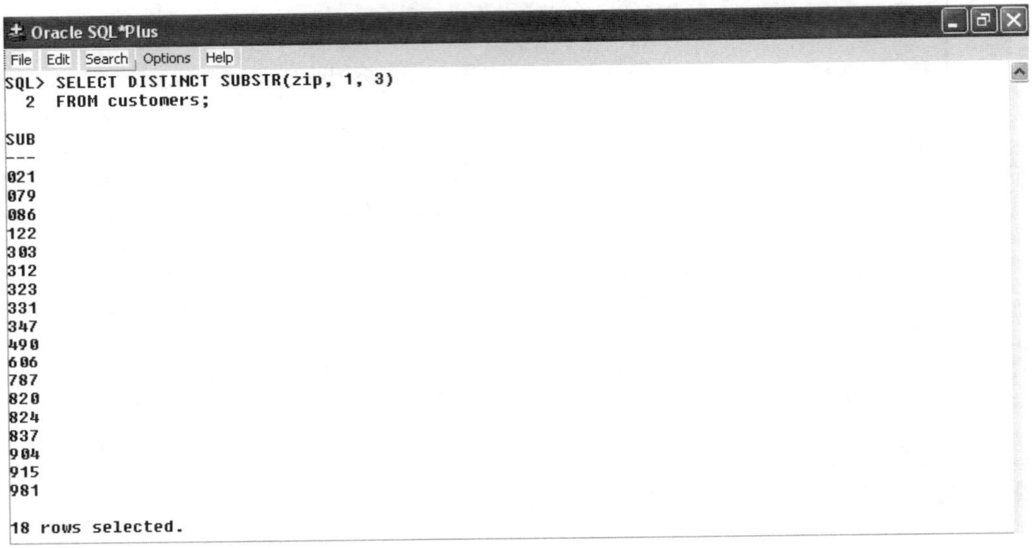

```
Oracle SQL*Plus
File Edit Search Options Help
SQL> SELECT DISTINCT SUBSTR(zip, 1, 3)
  2  FROM customers;

SUB
---
021
079
086
122
303
312
323
331
347
490
606
787
820
824
837
904
915
981

18 rows selected.
```

Figure 5-6 SUBSTR function

In Figure 5–6, the SELECT clause contains the DISTINCT keyword to eliminate duplication in the results. The arguments *(zip, 1, 3)* instruct the software to begin at the first character position of the Zip column and extract three characters as the substring, and then return them in the results.

The function can also extract substrings from the end of the data stored in the field. For example, if a negative 3 (-3) had been entered to indicate the beginning position, then the software would establish the beginning position by counting backward three positions from the end of the field. Similarly, if *(zip, -3, 2)* had been entered as the parameters for the SUBSTR function, then the third and fourth digits of the zip code would have been returned. Notice that in Figure 5–7, the entire zip code is displayed for each unique zip code stored in the CUSTOMERS table. The second column of the output contains just the first three digits of the zip code, and the third column contains just the third and fourth digits of the zip code as requested by the `SUBSTR(zip, -3, 2)` portion of the SELECT clause.

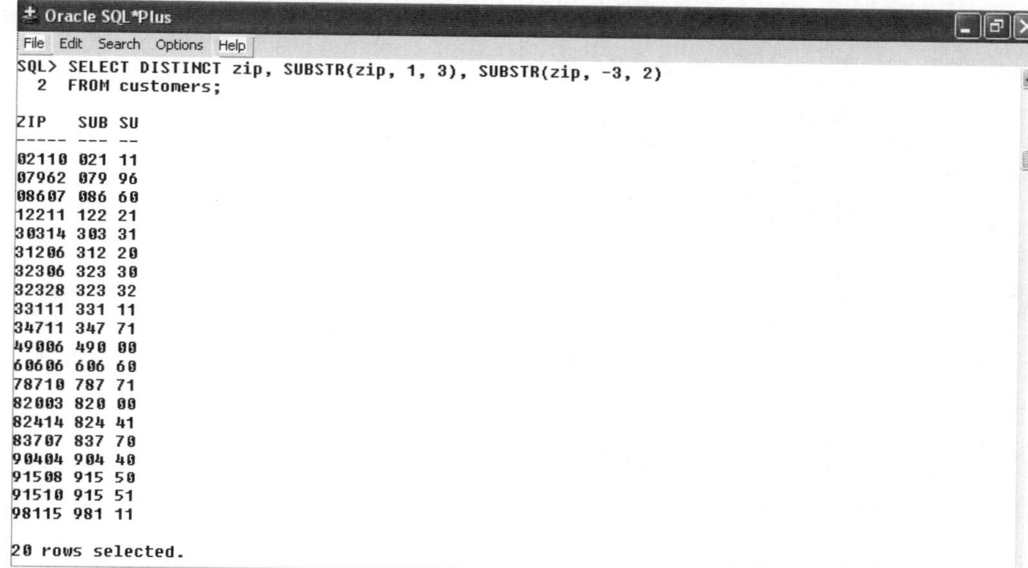

```
SQL> SELECT DISTINCT zip, SUBSTR(zip, 1, 3), SUBSTR(zip, -3, 2)
  2  FROM customers;

ZIP   SUB SU
----- --- --
02110 021 11
07962 079 96
08607 086 60
12211 122 21
30314 303 31
31206 312 20
32306 323 30
32328 323 32
33111 331 11
34711 347 71
49006 490 00
60606 606 60
78710 787 71
82003 820 00
82414 824 41
83707 837 70
90404 904 40
91508 915 50
91510 915 51
98115 981 11

20 rows selected.
```

Figure 5-7 Comparison of the SUBSTR arguments

LENGTH Function

When you plan the width of table columns, design text areas for forms, or determine the size of mailing labels, you might ask, "What is the greatest number of characters that will be entered on this line?" For example, suppose that you are creating mailing labels. You'll need to have labels that are wide enough to accommodate the longest mailing address. To determine the number of characters in a string, such as an address string, you can use the **LENGTH** function. The syntax of the LENGTH function is `LENGTH(c)`, where *c* represents the character string to be analyzed. Figure 5–8 shows the LENGTH function.

```
SQL> SELECT DISTINCT LENGTH(address)
  2  FROM customers;

LENGTH(ADDRESS)
---------------
             11
             12
             13
             16
             17
             18
             20

7 rows selected.
```

Figure 5-8 LENGTH function

The *(address)* argument of the LENGTH function in Figure 5–8 determines the number of characters, or the length of the data, contained in the Address field for each customer. The DISTINCT keyword instructs the software not to include duplicate values in the results. As shown by the output in Figure 5–8, a mailing label that will accommodate at least 20 characters will be required to send mail to current customers.

LPAD and RPAD Functions

Have you ever received a check in which the amount of the check is preceded by a series of asterisks? Many companies fill in the blank spaces on checks and forms with symbols to make it difficult for someone to alter the numbers listed. The **LPAD** function can be used to pad, or fill in, the area to the left of a character string with a specific character—or even a blank space.

The syntax of the LPAD function is `LPAD(c, l, s)`, where *c* represents the character string to be padded, *l* represents the length of the character string *after* being padded, and *s* represents the symbol or character to be used as padding, as shown in Figure 5–9.

```
SQL> SELECT firstname, LPAD(firstname, 12, ' ')
  2  FROM customers
  3  WHERE firstname LIKE '%E%';

FIRSTNAME  LPAD(FIRSTNA
---------- ------------
LEILA             LEILA
MESHIA           MESHIA
KENNETH         KENNETH
JORGE             JORGE
JAKE               JAKE
REESE             REESE
JASMINE         JASMINE
STEVE             STEVE
MICHELL         MICHELL
BECCA             BECCA
GREG               GREG
JENNIFER       JENNIFER
KENNETH         KENNETH

13 rows selected.
```

Figure 5-9 LPAD function

In the example in Figure 5–9, the LPAD function contains the argument `(firstname, 12, ' ')`.

- The first argument, `firstname`, informs the software that the Firstname column field will be padded.
- The second argument, *12*, means that the data contained in the Firstname column should be padded to a total length of 12 spaces—the total length includes both the data and the padding symbol.

- The third argument, `' '`, is an instruction to use a blank space as the padding symbol. Because the LPAD function is used, blank spaces are placed to the left of the customer's first name, until the total width of the data is 12 spaces. Notice that in Figure 5–9, because blank spaces are used to "left pad" the data, the customers' first names are right-aligned in the second column of output.

Oracle9*i* also provides an **RPAD** function that uses a symbol to pad the right side of a character string to a specific width. The syntax of the RPAD function is `RPAD(c, l, s)`, where `c` represents the character string to be padded, `l` represents the total length of the character string *after* being padded, and `s` represents the symbol or character to be used as padding.

5

LTRIM and RTRIM Functions

You can use the **LTRIM** function to remove a specific string of characters from the left side of a set of data. The syntax for the LTRIM function is `LTRIM(c, s)`, where `c` represents the data to be affected and `s` represents the string to be removed from the left of the data.

For example, suppose that some of the preprinted forms used by JustLee Books already contain the string "P.O. Box". However, some of your customers have "P.O. Box" as part of their address in the CUSTOMERS table. The LTRIM function in Figure 5–10 removes the character string 'P.O. Box' from each customer's address before it is displayed in the output. This prevents 'P.O. Box' from being displayed twice on the form.

Oracle SQL*Plus

File Edit Search Options Help

```
SQL> SELECT firstname, lastname, LTRIM(address, 'P.O. BOX')
  2  FROM customers
  3  WHERE address LIKE 'P.O. BOX%';

FIRSTNAME  LASTNAME   LTRIM(ADDRESS,'P.O.B
---------- ---------- --------------------
BONITA     MORALES    651
RYAN       THOMPSON   9835
LEILA      SMITH      66
CINDY      GIRARD     851
KENNETH    JONES      137
JORGE      PEREZ      8564
REESE      MCGOVERN   18
WILLIAM    MCKENZIE   971
JASMINE    LEE        2947
STEVE      SCHELL     677
BECCA      NELSON     563
JENNIFER   SMITH      1151
KENNETH    FALAH      335

13 rows selected.
```

Figure 5-10 LTRIM function

Oracle9*i* also supports the **RTRIM** function to remove specific characters from the right side of a set of data. The syntax for the RTRIM function is `RTRIM(c, s)`, where `c` represents the data to be affected and `s` represents the string to be removed from the right side of the data.

REPLACE Function

The **REPLACE** function is similar to the "search and replace" function used in some application programs. Basically, the REPLACE function looks for the occurrence of a specified string of characters and, if found, substitutes it with another set of characters. The syntax for the REPLACE function is **REPLACE(*c*, *s*, *r*)**, where *c* represents the data or column to be searched, *s* represents the string of characters to be found, and *r* represents the string of characters to be substituted for *s*. In Figure 5–11, every occurrence of "P.O." in a customer's address has been replaced with the words "POST OFFICE," using **REPLACE(address, 'P.O.', 'POST OFFICE')** to indicate that the character string POST OFFICE should be substituted in the display every time Oracle9*i* encounters the string P.O. in the address column of a customer.

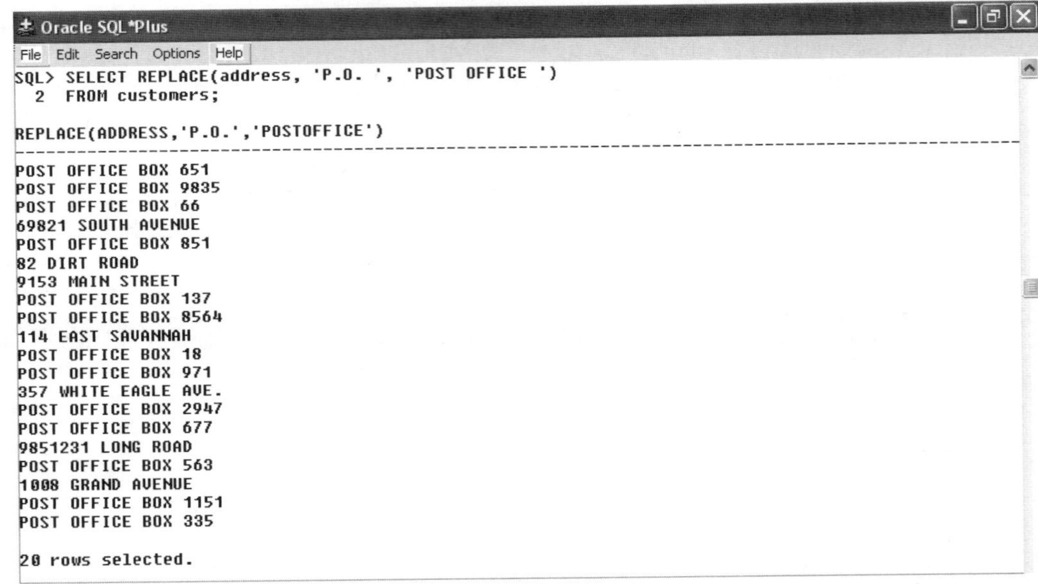

Figure 5-11 REPLACE function

CONCAT Function

Previously you learned how to use vertical bars (| |) to concatenate, or combine, the data from various columns with string literals. The **CONCAT** function can also be used to concatenate the data from two columns. The main difference between the concatenation operator and the CONCAT function is this: You can use the concatenation operator (| |) to combine a long list of columns and string literals; by contrast, you can use the CONCAT function only to combine *two items* (columns or string literals). If you need to combine more than two items using the CONCAT function, you must nest a CONCAT function inside another CONCAT function. The nesting of functions will be discussed later in this chapter.

The syntax for the CONCAT function is **CONCAT(*c1*, *c2*)**, where ***c1*** represents the first item to be included in the concatenation and ***c2*** represents the second item to be included in the operation. Both ***c1*** and ***c2*** can be either a column name or a string literal.

In Figure 5–12, a label has been added to each customer's customer number, so someone reading the output will know that the column is a customer number without having to look at the column heading. A column alias has also been added to the column to identify its contents.


```
SQL> SELECT firstname, lastname, CONCAT('Customer Number: ', customer#) "Number"
  2  FROM customers;

FIRSTNAME  LASTNAME   Number
---------- ---------- ----------------------------------------------------------
BONITA     MORALES    Customer Number: 1001
RYAN       THOMPSON   Customer Number: 1002
LEILA      SMITH      Customer Number: 1003
THOMAS     PIERSON    Customer Number: 1004
CINDY      GIRARD     Customer Number: 1005
MESHIA     CRUZ       Customer Number: 1006
TAMMY      GIANA      Customer Number: 1007
KENNETH    JONES      Customer Number: 1008
JORGE      PEREZ      Customer Number: 1009
JAKE       LUCAS      Customer Number: 1010
REESE      MCGOVERN   Customer Number: 1011
WILLIAM    MCKENZIE   Customer Number: 1012
NICHOLAS   NGUYEN     Customer Number: 1013
JASMINE    LEE        Customer Number: 1014
STEVE      SCHELL     Customer Number: 1015
MICHELL    DAUM       Customer Number: 1016
BECCA      NELSON     Customer Number: 1017
GREG       MONTIASA   Customer Number: 1018
JENNIFER   SMITH      Customer Number: 1019
KENNETH    FALAH      Customer Number: 1020

20 rows selected.
```

Figure 5-12 CONCAT function

NUMBER FUNCTIONS

Oracle9*i* provides a set of functions that are specifically designed to address the manipulation of numeric data. The majority of the number functions relate to trigonometry, such as COS for determining the cosine in radians. In organizations' daily operations, two of the most popular number functions are the ROUND and TRUNC functions.

ROUND Function

The **ROUND** function is used to round numeric fields to the stated precision. The syntax of the ROUND function is ROUND(*n*, *p*), where *n* represents the numeric data, or field, to be rounded and *p* represents the position of the digit to which the data should be rounded. If the value of *p* is a positive number, then the function refers to the right side of the decimal. However, if a negative value is entered, then Oracle9*i* will round to the left side of the decimal position.

In the third column of the results in Figure 5–13, the retail price of each book has been rounded to the nearest tenth, or dime. As shown in the second column of output, the actual retail price of most books ends with .95. In Oracle9*i* (as in most programs), values of five or more are rounded up, and values of less than five are rounded down. In Figure 5–13, the retail price for most of the books appears to be rounded to the nearest dollar. However, notice the book titled *The Wok Way to Cook*. The retail price of this book is $28.75, and when rounded to the nearest tenths' position, the price is $28.80.

```
Oracle SQL*Plus
File Edit Search Options Help
SQL> SELECT title, retail, ROUND(retail, 1), TRUNC(retail, 1)
  2  FROM books;

TITLE                              RETAIL ROUND(RETAIL,1) TRUNC(RETAIL,1)
------------------------------ ---------- --------------- ---------------
BODYBUILD IN 10 MINUTES A DAY       30.95              31            30.9
REVENGE OF MICKEY                      22              22              22
BUILDING A CAR WITH TOOTHPICKS      59.95              60            59.9
DATABASE IMPLEMENTATION             55.95              56            55.9
COOKING WITH MUSHROOMS              19.95              20            19.9
HOLY GRAIL OF ORACLE                75.95              76            75.9
HANDCRANKED COMPUTERS                  25              25              25
E-BUSINESS THE EASY WAY              54.5            54.5            54.5
PAINLESS CHILD-REARING              89.95              90            89.9
THE WOK WAY TO COOK                 28.75            28.8            28.7
BIG BEAR AND LITTLE DOVE             8.95               9             8.9
HOW TO GET FASTER PIZZA             29.95              30            29.9
HOW TO MANAGE THE MANAGER           31.95              32            31.9
SHORTEST POEMS                      39.95              40            39.9

14 rows selected.
```

Figure 5-13 ROUND and TRUNC functions

If the retail price of the books should have been rounded to the nearest dollar, the function entered would have to be changed to **`ROUND(retail, 0)`**, as shown in Figure 5–14. The "0" indicates that the retail price should be rounded to no decimal places. The last column displayed in Figure 5–14 shows the results of rounding the retail price to the nearest tens of dollars, using -1 to indicate the amount to the left of the decimal position that should be rounded.

```
Oracle SQL*Plus
File Edit Search Options Help
SQL> SELECT title, retail, ROUND(retail, 0), ROUND(retail, -1)
  2  FROM books;

TITLE                              RETAIL ROUND(RETAIL,0) ROUND(RETAIL,-1)
------------------------------ ---------- --------------- ----------------
BODYBUILD IN 10 MINUTES A DAY       30.95              31               30
REVENGE OF MICKEY                      22              22               20
BUILDING A CAR WITH TOOTHPICKS      59.95              60               60
DATABASE IMPLEMENTATION             55.95              56               60
COOKING WITH MUSHROOMS              19.95              20               20
HOLY GRAIL OF ORACLE                75.95              76               80
HANDCRANKED COMPUTERS                  25              25               30
E-BUSINESS THE EASY WAY              54.5              55               50
PAINLESS CHILD-REARING              89.95              90               90
THE WOK WAY TO COOK                 28.75              29               30
BIG BEAR AND LITTLE DOVE             8.95               9               10
HOW TO GET FASTER PIZZA             29.95              30               30
HOW TO MANAGE THE MANAGER           31.95              32               30
SHORTEST POEMS                      39.95              40               40

14 rows selected.
```

Figure 5-14 Rounding dollar amounts

TRUNC Function

There may be times when you need to truncate, rather than round, numeric data. You can use the **TRUNC** function to truncate a numeric value to a specific position. Any number(s) after that position is simply removed, or "dropped off." The syntax for the TRUNC function is `TRUNC(n, p)`, where *n* represents the numeric data or field to be truncated, and *p* represents the position of the digit from which the data should be removed or truncated. As with the ROUND function, entering a positive value for *p* indicates a position to the right of the decimal, whereas a negative number indicates a position to the left of the decimal.

The fourth column of the output in Figure 5–13 displays the results of **`TRUNC(retail, 1)`**. Compare the results of the TRUNC function with the results of the ROUND function. Unlike the ROUND function, the results of the TRUNC function did not depend on what value came after the tenths' position; the TRUNC function simply dropped any value beyond the first number after the decimal—without changing the value of the first number.

Again refer to the book *The Wok Way to Cook*. The retail price of the book is $28.75. After the retail price is truncated, notice the result is $28.7 rather than the $28.8 value received after the price was rounded. The value after the tenths' position had no effect on the result returned—it was simply removed from the retail price. If a zero had been included as the second argument of the TRUNC function, no decimal positions would have been displayed for the retail price in the output. However, remember that the dollar amount displayed would not be rounded; the decimals would simply have been dropped, or eliminated.

DATE FUNCTIONS

Oracle9*i*'s **DATE** function displays date values in a DD–MON–YY format that represents a two-digit day, a three-letter month abbreviation, and a two-digit year. For example, the date of February 2, 2004, would be stored as 02–FEB–04. Although users reference a date as a non-numeric field (i.e., a character string that must be enclosed in single quotation marks), it is actually stored internally in a numeric format that includes century, year, month, day, hours, minutes, and seconds. The valid range of dates that Oracle9*i* can reference is January 1, 4712, B.C. to December 31, 9999, A.D. Although dates appear as non-numeric fields, users can perform calculations with dates because they are stored internally as numeric data. The internal numeric version of a date used by Oracle9*i* is a Julian date. A **Julian date** represents the number of days that have passed between a specified date and January 1, 4712, B.C. For example, if you need to calculate the number of days between two dates, Oracle9*i* would first convert the dates to the Julian date numeric format and then determine the difference between the two dates. If Oracle9*i* did not have a numeric equivalent for a date, there would be a problem trying to derive the solution for the arithmetic expression '02–JAN–03' - '08–SEP–99'. Look at the calculation with date columns in Figure 5–15.

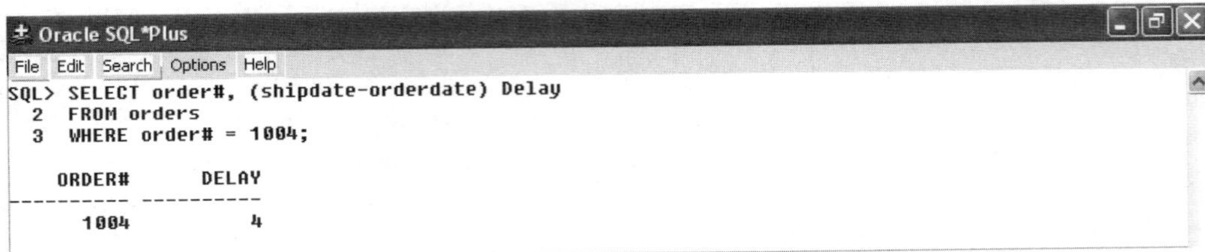

Figure 5–15 Calculation with date columns

In Figure 5–15, order 1004 did not ship the same day as it was ordered. To determine how many days shipment was delayed, the Orderdate column is subtracted from the Shipdate column. If calculations are performed between two date fields without the use of the DATE function, then the results are returned in terms of days because the dates are internally stored as numeric values.

By contrast, if you need the results returned in terms of weeks rather than days, simply divide the results by seven. For example, to have the results of the query in Figure 5–15 reported in terms of weeks (or portion of a week in this case), the equation in the SELECT clause would have been changed to `(shipdate-orderdate)/7`. However, there are times when it is difficult to convert date calculation results to a unit that you need. For example, suppose that you needed to know the number of months between two dates: Do you divide by 30 or by 31? What number would you use for February? As shown in the following sections, Oracle9*i* provides various functions to assist you with these kinds of calculations.

MONTHS_BETWEEN Function

Suppose that management would like to know whether customers are ordering books that have recently been released, or books that were published several months, or even years, ago. To find an answer to management's question, you might simply subtract the publication date (Pubdate) from the order date (Orderdate) for a book to determine how many days a book had been available to the public before it was ordered. However, as mentioned in the previous section, what number should be used to convert days to months? Oracle9*i* provides the **MONTHS_BETWEEN** function to determine the number of months between two dates. The syntax for the function is `MONTHS_BETWEEN(d1, d2)`, where *d1* and *d2* are the two dates in question and *d2* is subtracted from *d1*. Figure 5–16 shows this function.

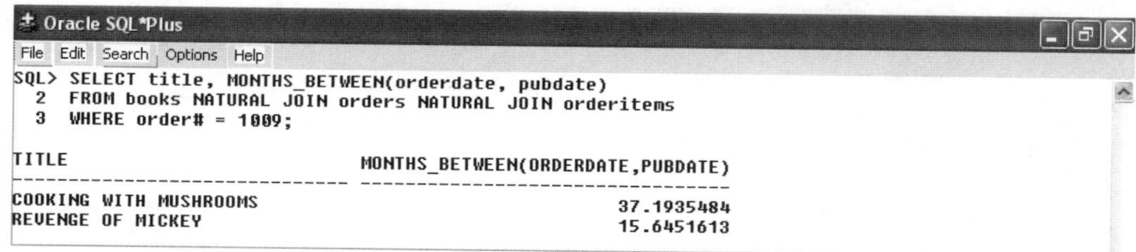


```
SQL> SELECT title, MONTHS_BETWEEN(orderdate, pubdate)
  2  FROM books NATURAL JOIN orders NATURAL JOIN orderitems
  3  WHERE order# = 1009;

TITLE                          MONTHS_BETWEEN(ORDERDATE,PUBDATE)
------------------------------ ---------------------------------
COOKING WITH MUSHROOMS                                37.1935484
REVENGE OF MICKEY                                     15.6451613
```

Figure 5-16 MONTHS_BETWEEN function

In Figure 5–16, the user wants to determine how many months the two books from order 1009 were available before this order was placed. Notice that the results of the query do not simply provide a whole number to indicate the number of months that have elapsed. Instead, the digits after the decimal indicate portions of a month. To remove the portions of a month, include the MONTHS_BETWEEN function inside a TRUNC function. This is nesting, and it will be demonstrated later in this chapter.

ADD_MONTHS Function

The management of JustLee Books has adopted a policy that states the company will stock a book for only five years after it is published. Management believes that after five years, sales for most books will decline to such a level that it is no longer profitable to keep them in inventory. Thus, management periodically requests a list of the current books in inventory and the date on which each book should be dropped from inventory. One method of calculating the "drop date" is to simply add 1825 (365*5) days to the publication date of each book. However, an even better approach is to use the **ADD_MONTHS** function, as shown in Figure 5–17.

```
Oracle SQL*Plus
File Edit Search Options Help
SQL> SELECT title, pubdate, ADD_MONTHS(pubdate, 60) "Drop Date"
  2  FROM books
  3  ORDER BY "Drop Date";

TITLE                          PUBDATE   Drop Date
------------------------------ --------- ---------
HOW TO MANAGE THE MANAGER      09-MAY-99 09-MAY-04
DATABASE IMPLEMENTATION        04-JUN-99 04-JUN-04
COOKING WITH MUSHROOMS         28-FEB-00 28-FEB-05
PAINLESS CHILD-REARING         17-JUL-00 17-JUL-05
THE WOK WAY TO COOK            11-SEP-00 11-SEP-05
BODYBUILD IN 10 MINUTES A DAY  21-JAN-01 21-JAN-06
HANDCRANKED COMPUTERS          21-JAN-01 21-JAN-06
SHORTEST POEMS                 01-MAY-01 01-MAY-06
BIG BEAR AND LITTLE DOVE       08-NOV-01 08-NOV-06
REVENGE OF MICKEY              14-DEC-01 14-DEC-06
HOLY GRAIL OF ORACLE           31-DEC-01 31-DEC-06
E-BUSINESS THE EASY WAY        01-MAR-02 01-MAR-07
BUILDING A CAR WITH TOOTHPICKS 18-MAR-02 18-MAR-07
HOW TO GET FASTER PIZZA        11-NOV-02 11-NOV-07

14 rows selected.
```

Figure 5-17 ADD_MONTHS function

In Figure 5–17, the ADD_MONTHS function is applied to the Pubdate column for each book to determine its "drop date." The syntax for the ADD_MONTHS function is `ADD_MONTHS(d, m)`, where *d* represents the beginning date for the calculation and *m* represents the number of months to add to the date. As shown in the output of the query, the result of the ADD_MONTHS function is a new date with the correct number of months added to the old date.

NEXT_DAY Function

JustLee Books' policy is that books must be shipped by the first Monday after they receive a customer's order. Whenever an order is received, the customer is informed of the latest date that order is expected to ship. The **NEXT_DAY** function can determine the next occurrence of a specific day of the week after a given date. The syntax for the NEXT_DAY function is `NEXT_DAY(d, DAY)`, where *d* represents the starting date and *DAY* represents the day of the week to be identified. Figure 5–18 shows the NEXT_DAY function.

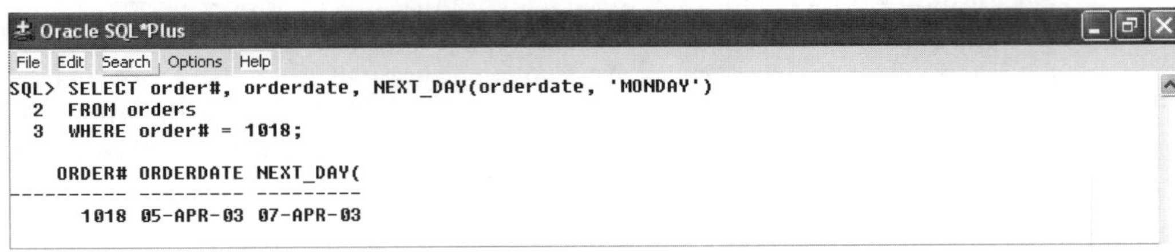

```
Oracle SQL*Plus
File Edit Search Options Help
SQL> SELECT order#, orderdate, NEXT_DAY(orderdate, 'MONDAY')
  2  FROM orders
  3  WHERE order# = 1018;

    ORDER# ORDERDATE NEXT_DAY(
---------- --------- ---------
      1018 05-APR-03 07-APR-03
```

Figure 5-18 NEXT_DAY function

In Figure 5–18, order 1018 was ordered on April 5, 2003. Because JustLee's policy is to inform a customer of the latest possible ship date, the query requests the date of the first Monday following the date of the order. As shown in the results, the customer can expect the order to be shipped by April 7—the following Monday.

TO_DATE Function

The **TO_DATE** function is of particular interest to application developers. Many database users may be uncomfortable entering a date in the format of DD–MON–YY and would prefer to enter a date as MM/DD/YY or Month DD,YYYY. The TO_DATE function allows users to enter a date in any format, and then it converts the entry into the default format used by Oracle9*i* of a two-digit day, three-letter month abbreviation, and two-digit year. The syntax for the TO_DATE function is `TO_DATE(d,f)`, where *d* represents the date being entered by the user and *f* is the format for the date that was entered. Figure 5–19 shows valid formats for entering a date in the TO_DATE function.

5

Date Formats		
Element	**Description**	**Example**
MONTH	Name of the month spelled out—padded with blank spaces to a total width of nine spaces	APRIL
MON	Three-letter abbreviation for the name of the month	APR
MM	Two-digit numeric value of the month	04
RM	Roman numeral month	IV
D	Numeric value for the day of the week	Wednesday = 4
DD	Numeric value for the day of the month	28
DDD	Numeric value for the day of the year	December 31 = 365
DAY	Name of the day of the week—padded with blank spaces to a length of nine characters	Wednesday
DY	Three-letter abbreviation for the day of the week	WED
YYYY	Displays the four-digit year	2004
YYY or YY or Y	Displays the last three, two, or single digit(s) of the year	2004 = 004; 2004 = 04; 2004 = 4
YEAR	Spells out the year	TWO THOUSAND FOUR
B.C. or A.D.	Indicates B.C. or A.D.	2004 A.D.

Figure 5-19 Data format elements

A more comprehensive list of the format model elements is provided later in this chapter, in the discussion of the TO_CHAR function.

When working with the TO_DATE function, the user enters the actual date specified as the first argument of the function. The second argument is a format model that allows Oracle9*i* to distinguish the different parts of the date. Because both the date entered and the format model are character strings, each argument must be enclosed in single quotation marks.

Suppose that a user needs a list of the orders placed on March 31, 2003, and shipped in April. The only March orders that will still be in the ORDERS table will be those that are still outstanding (not filled) at the beginning of April. However, for illustrative purposes, suppose that the user does not like the default date format used by Oracle9*i* and wants to use a more familiar format. In this case, the user prefers the format of Month DD, YYYY when referring to dates. As shown in Figure 5–20, using the TO_DATE function in a WHERE clause enables the user to enter the preferred format for the desired order date, and then includes the format model necessary for Oracle9*i* to interpret the order date.

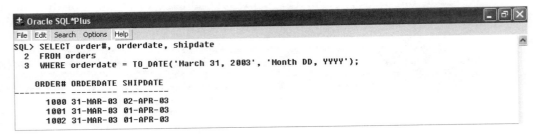

```
Oracle SQL*Plus
File Edit Search Options Help
SQL> SELECT order#, orderdate, shipdate
  2  FROM orders
  3  WHERE orderdate = TO_DATE('March 31, 2003', 'Month DD, YYYY');

    ORDER# ORDERDATE SHIPDATE
---------- --------- ---------
      1000 31-MAR-03 02-APR-03
      1001 31-MAR-03 01-APR-03
      1002 31-MAR-03 01-APR-03
```

Figure 5-20 TO_DATE function

The TO_DATE function is traditionally used for entering a date value into a table, using the INSERT command. The INSERT command will be discussed in Chapter 8.

OTHER FUNCTIONS

Some functions provided by Oracle9*i* do not fall neatly into a character, numeric, or date category. However, these functions are very important and are widely used in the work environment. There are five functions that will be discussed in this section: NVL, NVL2, TO_CHAR, DECODE, and SOUNDEX.

NVL Function

You can use the **NVL** function to address problems that can be caused when performing arithmetic operations with fields that may contain NULL values. (Recall that a NULL value is the absence of data.) In Oracle9*i*, a NULL value is not equivalent to a blank space or a zero. When a NULL value is used in a calculation, the result is a NULL value. The NVL function is used to substitute a value for the existing NULL. The syntax for the NVL function is `NVL(x, y)`, where *y* represents the value to be substituted if *x* is NULL. In many cases, the substitute for a NULL value in a calculation is zero (0).

In the real world, the NVL function is most commonly used to calculate an individual's gross pay as "salary + commission." But what happens when an individual's sales are not high enough, and he or she is not entitled to a commission? If you add the individual's salary to a NULL commission, the resulting gross pay is NULL—no paycheck. Unfortunately, you will probably realize that the error has occurred about the same time the individual storms into your office, wanting to know why he or she did not get a paycheck. To avoid this problem, rather than calculating gross pay as salary plus commission, use **`salary + NVL(commission, 0)`**. The NVL function simply substitutes a zero whenever the commission is NULL, and the person still gets paid, because "salary + 0" still equals "salary."

There are times, however, when substituting a simple zero for a NULL value will not return the desired results. In such cases, you may need to use a little more imagination. For example, suppose that management requests a report that reflects shipment delays for orders. Some of those orders may not yet have shipped, but they are expected to ship on the following Monday. If the Shipdate column for an order is left blank for the pending shipments, management may assume that there was no delay in the shipment of the order. However, if a zero is substituted for the missing ship date, weird results may be displayed in the query's output, as you can see in Figure 5–21.

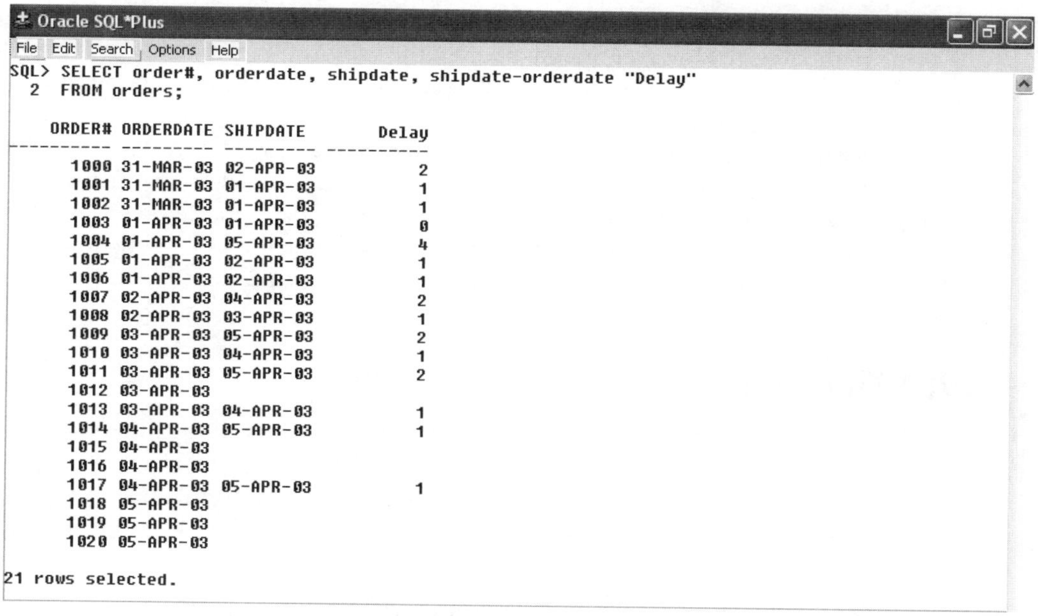

```
Oracle SQL*Plus
File Edit Search Options Help
SQL> SELECT order#, orderdate, shipdate, shipdate-orderdate "Delay"
  2  FROM orders;

    ORDER# ORDERDATE SHIPDATE       Delay
---------- --------- --------- ----------
      1000 31-MAR-03 02-APR-03          2
      1001 31-MAR-03 01-APR-03          1
      1002 31-MAR-03 01-APR-03          1
      1003 01-APR-03 01-APR-03          0
      1004 01-APR-03 05-APR-03          4
      1005 01-APR-03 02-APR-03          1
      1006 01-APR-03 02-APR-03          1
      1007 02-APR-03 04-APR-03          2
      1008 02-APR-03 03-APR-03          1
      1009 03-APR-03 05-APR-03          2
      1010 03-APR-03 04-APR-03          1
      1011 03-APR-03 05-APR-03          2
      1012 03-APR-03
      1013 03-APR-03 04-APR-03          1
      1014 04-APR-03 05-APR-03          1
      1015 04-APR-03
      1016 04-APR-03
      1017 04-APR-03 05-APR-03          1
      1018 05-APR-03
      1019 05-APR-03
      1020 05-APR-03

21 rows selected.
```

Figure 5-21 Calculation involving a NULL value

In Figure 5–21, the Delay between the Orderdate and Shipdate columns for each order is calculated. Since some orders have not yet been shipped, the actual ship date is a NULL value, and the Delay column is left blank. If management simply skims over the report, it would appear that the longest delay is four days (order 1004). However, this is not accurate. Because management's policy is that orders must be shipped by the first Monday after an order is received, you can substitute the expected Shipdate into the Delay calculation. See Figure 5–22.

```
SQL> SELECT order#, orderdate, NVL(shipdate, '07-APR-03'),
  2  NVL(shipdate, '07-APR-03')-orderdate "Delay"
  3  FROM orders
  4  WHERE order# = 1018;

    ORDER# ORDERDATE NVL(SHIPD      Delay
---------- --------- --------- ----------
      1018 05-APR-03 07-APR-03          2
```

Figure 5-22 NVL function to calculate shipping delay

Order 1018 in Figure 5–21 has not yet shipped. To calculate the anticipated shipping delay for that order, Figure 5–22 shows using the NVL function to substitute the date for the first Monday (07-APR-03) after the date of the order to estimate the shipping date and the resulting delay. The first NVL function given in the SELECT clause is **`NVL(shipdate, '07-APR-03')`**.

The purpose of this function is to substitute the anticipated shipping date for the NULL value in the results. However, simply because the substitution was made in the third column of the output does not mean that it will affect the calculation of the delay in the shipment of the order. The portion of the SELECT clause that is listed on line 2 of the query is what is actually calculating the shipment delay: **`NVL(shipdate, '07-APR-03')-orderdate "Delay"`**. This portion of the SELECT clause instructs Oracle9*i* that if the Shipdate column is NULL, substitute April 7, 2003, as the shipping date for the order, and then subtract the order date from the assigned date of shipment. After the NVL function has made the substitution, the subtraction is performed and the delay is calculated. The "Delay" portion is simply the column alias for that column.

NVL2 Function

The **NVL2** function is a variation of the NVL function that allows different options, based on whether a NULL value exists. The syntax for the NVL2 function is `NVL2(`*`x, y, z`*`)`, where *y* represents what should be substituted if *x* is not NULL, and *z* represents what should be substituted if *x* is NULL. This allows the user a little more flexibility when working with NULL values.

In reference to the gross pay calculation described in the previous section, rather than using the equation **`salary+NVL(commission, 0)`** to substitute a zero whenever the commission is NULL, the user could have used **`NVL2(commission, salary, salary+commission)`**. The NVL2 function would be read as, "If the commission is NULL, then the gross pay is simply salary. If the commission is not NULL, then calculate gross pay as salary plus commission."

Suppose that management needs a report describing the shipment status of orders. The report should list an order as being shipped or not shipped. You could use the NVL2 function to display the status of each order, based on whether the Shipdate column contains a NULL value. Because a date indicates that an order has been shipped, and a NULL value would indicate that an order has not yet shipped, this is the ideal situation for using the NVL2 function. See Figure 5–23.

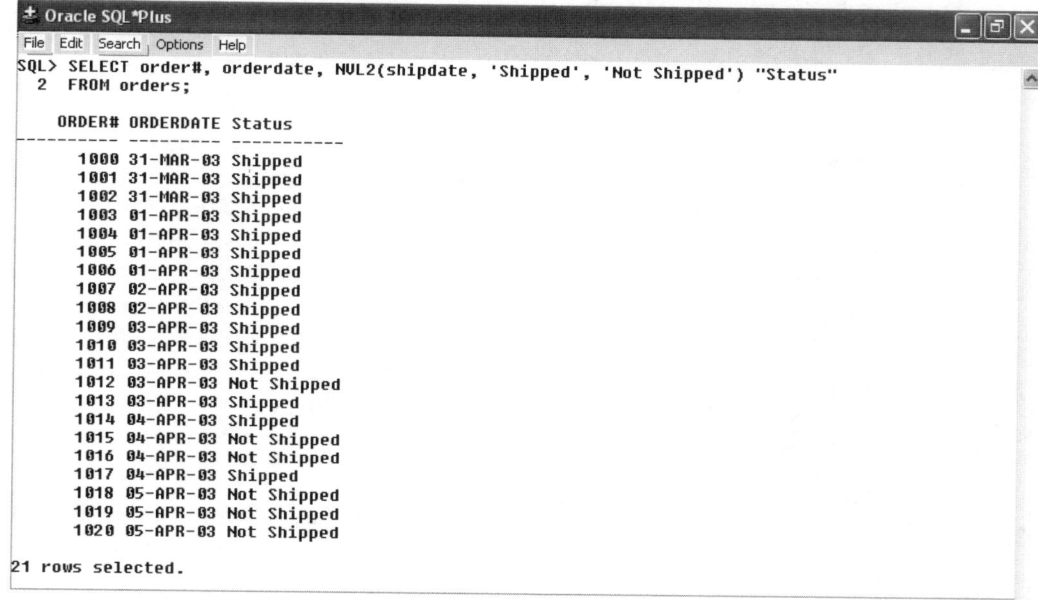


```
SQL> SELECT order#, orderdate, NVL2(shipdate, 'Shipped', 'Not Shipped') "Status"
  2  FROM orders;

    ORDER# ORDERDATE Status
---------- --------- -----------
      1000 31-MAR-03 Shipped
      1001 31-MAR-03 Shipped
      1002 31-MAR-03 Shipped
      1003 01-APR-03 Shipped
      1004 01-APR-03 Shipped
      1005 01-APR-03 Shipped
      1006 01-APR-03 Shipped
      1007 02-APR-03 Shipped
      1008 02-APR-03 Shipped
      1009 03-APR-03 Shipped
      1010 03-APR-03 Shipped
      1011 03-APR-03 Shipped
      1012 03-APR-03 Not Shipped
      1013 03-APR-03 Shipped
      1014 04-APR-03 Shipped
      1015 04-APR-03 Not Shipped
      1016 04-APR-03 Not Shipped
      1017 04-APR-03 Shipped
      1018 05-APR-03 Not Shipped
      1019 05-APR-03 Not Shipped
      1020 05-APR-03 Not Shipped

21 rows selected.
```

Figure 5-23 NVL2 function

In Figure 5–23, the NVL2 function has one of two character strings displayed, depending on whether the Shipdate column contains a NULL value. Because character strings are being used in the function, the strings must be enclosed in single quotation marks. If the strings are not enclosed in single quotation marks, Oracle9*i* assumes an existing column is being referenced and it returns an error because such columns do not exist in the table.

TO_CHAR Function

The **TO_CHAR** function is widely used to convert dates and numbers to a formatted character string. It is the opposite of the TO_DATE function discussed previously. The TO_DATE function allows a user to *enter* a date in any type of format, whereas the TO_CHAR function is used to have Oracle9*i* *display* dates in a particular format. The syntax of the TO_CHAR function is `TO_CHAR(n, 'f')`, where `n` is the date or number to be formatted and `f` is the format model to be used. A format model consists of a series of elements that represents exactly how the data should appear and must be entered within single quotation marks, as shown in Figure 5–24.

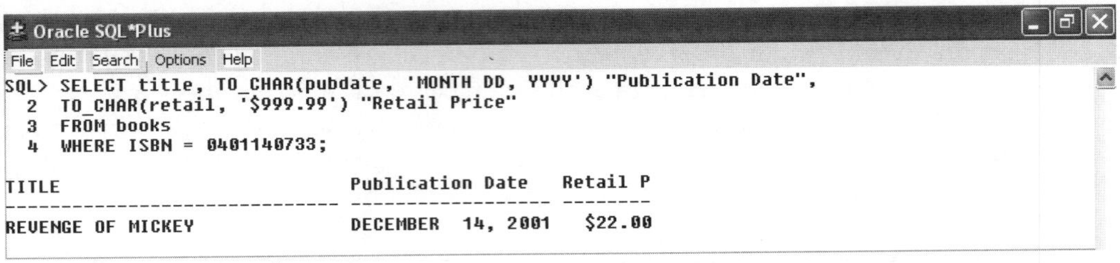

```
Oracle SQL*Plus
File Edit Search Options Help
SQL> SELECT title, TO_CHAR(pubdate, 'MONTH DD, YYYY') "Publication Date",
  2  TO_CHAR(retail, '$999.99') "Retail Price"
  3  FROM books
  4  WHERE ISBN = 0401140733;

TITLE                          Publication Date   Retail P
------------------------------ ------------------ --------
REVENGE OF MICKEY              DECEMBER  14, 2001   $22.00
```

Figure 5-24 TO_CHAR function

The query in Figure 5–24 contains two TO_CHAR functions.

The first TO_CHAR function is used to convert the publication date (Pubdate) to a specific date format (Month DD,YYYY) that spells out the month of the year, followed by the day of the month, a comma, and then the four-digit year. Notice that there is an unusual amount of space after the word DECEMBER. If you would like to eliminate insignificant spaces or zeros in the display, simply enter an *fm* at the beginning of the model to be used to display the data. The *fm* instructs Oracle9*i* to turn off the "fill mode," which is basically blank padding or added blank spaces, to create a fixed width for the name of the month. To eliminate the space from DECEMBER in the Pubdate column, the correct function and argument are **`TO_CHAR(pubdate, 'fmMONTH DD, YYYY')`**. If you would prefer to have the name of the month displayed in mixed case (e.g., December), simply use that case in the format model—`('fmMonth DD, YYYY')`.

The second TO_CHAR function in Figure 5–24 is used to format the "Retail Price" of the book to display a dollar sign and two decimal positions. Without the format model, Retail Price would have been displayed as 22—without the dollar sign or any decimals. Notice that the column heading for Retail Price is truncated. Although column headings for numeric fields are never truncated, in this case, Retail Price was converted to a character string so a format model could be applied. Therefore, Retail Price in this output is not considered a column of numeric data, but is actually considered a character string. Because the column heading for the character string is longer than the width of the column, it is truncated.

Oracle9*i* provides a wide variety of elements you can use to create format models for dates and numbers. The table in Figure 5–25 describes some commonly used format elements.

Formats		
Element	**Description**	**Example**
Date Elements		
MONTH	Name of the month spelled out—padded with blank spaces to a total width of nine spaces	APRIL
MON	Three-letter abbreviation for the name of the month	APR
MM	Two-digit numeric value of the month	04
RM	Roman numeral month	IV
D	Numeric value for the day of the week	Wednesday = 4
DD	Numeric value for the day of the month	28
DDD	Numeric value for the day of the year	December 31 = 365
DAY	Name of the day of the week—padded with blank spaces to a length of nine characters	Wednesday
DY	Three-letter abbreviation for the day of the week	WED
YYYY	Displays the four-digit year	2004
YYY or YY or Y	Displays the last three, two, or single digit(s) of the year	2004 = 004; 2004=04; 2004 = 4
YEAR	Spells out the year	TWO THOUSAND FOUR
BC or AD	Indicates B.C. or A.D.	2004 A.D.
Time Elements		
SS	Seconds	Value between 0–59
SSSS	Seconds past midnight	Value between 0–86399
MI	Minutes	Value between 0–59
HH or HH12	Hours	Value between 1–12
HH24	Hours	Value between 0–23
A.M. or P.M.	Indicates morning or evening hours	A.M. (before noon) or P.M. (after noon)
Number Elements		
9	Series of 9's indicates width of display (with insignificant leading zeros not displayed)	99999
0	Displays insignificant leading zeros	0009999
$	Displays a floating dollar sign	$99999
.	Indicates number of decimals to display	999.99
,	Displays a comma in the position indicated	9,999

Figure 5-25 Format model elements

Other Elements		
, . (punctuation symbols)	Display indicated punctuation	DD, YYYY = 24, 2001
"string"	The exact character string inside the double quotation marks is displayed	"of the year" YYYY = of the year 2001
TH	Ordinal number	DDTH = 8th
SP	Spell out number	DDSP = EIGHT
SPTH	Spell out ordinal number	DDSPTH = EIGHTH

Figure 5-25 Format model elements (continued)

An "RR" format was previously used to address potential problems raised by Y2K. However, most individuals in industry use the YYYY format model element to specify the exact century for a date.

DECODE Function

The **DECODE** function takes a specified value and compares it to values in a list. If a match is found, then the specified result is returned. If no match is found, then a default result is returned. If no default result is defined, a NULL is returned as the result. The DECODE function allows the user to specify different actions to be taken, depending on the circumstances (e.g., the exact value is or is not contained within a column). It basically saves the user from having to enter multiple statements for each possible scenario. The syntax for the DECODE function is `DECODE(V, L1, R1, L2, R2,..., D)`, where *V* is the value you are searching for, *L1* represents the first value in the list, *R1* represents the result to be returned if *L1* and *V* are equivalent, etc., and *D* is the default result to return if no match is found.

The DECODE function is similar to the CASE or IF...THEN...ELSE structures found in many programming languages.

JustLee Books is required to collect sales tax from customers who live in Florida and California. They are not required to collect sales tax on sales made to customers residing in other states. So, if a customer resides in California, 8% of the total order price must be collected as sales tax. However, if the customer lives in Florida, the customer must pay 7% sales tax.

To determine the sales tax rate that applies to each customer, the DECODE function can be used to compare the state in which each customer lives to a list of states. If a match occurs, then the sales tax rate that applies to that state is returned. However, if the customer lives in a state that is not listed, a default sales rate of 0 is applied, as shown in Figure 5–26.

```
Oracle SQL*Plus
File Edit Search Options Help
SQL> SELECT customer#, state,
  2         DECODE(state, 'CA', .08,
  3                       'FL', .07,
  4                             0) "Sales Tax Rate"
  5   FROM customers;

 CUSTOMER# ST Sales Tax Rate
---------- -- --------------
      1001 FL            .07
      1002 CA            .08
      1003 FL            .07
      1004 ID              0
      1005 WA              0
      1006 NY              0
      1007 TX              0
      1008 WY              0
      1009 CA            .08
      1010 GA              0
      1011 IL              0
      1012 MA              0
      1013 FL            .07
      1014 WY              0
      1015 FL            .07
      1016 CA            .08
      1017 MI              0
      1018 GA              0
      1019 NJ              0
      1020 NJ              0

20 rows selected.
```

Figure 5-26 DECODE function

In Figure 5–26, the DECODE function begins on line 2. The State (ST) column is identified as the value to be compared against the list. The state of California (CA) is the first item against which the value of the State column is compared. If the State column contains the value of CA, then .08 is returned as the sales tax rate, and the DECODE function is processed again for the next customer. If the value in the State column is not CA, then the value is compared against the next item in the list. (Note that in Figure 5–26, the second listed item has been placed on line 3 to improve the readability of the function.) If the value for the State column is equal to FL, then a sales tax rate of .07 is returned. If the value in the State column is not equal to the two items listed (CA or FL), then the default value is assigned, which in this case is zero.

SOUNDEX Function

Many government agencies and organizations perform searches for information based on the phonetic pronunciation of words rather than their actual spelling. For example, in many states, the first set of four characters and numbers of a driver's license number represents the phonetic sound of the individual's last name at the time the license was issued (it does not change if the person changes his or her last name). Oracle9*i* can reference the phonetic sound or representation of words using the **SOUNDEX** function. The syntax of the function is `SOUNDEX(c)`, where *c* is the character string being referenced. To demonstrate how the phonetic sound of a person's last name can be different from the actual spelling of the name, try sorting the last name of the customers in the CUSTOMERS table, based on the phonetic representation of the name as shown in Figure 5–27.

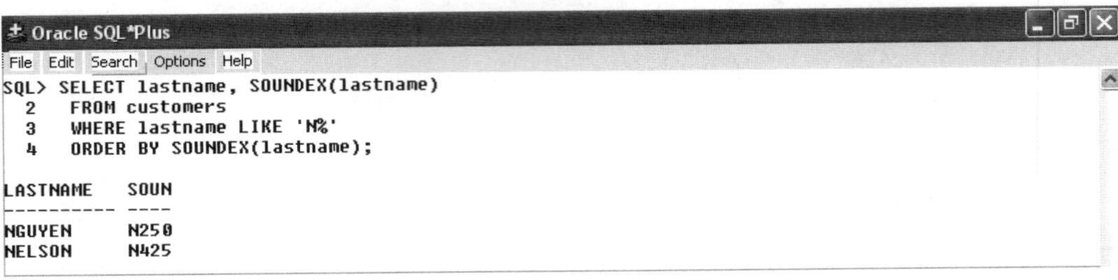

Figure 5-27 SOUNDEX function

Notice that in the results, the last name Nguyen appears before Nelson in the output. Why? Notice the second column of output. The letter "N" followed by three numbers is the phonetic representation of the last names listed. The letter indicates the actual first letter of the name, and the three numbers represent the sound produced by the remaining letters. The actual number assigned is based on an algorithm that assigns a numeric value based on certain letters that occur in the word. If either of these individuals receives a driver's license in a state that uses phonetics in deriving the license number, the letter "N" and the three numbers would represent the first part of their assigned driver's license number.

NESTING FUNCTIONS

Any of the single-row functions can be nested inside other single-row functions. **Nesting** functions simply means that one function is used as an argument inside of another function. There are some important rules to remember when nesting functions:

1. All arguments required for each function must be provided.
2. For every open parenthesis, there must be a corresponding closed parenthesis.
3. The nested, or inner, function is solved first. The result of the inner function is passed to the outer function, and then the outer function is executed.

The query previously presented in Figure 5–16 returned the number of months between the date of an order and the publication date of the book ordered. However, the output also included portions of a month. To eliminate the decimal portion of the output and return only the number of whole months between the two dates, you can nest the MONTHS_BETWEEN function inside the TRUNC function, as shown in Figure 5–28.

```
SQL> SELECT title, TRUNC(MONTHS_BETWEEN(orderdate, pubdate), 0)
  2  FROM books NATURAL JOIN orders NATURAL JOIN orderitems
  3  WHERE order# = 1009;

TITLE                          TRUNC(MONTHS_BETWEEN(ORDERDATE,PUBDATE),0)
------------------------------ ------------------------------------------
COOKING WITH MUSHROOMS                                                 37
REVENGE OF MICKEY                                                      15
```

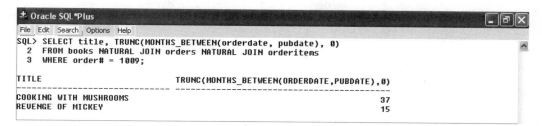

Figure 5-28 Nesting functions

Remember that the TRUNC function has two arguments—the value to be truncated followed by the position at which the truncation should occur. The value that you need truncated is the result of the MONTHS_BETWEEN function. Therefore, the MONTHS_BETWEEN function is entered as the first argument of the TRUNC function in Figure 5–28. The closing parenthesis after the Pubdate column name completes the MONTHS_BETWEEN function.

The nested MONTHS_BETWEEN function is followed by a comma, a zero, and a parenthesis. These serve to complete the TRUNC function. The zero indicates that there should be no decimal positions after the truncation occurs, and a parenthesis closes the TRUNC function. The value calculated by the inner function (MONTHS_BETWEEN) is used to complete the outer function (TRUNC), and the result is the whole number of months between the publication date of each book and the date the book was ordered.

DUAL TABLE

All the examples for the functions presented in this chapter have been based on data contained within a table. There may be times, however, when you need to use a function or to display data not contained in a table. For example, you might need to see the current date setting for your computer. The keyword **SYSDATE** returns the value of the date according to the computer. For example, if you need to determine how long ago a particular book was released, you can simply subtract the publication date from the SYSDATE. As shown in Figure 5–29, if the current date were February 19, 2004, you would find that the book titled *Handcranked Computers* is approximately 1125 days old.

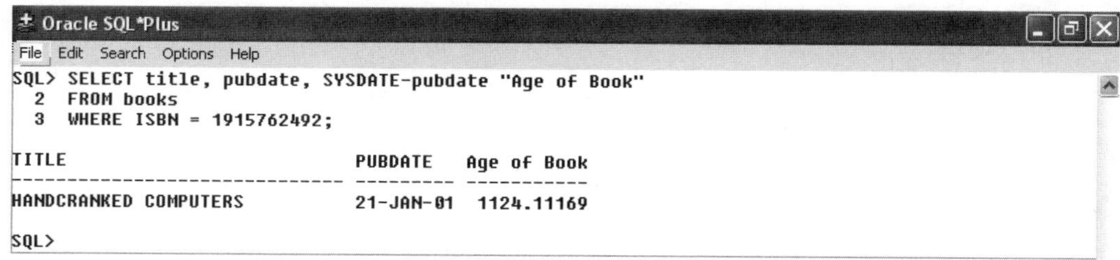

Figure 5-29 Calculation using SYSDATE

However, if you simply need to see the current date setting for your computer, you cannot use `SELECT SYSDATE;` and retrieve the date, because all SELECT statements must contain a SELECT clause and a FROM clause. Oracle9*i* has a dummy table that contains a single column and a single row to address this problem. The dummy table is called **DUAL**. If you need to display the current date, simply enter **`SELECT SYSDATE FROM dual;`** and the date setting for the computer will be displayed, as shown in Figure 5–30.

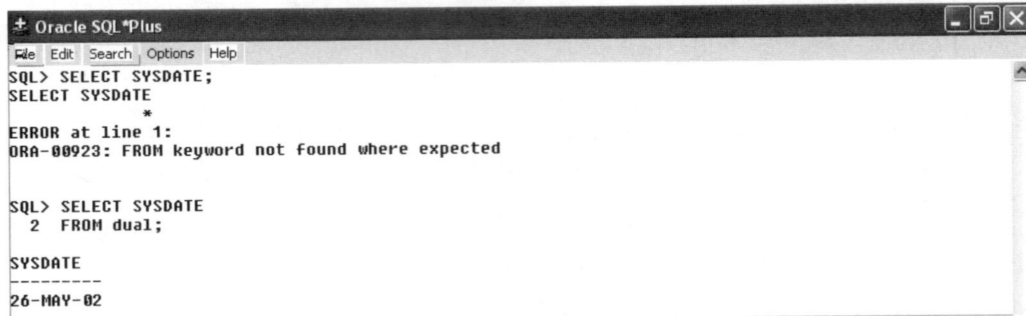

Figure 5-30 DUAL table reference

Any of the single-row functions presented in this chapter can be used with the DUAL table. Although the DUAL table is rarely used in industry for anything other than retrieving the system date and for certain programming procedures, it can be valuable for someone learning how to work with functions. For example, if you would like to practice rounding numbers or determining the length of character strings, you can enter a specific value in the appropriate function and reference the DUAL table in the FROM clause, as shown in Figure 5–31.

```
Oracle SQL*Plus
File Edit Search Options Help
SQL> SELECT ROUND(43769.43, -2)
  2  FROM dual;

ROUND(43769.43,-2)
------------------
             43800

SQL> SELECT LENGTH('Hello')
  2  FROM dual;

LENGTH('HELLO')
---------------
              5

SQL> SELECT SOUNDEX('Morris')
  2  FROM dual;

SOUN
----
M620
```

Figure 5-31 Function practice using the DUAL table

CHAPTER SUMMARY

- Oracle9*i* provides more than 150 predefined functions.
- Single-row functions return a result for each row or record processed.
- Character case conversion functions such as UPPER, LOWER, and INITCAP can be used to alter the case of character strings.
- Character manipulation functions can be used to extract substrings or portions of a string, replace occurrences of a string with another string, determine the length of a character string, and trim spaces or characters from strings.
- Simple number functions can round or truncate a number on both the left and right side of a decimal.
- Date functions can be used to perform calculations with dates or to change the format of a date entered by a user.
- The NVL and NVL2 functions are used to address problems encountered with NULL values.
- The TO_CHAR function lets a user present numeric data and dates in a specific format.
- The DECODE function allows an action to be taken to be determined by a specific value.
- The SOUNDEX function is based on the phonetic representation of characters.
- The ability to nest functions allows multiple operations to be performed on data.
- The DUAL table can be used as a dummy table when a SELECT statement must be issued for data that do not exist in a table.

CHAPTER 5 SYNTAX SUMMARY

The following table presents a summary of the syntax that you have learned in this chapter. You can use the table as a study guide and reference.

Syntax Guide		
Element	**Description**	**Example**
Case Conversion Functions		
LOWER	Converts characters to lower-case letters	`LOWER(c)` *c* = character string or field to be converted to lower case
UPPER	Converts characters to upper-case letters	`UPPER(c)` *c* = character string or field to be converted to upper case
INITCAP	Converts words to mixed-case, initial capital letters	`INITCAP(c)` *c* = character string or field to be converted to mixed case
Character Manipulation Functions		
SUBSTR	Returns a substring, or portion of a string, in output	`SUBSTR(c, p, l)` *c* = character string *p* = position (beginning) for the extraction *l* = length of output string
LENGTH	Returns the numbers of characters in a string	`LENGTH(c)` *c* = character string to be analyzed
LPAD / RPAD	Pads, or fills in, the area to the **L**eft (or **R**ight) of a character string, using a specific character—or even a blank space	`LPAD(c, l, s)` *c* = character string to be padded *l* = length of character string after being padded *s* = symbol or character to be used as padding
RTRIM / LTRIM	Trims, or removes, a specific string of characters from the **R**ight (or **L**eft) of a set of data	`LTRIM(c, s)` *c* = characters to be affected *s* = string to be removed from the left of the data
REPLACE	Used to perform a search and replace of displayed results	`REPLACE(c, s, r)` *c* = the data or column to be searched *s* = the string of characters to be found *r* = the string of characters to be substituted for *s*
CONCATE	Used to concatenate two data items	`CONCAT(c1, c2)` *c1*= first data item to be concatenated *c2*= second data item to be included in the concatenation

Syntax Guide		
Element	**Description**	**Example**
Number Functions		
ROUND	Rounds numeric fields	`ROUND(n, p)` *n* = numeric data, or a field, to be rounded *p* = position of the digit to which the data should be rounded
TRUNC	Truncates, or cuts, numbers to a specific position	`TRUNC(n, p)` *n* = numeric data, or a field, to be truncated *p* = position of the digit to which the data should be truncated
Date Functions		
MONTHS_ BETWEEN	Determines the number of months between two dates	`MONTHS_BETWEEN(d1, d2)` *d1* and *d2* = dates in question *d2* is subtracted from *d1*
ADD_MONTHS	Adds months to a date to signal a target date in the future	`ADD_MONTHS(d, m)` *d* = date (beginning) for the calculation *m*= months—the number of months to add to the date
NEXT_DAY	Determines the next day—a specific day of the week after a given date	`NEXT_DAY(d, DAY)` *d* = date (starting) *DAY* = the day of the week to be identified
TO_DATE	Converts a date in a specified format to the default date format	`TO_DATE(d,f)` *d* = date entered by the user *f* = format of the entered date
Other Functions		
NVL	Solves problems arising from arithmetic operations having fields that may contain NULL values. When a NULL value is used in a calculation, the result is a NULL value. The NVL function is used to substitute a value for the existing NULL.	`NVL(x, y)` *y* = the value to be substituted if *x* is NULL
NVL2	Provides options based on whether a NULL value exists	`NVL2(x, y, z)` *y* = what should be substituted if *x* is not NULL *z* = what should be substituted if *x* is NULL
TO_CHAR	Converts dates and numbers to a formatted character string	`TO_CHAR(n, 'f')` *n* = number or date to be formatted *f* = format model to be used

5

Syntax Guide		
Element	**Description**	**Example**
Other Functions (Continued)		
DECODE	Takes a given value and compares it to values in a list. If a match is found, then the specified result is returned. If no match is found, then a default result is returned. If no default result is defined, a NULL is returned as the result.	`DECODE(V, L1, R1, L2, R2,…, D)` *V* = value sought *L1* = the first value in the list *R1* = result to be returned if *L1* and *V* match *D* = default result to return if no match is found
SOUNDEX	Converts alphabetic characters to their phonetic representation, using an alphanumeric algorithm	`SOUNDEX(c)` *c* = characters to be phonetically represented

Review Questions

1. What is the purpose of the SOUNDEX function?
2. What is the difference between the NVL and NVL2 functions?
3. What is the difference between the TO_CHAR and TO_DATE functions when working with date values?
4. How is the TRUNC function different from the ROUND function?
5. How can padding be removed from a date that is displayed by the TO_CHAR function?
6. What is the difference between using the CONCAT function and the concatenation operator (| |) in a SELECT clause?
7. Which functions can be used to convert the case of character values?
8. When should you reference the DUAL table?
9. What format model would be used to display the date 25-DEC-04 as Dec. 25?
10. Why would the function NVL(shipdate, 'Not Shipped') return an error message?

MULTIPLE CHOICE

To answer the following questions, refer to the tables in Appendix A.

1. Which of the following is a valid SQL statement?
   a. `SELECT SYSDATE;`
   b. `SELECT UPPER(Hello) FROM dual;`
   c. `SELECT TO_CHAR(SYSDATE, 'fmMONTH DD, YYYY')`
      `FROM dual;`
   d. all of the above
   e. none of the above

2. Which of the following functions can be used to extract a portion of a character string?
   a. EXTRACT
   b. TRUNC
   c. SUBSTR
   d. INITCAP
3. Which of the following will determine how long ago orders that have not been shipped were received?
   a. ```
      SELECT order#, shipdate-orderdate delay
      FROM orders;
      ```
   b. ```
      SELECT order#, SYSDATE - orderdate
      FROM orders
      WHERE shipdate IS NULL;
      ```
   c. ```
      SELECT order#, NVL(shipdate, 0)
      FROM orders
      WHERE orderdate is NULL;
      ```
   d. ```
      SELECT order#, NULL(shipdate)
      FROM orders;
      ```
4. Which of the following SQL statements will produce Hello World as the output?
   a. ```
      SELECT "Hello World"
      FROM dual;
      ```
   b. ```
      SELECT INITCAP('HELLO WORLD')
      FROM dual;
      ```
   c. ```
      SELECT LOWER('HELLO WORLD')
      FROM dual;
      ```
   d. both A and B
   e. none of the above
      ```

5. Which of the following functions can be used to substitute a value for a NULL value?
   a. NVL
   b. TRUNC
   c. NVL2
   d. SUBSTR
   e. both A and D
   f. both A and C
6. Which of the following is not a valid format model for displaying the current time?
   a. 'HH:MM:SS'
   b. 'HH24:SS'
   c. 'HH12:MI:SS'
   d. All of the above are valid.
7. Which of the following will list only the last four digits of the contact person's phone number at American Publishing?
   a.
   ```
   SELECT EXTRACT(phone, -4, 1)
   FROM publisher
   WHERE name = 'AMERICAN PUBLISHING';
   ```
   b.
   ```
   SELECT SUBSTR(phone, -4, 1)
   FROM publisher
   WHERE name = 'AMERICAN PUBLISHING';
   ```
   c.
   ```
   SELECT EXTRACT(phone, -1, 4)
   FROM publisher
   WHERE name = 'AMERICAN PUBLISHING';
   ```
   d.
   ```
   SELECT SUBSTR(phone, -4, 4)
   FROM publisher
   WHERE name = 'AMERICAN PUBLISHING';
   ```
8. Which of the following functions can be used to determine how many months a book has been available?
   a. MONTH
   b. MON
   c. MONTH_BETWEEN
   d. none of the above

9. Which of the following will display the order date for order 1000 as 03/31?
   a. ```
      SELECT TO_CHAR(orderdate, 'MM/DD')
      FROM orders
      WHERE order# = 1000;
      ```
   b. ```
      SELECT TO_CHAR(orderdate, 'fmMM/DD')
      FROM orders
      WHERE order# = 1000;
      ```
   c. ```
      SELECT TO_CHAR(orderdate, 'fmMONTH/YY')
      FROM orders
      WHERE order# = 1000;
      ```
   d. both A and B
   e. none of the above



10. Which of the following functions includes different options depending on the value of a specified column?
    a. NVL
    b. DECODE
    c. UPPER
    d. SUBSTR
11. Which of the following SQL statements is not valid?
    a. ```
       SELECT TO_CHAR(orderdate, '99/9999')
       FROM orders;
       ```
    b. ```
       SELECT INITCAP(firstname), UPPER(lastname)
       FROM customers;
       ```
    c. ```
       SELECT cost, retail, TO_CHAR(retail-cost, '$999.99') profit
       FROM books;
       ```
    d. All of the above are valid.
12. Which function can be used to add spaces to a column until it is a specific width?
    a. TRIML
    b. PADL
    c. LWIDTH
    d. none of the above
13. Which of the following SELECT statements will return 30 as the result?
    a. `SELECT ROUND(24.37, 2) FROM dual;`
    b. `SELECT TRUNC(29.99, 2) FROM dual;`
    c. `SELECT ROUND(29.01, -1) FROM dual;`
    d. `SELECT TRUNC(29.99, -1) FROM dual;`

14. Which of the following is a valid SQL statement?
    a. `SELECT TRUNC(ROUND(125.38, 1), 0) from dual;`
    b. `SELECT ROUND(TRUNC(125.38, 0) from dual;`
    c. `SELECT LTRIM(LPAD(state, 5, ' '), 4, -3, "*") from dual;`
    d. `SELECT SUBSTR(ROUND(14.87, 2, 1), -4, 1) from dual;`
15. Which of the following functions cannot be used to convert the case of a character string?
    a. UPPER
    b. LOWER
    c. INITIALCAP
    d. All of the above can be used for case conversion.
16. Which of the following format elements will cause months to be displayed in a three-letter abbreviated format?
    a. MMM
    b. fmMONTH
    c. MON
    d. none of the above
17. Which of the following SQL statements will display a customer's name in all upper-case characters?
    a. `SELECT UPPER('firstname', 'lastname') FROM customers;`
    b. `SELECT UPPER(firstname, lastname) FROM customers;`
    c. `SELECT UPPER(lastname, ',' firstname) FROM customers;`
    d. none of the above
18. Which of the following functions can be used to display the character string FLORIDA in the results of a query whenever FL is encountered in the State field?
    a. SUBSTR
    b. NVL2
    c. REPLACE
    d. TRUNC
    e. none of the above
19. The name of the dummy table provided by Oracle9*i* is:
    a. DUMDUM
    b. DUAL
    c. ORAC
    d. SYS

20. If an integer is multiplied by a NULL value, the result will be:
    a. an integer
    b. a whole number
    c. a NULL value
    d. None of the above—a syntax error message is returned.

HANDS-ON ASSIGNMENTS

To perform the following activities, refer to the tables in Appendix A.

1. Obtain a list of all customer names that displays the first letter of the first and last names in upper-case letters and the rest in lower-case letters.
2. Create a list of all customers that will display the characters 'NOT REFERRED' if the customer was not referred to JustLee Books by another customer.
3. Determine the amount of profit generated by the book purchased on order 1002. The profit should be formatted to display a dollar sign and two decimal places.
4. Display a list of all books and the percentage of markup for each book. The percentage of markup should be displayed as a whole number (i.e., multiplied by 100) with no decimal position, followed by a percent sign (e.g., .2793 = 28%).
5. Display the current day of the week, hour, minutes, and seconds of the current date setting on the computer you are using.
6. Create a list of JustLee's books and precede the cost of each book with asterisks so that the width of the displayed cost field is 12.
7. Determine the length of the data stored in the ISBN field of the BOOKS table. Make certain to have the length only display once (not once for each book).
8. Using today's date, determine the age (in months) of each book that JustLee sells. Make certain that only whole months are displayed, rather than portions of months.
9. Determine the calendar date of the next occurrence of Wednesday, based on today's date.
10. List the phonetic representation of all authors contained in JustLee's database.

A Case for Oracle9*i*

To perform this activity, refer to the tables in Appendix A.

Management is proposing to increase the price of each book. The amount of the increase will be based on each book's category, according to the following scale: computer books, 10%; fitness books, 15%; self-help books, 25%; all other categories, 3%. Create a list that displays each book's title, category, current retail price, and revised retail price. The prices should be displayed with two decimal places. The column headings for the output should be as follows: Title, Category, Current Price, Revised Price. Make certain the results are sorted in order of category. If there is more than one book in a category, a secondary sort should be performed on the book's title.

Create a memo that provides management with the SELECT statement used to generate the results as well as the result of the statement.

CHAPTER

6

GROUP FUNCTIONS

Objectives

After completing this chapter, you should be able to do the following:

- Differentiate between single-row and multiple-row functions
- Use the SUM and AVG functions for numeric calculations
- Use the COUNT function to return the number of records containing non-NULL values
- Use COUNT(*) to include records containing NULL values
- Use the MIN and MAX functions with non-numeric fields
- Determine when to use the GROUP BY clause to group data
- Identify when the HAVING clause should be used
- List the order of precedence for evaluating WHERE, GROUP BY, and HAVING clauses
- State the maximum depth for nesting group functions
- Nest a group function inside a single-row function
- Calculate the standard deviation and variance of a set of data, using the STDDEV and VARIANCE functions

Group functions, also called **multiple-row functions**, return one result per group of rows processed. Multiple-row functions to be discussed in this chapter include SUM, AVG, COUNT, MIN, MAX, STDDEV, and VARIANCE. This chapter will present the GROUP BY clause to identify the group(s) of records to be processed and the HAVING clause to restrict the groups returned in the query results. A discussion of statistical functions concludes the chapter.

Figure 6–1 provides an overview of this chapter's contents.

Group (Multiple-Row) Functions		
Function (and syntax)	**Description**	**Example**
`SUM([DISTINCT|ALL] n)`	Returns the sum or total value of the selected numeric field. Ignores NULL values.	`SELECT SUM (retail-cost) FROM books;`
`AVG([DISTINCT|ALL] n)`	Returns the average value of the selected numeric field. Ignores NULL values.	`SELECT AVG(cost) FROM books;`
`COUNT(*|[|DISTINCT| ALL] c)`	Returns the number of rows that contain a value in the identified field. Rows containing NULL values in the field will not be included in the results. To count rows containing NULL values, use an * rather than a field name.	`SELECT COUNT(*) FROM books;` *or* `SELECT COUNT (shipdate) FROM orders;`
`MAX([DISTINCT|ALL] c)`	Returns the highest (maximum) value from the selected field. Ignores NULL values.	`SELECT MAX (customer#) FROM customers;`
`MIN([DISTINCT|ALL] c)`	Returns the lowest (minimum) value from the selected field. Ignores NULL values.	`SELECT MIN (retail-cost) FROM books;`
`STDDEV([DISTINCT| ALL] n)`	Returns the standard deviation of the numeric field selected. Ignores NULL values.	`SELECT STDDEV (retail) FROM books;`
`VARIANCE([DISTINCT| ALL] n)`	Returns the variance of the numeric field selected. Ignores NULL values.	`SELECT VARIANCE (retail) FROM books;`

Figure 6-1 Overview of chapter contents

Go to the JustLee Database folder in your Data Files. If you run the **Bookscript.sql** file, you will be able to work through the queries shown in this chapter. Your output should match the output shown.

Group Function Concepts

Multiple-row functions are commonly referred to as group functions because they process groups of rows. Because these functions return only one result per group of data, they are also known as **aggregate functions**. Figure 6–2 shows the position of these clauses in the SELECT statement.

```
SELECT *|columnname, columnname…
FROM tablename
[WHERE condition]
[GROUP BY columnname, columnname…] ←
[HAVING group condition]; ←
```

Figure 6-2 Location of multiple-row functions in the SELECT statement

Follow these rules when working with group functions:

1. Use the DISTINCT keyword to include only unique values. *The ALL keyword is the default*, and it instructs Oracle9*i* to include all values (except nulls).
2. All group functions ignore NULL values except COUNT(*). To include NULL values, nest the NVL function within the group function. For example: **`SELECT MAX(NVL(shipdate, SYSDATE) - orderdate) FROM ORDERs;`** will substitute the system date for the shipping date of any order that has not yet shipped.

6

SUM Function

The **SUM** function is used to calculate the total amount stored in a numeric field for a group of records. The syntax of the SUM function is `SUM([DISTINCT|ALL] n)`, where *n* is a column containing numeric data. The optional DISTINCT keyword instructs Oracle9*i* to include only *unique* numeric values in its calculation. The ALL keyword instructs Oracle9*i* to include *multiple* occurrences of numeric values when totaling a field. If the DISTINCT or ALL keywords are not included when the SUM function is used in a query, Oracle9*i* will assume the ALL keyword by default and use all the numeric values that exist when the query is executed, as shown in Figure 6–3.

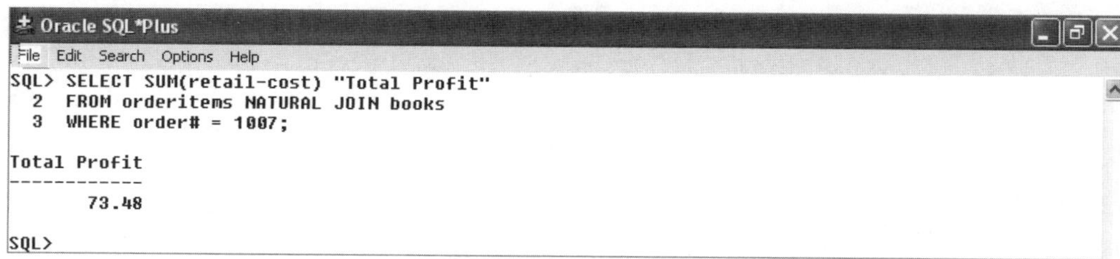

Figure 6-3 The SUM function

In Figure 6–3, the query calculates the total profit from books sold on order 1007. The SUM function in the SELECT clause uses the argument of **`retail - cost`** to instruct Oracle9*i* to calculate the profit generated by each book before totaling the profit. Notice that the SELECT clause also includes a column alias to describe the output. As with single-row functions, if a group function is used in a SELECT clause, the actual function will be displayed as the column heading unless a column alias is assigned.

The WHERE clause in Figure 6–3 restricts the rows used in the calculation to only those books that appear on order 1007. Because the books ordered are identified in the ORDERITEMS table, and the cost and retail price of the books are stored in the BOOKS table, the two tables are joined in the FROM clause. The difference between the cost and retail price for each book is calculated. Then, individually calculated profits are totaled, and the total profit for the order is presented as a single output for the query, as shown in Figure 6–3.

Suppose that management wants to determine total sales for one day—April 2, 2003. You might assume that you could simply query the ORDERS table; however, this table does not include the total "amount due" for an order. For example, one customer's order might have been for two copies of *Revenge of Mickey* and one copy of *Handcranked Computers*. The only way to calculate how much that customer owes for his or her order is to multiply the quantity of books purchased by the retail price of each book (quantity * retail). This is commonly referred to as an "extended price". The extended prices would then be totaled to yield a total amount due for the customer. (You will learn how to determine a customer's amount due later in the chapter.) However, in this case, management wants to know just the total sales for April 2, 2003. To calculate that total, simply add the extended prices for all orders placed on that day. The query in Figure 6–4 is used to determine the day's total sales.

```
SELECT SUM(quantity*retail) "Total Sales"
FROM orders JOIN orderitems
ON orders.order# = orderitems.order#
JOIN books ON orderitems.ISBN  = books.ISBN
WHERE orderdate = '02-APR-03';
```

Figure 6-4 SELECT statement to calculate total sales for a specified date

Let's look at each of the clauses in Figure 6–4. The SELECT clause uses the SUM function to calculate total sales by adding the extended price for each item ordered. The argument of the SUM function calculates the extended price for each book ordered. Notice that in the FROM clause, three tables must be joined. WHY? The date of the order is stored in the ORDERS table, the actual books and quantity ordered are stored in the ORDERITEMS table, and the retail price of each book is stored in the BOOKS table. To calculate the day's sales, each of these tables must be included in the FROM clause. The WHERE clause is included to restrict the calculation to only the orders that were placed on April 2, 2003. As shown in Figure 6–5, total sales for that date amounted to $387.15.

```
Oracle SQL*Plus
File  Edit  Search  Options  Help
SQL> SELECT SUM(quantity*retail) "Total Sales"
  2  FROM orders JOIN orderitems ON
  3  orders.order# = orderitems.order#
  4  JOIN books ON orderitems.ISBN = books.ISBN
  5  WHERE orderdate = '02-APR-03';

Total Sales
-----------
     387.15

SQL>
```

Figure 6-5 Results of the total sales calculation

Remember that the ON keyword is required for each JOIN keyword. If the ON keyword was not included at the end of the second line of the SELECT statement in Figure 6–5, edit the statement and either add the ON keyword to the end of line 2 or the beginning of line 3, and execute the query again.

AVG Function

The **AVG** function calculates the average of the numeric values in a specified column. The syntax of the AVG function is `AVG([DISTINCT|ALL] n)`, where *n* is a column containing numeric data.

For example, if the management of JustLee Books wants to know the average profit generated by all books in the Computer Category, the WHERE clause would be used to restrict the rows processed to those containing the value "COMPUTER" in the Category column. As with the SUM function, first the profit for each book is calculated and then totaled. That total is then divided by the number of records that contain non-NULL values in the specified field, as shown in Figure 6–6.

```
Oracle SQL*Plus
File  Edit  Search  Options  Help
SQL> SELECT AVG(retail-cost) "Average Profit"
  2  FROM books
  3  WHERE category = 'COMPUTER';

Average Profit
--------------
       18.2625

SQL> SELECT TO_CHAR(AVG(retail-cost),'999.99') "Average Profit"
  2  FROM books
  3  WHERE category = 'COMPUTER';

Average
-------
  18.26

SQL>
```

Figure 6-6 The AVG function

As shown in the first query in Figure 6–6, the average profit returned by books in the Computer Category is $18.2625. When a query includes division, the resulting display may include more than two decimal positions. Because management prefers the results to be displayed with only two decimal positions, the TO_CHAR function is included in the second query in Figure 6–6 to specify that format. When the TO_CHAR function is used on numeric data, any excess decimals are rounded (not truncated) to the specified number of digits. As shown in this example, group functions can be nested inside single-row functions.

Notice that in the output of the second query, the entire column alias is not displayed. Why? When the TO_CHAR function is included to apply a format model to the average profit, the numeric values are converted to characters for display purposes. Since the average profit is now considered to be a character string, Oracle9*i* truncates the column alias to match the width of the displayed column.

COUNT Function

Depending on the argument used, the **COUNT** function can either (1) count the records that have non-NULL values in a specified field or (2) count the total records that meet a specific condition, including those containing NULL values. The syntax of the COUNT function is `COUNT(*|[|DISTINCT|ALL] c)`, where *c* represents any type of column, numeric or non-numeric.

The query in Figure 6–7 tells Oracle9*i* to use the COUNT function to return the number of distinct categories represented by the titles currently stored in the BOOKS table.

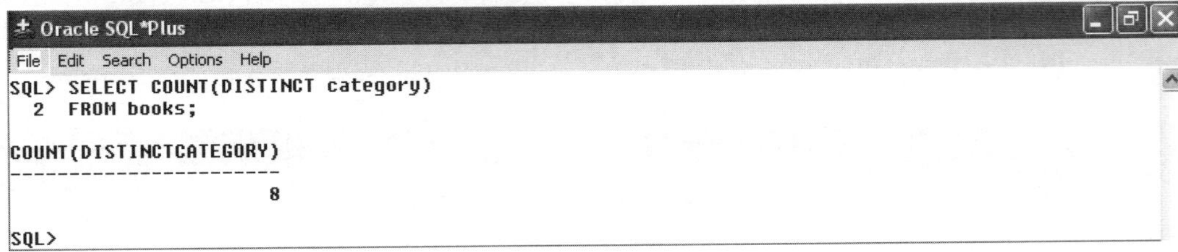

Figure 6-7 The COUNT function

The column heading for the results displayed in Figure 6–7 is the actual COUNT function used in the SELECT statement. To create a more "attractive" column heading, simply add a column alias in the SELECT clause.

Notice that in Figure 6–7, the DISTINCT keyword precedes the column name in the argument of the COUNT function, rather than appearing directly after the SELECT keyword in the SELECT clause. This instructs Oracle9*i* to count each different value found in the Category column. If the DISTINCT keyword were entered directly after

SELECT, it would apply to the entire COUNT function and would have been interpreted to mean that only duplicate rows, not duplicate category values, should be suppressed.

As shown in Figure 6–8, if the DISTINCT keyword is included as part of the SELECT clause, Oracle9*i* will return an actual count of how many rows are in the BOOKS table. Why? Because the DISTINCT keyword applies to the results of the COUNT function—after all the rows containing a value in the Category column have been counted.

Figure 6-8 Flawed query: DISTINCT after SELECT keyword returns a count of the number of rows in the BOOKS table, not the distinct categories.

However, in this case, management wants to know how many different categories are represented by the books in the BOOKS table. Because the DISTINCT keyword should apply to the contents of the Category column, the keyword must be listed immediately before the column name, inside the argument of the COUNT function. As was shown in Figure 6–7, the 14 titles in the BOOKS table represent 8 different categories.

Suppose that management asks how many orders are currently outstanding; that is, they have not been shipped to customers. One solution is to print a list of all orders that have a NULL value for the date shipped. However, you would still need to count the records returned in the results. There is a simpler solution, as shown in Figure 6–9.

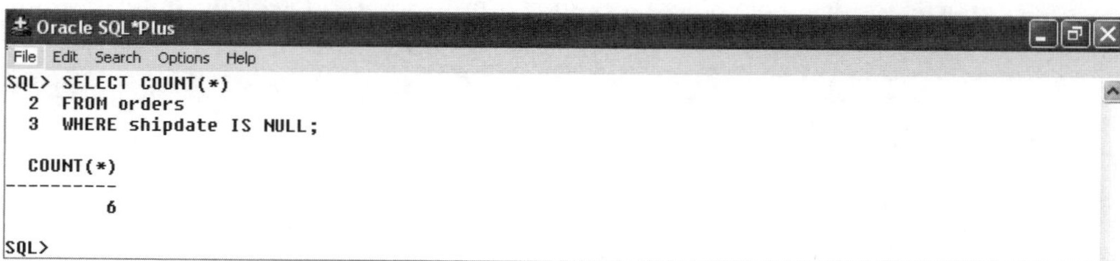

Figure 6-9 The COUNT(*) function

Remember that the equal sign (=) cannot be used when searching for a NULL value. Use IS NULL, the appropriate comparison operator, for finding rows containing a NULL value in a specified column.

When the argument supplied in the COUNT function is an asterisk (*), the existence of the entire record is counted. By counting the entire record, a NULL value will not be discarded by the COUNT function. As shown by the query in Figure 6–9, the WHERE clause restricts the rows that should be counted to only those that do not have a value stored in the Shipdate column. By contrast, look at the flawed query in Figure 6–10.

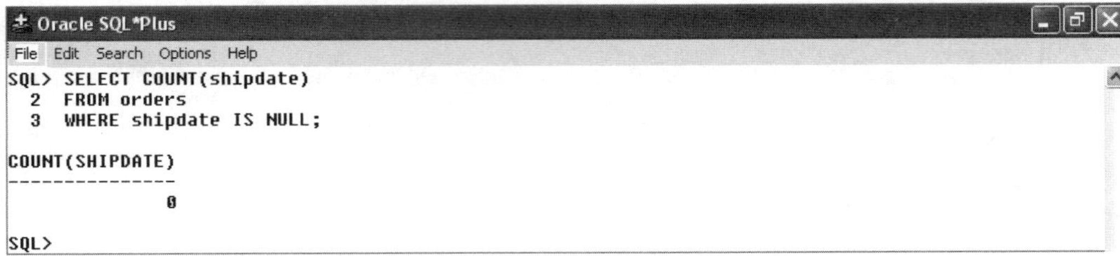

Figure 6-10 Flawed query: the COUNT function with NULL values—and no asterisk (*)

The query in Figure 6–10 is a modification of the example shown in Figure 6–9 and illustrates a common query error—the asterisk has been replaced with the Shipdate column in the COUNT argument. Because the WHERE clause restricts the records to only those having a NULL value in the Shipdate column, the function returned a count of zero. Basically, since the specified column contained no value, there was nothing to count. Therefore, *whenever NULL values may affect the COUNT function, you should use an asterisk as the argument rather than a column name.*

MAX Function

The **MAX** function returns the largest value stored in the specified column. The syntax for the MAX function is `MAX([DISTINCT|ALL] c)`, where *c* can represent any numeric, character, or date field. The query shown in Figure 6–11 requests numeric output.

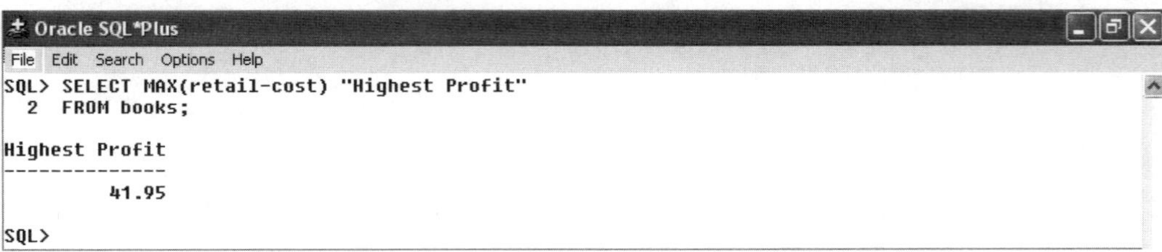

Figure 6-11 The MAX function

In Figure 6–11, the maximum profit generated by a book is returned from the **`MAX(retail-cost)`** function. As shown in the results, the largest profit earned by a single book is $41.95.

The problem with the query in Figure 6–11 is that you cannot tell which book is actually generating the profit. Therefore, the result of the query would not be very helpful to management. They would also need at least the title of the book to identify which book is the most profitable. Because the SELECT clause cannot contain the column name without grouping the books by title (and thereby displaying the profit generated by each book), a subquery would need to be used to identify both the title and profit of the most profitable book in inventory. You will learn how to work with subqueries in Chapter 7.

The MAX function can also be used with non-numeric data. Then, the output shows the first value that occurs when a column is sorted in descending order. Similarly, if a column contains dates, the most recent date is considered to have the highest value (based on its Julian date, discussed in Chapter 5). If the MAX function is applied to a character column, the letter *Z* would have a higher, or larger, "value" than the letter *A*.

For illustrative purposes, suppose that you needed to find the book title that alphabetically appears at the end of an ascending list (A–Z) of book titles. In other words, the book title has the largest value of all books in the BOOKS table. As shown in Figure 6–12, the MAX function is applied to the Title column, and the book title with the largest value is displayed.

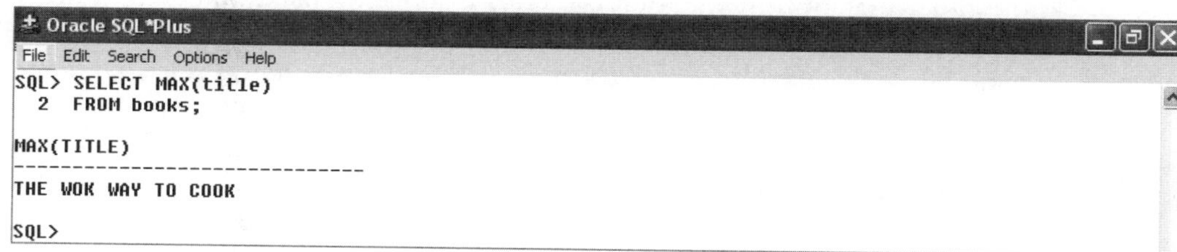

Figure 6-12 MAX function applied to character data

MIN Function

In contrast to the MAX function, the **MIN** function returns the smallest value in a specified column. As with the MAX function, the MIN function works with any numeric, character, or date column. The syntax for the MIN function is `MIN([DISTINCT|ALL] c)`, where *c* represents a character, numeric, or date column. Figure 6–13 shows the MIN function.

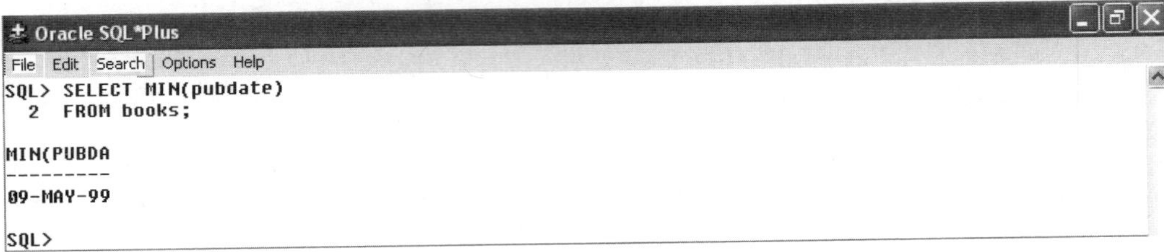

```
SQL> SELECT MIN(pubdate)
  2  FROM books;

MIN(PUBDA
---------
09-MAY-99

SQL>
```

Figure 6-13 The MIN function

The SELECT statement in Figure 6–13 instructs Oracle9*i* to find the earliest (MIN) publication date (pubdate) of all books stored in the BOOKS table. In this case, the earliest publication date is May 9, 1999. Of course, if you wanted the title of the most recently published book, you would substitute the MIN function with the MAX function. The MIN function uses the same logic as the MAX function for numeric and character data—except that it returns the smallest value rather than the largest value.

GROUP BY Clause

In Figure 6–6, the SELECT query returned the average profit of all books in the Computer Category. However, suppose that the management of JustLee Books wants to know the average profit for each category of books. The simplest solution is to reissue the query in Figure 6–6, once for each category, using the WHERE clause to restrict the query to a specific category each time. An alternative solution is to divide the records in the BOOKS table into groups, and then calculate the average for each group—preferably all in one query. This can be done with the GROUP BY clause. The syntax for the GROUP BY clause is `GROUP BY` ***`columnname`*** `[,` ***`columnname`***`,...]`, where the ***`columnname`*** is the column(s) to be used to create the groups or sets of data.

What happens when you attempt to create this query without the GROUP BY clause? In Figure 6–14, the SELECT clause includes both the single column named Category and the AVG group function. Because the SELECT statement did not include a GROUP BY clause, an error message is returned.

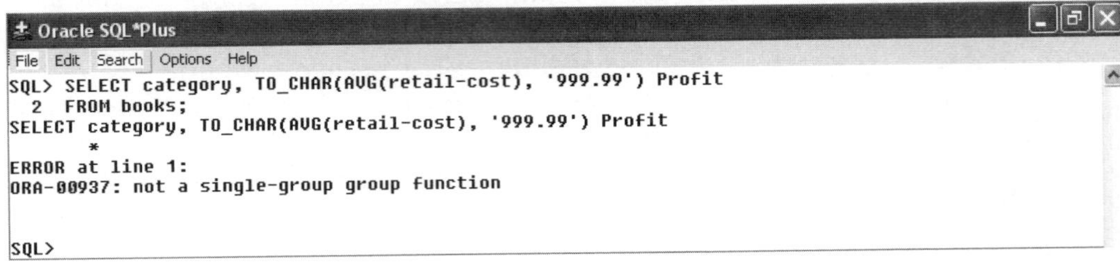

```
SQL> SELECT category, TO_CHAR(AVG(retail-cost), '999.99') Profit
  2  FROM books;
SELECT category, TO_CHAR(AVG(retail-cost), '999.99') Profit
       *
ERROR at line 1:
ORA-00937: not a single-group group function

SQL>
```

Figure 6-14 Flawed query: Group function and individual field—incorrectly executed without a GROUP BY clause

To specify that groups should be created, add the GROUP BY clause to the SELECT statement. When using the GROUP BY clause, remember the following:

1. If a group function is used in the SELECT clause, then any individual columns listed in the SELECT clause must also be listed in the GROUP BY clause.
2. Columns used to group data in the GROUP BY clause do not have to be listed in the SELECT clause. They are only included in the SELECT clause to have the groups identified in the output.
3. Column aliases cannot be used in the GROUP BY clause.
4. Results returned from a SELECT statement that include a GROUP BY clause will present the results in ascending order of the column(s) listed in the GROUP BY clause. To present the results in a different order, use the ORDER BY clause.

If you think of how results would appear if the statement in Figure 6–14 actually executed, then it becomes obvious why the GROUP BY clause is required: The SELECT clause would require a display with the category for each record in the first column, and then the average profit generated by all books in the BOOKS table (not for each category) in the second column. In other words, the category displayed may change, but all the rows returned would display the same average profit, which might lead the user to conclude that each category is generating the same average profit.

The required GROUP BY clause is added in Figure 6–15 to correct the error in the query in Figure 6–14.

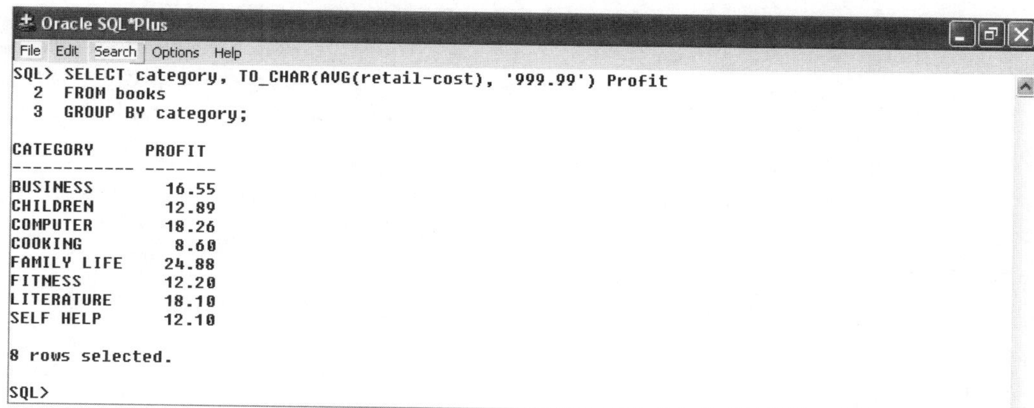

Figure 6-15 The GROUP BY clause, correctly executed

In the example in Figure 6–15, the single column name listed in the SELECT clause is included in the GROUP BY clause. Then, when the query is executed, the records in the BOOKS table are first grouped by category, and then the average profit for each category is calculated. Because the Category column is listed in the SELECT clause, the name of each category is displayed with the average profit generated by each category.

In Figure 6–5, the SUM function was used to calculate the total sales for a particular day. But how can you determine how much each customer owes for an order? The SUM function is a group function and, therefore, returns one total for all rows processed, but how do you display a list of all orders and the total amount due for *each* order? This is the perfect scenario for the GROUP BY clause.

For example, if the Billing Department requests a list of the amount due from each customer for each order, use the GROUP BY function to group the rows for each order, and then use the SUM function to calculate the extended price for the items ordered and return the total amount due for each order. The SQL statement to create this list is shown in Figure 6–16.

```
SELECT customer#, order#,
     SUM(quantity*retail) "Order Total"
FROM orders NATURAL JOIN orderitems
     NATURAL JOIN books
GROUP BY customer#, order#;
```

Figure 6-16 Query to calculate the total amount due for each order

Because the SELECT clause in Figure 6–16 includes the individual Customer# and Order# columns, these columns must also be listed in the GROUP BY clause. You might ask whether it is necessary to include the order number in the query. Suppose, however, that a customer recently placed two orders. If the order number had not been included in the query, then it is possible the SQL statement would return the total amount due from each customer, not the amount due from a customer for each order. The customer number was included to identify who placed each order.

As shown in Figure 6–17, when the statement is executed, Oracle9*i* displays each order, the number of the customer who placed the order, and the total amount due for each order.

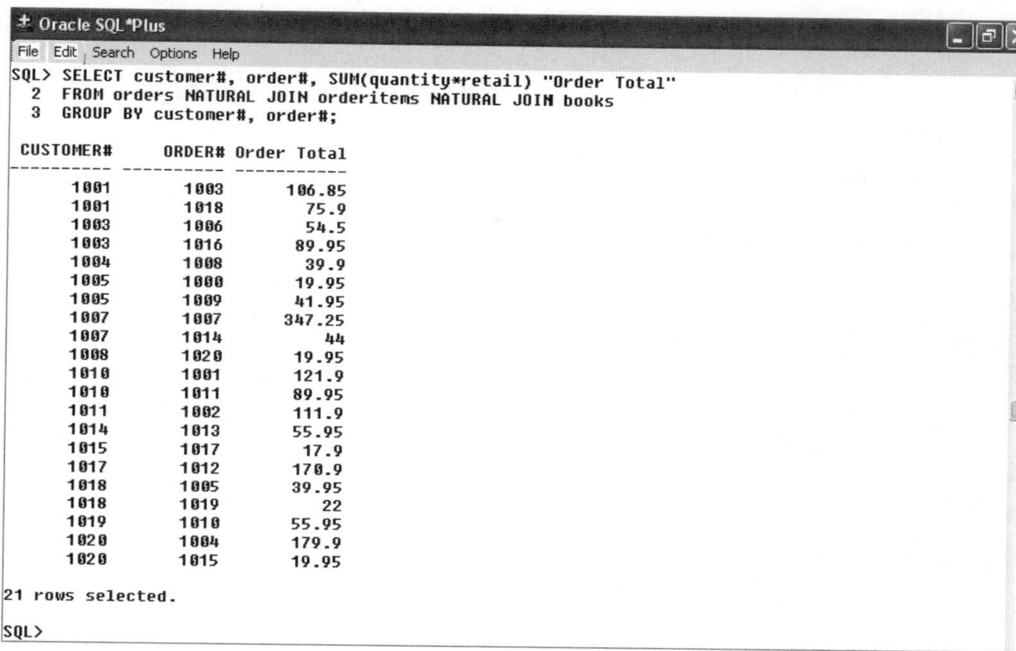

```
Oracle SQL*Plus
File Edit Search Options Help
SQL> SELECT customer#, order#, SUM(quantity*retail) "Order Total"
  2  FROM orders NATURAL JOIN orderitems NATURAL JOIN books
  3  GROUP BY customer#, order#;

 CUSTOMER#     ORDER# Order Total
---------- ---------- -----------
      1001       1003      106.85
      1001       1018        75.9
      1003       1006        54.5
      1003       1016       89.95
      1004       1008        39.9
      1005       1000       19.95
      1005       1009       41.95
      1007       1007      347.25
      1007       1014          44
      1008       1020       19.95
      1010       1001       121.9
      1010       1011       89.95
      1011       1002       111.9
      1014       1013       55.95
      1015       1017        17.9
      1017       1012       170.9
      1018       1005       39.95
      1018       1019          22
      1019       1010       55.95
      1020       1004       179.9
      1020       1015       19.95

21 rows selected.

SQL>
```

Figure 6-17 Total amount due for each order in the ORDERS table

HAVING CLAUSE

The **HAVING** clause is used to restrict the groups returned by a query. A general rule of thumb is that if you need to use a group function to restrict groups, then you must use the HAVING clause because *the WHERE clause cannot contain group functions*. Although the WHERE clause restricts the records that enter the query for processing, the HAVING clause specifies which groups will be displayed in the results, as shown in Figure 6–18. In other words, the HAVING clause serves as the WHERE clause for groups.

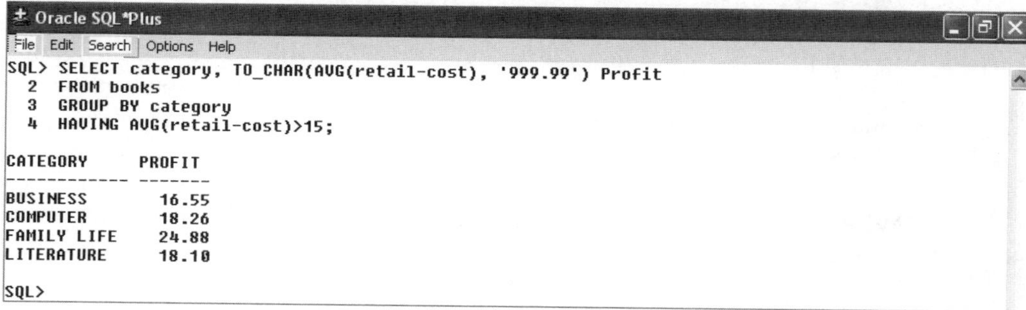

```
Oracle SQL*Plus
File Edit Search Options Help
SQL> SELECT category, TO_CHAR(AVG(retail-cost), '999.99') Profit
  2  FROM books
  3  GROUP BY category
  4  HAVING AVG(retail-cost)>15;

CATEGORY     PROFIT
------------ -------
BUSINESS       16.55
COMPUTER       18.26
FAMILY LIFE    24.88
LITERATURE     18.10

SQL>
```

Figure 6-18 The HAVING clause

The syntax for the HAVING clause is the keyword HAVING followed by the group condition(s): **HAVING** ***groupfunction comparisonoperator value***. In addition, the logical operators NOT, AND, and OR can be used to join group conditions in the HAVING clause, if necessary.

In Figure 6–18, the HAVING clause is used to limit the groups displayed to those categories with an average profit of more than $15.00. In this case, only four categories return an average profit of more than $15.00.

In Figure 6–19, the WHERE clause restricts the records to be processed to only those having a publication date after January 1, 2002. The GROUP BY clause groups the records that met the publication date restriction by the category to which each book is assigned. The HAVING clause is used to restrict the group data displayed to only those with an average profit greater than $15.00. Whenever a SELECT statement includes all three clauses, the order in which they are evaluated is as follows:

1. The WHERE clause
2. The GROUP BY clause
3. The HAVING clause

In essence, the WHERE clause filters the data *before* grouping, whereas the HAVING clause filters the groups *after* the grouping occurs.

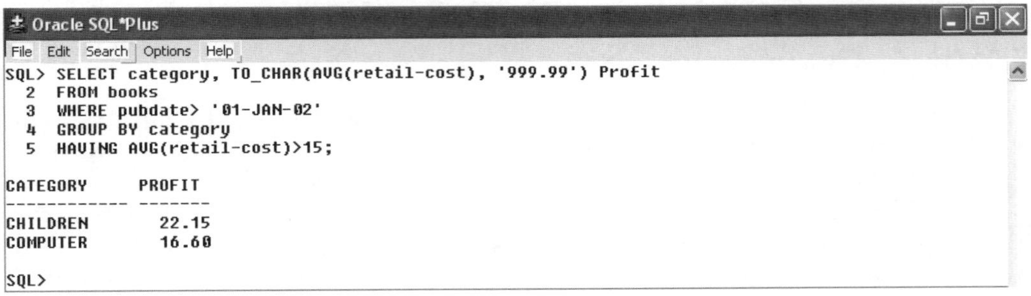

```
Oracle SQL*Plus
File Edit Search Options Help
SQL> SELECT category, TO_CHAR(AVG(retail-cost), '999.99') Profit
  2  FROM books
  3  WHERE pubdate> '01-JAN-02'
  4  GROUP BY category
  5  HAVING AVG(retail-cost)>15;

CATEGORY     PROFIT
------------ -------
CHILDREN       22.15
COMPUTER       16.60

SQL>
```

Figure 6-19 Query with WHERE and HAVING clauses

Suppose that after the Billing Department receives the list created in Figure 6–16, the department manager asks for another list of the amount due—but only for orders with a total amount due greater than $100.00. Since output is restricted (based on the results of the SUM function, which is a group function), a HAVING clause will be required. The query previously given in Figure 6–16 could be modified by adding **HAVING SUM(quantity*retail)>100**. As shown in Figure 6–20, only six orders have a total amount due greater than $100.

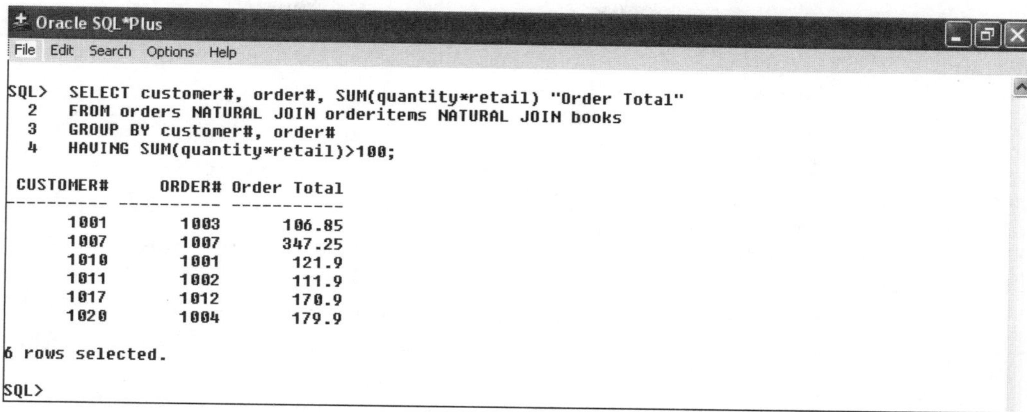

Figure 6-20 Using the HAVING clause to restrict output results based on a group function

Nesting Functions

As with single-row functions, when group functions are nested, *the inner function is resolved first.* The result of the inner function is then passed back as input for the outer function. Unlike single-row functions that have no restriction on how many nesting levels can occur, *group functions can only be nested to a depth of two.* As was shown in Figure 6–18, group functions can be nested inside single-row functions. In addition, single-row functions can be nested inside group functions.

In Figure 6–21, the SUM function is nested inside the AVG function to determine the average total amount for an order. The GROUP BY clause first groups all the records, based on the Order# column. Then, the total order amount is calculated for each order by the SUM function. The AVG function is used to calculate the average of the total order amounts calculated by the SUM function. The resulting output is the average total amount due for orders currently stored in the ORDERS table.

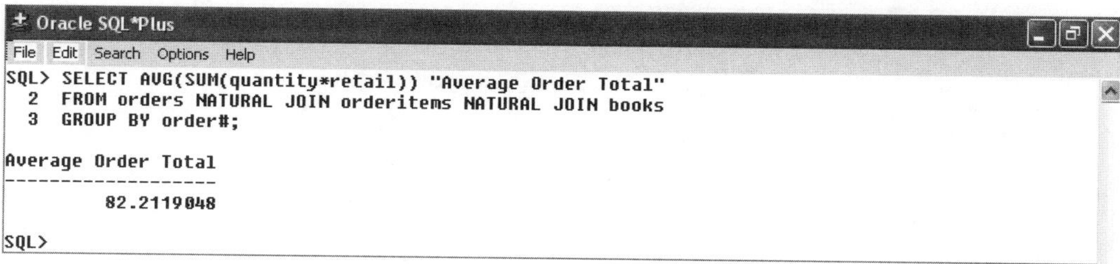

Figure 6-21 Nested group functions

Don't forget to include two closing parentheses at the end of the function in the SELECT clause. The first parenthesis closes the SUM function, and the second closes the AVG function.

STATISTICAL GROUP FUNCTIONS

Oracle9*i* uses **statistical group functions** to perform calculations for data analysis. In most organizations, functional areas, such as Marketing and Accounting, need to perform data analyses to detect sales trends, price fluctuations, etc. Oracle9*i* provides functions to support basic statistical calculations, such as standard deviation and variance. Although these calculations are easy to perform in Oracle9*i*, most people need training in statistical analysis to interpret the results of the calculations. This chapter's discussion of the results of those calculations is intended only to give you an overview of the calculations' purpose, not to train you in statistical analysis. The statistical functions to be covered in this chapter are STDDEV and VARIANCE.

STDDEV Function

The **STDDEV** function calculates the standard deviation for a specified field. A **standard deviation** calculation is used to determine how close individual values are to the mean, or average, of a group of numbers. The syntax for the STDDEV function is `STDDEV([DISTINCT|ALL] n)`, where *n* represents a numeric column.

The SELECT statement shown in Figure 6–22 displays each book category in JustLee Books' database, the average profit for each category, and the standard deviation of the profit for each category. Let's look at how to interpret these results.

```
Oracle SQL*Plus
File Edit Search Options Help
SQL> SELECT category, AVG(retail-cost), STDDEV(retail-cost)
  2  FROM books
  3  GROUP BY category;

CATEGORY     AVG(RETAIL-COST) STDDEV(RETAIL-COST)
------------ ---------------- -------------------
BUSINESS                16.55                   0
CHILDREN                12.89          13.0956176
COMPUTER              18.2625          11.2267074
COOKING                   8.6           1.6263456
FAMILY LIFE            24.875          24.1476966
FITNESS                  12.2                   0
LITERATURE               18.1                   0
SELF HELP                12.1                   0

8 rows selected.

SQL>
```

Figure 6-22 The STDDEV function

For the value calculated by the standard deviation to be useful, it must be compared to the calculated "average profit" for each category. For example, look at the Cooking Category in Figure 6–22. The average profit for books in the Cooking Category is $8.60. However, are most books close to that average, or do the majority of the books generate a profit of only $1, and one book generates a profit of $20—which would inflate the average? The standard deviation is a statistical approximation of how many books are within a certain range around the average.

The STDDEV function is based on the concept of normal distribution. A **normal distribution** means that if you input a large number of data values, those values tend to cluster around some average value. The basic assumption is that as you move closer to the average value, you will find the majority of data values clustered around that average value. However, some values may be extreme, and thus they may be much larger or smaller than the average value. Of course, each extreme data value affects the group's average. For example, calculating the average of 5, 6, 7, and 100 would result in a larger group average than if the 100 had not been included. When performing statistical analysis, the standard deviation is calculated to determine how closely the data match the average value for the group.

In a normal distribution, you can expect to find 68 percent of the books within one standard deviation (plus or minus) of the average, and 95 percent of the books within two standard deviations (plus or minus) of the average. In simpler terms, basically 68 percent of the books will have a profit between $6.97 ($8.60 - $1.63) and $10.23 ($8.60 + $1.63), and 95 percent of the books will have a profit between $5.34 ($8.60 - $3.26) and $11.86 ($8.60 + $3.26). The standard deviation can give management a quick picture of the average profit for books in a particular category, without having to examine each book—a very time-consuming task if the category includes thousands of books.

Notice that some of the categories in Figure 6–22 have a standard deviation of zero. If the STDDEV function is processing only one record per group, the result will always be zero due to the mathematical formula used to calculate the standard deviation.

VARIANCE Function

The **VARIANCE** function is used to determine how widely data are spread in a group. The variance of a group of records is calculated based on the minimum and maximum values for a specified field. If the data values are closely clustered together, the variance will be small. However, if the data contain extreme values (unusually high or low values), the variance will be larger. The syntax for the VARIANCE function is `VARIANCE([DISTINCT|ALL] n)`, where *n* represents a numeric field. Figure 6–23 presents the VARIANCE function.

```
Oracle SQL*Plus
File Edit Search Options Help
SQL> SELECT category, VARIANCE(retail-cost), MIN(retail-cost), MAX(retail-cost)
  2  FROM books
  3  GROUP BY category;

CATEGORY     VARIANCE(RETAIL-COST) MIN(RETAIL-COST) MAX(RETAIL-COST)
------------ --------------------- ---------------- ----------------
BUSINESS                         0            16.55            16.55
CHILDREN                  171.4952             3.63            22.15
COMPUTER                126.038958              3.2             28.7
COOKING                      2.645             7.45             9.75
FAMILY LIFE              583.11125              7.8            41.95
FITNESS                          0             12.2             12.2
LITERATURE                       0             18.1             18.1
SELF HELP                        0             12.1             12.1

8 rows selected.

SQL>
```

Figure 6-23 The VARIANCE function

The query in Figure 6–23 lists the categories for all books in the BOOKS table, the profit variance of each category, and (for comparison purposes) the lowest and highest profit within each category. As with the standard deviation, if a group of data consists of only one value (e.g., Business, Fitness, Literature, and Self Help categories), the calculated variance is zero. However, unlike standard deviation, variance is not measured with the same units (e.g., dollars) as the source data used for the calculation.

To interpret the results of a VARIANCE function, you must look at how large or small the value is. For example, the Cooking Category has a smaller variance than the other categories. This means that the profits for books in the Cooking Category are clustered tightly together (i.e., the profit does not cover a wide range). Look at the minimum and maximum profit for all books in the Cooking Category, and notice that the profit range is $2.30 ($9.75 - $7.45). On the other hand, look at the Family Life Category. This category has the largest variance, and if you compare the minimum and maximum profit, it has the largest profit range of all the categories presented. This should throw up a warning flag to management that some books may generate very little profit, whereas others may return a very large profit, and that considering only the average profit for books in the Family Life Category should not be used as the basis for decision-making.

Chapter Summary

- The AVG, SUM, STDDEV, and VARIANCE functions are used only with numeric fields.
- The COUNT, MAX, and MIN functions can be applied to any datatype.
- The AVG, SUM, MAX, MIN, STDDEV, and VARIANCE functions all ignore NULL values. COUNT(*) counts records containing NULL values. To include NULL values in other group functions, the NVL function is required.
- By default, the AVG, SUM, MAX, MIN, COUNT, STDDEV, and VARIANCE functions include duplicate values. To only include unique values, the DISTINCT keyword must be used.
- The GROUP BY clause is used to divide table data into groups.
- If a SELECT clause contains both an individual field name and a group function, the field name must also be included in a GROUP BY clause.
- The HAVING clause is used to restrict groups in a group function.
- Group functions can be nested to a depth of two. The inner function is always solved first. The results of the nesting function are used as input for the outer function.
- The functions STDDEV and VARIANCE are used to perform statistical analyses on a set of data.

Chapter 6 Syntax Summary

The following table presents a summary of the syntax that you have learned in this chapter. You can use the table as a study guide and reference.

Syntax Guide		
Group (Multiple-Row) Functions		
Function (and syntax)	**Description**	**Example**
`SUM([DISTINCT|ALL] n)`	Returns the sum or total value of the selected numeric field. Ignores NULL values.	`SELECT SUM (retail-cost) FROM books;`
`AVG([DISTINCT|ALL] n)`	Returns the average value of the selected numeric field. Ignores NULL values.	`SELECT AVG(cost) FROM books;`
`COUNT(*|[|DISTINCT| ALL] c)`	Returns the number of rows that contain a value in the identified field. Rows containing NULL values in the field will not be included in the results. To count all rows, including those with NULL values, use an * rather than a field name.	`SELECT COUNT(*) FROM books;` *or* `SELECT COUNT (shipdate) FROM orders;`
`MAX([DISTINCT|ALL] c)`	Returns the highest (maximum) value from the selected field. Ignores NULL values.	`SELECT MAX (customer#) FROM customers;`
`MIN([DISTINCT|ALL] c)`	Returns the lowest (minimum) value from the selected field. Ignores NULL values.	`SELECT MIN (retail-cost) FROM books;`
`STDDEV([DISTINCT| ALL] n)`	Returns the standard deviation of the selected numeric field. Ignores NULL values.	`SELECT STDDEV (retail) FROM books;`
`VARIANCE([DISTINCT| ALL] n)`	Returns the variance of the selected numeric field. Ignores NULL values.	`SELECT VARIANCE (retail) FROM books;`
Clauses		
Clause	**Description**	**Example**
`GROUP BY columnname [,columnname, ...]`	Divides data into sets or groups based on the contents of the specified column(s).	`SELECT AVG(cost) FROM books GROUP BY category;`
`HAVING groupfunction comparisonoperator value`	Restricts the groups displayed in the results of a query.	`SELECT AVG(cost) FROM books GROUP BY category HAVING AVG (cost)>21;`

REVIEW QUESTIONS

To answer these questions, refer to the tables in Appendix A.

1. Explain the difference between single-row and group functions.
2. Which group function can be used to count NULL values?
3. Which clause can be used to restrict the groups returned by a query based on a group function?
4. Under what circumstances *must* you include a GROUP BY clause in a query?
5. In which clause should you include the condition "pubid=4" to restrict the rows processed by a query?
6. In which clause should you include the condition **`MAX(cost)>39`** to restrict the groups displayed in the results of a query?
7. Which function returns the lowest numeric or character value stored in a specified column?
8. What is the maximum depth allowed when nesting group functions?
9. In what order will output results be presented if a SELECT statement contains a GROUP BY clause and no ORDER BY clause?
10. Which clause is used to restrict the records retrieved from a table? Which clause restricts the groups displayed in the results of a query?

MULTIPLE CHOICE

To answer these questions, refer to the tables in Appendix A.

1. Which of the following statements is true?
   a. The MIN function can only be used with numeric data.
   b. The MAX function can only be used with date values.
   c. The AVG function can only be used with numeric data.
   d. The SUM function cannot be part of a nested function.
2. Which of the following is a valid SELECT statement?
   a. `SELECT AVG(retail-cost) FROM books GROUP BY category;`
   b. `SELECT category, AVG(retail-cost) FROM books;`
   c.
   ```
   SELECT category, AVG(retail-cost) FROM books
   WHERE AVG(retail-cost) > 8.56
   GROUP BY category;
   ```
   d.
   ```
   SELECT category, AVG(retail-cost) profit
   FROM books
   GROUP BY category
   HAVING profit > 8.56;
   ```

3. Which of the following statements is correct?
   a. The WHERE clause can only contain a group function if that group function is not also listed in the SELECT clause.
   b. Group functions cannot be used in the SELECT, FROM, or WHERE clauses.
   c. The HAVING clause is always processed before the WHERE clause.
   d. The GROUP BY clause is always processed before the HAVING clause.
4. Which of the following is not a valid SQL statement?
   a.
   ```
   SELECT MIN(pubdate)
   FROM books
   GROUP BY category
   HAVING pubid = 4;
   ```
   b.
   ```
   SELECT MIN(pubdate)
   FROM books
   WHERE category = 'COOKING';
   ```
   c.
   ```
   SELECT COUNT(*)
   FROM orders
   WHERE customer# = 1005;
   ```
   d.
   ```
   SELECT MAX(COUNT(customer#))
   FROM orders
   GROUP BY customer#;
   ```
5. Which of the following statements is correct?
   a. The COUNT function can be used to determine how many rows contain a NULL value.
   b. Only distinct values are included in group functions, unless the ALL keyword is included in the SELECT clause.
   c. The HAVING clause restricts rows to be processed.
   d. The WHERE clause determines which groups will be displayed in the query results.
   e. None of the statements are correct.

6. Which of the following is a valid SQL statement?
   a. ```
      SELECT customer#, order#,
      MAX(shipdate-orderdate)
      FROM orders
      GROUP BY customer#
      WHERE customer# = 1001;
      ```
   b. ```
      SELECT customer#, COUNT(order#)
      FROM orders
      GROUP BY customer#;
      ```
   c. ```
      SELECT customer#, COUNT(order#)
      FROM orders
      GROUP BY COUNT(order#);
      ```
   d. ```
      SELECT customer#, COUNT(order#)
      FROM orders
      GROUP BY order#;
      ```
7. Which of the following SELECT statements will list only the book with the largest profit?
   a. ```
      SELECT title, MAX(retail-cost)
      FROM books
      GROUP BY title;
      ```
   b. ```
      SELECT title, MAX(retail-cost)
      FROM books
      GROUP BY title
      HAVING MAX(retail-cost);
      ```
   c. ```
      SELECT title, MAX(retail-cost)
      FROM books;
      ```
   d. none of the above
8. Which of the following is correct?
   a. A group function can be nested inside a group function.
   b. A group function can be nested inside a single-row function.
   c. A single-row function can be nested inside a group function.
   d. A and B
   e. A, B, and C
9. Which of the following functions is used to calculate the total value contained in a specified column?
   a. COUNT
   b. MIN
   c. TOTAL
   d. SUM
   e. ADD
      ```

10. Which of the following SELECT statements will list the highest retail price of all books in the Family Category?
    a.
    ```
    SELECT MAX(retail)
    FROM books
    WHERE category = 'FAMILY';
    ```
    b.
    ```
    SELECT MAX(retail)
    FROM books
    HAVING category = 'FAMILY';
    ```
    c.
    ```
    SELECT retail
    FROM books
    WHERE category = 'FAMILY'
    HAVING MAX(retail);
    ```
    d. none of the above
11. Which of the following functions can be used to include NULL values in calculations?
    a. SUM
    b. NVL
    c. MAX
    d. MIN
12. Which of the following is not a valid statement?
    a. You must enter the ALL keyword in a function to include all non-unique values.
    b. The AVG function can be used to find the average calculated difference between two dates.
    c. The MIN and MAX functions can be used on any type of data.
    d. All of the above are valid statements.
    e. None of the above are valid statements.
13. Which of the following SQL statements will determine how many total customers were referred by other customers?
    a.
    ```
    SELECT customer#, SUM(referred)
    FROM customers
    GROUP BY customer#;
    ```
    b.
    ```
    SELECT COUNT(referred)
    FROM customers;
    ```
    c.
    ```
    SELECT COUNT(*)
    FROM customers;
    ```
    d.
    ```
    SELECT COUNT(*)
    FROM customers
    WHERE referred IS NULL;
    ```

Use the following SELECT statement to answer questions 14–18.

```
1 SELECT customer#, COUNT(*)
2 FROM customers NATURAL JOIN orders
3 NATURAL JOIN orderitems
4 WHERE orderdate > '02-APR-02'
5 GROUP BY customer#
6 HAVING COUNT(*) > 2;
```

14. Which line of the SELECT statement is used to restrict the number of records to be processed by the query?
    a. 1
    b. 4
    c. 5
    d. 6

15. Which line of the SELECT statement is used to restrict the groups displayed in the results of the query?
    a. 1
    b. 4
    c. 5
    d. 6
16. Which line of the SELECT statement is used to group the data contained in the database?
    a. 1
    b. 4
    c. 5
    d. 6
17. Which clause must be included for the query to execute since the SELECT clause contains the Customer# column?
    a. 1
    b. 4
    c. 5
    d. 6
18. The COUNT(*) function in the SELECT clause is used to return:
    a. the number of records in the specified tables
    b. the number of orders placed by each customer
    c. the number of NULL values in the specified tables
    d. the number of customers who have placed an order

19. Which of the following functions can be used to determine the earliest ship date for all orders recently processed by JustLee Books?
    a. COUNT function
    b. MAX function
    c. MIN function
    d. STDDEV function
    e. VARIANCE function
20. Which of the following is not a valid SELECT statement?
    a. ```
       SELECT STDDEV(retail)
       FROM books;
       ```
    b. ```
       SELECT AVG(SUM(retail))
       FROM orders NATURAL JOIN orderitems NATURAL JOIN books
       GROUP BY customer#;
       ```
    c. ```
       SELECT order#, TO_CHAR(SUM(retail, '999.99')
       FROM orderitems NATURAL JOIN books
       GROUP BY order#;
       ```
    d. ```
       SELECT title, VARIANCE(retail-cost)
       FROM books
       GROUP BY pubid;
       ```

Hands-on Assignments

To perform these activities, refer to the tables in Appendix A.

1. Determine how many books are in the Cooking Category.
2. Display the number of books that have a retail price of more than $30.00.
3. Display the date of the most recently published book.
4. Determine the total profit generated by sales to customer 1017.
5. List the most expensive book purchased by customer 1017.
6. List the least expensive book in the Computer Category.
7. Determine the average profit generated by orders contained in the ORDERS table.
8. Determine how many orders have been placed by each customer in the CUSTOMERS table. Do not include any customer in the results who has not recently placed an order with JustLee Books.
9. List the customers living in Georgia or Florida who have recently placed an order totaling more than $80.
10. What is the retail price of the most expensive book written by Lisa White?

A Case for Oracle9*i*

To perform this activity, refer to the tables in Appendix A.

JustLee Books has a problem: Their book storage space is becoming very limited. As a solution, management is considering limiting the inventory to only those books that return at least a 55 percent profit. Any book that returns less than a 55 percent profit would be dropped from inventory and not reordered.

This plan could, however, have a negative impact on overall sales. Management fears that if JustLee stops carrying the less-profitable books, the company might lose repeat business from its customers. As part of management's decision-making process, they want to know whether less-profitable books are frequently purchased by current customers. Therefore, management would like to know how many times these less-profitable books have been purchased recently.

Determine which books generate less than a 55 percent profit and how many copies of those books have been sold recently. Summarize your findings for management, and include a copy of the query necessary to retrieve the data from the database tables.

CHAPTER

7

SUBQUERIES

Objectives

After completing this chapter, you should be able to do the following:

- Determine when it is appropriate to use a subquery
- Identify which clauses can contain subqueries
- Distinguish between an outer query and a subquery
- Use a single-row subquery in a WHERE clause
- Use a single-row subquery in a HAVING clause
- Use a single-row subquery in a SELECT clause
- Distinguish between single-row and multiple-row comparison operators
- Use a multiple-row subquery in a WHERE clause
- Use a multiple-row subquery in a HAVING clause
- Use a multiple-column subquery in a WHERE clause
- Create an inline view, using a multiple-column subquery in a FROM clause
- Compensate for NULL values in subqueries
- Distinguish between correlated and uncorrelated subqueries
- Nest a subquery inside another subquery

Suppose that the management of JustLee Books requests a list of every computer book that has a higher retail price than *Database Implementation*. In previous chapters, you would have followed this procedure: (1) Query the database to determine the retail price of *Database Implementation*, and then (2) create a second SELECT statement to find the titles of all books retailing for more than *Database Implementation*. In this chapter, you will learn how to use an alternative approach, a subquery, to get the same output. A **subquery** is a nested query—one complete query inside another query.

The output of the subquery can consist of a single value (a single-row subquery), several rows of values (a multiple-row subquery), or even multiple columns of data (a multiple-column subquery). Figure 7–1 provides an overview of this chapter's contents.

Subquery	Description
Single-Row Subquery	Returns one row of results that consists of one column to the outer query
Multiple-Row Subquery	Returns more than one row of results to the outer query
Multiple-Column Subquery	Returns more than one column of results to the outer query
Correlated Subquery	References a column in the outer query. Executes the subquery once for every row in the outer query
Uncorrelated Subquery	Executes the subquery first and passes the value to the outer query

Figure 7-1 Overview of chapter contents

Go to the JustLee Database folder in your Data Files. If you run the **Bookscript.sql** file, you will be able to work through the queries shown in this chapter. Your output should match the output shown.

SUBQUERIES AND THEIR USES

Sometimes, getting an answer to a query requires a multistep operation. First, a query must be created to determine a value that is unknown to the user but is contained within the database. That first query is the subquery. The results of the subquery are passed back as input to the **parent query**, or **outer query**. The parent query incorporates that value into its calculations to determine the final output.

Although subqueries are most commonly used in the WHERE or HAVING clause of a SELECT statement, there may be times when it is appropriate to use a subquery in the SELECT or FROM clause. When the subquery is nested in a WHERE or HAVING clause, the results returned from the subquery are used as a condition in the outer query. Any type of subquery (single-row, multiple-row, or multiple-column) can be used in the WHERE, HAVING, or FROM clause of a SELECT statement. As will be discussed, the only type of subquery that can be used in a SELECT clause is a single-row subquery.

Using a subquery in a FROM clause has a specific purpose and will be discussed in a separate section in this chapter.

Keep the following rules in mind when working with any type of subquery:

- A subquery must be *a complete query in itself*—i.e., it must have at least a SELECT and a FROM clause.
- A subquery cannot have an ORDER BY clause. If the displayed output needs to be presented in a specific order, an ORDER BY clause should be listed as the last clause of the outer query.
- A subquery *must be enclosed within a set of parentheses* to separate it from the outer query.
- If the subquery is placed in the WHERE or HAVING clause of an outer query, the subquery can be placed only on the *right side* of the comparison operator.

SINGLE-ROW SUBQUERIES

A **single-row subquery** is used when the results of the outer query are based on a single, unknown value. Although it is formally called "single-row," this implies that the query will return multiple columns—but only one row—of results. However, *a single-row subquery can return to the outer query only one row of results that consists of only one column.* Therefore, this text will refer to the output of a single-row subquery as a **single value**.

Single-Row Subquery in a WHERE Clause

Let's compare creating multiple queries, which you've studied in previous chapters, with creating one query containing a subquery. In this chapter's introduction, management requested a list of all computer books that have a higher retail price than *Database Implementation*. As shown in Figure 7–2, the first step is to create a query to determine the retail price of that book, which is $31.40.

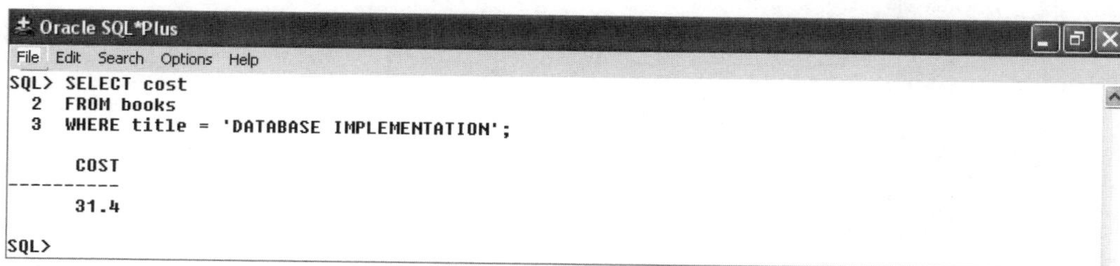

Figure 7-2 Query to determine the retail price of *Database Implementation*

To determine which computer books retail for more than $31.40, a second query must be issued that explicitly states the cost of *Database Implementation*. That second query is issued in Figure 7–3.

```
SQL> SELECT title, cost
     FROM books
     WHERE cost > 31.4
     AND category = 'COMPUTER';

TITLE                               COST
------------------------------ ----------
HOLY GRAIL OF ORACLE                47.25
E-BUSINESS THE EASY WAY              37.9

SQL>
```

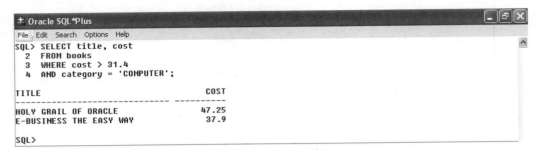

Figure 7-3 Query for computer books costing more than $31.40

The WHERE clause in Figure 7–3 explicitly includes the retail price of *Database Implementation* found by the first query in Figure 7–2, and the category condition is to restrict records only to those in the Computer Category.

However, these same results could be obtained through the use of a single-row subquery. A single-row subquery is appropriate in this example because (1) to obtain the desired results, an unknown value must be obtained, and that value is contained in the database, and (2) only one value should be returned from the inner query (i.e., the retail price of *Database Implementation*).

In Figure 7–4, a single-row subquery is substituted for the Cost condition of the SELECT statement given in Figure 7–3. The subquery is enclosed in parentheses to distinguish it from the clauses of the parent query.

```
SQL> SELECT title, cost
     FROM books
     WHERE cost >
                  (SELECT cost
                   FROM books
                   WHERE title = 'DATABASE IMPLEMENTATION')
     AND category = 'COMPUTER';

TITLE                               COST
------------------------------ ----------
HOLY GRAIL OF ORACLE                47.25
E-BUSINESS THE EASY WAY              37.9

SQL>
```

Figure 7-4 A single-row subquery

In Figure 7–4, the inner query is executed first, and the result of the query, a single value of 31.4, is passed back to the outer query. The outer query is then executed, and all books having a retail price greater than $31.40 and belonging to the Computer Category are listed in the output.

The indention used for the subquery in Figure 7–4 is merely to improve readability and is not required by Oracle9*i*.

Operators indicate to Oracle9*i* whether a user is creating a single-row subquery or a multiple-row subquery. The single-row operators are =, >, <, >=, <=, and <>. Although other operators, such as IN, are allowed, single-row operators instruct Oracle9*i* that only one value is expected from the subquery. If more than one value is returned, the SELECT statement will fail, and you will receive an error message.

Suppose that management makes another request. They want the title of the most expensive book sold by JustLee Books. Because you learned how to use the MAX function in Chapter 6, this should be simple. You might create the query shown in Figure 7–5.

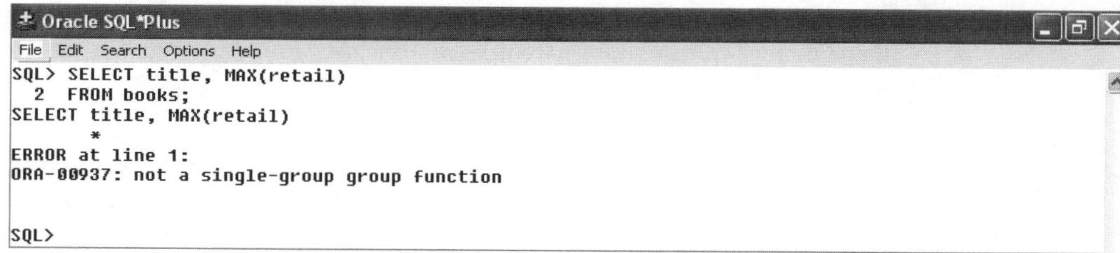

Figure 7-5 Flawed query: attempt to determine the book with the highest retail price

Perhaps it is not quite that simple. Remember the rule when working with group functions: If an individual field with a group function is listed in the SELECT clause, the individual field must also be listed in a GROUP BY clause. In this instance, adding a GROUP BY clause would not make sense: If a GROUP BY clause is added that contains the Title column, then each book would be its own group because each title is different. In other words, the results would be the same as using **`SELECT title, retail`** in the query.

Thus, to retrieve the title of the most expensive book, you can use a subquery to determine the highest retail price of any book. That retail price can then be returned to an outer query and displayed in the results.

As shown in Figure 7–6, the most expensive book sold by JustLee Books is *Painless Child-Rearing*.

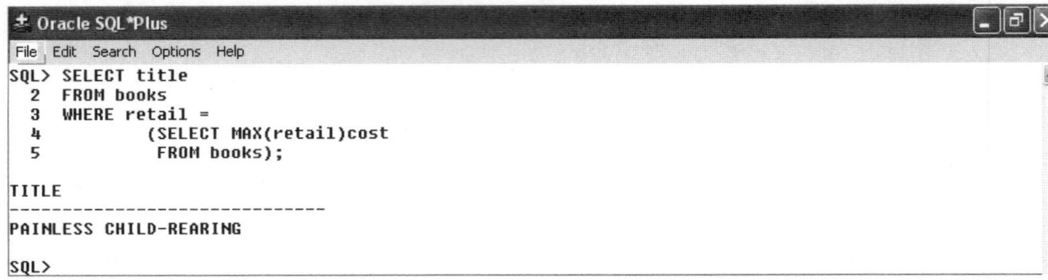

```
Oracle SQL*Plus
File Edit Search Options Help
SQL> SELECT title
FROM books
WHERE retail =
        (SELECT MAX(retail)cost
         FROM books);

TITLE
------------------------------
PAINLESS CHILD-REARING

SQL>
```

Figure 7-6 Query to determine the title of the most expensive book

If management also requests the actual retail price of the book, the Retail field can simply be listed in the SELECT clause of the query and then displayed in the results.

If an error message is returned for the query in Figure 7–6, make certain that the subquery contains four parentheses—one set around the *retail* argument for the MAX function and one set around the subquery.

Multiple subqueries can also be included in a SELECT statement. For example, suppose that management needs to know the title of all books published by the publisher of *Big Bear and Little Dove* that generate more than the average profit returned by all books sold through JustLee Books. In this case, two values are unknown: (1) the identity of the publisher of *Big Bear and Little Dove* and (2) the average profit of all books. How might you create a query that extracts those values? The SELECT statement in Figure 7–7 uses two separate subqueries in the WHERE clause to obtain the information.

```
Oracle SQL*Plus
File Edit Search Options Help
SQL> SELECT isbn, title
FROM books
WHERE pubid =
               (SELECT pubid
                FROM books
                WHERE title = 'BIG BEAR AND LITTLE DOVE')
AND retail-cost >
                (SELECT AVG(retail-cost)
                 FROM books);

ISBN       TITLE
---------- ------------------------------
2491748320 PAINLESS CHILD-REARING
2147428890 SHORTEST POEMS

SQL>
```

Figure 7-7 SELECT statement with two single-row subqueries

Notice that each subquery in Figure 7–7 is complete—both contain a minimum of one SELECT clause and one FROM clause. Because they are subqueries, each is enclosed in parentheses. The first subquery determines the publisher of *Big Bear and Little Dove* and returns that result to the first condition of the WHERE clause (line 3). The second subquery finds the average profit of all books sold by JustLee Books by using the AVG function and then passes that value back to the second condition of the WHERE clause (line 7) to be compared against the profit for each book. Because the two conditions of the WHERE clause in the outer query are combined with the AND logical operator, both values returned by the subqueries must be met for a book to be listed in the output of the outer query. In this example, two books are found that are published by the publisher of *Big Bear and Little Dove* and that return more than the average profit.

Single-Row Subquery in a HAVING Clause

As previously mentioned, a subquery can also be included in a HAVING clause. A HAVING clause is used when the group results of a query need to be restricted, based on some condition. If the result returned from a subquery must be compared to a group function, then the inner query must be nested in the outer query's HAVING clause.

For example, suppose that management needs a list of all book categories that return a higher average profit than the Literature Category does. You would follow these steps:

1. Determine the average profit for all literature books.
2. Calculate the average profit for every category.
3. Compare the average profit for every category with the average profit for the Literature Category.

Try writing a query that accomplishes this goal, and then look at the query and output in Figure 7–8.

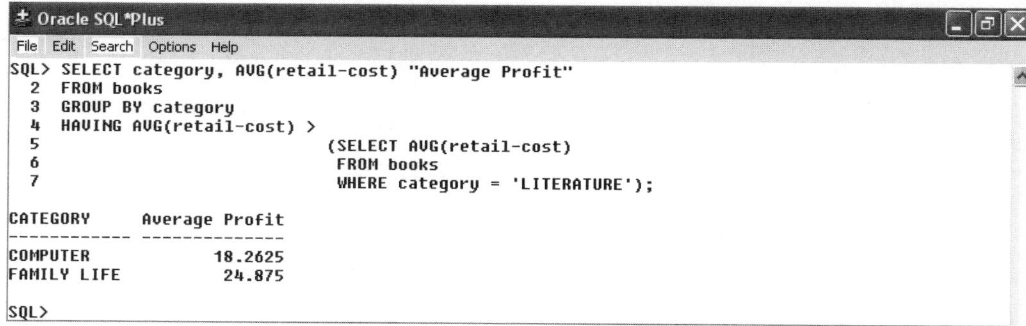

```
Oracle SQL*Plus
File Edit Search Options Help
SQL> SELECT category, AVG(retail-cost) "Average Profit"
  2  FROM books
  3  GROUP BY category
  4  HAVING AVG(retail-cost) >
  5                              (SELECT AVG(retail-cost)
  6                               FROM books
  7                               WHERE category = 'LITERATURE');

CATEGORY     Average Profit
------------ --------------
COMPUTER            18.2625
FAMILY LIFE          24.875

SQL>
```

Figure 7-8 Single-row subquery nested in a HAVING clause

As Figure 7–8 shows, the results are restricted to groups that have a higher average profit than the Literature category does. Getting these results requires the HAVING clause. Because the results of the subquery are applied to groups of data, it is necessary to nest the subquery in the HAVING clause.

As shown in Figure 7–8, the subquery in a HAVING clause must follow the same guidelines as those used in WHERE clauses; that is, it must include at least a SELECT clause and a FROM clause—and be enclosed in parentheses.

Single-Row Subquery in a SELECT Clause

A single-row subquery can also be nested in the SELECT clause of an outer query. However, this approach is rarely used because when the subquery is listed in the SELECT clause, this means the value returned by the subquery will be displayed for every row of output generated by the parent query. For illustrative purposes, suppose that management would like to compare the price of each book in inventory against the average price of all books in inventory. One approach would be to calculate the average price of all books and give management that figure and a separate list of all books with their current retail price.

On the other hand, you could use a subquery in a SELECT clause that calculates the average retail price of all books. When a single-row subquery is included in a SELECT clause, the result of the subquery is displayed in the output of the parent query. To include a subquery in a SELECT clause, the subquery is simply separated from the table columns by a comma, just as if you were listing another column. In fact, the results of the subquery can even be given a column alias. Look at the output of the query issued in Figure 7–9.

```
Oracle SQL*Plus
File Edit Search Options Help
SQL> SELECT title, retail,
                          (SELECT AVG(retail)
                           FROM books) "Overall Average"
     FROM books;

TITLE                              RETAIL Overall Average
------------------------------ ---------- ---------------
BODYBUILD IN 10 MINUTES A DAY       30.95      40.9821429
REVENGE OF MICKEY                      22      40.9821429
BUILDING A CAR WITH TOOTHPICKS      59.95      40.9821429
DATABASE IMPLEMENTATION             55.95      40.9821429
COOKING WITH MUSHROOMS              19.95      40.9821429
HOLY GRAIL OF ORACLE                75.95      40.9821429
HANDCRANKED COMPUTERS                  25      40.9821429
E-BUSINESS THE EASY WAY              54.5      40.9821429
PAINLESS CHILD-REARING              89.95      40.9821429
THE WOK WAY TO COOK                 28.75      40.9821429
BIG BEAR AND LITTLE DOVE             8.95      40.9821429
HOW TO GET FASTER PIZZA             29.95      40.9821429
HOW TO MANAGE THE MANAGER           31.95      40.9821429
SHORTEST POEMS                      39.95      40.9821429

14 rows selected.

SQL>
```

Figure 7-9 Single-row subquery in a SELECT clause

As shown in Figure 7–9, to calculate the average price of all books in inventory, the SELECT clause of the outer query includes the Title and Retail columns in the column list as well as the subquery. The average calculated by the subquery is displayed for every book included in the output. The column alias, Overall Average, is assigned to the results of the subquery to indicate the contents of that column. If a column alias had not been used, the actual subquery would have been shown as the column heading—a somewhat unattractive column heading. The overall result of having the subquery in the SELECT clause enables management to look at one list and compare each book's retail price to the average retail price for all books.

Try using the subquery in Figure 7–9 to calculate the difference between the retail price and the average price. Simply restructure the SELECT clause and have `retail -` precede the subquery, just as if you were calculating profit using the expression `(retail - cost)`.

In the "real world," users who are just learning to work with subqueries often do not have much confidence in the output when a subquery is contained in a WHERE clause because the actual value generated by the subquery is not displayed. However, if the subquery being used in the WHERE clause is also included in the SELECT clause, the value generated by the single-row subquery can be compared against the final output of the parent query. This allows a novice to validate the output. The subquery is removed from the SELECT clause after validation to generate the requested results. It gives the user more confidence in the final results and reduces the risk of distributing erroneous data. However, this will only work for single-row subqueries. For other types of subqueries, you must execute the subquery as a separate SELECT statement to determine the values it will generate because a SELECT clause can only process one data value.

MULTIPLE-ROW SUBQUERIES

Multiple-row subqueries are nested queries that can return more than one row of results to the parent query. Multiple-row subqueries are most commonly used in WHERE and HAVING clauses. The main rule to keep in mind when working with multiple-row subqueries is that *you must use multiple-row operators*. If a single-row operator is used with a subquery that returns more than one row of results, Oracle9*i* will return an error message, and the SELECT statement will fail. Valid multiple-row operators include IN, ALL, and ANY. Let's consider each of those operators.

IN Operator

Of the three multiple-row operators, the IN operator is the most commonly used. Figure 7–10 shows a multiple-row subquery using the IN operator.

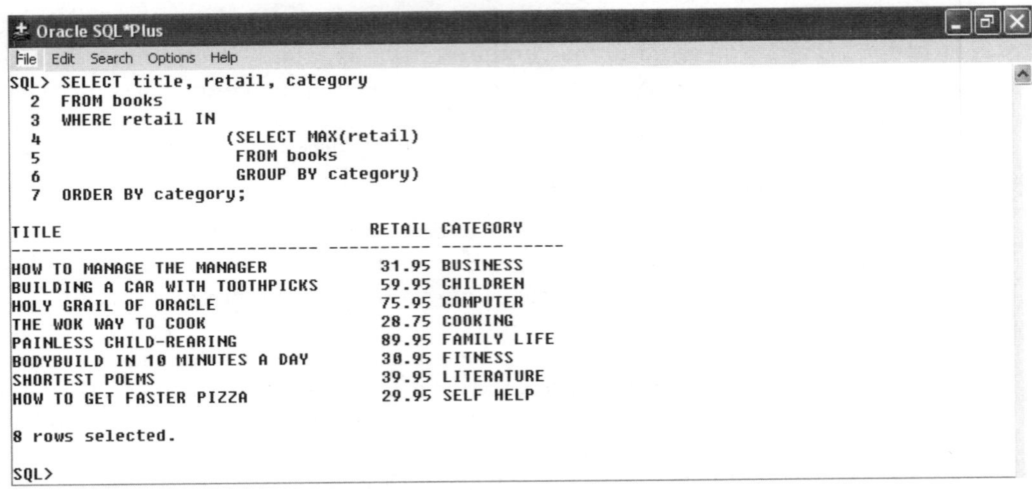

Figure 7-10 Multiple-row subquery using the IN operator

The IN operator in Figure 7–10 indicates that the records processed by the outer query must match one of the values returned by the subquery (i.e., it creates an OR condition). The order of execution in Figure 7–10 is as follows:

1. The subquery determines the price of the most expensive book in each category.
2. The maximum retail price in each category is passed to the WHERE clause of the outer query.
3. The outer query compares the price of each book to the prices generated by the subquery.
4. If the retail price of a book matches one of the prices returned by the subquery, then the title, retail price, and category of the book are displayed in the query's output.

ALL and ANY Operators

The ALL and ANY operators can be combined with other comparison operators to treat the results of a subquery as a set of values, rather than as individual values. Figure 7–11 summarizes the use of the ALL and ANY operators in conjunction with other comparison operators.

Operator	Description
>ALL	More than the highest value returned by the subquery
<ALL	Less than the lowest value returned by the subquery
<ANY	Less than the highest value returned by the subquery
>ANY	More than the lowest value returned by the subquery
=ANY	Equal to any value returned by the subquery (same as IN)

Figure 7-11 Descriptions of ALL and ANY operator combinations

The ALL operator is fairly straightforward:

- If the ALL operator is combined with the "greater than" symbol (>), then the outer query is searching for all records with a value higher than the highest value returned by the subquery (i.e., more than ALL the values returned).
- If the ALL operator is combined with the "less than" symbol (<), then the outer query is searching for all records with a value lower than the lowest value returned by the subquery (i.e., less than ALL the values returned).

To examine the impact of using the ALL comparison operator, the query given in Figure 7–12 will be used as a subquery.

Figure 7-12 Retail price of the books in the Cooking Category

The query in Figure 7–12 returns the retail prices for two books in the Cooking Category. The lowest value returned is $19.95, and the highest value is $28.75.

Suppose that you want to know the titles of all books having a retail price greater than the most expensive book in the Cooking Category. One approach is to use the MAX function in a subquery to find the highest retail price. Another approach is to use the >ALL operator, as shown in Figure 7–13.

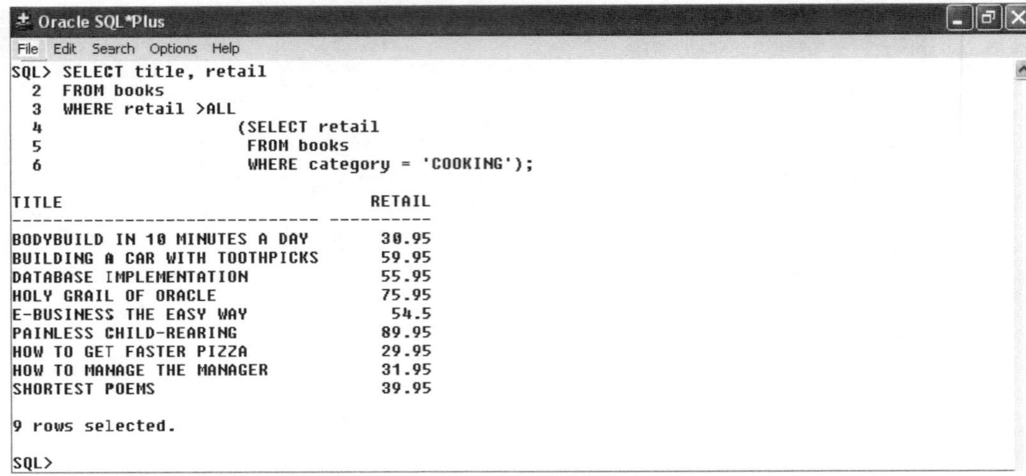

Figure 7-13 The >ALL operator

This is the Oracle9*i* "thinking" that goes into processing the SELECT command in Figure 7–13:

- The subquery shown in Figure 7–13 passes the retail prices of the two books in the Cooking Category ($19.95 and $28.75) to the outer query.
- Since the >ALL operator is used in the outer query, Oracle9*i* is instructed to list all the books with a retail price higher than the largest value returned by the subquery ($28.75).
- In this example, there are nine books that have a higher price than the most expensive book in the Cooking Category.

Similarly, the <ALL operator is used to determine the records that have a value less than the lowest value returned by a subquery. Thus, if you need to find books that are priced less than the least expensive book in the Cooking Category, first formulate a subquery that identifies the books in the Cooking Category. Then you can compare the retail price of the books in the BOOKS table against the values returned by a subquery, using the <ALL operator.

As in the previous query, the subquery shown in Figure 7–14 first finds the two books in the Cooking Category (Figure 7–12). The retail prices of those books ($19.95 and $28.75) are then passed to the outer query. Because $19.95 is the lowest retail price of all books in the Cooking Category, only those books having a retail price less than $19.95 will be displayed in the output. In this case, Oracle9*i* found only one book having a retail price lower than the least expensive book in the Cooking Category: *Big Bear and Little Dove*.

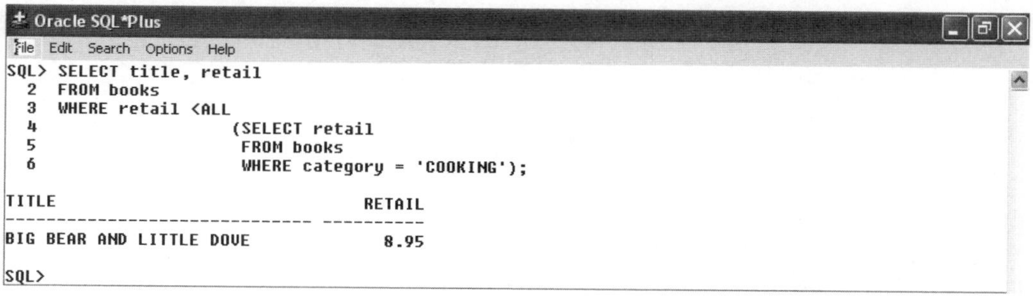

```
Oracle SQL*Plus
File Edit Search Options Help
SQL> SELECT title, retail
     FROM books
     WHERE retail <ALL
                      (SELECT retail
                       FROM books
                       WHERE category = 'COOKING');

TITLE                              RETAIL
------------------------------ ----------
BIG BEAR AND LITTLE DOVE             8.95

SQL>
```

Figure 7-14 The <ALL operator

By contrast, the <ANY operator is used to find records that have a value less than the highest value returned by a subquery. To determine which books cost less than the most expensive book in the Cooking Category, simply evaluate the results of the subquery, using the <ANY operator, as shown in Figure 7–15.

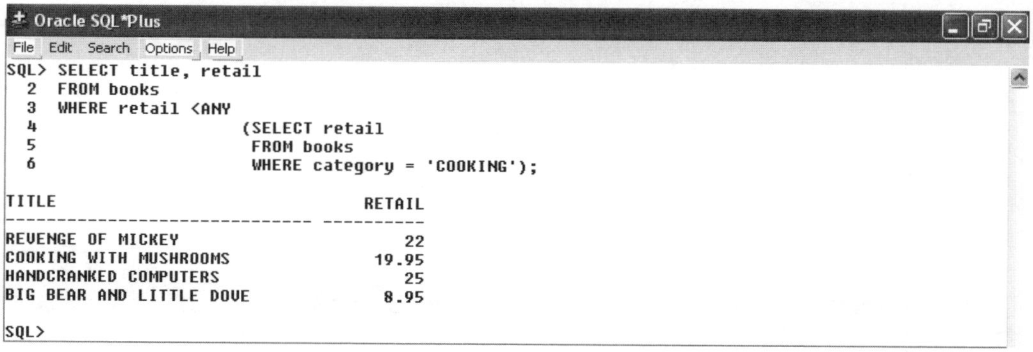

```
Oracle SQL*Plus
File Edit Search Options Help
SQL> SELECT title, retail
     FROM books
     WHERE retail <ANY
                      (SELECT retail
                       FROM books
                       WHERE category = 'COOKING');

TITLE                              RETAIL
------------------------------ ----------
REVENGE OF MICKEY                      22
COOKING WITH MUSHROOMS              19.95
HANDCRANKED COMPUTERS                  25
BIG BEAR AND LITTLE DOVE             8.95

SQL>
```

Figure 7-15 The <ANY operator

In Figure 7–15, the outer query found four books with a retail price less than the most expensive book in the Cooking Category. Notice, however, that the results also include *Cooking With Mushrooms*, which is a book found in the Cooking Category. Since the outer query compares the records to the highest value in the Cooking Category, then any other book in the Cooking Category will also be displayed in the query results. To eliminate any book in the Cooking Category from appearing in the output, simply add the condition **AND category <> 'COOKING'** to the WHERE clause of the outer query.

The >ANY operator is used to return records that have a value greater than the lowest value returned by the subquery. In Figure 7–16, there are 12 records that have a retail price greater than the lowest retail price returned by the subquery ($19.95).

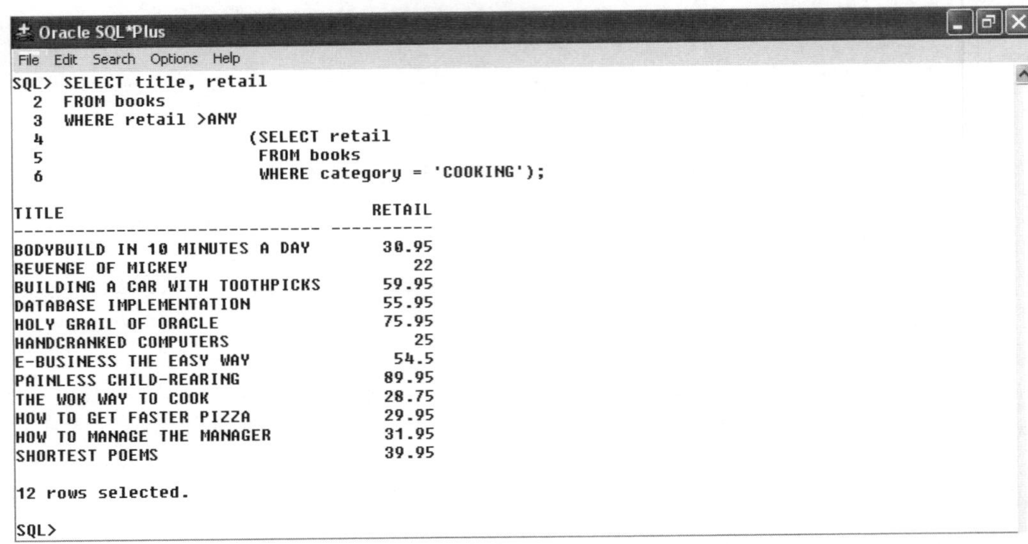

```
SQL> SELECT title, retail
FROM books
WHERE retail >ANY
                    (SELECT retail
                     FROM books
                     WHERE category = 'COOKING');

TITLE                           RETAIL
------------------------------ ----------
BODYBUILD IN 10 MINUTES A DAY      30.95
REVENGE OF MICKEY                     22
BUILDING A CAR WITH TOOTHPICKS     59.95
DATABASE IMPLEMENTATION            55.95
HOLY GRAIL OF ORACLE               75.95
HANDCRANKED COMPUTERS                 25
E-BUSINESS THE EASY WAY             54.5
PAINLESS CHILD-REARING             89.95
THE WOK WAY TO COOK                28.75
HOW TO GET FASTER PIZZA            29.95
HOW TO MANAGE THE MANAGER          31.95
SHORTEST POEMS                     39.95

12 rows selected.

SQL>
```

Figure 7-16 The >ANY operator

The =ANY operator works the same way as the IN comparison operator does. For example, in the query given in Figure 7–17, the user is searching for the title of books that were purchased by customers who also purchased the book with the ISBN of 0401140733. Since that book could have appeared on more than one order, and the user wants to identify all those orders, the =ANY operator is used.

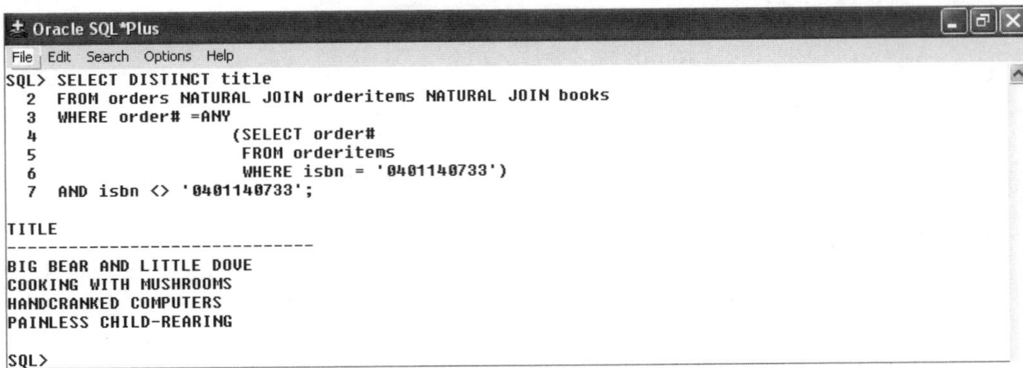

```
SQL> SELECT DISTINCT title
FROM orders NATURAL JOIN orderitems NATURAL JOIN books
WHERE order# =ANY
                   (SELECT order#
                    FROM orderitems
                    WHERE isbn = '0401140733')
AND isbn <> '0401140733';

TITLE
------------------------------
BIG BEAR AND LITTLE DOVE
COOKING WITH MUSHROOMS
HANDCRANKED COMPUTERS
PAINLESS CHILD-REARING

SQL>
```

Figure 7-17 The =ANY operator

The query in Figure 7–17 would have yielded the same results if the IN operator had been used instead of the =ANY operator. The DISTINCT keyword in the SELECT clause of the outer query is included because, as previously mentioned, a title could have been ordered by more than one customer and would have had multiple listings in the output.

If you do not get the same results as those shown in Figure 7–17, make certain the closing parenthesis for the subquery is after the condition on line 6 of the display and not after the condition on line 7, which is part of the outer query.

Also notice in Figure 7–17 that the columns needed to complete the query are in three different tables: ORDERS, ORDERITEMS, and BOOKS. Because the columns necessary to perform the inner query are only contained in the ORDERITEMS table, joins among the tables are not required in the subquery. However, the columns referenced by the outer query are contained in the BOOKS table (ISBN and Title) and the ORDERITEMS table (Order#). To complete the access path, or link, between the BOOKS and ORDERITEMS tables, the ORDERS table must be included in the FROM clause of the outer query.

EXISTS Operator

The **EXISTS** operator is used to determine whether a condition is present in a subquery. The results of the operator are Boolean—it is TRUE if the condition exists and FALSE if it does not. If the results are TRUE, then the records meeting the condition are displayed.

To better understand how the EXISTS operator works, let's look at an example. Suppose that management requests a list of all recently ordered books. To provide this list, you can take one of two approaches. One approach is to join the BOOKS and ORDERITEMS tables and display the title for each book. The other approach is to use the EXISTS operator to determine which books have been ordered recently, using the query shown in Figure 7–18.

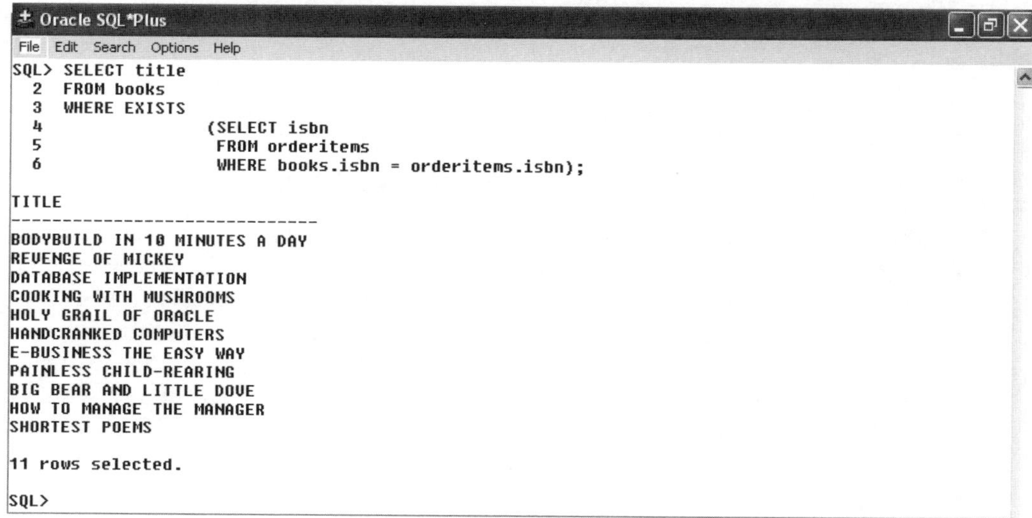

Oracle SQL*Plus

File Edit Search Options Help

```
SQL> SELECT title
  2  FROM books
  3  WHERE EXISTS
  4                (SELECT isbn
  5                 FROM orderitems
  6                 WHERE books.isbn = orderitems.isbn);

TITLE
------------------------------
BODYBUILD IN 10 MINUTES A DAY
REVENGE OF MICKEY
DATABASE IMPLEMENTATION
COOKING WITH MUSHROOMS
HOLY GRAIL OF ORACLE
HANDCRANKED COMPUTERS
E-BUSINESS THE EASY WAY
PAINLESS CHILD-REARING
BIG BEAR AND LITTLE DOVE
HOW TO MANAGE THE MANAGER
SHORTEST POEMS

11 rows selected.

SQL>
```

Figure 7-18 Subquery with the EXISTS operator

As shown in Figure 7–18, the SELECT and FROM clauses of the outer query indicate that the titles of the books in the BOOKS table should be displayed in the output. However, which titles should be displayed? The WHERE clause in the outer query uses the EXISTS operator, which can be interpreted to mean "include only those books that exist in, or are identified by, the subquery." In this example, the subquery simply identifies all book ISBNs that are contained in the ORDERITEMS table—which happen to be all the books that have been ordered lately. Because the BOOKS table also contains the ISBN column, the outer query lists the title of every book with a matching ISBN in the ORDERITEMS table (i.e., it also exists in the ORDERITEMS table) in the output.

However, what if you had wanted to know the title of every book that had *not* been ordered recently? In that case, you would be searching for books that do not exist in the ORDERITEMS table. As shown in Figure 7–19, adding the NOT logical operator before the EXISTS operator in the WHERE clause of the outer query specifies that a title should be displayed only if the book's ISBN is not contained in the ORDERITEMS table (i.e., the result of the EXISTS condition is FALSE).

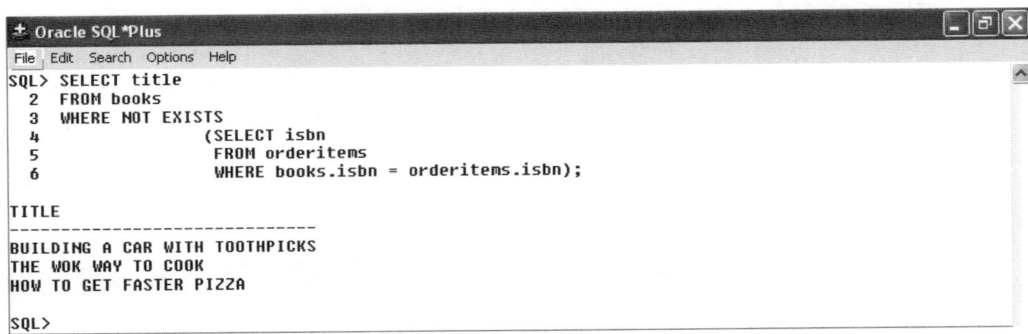

Figure 7-19 Subquery with the NOT and EXISTS operators

Multiple-Row Subquery in a HAVING Clause

Thus far, you have seen multiple-row subqueries in a WHERE clause, but they can also be included in a HAVING clause. When the results of the subquery are being compared to grouped data in the outer query, the subquery *must* be nested in a HAVING clause in the parent query. For illustrative purposes, suppose that you needed to determine whether any customer's recently placed order has a total "amount due" that is greater than the average "amount due" for all orders originating from that customer's state. Getting this output requires that you first determine the average "amount due" from all books ordered from each state, and then compare the state averages with each customer's order. The state averages can be calculated in a subquery, but because one value will be returned for each state contained in the ORDERS table, this requires a multiple-row subquery. The average "amount due" for the orders from each state will need to be compared with the total "amount due" for each order; this requires the outer query to group all the items in the ORDERITEMS table by the Order# column. Therefore, the outer query will require a HAVING clause since the comparison is based on grouped data.

As shown in Figure 7–20, the structure for using a multiple-row subquery in a HAVING clause is the same as using the subquery in a WHERE clause.

```
Oracle SQL*Plus
File Edit Search Options Help
SQL> SELECT order#, SUM(retail*quantity)
  FROM orders NATURAL JOIN orderitems NATURAL JOIN books
  HAVING SUM(retail*quantity) >ANY
                              (SELECT AVG(SUM(quantity*retail))
                               FROM orders NATURAL JOIN orderitems
                               NATURAL JOIN books
                               GROUP BY shipstate)
  GROUP BY order#;

    ORDER# SUM(RETAIL*QUANTITY)
---------- --------------------
      1007               347.25

SQL>
```


Figure 7-20 Multiple-row subquery in a HAVING clause

Single-row and multiple-row subqueries may look the same in terms of the subqueries themselves; however, they are distinctly different. A single-row subquery can only return *one* data value, whereas a multiple-row subquery can return *several* values. Thus, if you execute a subquery that returns more than one data value and the comparison operator is intended to be used only with single-row subqueries, you will receive an error message, and the query will not be executed, as shown in Figure 7–21.

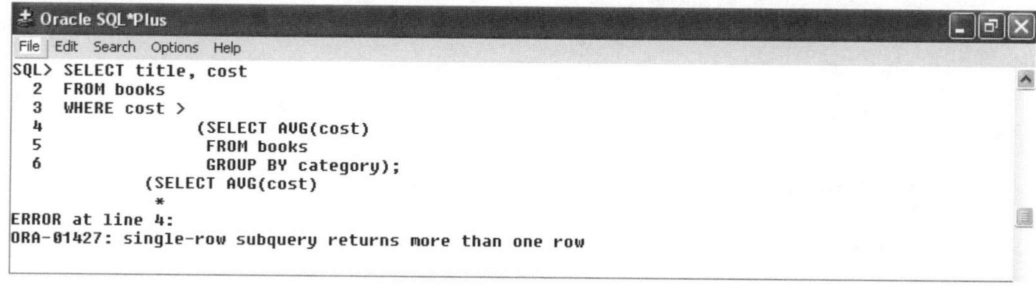

```
Oracle SQL*Plus
File Edit Search Options Help
SQL> SELECT title, cost
  FROM books
  WHERE cost >
               (SELECT AVG(cost)
                FROM books
                GROUP BY category);
             (SELECT AVG(cost)
              *
ERROR at line 4:
ORA-01427: single-row subquery returns more than one row
```

Figure 7-21 Flawed query: using a single-row operator for a multiple-row subquery

MULTIPLE-COLUMN SUBQUERIES

Now that you've examined multiple-row subqueries, let's look at multiple-column subqueries. A **multiple-column subquery** returns more than one column to the outer query. A multiple-column subquery can be listed in the FROM, WHERE, or HAVING clause of a query. If the multiple-column subquery is included in the FROM clause of the outer query, the subquery actually generates a temporary table that can be referenced by other clauses of the outer query.

Multiple-Column Subquery in a FROM Clause

When a multiple-column subquery is used in the FROM clause of an outer query, it basically creates a temporary table that can be referenced by other clauses of the outer query. This temporary table is more formally called an **inline view**. If the temporary table generated by the subquery contains grouped data, these data can be referenced or used just like individual data values.

Suppose that you need a list of all books in the BOOKS table that have a higher-than-average selling price than other books in their category. You need to display each book's title, retail price, category, and the average selling price of books in that category. Since the average selling price is based on grouped data, this presents a problem. How might you solve it? Look at Figure 7–22.

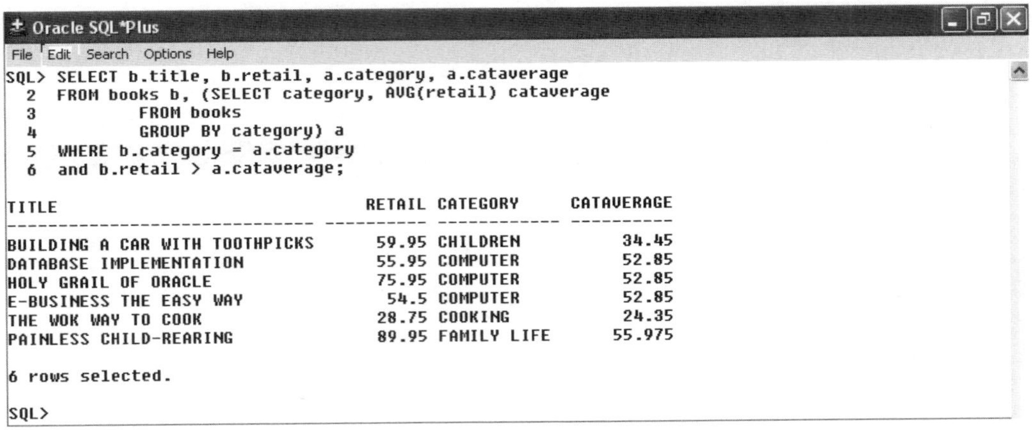

```
Oracle SQL*Plus
File Edit Search Options Help
SQL> SELECT b.title, b.retail, a.category, a.cataverage
     FROM books b, (SELECT category, AVG(retail) cataverage
             FROM books
             GROUP BY category) a
     WHERE b.category = a.category
     and b.retail > a.cataverage;

TITLE                              RETAIL CATEGORY     CATAVERAGE
------------------------------ ---------- ------------ ----------
BUILDING A CAR WITH TOOTHPICKS      59.95 CHILDREN          34.45
DATABASE IMPLEMENTATION             55.95 COMPUTER          52.85
HOLY GRAIL OF ORACLE                75.95 COMPUTER          52.85
E-BUSINESS THE EASY WAY              54.5 COMPUTER          52.85
THE WOK WAY TO COOK                 28.75 COOKING           24.35
PAINLESS CHILD-REARING              89.95 FAMILY LIFE      55.975

6 rows selected.

SQL>
```

Figure 7-22 Multiple-column subquery in a FROM clause

In Figure 7–22, a multiple-column subquery is nested in the FROM clause of the outer table. The subquery creates a temporary table. The subquery determines the categories that exist in the BOOKS table and the average selling price of every book in that particular category. However, how do you display the title of each book in the BOOKS table, its retail price, the category of the book, and the average price of all books in that same category?

The BOOKS table contains the individual data for each book, and the subquery has created a temporary table that stores the grouped data. Notice that on line 4 of the query in Figure 7–22 that a table alias has been assigned to the results of the subquery, so the columns contained in the subquery (Category and Cataverage) can be referenced by other clauses in the outer SELECT statement.

In essence, the query in Figure 7–22 is referencing, or obtaining, data from two different tables. The tables have been joined using the traditional approach—through the WHERE clause of the outer query. The problem with using the traditional approach is that both tables contain a column called Category—this will create a problem with ambiguity if the Category column is referenced anywhere in the outer query. To avoid ambiguity, the Category column needs a column qualifier to identify which table contains the category data to be displayed. Therefore, table aliases are used in the SELECT and WHERE clauses to identify the table containing the column being referenced.

As shown in Figure 7–23, the query could have also been created using the NATURAL JOIN keywords supported by Oracle9*i*.

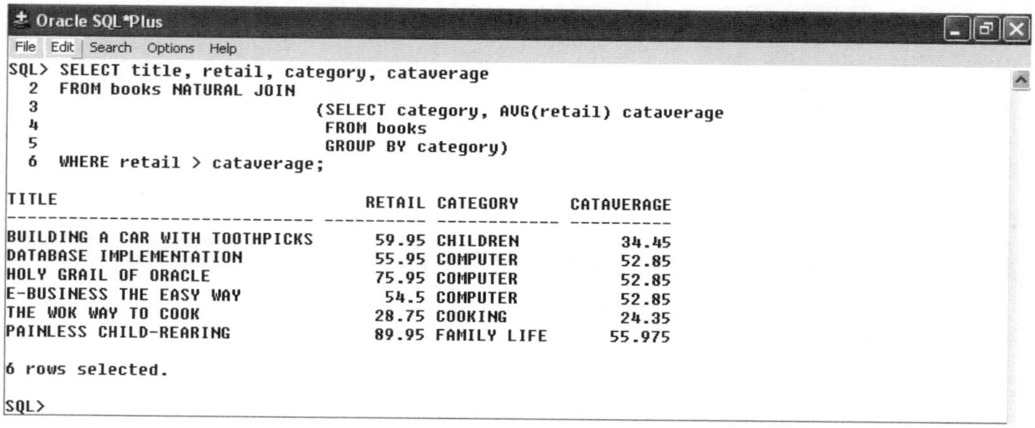

```
Oracle SQL*Plus
File Edit Search Options Help
SQL> SELECT title, retail, category, cataverage
  2  FROM books NATURAL JOIN
  3                          (SELECT category, AVG(retail) cataverage
  4                           FROM books
  5                           GROUP BY category)
  6  WHERE retail > cataverage;

TITLE                              RETAIL CATEGORY     CATAVERAGE
------------------------------ ---------- ------------ ----------
BUILDING A CAR WITH TOOTHPICKS      59.95 CHILDREN          34.45
DATABASE IMPLEMENTATION             55.95 COMPUTER          52.85
HOLY GRAIL OF ORACLE                75.95 COMPUTER          52.85
E-BUSINESS THE EASY WAY              54.5 COMPUTER          52.85
THE WOK WAY TO COOK                 28.75 COOKING           24.35
PAINLESS CHILD-REARING              89.95 FAMILY LIFE      55.975

6 rows selected.

SQL>
```

Figure 7-23 Multiple-column subquery in a FROM clause with the NATURAL JOIN keywords

Because both tables (BOOKS and the temporary table created by the subquery) in Figure 7–23 contain a column named Category, the tables are linked in the FROM clause with a NATURAL JOIN through the Category field. Because column qualifiers are not allowed with the NATURAL JOIN method, the temporary table created by the subquery is not assigned a table alias.

Multiple-Column Subquery in a WHERE Clause

When a multiple-column subquery is included in the WHERE or HAVING clause of the outer query, the IN operator is used by the outer query to evaluate the results of the subquery. The results of the subquery consist of more than one column of results.

The syntax for the outer WHERE clause is `WHERE (`*`columnname, columnname,..`*`)` `IN` *`subquery`*. Keep these rules in mind:

- Since the WHERE clause contains more than one column name, the column list must be enclosed within parentheses.
- Column names listed in the WHERE clause must be in the same order as they are listed in the SELECT clause of the subquery.

Double-check that the field list presented in the WHERE clause of the outer query is enclosed in parentheses and is in the same order as the list of fields given in the SELECT clause of the subquery.

Previously in Figure 7–10, the subquery returned the price of the most expensive book in each category, and the outer query generated a list of the title, retail price, and category of books matching the retail price returned by the subquery. The overall result of the outer query was to display the title, retail price, and category for the most expensive book in each category. However, that query will only work if two books do not have the same retail price. For example, suppose that books in two different categories both have the same price as one of the values returned by the subquery. Then one (or possibly both) would not be the most expensive book in its category, and you would receive erroneous results. To create a query specifically to create a list of the most expensive books in each category, a multiple-column subquery is more appropriate. Look at the example in Figure 7–24.

Oracle SQL*Plus

File Edit Search Options Help

```
SQL> SELECT title, retail, category
  2  FROM books
  3  WHERE (category, retail) IN
  4                              (SELECT category, MAX(retail)
  5                               FROM books
  6                               GROUP BY category)
  7  ORDER BY category;

TITLE                              RETAIL CATEGORY
------------------------------ ---------- ------------
HOW TO MANAGE THE MANAGER           31.95 BUSINESS
BUILDING A CAR WITH TOOTHPICKS      59.95 CHILDREN
HOLY GRAIL OF ORACLE                75.95 COMPUTER
THE WOK WAY TO COOK                 28.75 COOKING
PAINLESS CHILD-REARING              89.95 FAMILY LIFE
BODYBUILD IN 10 MINUTES A DAY       30.95 FITNESS
SHORTEST POEMS                      39.95 LITERATURE
HOW TO GET FASTER PIZZA             29.95 SELF HELP

8 rows selected.

SQL>
```

Figure 7-24 Multiple-column subquery in a WHERE clause

In Figure 7–24, the subquery finds the highest retail value in each category and passes both the category names and the retail prices back to the outer query.

Although a multiple-column subquery can be used in the HAVING clause of an outer query, it is generally only used when analyzing extremely large sets of numeric data that have been grouped and is generally presented in more advanced courses focusing upon quantitative methods.

NULL VALUES

As with everything else, NULL values also present a problem when using subqueries. Because a NULL value is the same as the absence of data, a NULL cannot be returned to an outer query for comparison purposes—it is not equal to anything, not even another NULL. Therefore, if a NULL value is passed from a subquery, the results of the outer query will be "no rows selected." Although the statement will not fail (generate an Oracle9*i* error message), you will not get the expected results, as you can see from the example in Figure 7–25.

```
Oracle SQL*Plus
File Edit Search Options Help
SQL> SELECT customer#
  2  FROM customers
  3  WHERE referred =
  4                  (SELECT referred
  5                   FROM customers
  6                   WHERE customer# = 1005);

no rows selected

SQL>
```

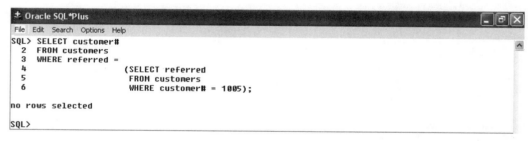

Figure 7-25 Flawed query: NULL results from a subquery

In Figure 7–25, the user is trying to determine whether the customer who referred customer 1005 has referred any other customers to JustLee Books. The problem is that no rows are listed as output from the outer query. The question is this: Are no rows listed because the customer who referred customer 1005 has not referred any other customers, or because customer 1005 was not originally referred to JustLee Books—i.e., the Referred column is NULL? If no one referred customer 1005, should the output of the outer query be a list of all customers who were not referred by other customers?

In this case, customer 1005 was not referred by any other customer; thus, the Referred column is NULL. Because a NULL value is passed to the outer query, no record processed by the outer query has an equivalent value in the Referred column. However, what if customer 1005 was not referred by another customer, and you wanted a list of all customers who were also not referred by other customers? As always, it is the NVL function to the rescue.

NVL in Subqueries

If it is possible for a subquery to return a NULL value to the outer query for comparison, the NVL function should be used to substitute an actual value for the NULL. However, you must keep two things in mind:

1. The substitution of the NULL value must occur for the NULL value both in the subquery and in the outer query.
2. The value substituted for the NULL value must be one that could not possibly exist anywhere else in that column.

Figure 7–26 uses the same premise as Figure 7–25 and provides an example of these two rules.

```
SQL> SELECT customer#
  2  FROM customers
  3  WHERE NVL(referred, 0) =
  4                              (SELECT NVL(referred, 0)
  5                               FROM customers
  6                               WHERE customer# = 1005);

 CUSTOMER#
----------
      1001
      1002
      1003
      1004
      1005
      1006
      1008
      1010
      1011
      1012
      1014
      1015
      1017
      1018
      1020

15 rows selected.

SQL>
```

Figure 7-26 The NVL function in a subquery

In the query presented in Figure 7–26, the NVL function is included whenever the Referred column is referenced—in both the subquery and the outer query. In this example, a zero is substituted for the NULL value. Because the value contained in the Referred column is actually the customer number of a customer in the CUSTOMERS table, and because no customer has the customer number of zero, substituting a zero for the NULL value will not accidentally make a NULL record equivalent to a non-NULL record.

Whenever you substitute a value for a NULL, make certain no other record contains the substituted value. For example, use ZZZ for a customer name; in the case of a date field, use a date that absolutely would not exist in the database.

IS NULL in Subqueries

Although problems exist when passing a NULL value from a subquery to an outer query, searches for NULL values are allowed in a subquery. As with regular queries, you can still search for NULL values, using the IS NULL comparison operator.

For example, suppose that you need to find the title of all books that have been ordered but not yet shipped to the customers. The subquery presented in Figure 7–27 identifies the orders that have not yet shipped—the ship date is NULL.

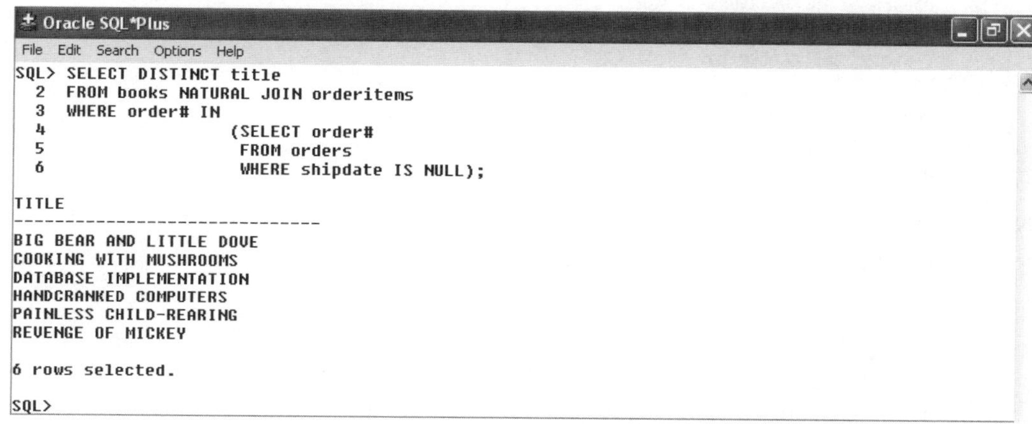

```
SQL> SELECT DISTINCT title
  2  FROM books NATURAL JOIN orderitems
  3  WHERE order# IN
  4                 (SELECT order#
  5                  FROM orders
  6                  WHERE shipdate IS NULL);

TITLE
------------------------------
BIG BEAR AND LITTLE DOVE
COOKING WITH MUSHROOMS
DATABASE IMPLEMENTATION
HANDCRANKED COMPUTERS
PAINLESS CHILD-REARING
REVENGE OF MICKEY

6 rows selected.

SQL>
```

Figure 7-27 Using IS NULL in a subquery

As shown in Figure 7–27, the order number for each order is returned to the outer query, and the title for each book is displayed. The DISTINCT keyword is used to suppress duplicate titles from being listed. Although the subquery searches for records containing NULL values, it is the Order# column that is passed back to the outer query. Because the Order# column is the primary key for the ORDERS table, no NULL values can exist in that field. Therefore, there is no need to use the NVL function in this example.

Correlated Subqueries

Thus far you have, for the most part, studied **uncorrelated subqueries**: The subquery is executed first, the results of the subquery are passed to the outer query, and then the outer query is executed. By contrast, a **correlated subquery**, which refers to the method of processing that Oracle9*i* uses to execute a query, uses a different procedure. The query using the EXISTS operator in Figure 7–18 is a correlated subquery and uses that different procedure. The query is displayed again in Figure 7–28 for easy reference.

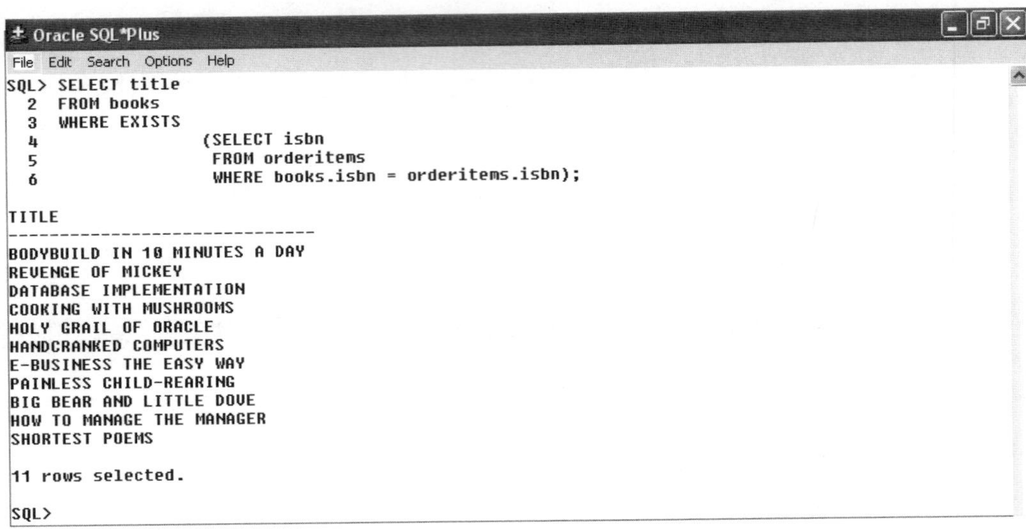

Oracle SQL*Plus
File Edit Search Options Help
SQL> SELECT title
2 FROM books
3 WHERE EXISTS
4 (SELECT isbn
5 FROM orderitems
6 WHERE books.isbn = orderitems.isbn);
TITLE

BODYBUILD IN 10 MINUTES A DAY
REVENGE OF MICKEY
DATABASE IMPLEMENTATION
COOKING WITH MUSHROOMS
HOLY GRAIL OF ORACLE
HANDCRANKED COMPUTERS
E-BUSINESS THE EASY WAY
PAINLESS CHILD-REARING
BIG BEAR AND LITTLE DOVE
HOW TO MANAGE THE MANAGER
SHORTEST POEMS
11 rows selected.
SQL>

Figure 7-28 Correlated subquery

Although the query in Figure 7–28 is a multiple-row subquery, the execution of the entire query requires that each row in the BOOKS table be processed to determine whether it also exists in the ORDERITEMS table. In other words, Oracle9*i* executes the outer query first, and when it encounters the WHERE clause of the outer query, it is evaluated to determine whether that row is TRUE (i.e., whether it exists in the ORDERITEMS table). If it is TRUE, then the book's title is displayed in the results. The outer query is executed again for the next book in the BOOKS table and compared to the contents of the ORDERITEMS table, and so on, for each row of the BOOKS table. In other words, *a correlated subquery is a subquery that is processed, or executed, once for each row in the outer query.*

How does Oracle9*i* distinguish between an uncorrelated and a correlated subquery? Simply speaking, *if a subquery references a column from the outer query, then it is a correlated subquery*. Notice that in the subquery, the WHERE clause specifies the ISBN column of the BOOKS table. Since the BOOKS table is not included in the FROM clause of the subquery, it is forced to use data that are processed by the outer query (i.e., the ISBN of the books being processed during that execution of the outer query).

If the comparison of the ISBNs from each table had been performed in the outer query (i.e., the two tables had been joined in the outer clause and not in the subquery), the results would have been completely different. Since the subquery would no longer reference a column contained in the outer query, it would have been considered an uncorrelated subquery. With an uncorrelated subquery, the subquery would have been executed first, and then the results would have been passed back to the outer query. Since the subquery is used to identify every ISBN contained in the ORDERITEMS table, each ISBN listed in the table would have been returned to the outer query. Then, the outer query would find the title for that book and display that title.

Notice that in Figure 7–29, the subquery no longer specifies the column ISBN from the BOOKS table—the join is created in the outer query. Therefore, the subquery is executed first, and for each value returned, it is matched to the contents of the BOOKS table. As shown in Figure 7–29, several titles are listed more than once. Why? Because their ISBN is listed more than once in the ORDERITEMS table and is returned from the subquery each time it is encountered.

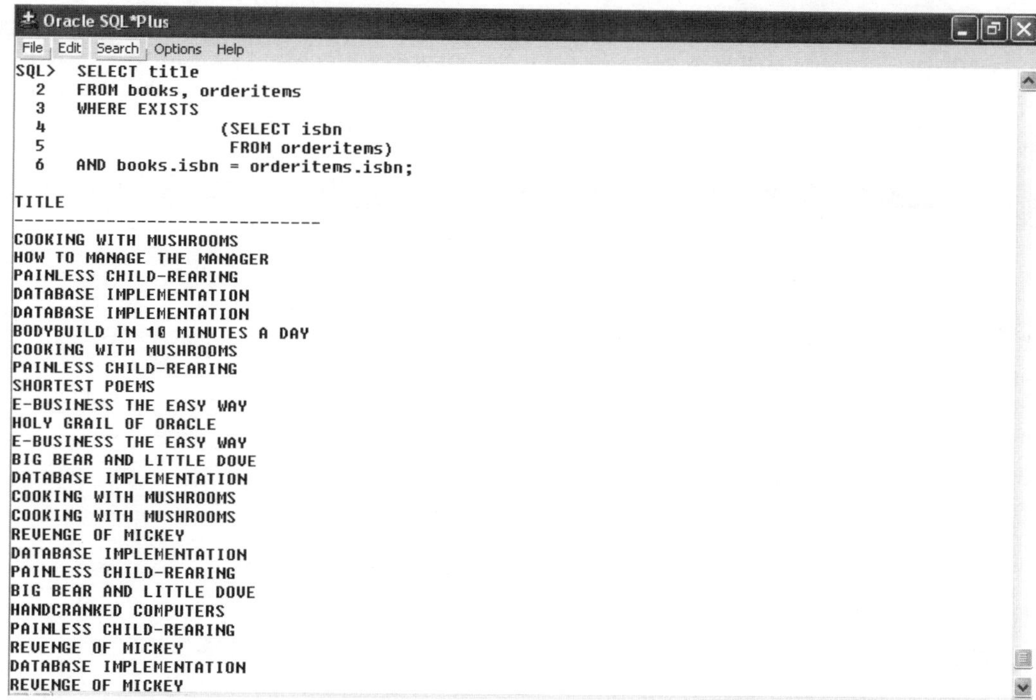

Figure 7-29 Uncorrelated subquery (partial output shown)

NESTED SUBQUERIES

Subqueries can be nested inside the FROM, WHERE, or HAVING clauses of other subqueries. In Oracle9*i*, subqueries in a WHERE clause can be nested to a depth of 255 subqueries, and there is no depth limit when the subqueries are nested in a FROM clause. When nesting subqueries, you may wish to use the following strategy:

- Determine exactly what you are trying to find. This is the goal of the query.
- Write the innermost subquery first.

- After you write the innermost query, look at the value you are able to pass back to the outer query. If this is not the value needed by the outer query (e.g., it references the wrong column), analyze how the data need to be converted to get the correct rows, and use another subquery between the outer query and the nested subquery. In some cases, you may need to create several layers of subqueries to link the value returned by the innermost subquery to the value needed by the outer query.

The most common reason for nesting subqueries is to create a chain of data. For example, suppose that you need to find the name of the customer who has ordered the most books from JustLee Books (not including multiple quantities of the same book) on *one* order. Figure 7–30 shows a query that provides these results.

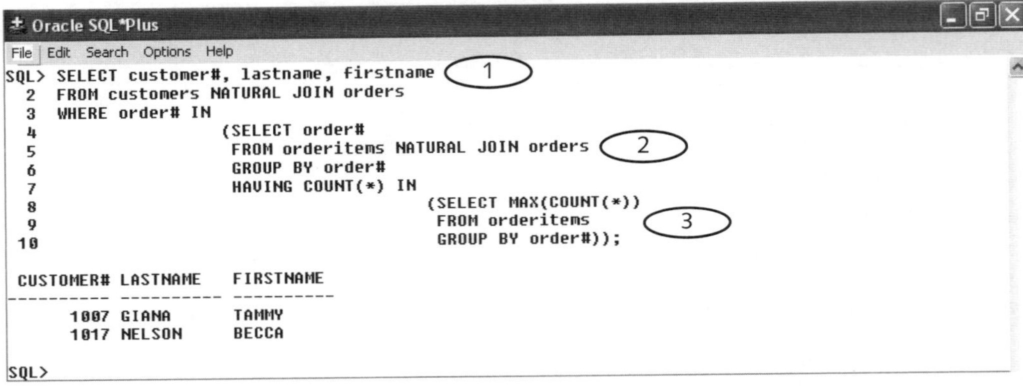

Figure 7-30 Nested subqueries

Here are the steps required to create the query shown in Figure 7–30:

1. The goal of the query is to count the number of items placed on each order and identify the order—or orders, in case of a tie—with the most items. The nested subquery identified by ③ in Figure 7–30 finds the order that has the most books.
2. The value of the highest count of the items ordered is then passed back to the outer subquery, ②.
3. The outer subquery, ②, is then used to identify which order(s) has the same number of items as the highest number of items previously found by the nested subquery.
4. Once the order number(s) has been identified, it is then passed back to the outer query, ①, which determines the customer number and name of the person who placed the order. In this case, two customers tied for placing an order that has the most items.

If the first subquery, ②, had not been included, the second subquery would have tried to return the value from the MAX(COUNT(*)) functions directly to the outer query for comparison with the Order# column. By including the first subquery, ②, Oracle9*i* is able to determine which orders have the same COUNT(*) value as the value returned by the second subquery, and can then pass the order number back to the outer query to determine the customer's information. In this case, there are two customers who meet the criteria—fortunately, the user included the IN operator in the query, thus avoiding an error message.

Don't forget to include the extra set of parentheses for the nested function on line 8, or you will receive an error message.

Chapter Summary

- A subquery is a complete query nested in the SELECT, FROM, HAVING, or WHERE clause of another query. The subquery must be enclosed in parentheses and have a SELECT and a FROM clause, at a minimum.
- Subqueries are completed first. The result of the subquery is used as input for the outer query.
- A single-row subquery can return a maximum of one value.
- Single-row operators include =, >, <, >=, <=, and <>.
- Multiple-row subqueries return more than one row of results.
- Operators that can be used with multiple-row subqueries include IN, ALL, ANY, and EXISTS.
- Multiple-column subqueries return more than one column to the outer query. The columns of data are passed back to the outer query in the same order in which they are listed in the SELECT clause of the subquery.
- NULL values returned by a multiple-row or multiple-column subquery will not present a problem if the IN or =ANY operator is used. The NVL function can be used to substitute a value for a NULL value when working with subqueries.
- Correlated subqueries reference a column contained in the outer query. When using correlated subqueries, the subquery is executed once for each row processed by the outer query.
- Subqueries can be nested to a maximum depth of 255 subqueries in the WHERE clause of the parent query. The depth is unlimited for subqueries nested in the FROM clause of the parent query.
- With nested subqueries, the innermost subquery is executed first, then the next highest level subquery is executed, and so on, until the outermost query is reached.

CHAPTER 7 SYNTAX SUMMARY

The following tables present a summary of the syntax that you have learned in this chapter. You can use the tables as a study guide and reference.

Syntax Guide	
Subquery Types	
Subquery Processing	**Example**

Correlated Subquery:
References a column in the outer query.
Executes the subquery once for every row in the outer query
	`SELECT title`
	`FROM books b`
	`WHERE b.isbn IN`
	`   (SELECT isbn`
	`    FROM orderitems o`
	`    WHERE b.isbn = o.isbn);`

Uncorrelated Subquery:
Executes the subquery first and passes the value to the outer query
	`SELECT title`
	`FROM books b, orderitems o`
	`WHERE books isbn IN`
	`   (SELECT isbn`
	`    FROM orderitems)`
	`AND b.isbn = o.isbn;`

Multiple-Row Comparison Operators	
Operator	**Description**
>ALL	More than the highest value returned by the subquery
<ALL	Less than the lowest value returned by the subquery
<ANY	Less than the highest value returned by the subquery
>ANY	More than the lowest value returned by the subquery
=ANY	Equal to any value returned by the subquery (same as IN)
[NOT] EXISTS	Row must match a value in the subquery

Subquery Processing Classifications	
Subquery Processing	**Example**

Correlated Subquery:
References a column in the outer query. Executes the subquery once for every row in the outer query
	SELECT title
	FROM books b
	WHERE b.isbn IN
	   (SELECT isbn
	    FROM orderitems o
	    WHERE b.isbn = o.isbn);

Uncorrelated Subquery:
Executes the subquery first and passes the value to the outer query
	SELECT title
	FROM books b, orderitems o
	WHERE books isbn IN
	   (SELECT isbn
	    FROM orderitems)
	AND b.isbn = o.isbn;

Multiple-Row Comparison Operators	
Operator	**Description**
>ALL	More than the highest value returned by the subquery
<ALL	Less than the lowest value returned by the subquery
<ANY	Less than the highest value returned by the subquery
>ANY	More than the lowest value returned by the subquery
=ANY	Equal to any value returned by the subquery (same as IN)
[NOT] EXISTS	Row must match a value in the subquery

Review Questions

1. What is the difference between a single-row subquery and a multiple-row subquery?
2. What comparison operators are required for multiple-row subqueries?
3. What happens if a single-row subquery returns more than one row of results?
4. Which clause(s) cannot be used in a subquery?
5. If a subquery is used in the FROM clause of an outer query, how are the results of the subquery referenced in other clauses of the outer query?
6. Which multiple-row subquery comparison operator is the same as the IN comparison operator?
7. How can Oracle9*i* determine whether clauses of a SELECT statement belong to an outer query or a subquery?

8. When should a subquery be nested in a HAVING clause?
9. What is the difference between correlated and uncorrelated subqueries?
10. What type of situation requires the use of a subquery?

MULTIPLE CHOICE

To answer these questions, refer to the tables in Appendix A.

1. Which query will identify the customers living in the same state as the customer named Leila Smith?

   a.
   ```
   SELECT customer# FROM customers WHERE state =
   (SELECT state FROM customers WHERE lastname = 'SMITH');
   ```
   b.
   ```
   SELECT customer# FROM customers WHERE state =
   (SELECT state FROM customers WHERE lastname = 'SMITH' OR
   firstname = 'LEILA');
   ```
   c.
   ```
   SELECT customer# FROM customers WHERE state =
   (SELECT state FROM customers
   WHERE lastname = 'SMITH' AND firstname = 'LEILA'
   ORDER BY customer);
   ```
   d.
   ```
   SELECT customer# FROM customers WHERE state =
   (SELECT state FROM customers
   WHERE lastname = 'SMITH' AND firstname = 'LEILA');
   ```

2. Which of the following is a valid SELECT statement?

   a.
   ```
   SELECT order# FROM orders WHERE shipdate =
   SELECT shipdate FROM orders WHERE order# = 1010;
   ```
   b.
   ```
   SELECT order# FROM orders WHERE shipdate =
   (SELECT shipdate FROM orders)
   AND order# = 1010;
   ```
   c.
   ```
   SELECT order# FROM orders WHERE shipdate =
   (SELECT shipdate FROM orders WHERE order# = 1010);
   ```
   d.
   ```
   SELECT order# FROM orders HAVING shipdate =
   (SELECT shipdate FROM orders WHERE order# = 1010);
   ```

3. Which of the following operators is considered a single-row operator?

   a. IN

   b. ALL

   c. <>

   d. <>ALL

4. Which of the following queries will determine which customers have ordered the same books as customer 1017?
   a. `SELECT order# FROM orders WHERE customer# = 1017;`
   b. ```
      SELECT customer# FROM orders NATURAL JOIN orderitems
      WHERE isbn = (SELECT isbn FROM orderitems
      WHERE customer# = 1017);
      ```
   c. ```
      SELECT customer# FROM orders WHERE order# =
      (SELECT order# FROM orderitems WHERE customer# = 1017);
      ```
   d. ```
      SELECT customer# FROM orders NATURAL JOIN orderitems
      WHERE isbn IN (SELECT isbn FROM orderitems
      NATURAL JOIN orders WHERE customer# = 1017);
      ```
5. Which of the following statements is valid?
   a. ```
      SELECT title FROM books WHERE retail <
      (SELECT cost FROM books WHERE isbn = '9959789321');
      ```
   b. ```
      SELECT title FROM books WHERE retail =
      (SELECT cost FROM books WHERE isbn = '9959789321'
      ORDER BY cost);
      ```
   c. ```
      SELECT title FROM books WHERE category IN
      (SELECT cost FROM orderitems WHERE isbn = '9959789321');
      ```
   d. none of the above statements
6. Which of the following statements is correct?
   a. If a subquery is used in the FROM clause of the outer query, the data contained in the temporary table cannot be referenced by clauses used in the outer query.
   b. The temporary table created by a subquery in the FROM clause of the outer query must be assigned a table alias or it cannot be joined with another table using the NATURAL JOIN keywords.
   c. If a temporary table is created through a subquery in the FROM clause of an outer query, the data in the temporary table can be referenced by another clause in the outer query.
   d. none of the above

7. Which of the following queries identifies other customers who were referred to JustLee Books by the same individual that referred Jorge Perez?
   a. ```
      SELECT customer# FROM customers WHERE referred =
      (SELECT referred FROM customers
      WHERE firstname = 'JORGE' AND lastname = 'PEREZ');
      ```
   b. ```
      SELECT referred FROM customers
      WHERE (customer#, referred) =
      (SELECT customer# FROM customers
      WHERE firstname = 'JORGE' AND lastname = 'PEREZ');
      ```
   c. ```
      SELECT referred FROM customers
      WHERE (customer#, referred) IN
      (SELECT customer# FROM customers
      WHERE firstname = 'JORGE' AND lastname = 'PEREZ');
      ```
   d. ```
      SELECT customer# FROM customers
      WHERE customer# =
      (SELECT customer# FROM customers
      WHERE firstname = 'JORGE' AND lastname = 'PEREZ');
      ```
8. In which of the following situations would it be appropriate to use a subquery?
   a. when you need to find all customers living in a particular region of the country
   b. when you need to find all publishers who have toll-free telephone numbers
   c. when you need to find the name of all books that were shipped on the same date as an order placed by a particular customer
   d. when you need to find all books published by Publisher 4
9. Which of the following customers have ordered the same books as customers 1001 and 1005?
   a. ```
      SELECT customer# FROM orders NATURAL JOIN books
      WHERE isbn =
      (SELECT isbn FROM orderitems NATURAL JOIN books
      WHERE customer# = 1001 OR customer# = 1005);
      ```
   b. ```
      SELECT customer# FROM orders NATURAL JOIN books
      WHERE isbn <ANY
      (SELECT isbn FROM orderitems NATURAL JOIN books
      WHERE customer# = 1001 OR customer# = 1005);
      ```
   c. ```
      SELECT customer# FROM orders NATURAL JOIN books
      WHERE isbn =
      (SELECT isbn FROM orderitems NATURAL JOIN books
      WHERE customer# = 1001 OR 1005);
      ```
   d. ```
      SELECT customer# FROM orders NATURAL JOIN books
      WHERE isbn IN
      (SELECT isbn FROM orderitems NATURAL JOIN orders
      NATURAL JOIN books
      WHERE customer# = 1001 OR customer# = 1005);
      ```

10. Which of the following operators is used to find all values that are greater than the highest value returned by a subquery?

    a. >ALL

    b. <ALL

    c. >ANY

    d. <ANY

    e. IN

11. Which query will determine the customer who has ordered the most books from JustLee Books?

    a.
    ```
    SELECT customer# FROM orders NATURAL JOIN orderitems
    NATURAL JOIN customers
    HAVING SUM(quantity) =
    (SELECT MAX(SUM(quantity)) FROM orders NATURAL JOIN
    orderitems GROUP BY customer#)
    GROUP BY customer#;
    ```

    b.
    ```
    SELECT customer# FROM orders NATURAL JOIN orderitems WHERE
    SUM(quantity) =
    (SELECT MAX(SUM(quantity)) FROM orderitems GROUP BY
    customer#);
    ```

    c.
    ```
    SELECT customer# FROM orders WHERE MAX(SUM(quantity)) =
    (SELECT MAX(SUM(quantity) FROM orderitems GROUP BY order#);
    ```

    d.
    ```
    SELECT customer# FROM orders WHERE quantity =
    (SELECT MAX(SUM(quantity)) FROM orderitems GROUP BY
    customer#);
    ```

12. Which of the following statements is correct?

    a. The IN comparison operator cannot be used with a subquery that returns only one row of results.

    b. The equals (=) comparison operator cannot be used with a subquery that returns more than one row of results.

    c. In an uncorrelated subquery, the statements in the outer query are executed first, and then the statements in the subquery are executed.

    d. A subquery can only be nested in the SELECT clause of an outer query.

13. What is the purpose of the following query?

```
SELECT isbn, title
FROM books
WHERE (pubid, category) IN
                    (SELECT pubid, category
                    FROM books
                    WHERE title LIKE '%ORACLE%');
```

   a. It determines which publisher published a book belonging to the Oracle Category and then lists all other books published by that same publisher.
   b. It lists all the publishers and categories containing the value ORACLE.
   c. It lists the ISBN and title of all books belonging to the same category and having the same publisher as any book with the letters ORACLE in its title.
   d. None of the above—the query contains a multiple-row operator, and since the inner query only returns one value, the SELECT statement will fail and return an error message.

14. A subquery must be placed in the HAVING clause of the outer query if:
   a. The inner query needs to reference the value returned to the outer query.
   b. The value returned by the inner query is to be compared to grouped data in the outer query.
   c. The subquery returns more than one value to the outer query.
   d. None of the above—subqueries cannot be used in the HAVING clause of an outer query.

15. Which of the following SQL statements will list all books written by the same author who wrote *The Wok Way to Cook?*

   a.
   ```
   SELECT title FROM books WHERE isbn IN
   (SELECT isbn FROM bookauthor
   HAVING authorid IN 'THE WOK WAY TO COOK);
   ```
   b.
   ```
   SELECT isbn FROM bookauthor WHERE authorid IN
   (SELECT authorid FROM books NATURAL JOIN bookauthor
   WHERE title = 'THE WOK WAY TO COOK');
   ```
   c.
   ```
   SELECT title FROM bookauthor WHERE authorid IN
   (SELECT authorid FROM books NATURAL JOIN bookauthor
   WHERE title = 'THE WOK WAY TO COOK);
   ```
   d.
   ```
   SELECT isbn FROM bookauthor HAVING authorid =
   SELECT authorid FROM books NATURAL JOIN bookauthor
   WHERE title = 'THE WOK WAY TO COOK';
   ```

16. Which of the following statements is correct?
    a. If the subquery only returns a NULL value, the only records returned by an outer query are those that contain an equivalent NULL value.
    b. A multiple-column subquery can only be used in the FROM clause of an outer query.
    c. A subquery can only contain one condition in its WHERE clause.
    d. The order of the columns listed in the SELECT clause of a multiple-column subquery must be in the same order as the corresponding columns listed in the WHERE clause of the outer query.
17. Subqueries in a WHERE clause can be nested to a depth of:
    a. 2
    b. 25
    c. 225
    d. 255
    e. unlimited

18. Given the following query, which statement is correct?

```
SELECT order# FROM orders WHERE order# IN
(SELECT order# FROM orderitems
WHERE isbn = '9959789321');
```

    a. The statement will not execute since the subquery and the outer query do not reference the same table.
    b. The outer query is not necessary since it has no effect on the results displayed.
    c. The query will fail if only one result is returned to the outer query since the WHERE clause of the outer query uses the IN comparison operator.
    d. No rows will be displayed since the ISBN in the WHERE clause is enclosed in single quotation marks.
19. Given the following SQL statement, which statement is most accurate?

```
SELECT customer#
FROM customers NATURAL JOIN orders
WHERE shipdate-orderdate IN
                    (SELECT MAX(shipdate-orderdate)
                    FROM orders
                    WHERE shipdate IS NULL);
```

    a. The SELECT statement will fail and return an Oracle error message.
    b. The outer query will display no rows in its results since the subquery passes a NULL value to the outer query.
    c. The customer number will be displayed for those customers whose orders have not yet shipped.
    d. The customer number of all customers who have not placed an order will be displayed.

20. Which of the following statements is correct?
    a. In a correlated subquery, the outer query is executed once for every row processed by the inner query.
    b. In an uncorrelated subquery, the outer query is executed once for every row processed by the inner query.
    c. In a correlated subquery, the inner query is executed once for every row processed by the outer query.
    d. In an uncorrelated subquery, the inner query is executed once for every row returned by the outer query.

HANDS-ON ASSIGNMENTS

To perform these activities, refer to the tables in Appendix A.

1. Determine which books have a retail price that is less than the average retail price of all books sold by JustLee Books.
2. Determine which books cost less than the average cost of other books in the same category.
3. Determine which orders were shipped to the same state as order 1014.
4. Determine which orders had a higher total amount due than order 1008.
5. Determine which author or authors wrote the book(s) most frequently purchased by customers of JustLee Books.
6. List the title of all books in the same category as books previously purchased by customer 1007. Do not include books already purchased by this customer.
7. List the city and state for the customer(s) who experienced the longest shipping delay.
8. Determine which customers placed orders for the least expensive book (in terms of retail price) carried by JustLee Books.
9. Determine how many different customers have placed an order for books written or co-written by James Austin.
10. Determine which books were published by the publisher of *The Wok Way to Cook*.

A CASE FOR ORACLE9*i*

To perform this activity, refer to the tables in Appendix A.

Currently, JustLee Books bills customers for orders by enclosing an invoice with each order when it is shipped. A customer then has 10 days to send in the payment. Of course, this has resulted in the company having to list some debts as "uncollectible." By contrast, most other online booksellers receive payment through a customer's credit card at the time of purchase. Although payment would be deposited within 24 hours into JustLee

Books' bank account by accepting credit cards, there is a downside to this alternative. When a merchant accepts credit cards for payment, the company that processes the credit card sales (usually called a "credit card clearinghouse") deducts a 1.5 percent processing fee from the total amount of the credit card sale.

The management of JustLee Books is trying to determine whether the surcharge for credit card processing is more than the amount usually deemed "uncollectible" when customers are sent an invoice. Historically, the average amount that has been lost by JustLee Books is about 4 percent of the total amount due from orders that have a higher-than-average amount due. In other words, it is usually customers who have an order with a larger-than-average invoice total who default on their payment.

To determine how much money would be lost, or gained, by accepting credit card payments, management has requested that you do the following:

1. Determine how much the surcharge would be for all recently placed orders if payment had been made by a credit card.
2. Determine the total amount that can be expected to be written off as "uncollectible" based on recently placed orders having an invoice total that is more than the average of all recently placed orders.

Based on the results of these two calculations, you should determine whether the company will lose money by accepting payment via credit card. State your findings in a memo to management. Include the SQL statement(s) necessary to calculate the expected surcharge and the expected amount of "uncollectible" payments.

CHAPTER

8

TABLE CREATION AND MANAGEMENT

Objectives

After completing this chapter, you should be able to do the following:

- Create a new table, using the CREATE TABLE command
- Name a new column or table
- Use a subquery to create a new table
- Add a column to an existing table
- Modify the size of a column in an existing table
- Drop a column from an existing table
- Mark a column as unused, then delete it at a later time
- Rename a table
- Truncate a table
- Drop a table

The management of JustLee Books has decided to implement a sales commission program for all account managers who have been employed by the company for more than six months. Beginning with orders placed after March 31, 2003, account managers will receive a commission for each order received from the geographical region they supervise. The commission rate will be based on the retail price of books sold in a manager's region.

To implement this new policy, several changes must be made to the database because it does not contain any information about the account managers, nor does it identify geographical regions in which customers live. Because the account managers represent a new entity, at least one new table must be added to the database.

Now that you have become skilled at using SQL to retrieve existing data, Chapters 8–10 will demonstrate SQL commands to create and modify tables, add data to tables, and edit existing data. This chapter will address methods for creating tables and modifying existing tables. Commands that are used to create or modify database tables are called **data definition language (DDL)** commands. These commands are basically SQL commands that are used specifically to create or modify database objects. A **database object** is a defined, self-contained structure in Oracle9*i*. In this chapter, you will create database tables, which are considered database objects. Later, in Chapter 12, you will learn how to create and modify other types of database objects. Figure 8–1 provides an overview of this chapter's topics.

Creating Tables	
Commands and Clauses	**Description**
CREATE TABLE	Creates a new table in the database. The user names the columns and identifies the type of data to be stored. To view a table, use the SQL*PLUS command DESCRIBE.
CREATE TABLE...AS	Creates a table from existing database tables, using the AS clause and subqueries
Modifying Tables	
ALTER TABLE... ADD	Adds a column to a table
ALTER TABLE... MODIFY	Changes a column size, datatype, or default value
ALTER TABLE... DROP COLUMN	Deletes one column from a table
ALTER TABLE... SET UNUSED *or* SET UNUSED COLUMN	Marks a column for deletion at a later time
DROP UNUSED COLUMNS	Completes the deletion of a column previously marked with SET UNUSED
RENAME...TO	Changes a table name
TRUNCATE TABLE	Deletes all table rows, but table name and column structure remain
Deleting Tables	
DROP TABLE	Removes an entire table from the Oracle9*i* database

Figure 8-1 Overview of chapter contents

Go to the JustLee Database folder in your Data Files. If you run the **Bookscript.sql** file, you will be able to work through the queries shown in this chapter. Your output should match the output shown.

TABLE DESIGN

Before actually issuing a SQL command to create a table, you must first choose the table's name and determine its structure—that is, you must determine what columns will be included in the table. In addition, you will need to determine the necessary width of any character or numeric columns.

Let's look at these requirements in more depth. Oracle9*i* has the following rules for naming both tables and columns:

- The names of tables and columns can be up to 30 characters in length and must begin with a letter. This limitation only applies to the name of a table or column, not to the amount of data in a column.
- Numbers, the underscore symbol (_), and the number sign (#) are allowed in table and column names; however, *you cannot include any blank spaces in table or column names.*
- Each table owned by a user should have a unique table name, and the column names within each table should be unique.
- Oracle9*i* "reserved words," such as SELECT, DISTINCT, CHAR, and NUMBER, cannot be used.

Because the new table will contain data about account managers, the name of the table will be ACCTMANAGER. The table will need to contain each account manager's name, employment date, and assigned region. In addition, the ACCTMANAGER table should also contain an ID code to act as the table's primary key and uniquely identify each account manager. Although it is unlikely that JustLee Books will ever have two account managers with the same name, the ID code can reduce data-entry errors because users will only need to type a short code instead of a manager's entire name in SQL commands.

Now that the contents of the table have been determined, the columns can be designed. When a table is created in Oracle9*i*, the columns must also be created. Before you can create the columns, you must do the following:

1. Choose a name for each column.
2. Determine the type of data each column will store.
3. Determine (in some cases) the maximum width of the column.

Before choosing column names, let's look at some datatypes and their default values. Valid datatypes are shown in Figure 8–2.

Oracle9*i* Datatypes	
Datatype	**Description**
VARCHAR2(*n*)	**Var**iable-length **char**acter data, where *n* represents the maximum length of the column. Maximum size is 4,000 characters. There is no default size for this datatype; a minimum value must be specified. *Example:* VARCHAR2(9) can contain up to 9 letters, numbers, or symbols.
CHAR(*n*)	Fixed-length character column, where *n* represents the length of the column. Default size is 1. Maximum size is 2,000 characters. *Example:* CHAR(9) can contain 9 letters, numbers, or symbols. However, if fewer than 9 are entered, spaces will be added to the right to force the data to reach a length of 9.
NUMBER(*p*,*s*)	Numeric column, where *p* indicates **precision**, or the total number of digits to the left and right of the decimal position, to a maximum of 38 digits; and *s*, or **scale**, indicates the number of positions to the right of the decimal. *Example:* NUMBER(7, 2) can store a numeric value up to 99999.99. If precision or scale is not specified, the column will default to a precision of 38 digits.
DATE	Stores date and time between January 1, 4712 B.C. and December 31, 9999 A.D. Seven bytes are allocated to the column to store the century, year, month, day, hour, minute, and second of a date. Oracle9*i* displays the date in the format DD-MON-YY. Other aspects of a date can be displayed by using the TO_CHAR format. The width of the field is predefined by Oracle9*i* as 7 bytes.

Figure 8-2 Oracle9*i* datatypes

The datatypes of LONG, CLOB, RAW(*n*), LONG RAW, BLOB, BFILE, TIMESTAMP, and INTERVAL are also available. LONG stores variable-length character data up to 2 gigabytes. CLOB is used for single-byte character data up to 4 gigabytes. RAW(*n*) stores raw binary data up to 2000 bytes. LONG RAW can contain up to 2 gigabytes of unstructured data, while BLOB can store up to 4 gigabytes of unstructured binary object data. A BFILE column stores a file locator to a binary file stored by the operating system. TIMESTAMP and INTERVAL are new datatypes supported by Oracle9*i*. They are extensions of the DATE datatype, where TIMESTAMP refers to the time value (hour, minute, and second can be referenced without the TO_CHAR function), and INTERVAL is used to identify a specific interval, or amount, of time.

A **datatype** simply identifies the type of data Oracle9*i* will be expected to store in a column. For example, previously you performed calculations using the Orderdate and Shipdate columns from the ORDERS table. The values stored in those columns were dates. How was Oracle9*i* able to determine that the values were dates and, therefore, able to compute the difference between the dates in the two columns? When the ORDERS table was originally created, the two columns were assigned a datatype of DATE. By having been defined as DATE columns, calendar dates were the only possible kind of data that could be stored in the columns.

As previously noted, the ACCTMANAGER table will consist of four columns. The first column will serve to uniquely identify each account manager and will be named AmID. The ID code assigned to each account manager will consist of the first letter of a manager's last name, followed by a three-digit number. Because the column will consist of both letters and numbers, the column will be defined to store the datatype of VARCHAR2 and have a width of four. The CHAR datatype could have been used; however, many individuals in industry avoid using that datatype because Oracle9*i* will pad the column to the specified length if an entry does not fill the entire width of the column. This can result in wasted storage space if the majority of the values stored by Oracle9*i* must include blank spaces to force the contents of a column to a specific length.

The second column of the ACCTMANAGER table will be used to store the first and last name of each account manager. If you store first names and last names in individual columns, you can perform simple searches on each part of a manager's name. However, the table is not expected to contain a massive number of records. In addition, the names contained in the ACCTMANAGER table will not be used as direct input for other types of application programs (e.g., those used by Human Resources or Accounting), so there is no need to store each portion of the name separately.

Because each account manager's name will consist of characters, this column will also be assigned the datatype of VARCHAR2. On the other hand, there is no clear choice for the width of the column. Generally, you will define the width so it can hold the largest value that could ever be entered into that column. However, the possibility exists that an account manager will be hired in the future who has an extremely long first and/or last name. If necessary, it's relatively easy to increase a column's width at a later time. Therefore, the assumption is that a column width of 20 characters is sufficient to store the names of the current account managers. The column will be named Amname.

The actual names of the account managers will be provided in Chapter 10.

The third column of the table will be used to store the employment date of each account manager. Because the datatype will be DATE, you will not have to worry about the length of the column—the length is predetermined by Oracle9*i*. All that remains is to choose a name for the column. In this case, Amedate, for "account manager employment date," will be appropriate.

The fourth and final column for the ACCTMANAGER table is the region to which the account manager is assigned. The United States will be divided into six geographical regions, each identified by a two-letter code (NE will represent the Northeast region, etc.). The name of the column will be Region and will be defined as a CHAR datatype with a width of two characters. The VARCHAR2 datatype could also be used; however, because the values stored in the Region column will always consist of two characters, the CHAR was selected for illustrative purposes.

TABLE CREATION

The basic syntax of the SQL command to create a table in Oracle9*i* is shown in Figure 8–3.

```
CREATE TABLE [schema] tablename
     (columnname datatype [DEFAULT value],
     [columnname datatype [DEFAULT value], …);
```

Figure 8-3 CREATE TABLE syntax

The keywords **CREATE TABLE** instruct Oracle9*i* to create a table. The optional **schema** can be included to indicate who will "own" the table that is about to be created. For example, if the person creating the table is also the person who will own the table, then the schema can be omitted, and the current user name will be assumed by default. On the other hand, if you were creating the ACCTMANAGER table for someone with the user name of DRAKE, the schema and table name would be entered as **DRAKE.ACCTMANAGER** to inform Oracle9*i* that the table ACCTMANAGER will belong to DRAKE's schema, not yours. The owner of a database object has the right to perform certain operations on that object. In the case of a table, the only way another database user can query or manipulate the data contained in the table is to be given permission from the table's owner or the database administrator. The table name, of course, is the name that will be used to identify the table being created.

To create a table for someone else's schema (i.e., a table owned by someone else), you must be granted permission, or the privilege, to use the CREATE TABLE command for that user's schema. The different privileges available in Oracle9*i* will be presented in Chapter 13.

Defining Columns

Once the table name has been entered, the columns to be included in the table must be defined. A table can contain a maximum of 1,000 columns. *The syntax of the CREATE TABLE command requires the column list to be enclosed within parentheses.* If more than one column will exist in the table, then the name, datatype, and appropriate width are listed for the first column before the next column is defined. Commas separate columns in the list. The CREATE TABLE command also allows a default value to be assigned to a column. The default value is the value that will automatically be stored by Oracle9*i* if the user makes no entry into that column.

Using the syntax presented in Figure 8–3, the SQL command in Figure 8–4 will create the ACCTMANAGER table.

```
CREATE TABLE acctmanager
     (amid         VARCHAR2(4),
     amname        VARCHAR2(20),
     amedate       DATE       DEFAULT SYSDATE,
     region        CHAR(2));
```

Figure 8-4 CREATE TABLE command to create the ACCTMANAGER table

In the command given in Figure 8–4, the name of the table is ACCTMANAGER. Although it is entered in lower-case letters, Oracle9*i* will automatically convert the table name to upper-case letters when the command is processed. It is entered here in lower-case letters to distinguish it from the CREATE TABLE keywords. Because the user who creates the table will also be the owner of the table, the schema has been omitted.

The four columns to be created are listed next, within parentheses. Each column is defined on a separate line simply to improve readability; this is not an Oracle9*i* requirement. Notice the definition for the Amedate column; it has been assigned a default value of SYSDATE. This will instruct Oracle9*i* to insert the current date, according to the Oracle9*i* server, if the user enters a new account manager without including the individual's date of employment. Of course, this will only be beneficial if the account manager's record is created on the same date the person is hired; otherwise, this will make an incorrect entry if the Amedate column is left blank. In other words, assign defaults with caution!

Figure 8–5 shows the creation of the ACCTMANAGER table.

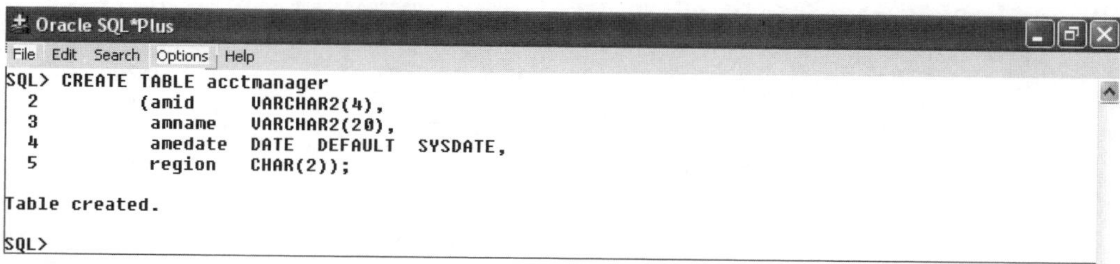

Figure 8-5 Creating the ACCTMANAGER table

A user cannot have two tables with the same name. If you attempt to create a table that with same name as another table in your schema, Oracle9*i* will display an ORA-00955 error message. Similarly, if you create a table and then want to create the same table a second time, using the same table name, you must first delete the existing table using the command `DROP TABLE tablename;`. (The DROP TABLE command will be shown later in this chapter.)

After the command has been entered, Oracle9*i* will return the message "Table created" to let the user know the table was created successfully. Notice that the message does not contain any reference to rows being created. At this point, the table does not contain any data; only the structure of the table has been created (i.e., the table has been defined in terms of a table name and the type of data it will contain). The data, or rows, must be entered into the table as a separate step using the INSERT command. You will enter all the data for the account managers in Chapter 10.

If you receive an error message, such as an ORA-00922 error message, when executing the CREATE TABLE command, it could be a result of either (1) not including a second closing parenthesis at the end of the command to close the column list, or (2) not separating each complete column definition with a comma. If an error message appears stating that you do not have sufficient privileges, ask the database administrator to grant you the CREATE TABLE privilege.

Viewing Table Structures: DESCRIBE

To determine whether the structure of the table was created correctly, you can use the SQL*Plus command `DESCRIBE` *`tablename`* to display the structure of the table, as shown in Figure 8–6. Because the DESCRIBE command is a SQL*Plus command rather than a SQL command, it can be abbreviated. The abbreviation for the DESCRIBE command is DESC.

```
Oracle SQL*Plus
File  Edit  Search  Options  Help
SQL> DESCRIBE acctmanager
 Name                                      Null?    Type
 ----------------------------------------- -------- ----------------------------
 AMID                                               VARCHAR2(4)
 AMNAME                                             VARCHAR2(20)
 AMEDATE                                            DATE
 REGION                                             CHAR(2)

SQL>
```

Figure 8-6 The DESCRIBE command

When you issue the **DESCRIBE** command, all columns defined for the ACCTMANAGER table are listed. With each column name, the results also display whether the column will allow NULL values and the column's datatype. However, notice that the results do not display a "NOT NULL" requirement for the AmId column—it is blank. Because this column will be the primary key for the table, it should not be allowed to contain any blank entries. (This problem will be corrected in the next chapter.) Thus, if the four columns have the correct name, datatype, and width—*and* your CREATE TABLE command executed—you now have a table ready to accept account managers' data.

Table Creation Through Subqueries

In the previous section, a table was created "from scratch." However, sometimes you might need to create a table based on data contained in existing tables. For example, suppose that management wants each account manager to have a table containing his or her customers' names. In addition, the table should include the book titles ordered by each customer. A word of caution is in order, however. You generally would not want multiple copies of data within the same database—it could lead to data redundancy. However, in this case, account managers would have the flexibility of adding columns for making comments or changes that should not be reflected in the main database tables. Because the account managers' tables would not be relevant to the transaction-processing activities for JustLee Books (e.g., order processing), allowing account managers to have tables containing data that already exist in other tables would not be a cause for alarm.

When you create the account managers' tables, bear in mind that the current tables' design is based on one entity type—i.e., the CUSTOMERS table contains information about all customers, and the ORDERS table contains information about all orders. Now, management wants a new table that contains not only customers' names but also data about orders that customers have placed. Before you tell management that creating the tables for the account managers will not be a problem, you should determine how difficult this task will be—you don't want to say you can have it completed in a few hours and then discover that it will require someone to create the tables and re-enter the data manually, which could take a day or longer.

CREATE TABLE...AS

To create a table that will contain data from existing tables, you can use the CREATE TABLE command with an AS clause that contains a subquery. The syntax is shown in Figure 8-7.

```
CREATE TABLE tablename [(columnname, …)]
AS (subquery);
```

Figure 8-7 CREATE TABLE...AS command syntax

The command syntax given in Figure 8-7 uses the CREATE TABLE keywords to instruct Oracle9*i* to create a table. The name of the new table is then provided.

If you need to give the columns in the new table names that are different from the column names in the existing table, then the new column names must be listed after the table name, inside parentheses. However, if you do not want to change any of the column names, the column list in the CREATE TABLE clause is omitted. If you do provide a column list in the CREATE TABLE clause, it must contain a name for *every* column to be returned by the query—including those names that will remain the same. In other words, if five columns are going to be returned from the subquery, then five columns must be listed in the CREATE TABLE clause, or Oracle9*i* will return an error message, and the statement will fail. In addition, the column list must be in the *same order* as the columns listed in the SELECT clause of the subquery, so Oracle9*i* will know which column from the subquery is assigned to which column in the new table.

The subquery portion of the command follows the same guidelines as those previously described in Chapter 7. The AS keyword is used to instruct Oracle9*i* that the columns in the new table will be based on the columns returned by the subquery and must precede the subquery. The columns in the new table will be created based on the same datatype and width the columns had in the existing table(s). To distinguish the subquery from the rest of the CREATE TABLE command, *the subquery must be enclosed within parentheses.*

Management has specified that for each customer, tables should contain the customer's number and state of residence plus the ISBN, category, quantity, cost, and retail price of each book sold to that customer. As a test run, try to create this table, using only the customers in Florida, Georgia, and Alabama to represent the Southeast (SE) region. The subquery used to identify the data to be included in the new table is shown in Figure 8–8.

```
SELECT customer#, state, ISBN, category,
       quantity, cost, retail
FROM customers NATURAL JOIN orders NATURAL JOIN
     orderitems NATURAL JOIN books
WHERE state IN ('FL', 'GA', 'AL');
```

Figure 8-8 Subquery for the CREATE TABLE...AS command

Although customer data are stored in the CUSTOMERS table and book data are stored in the BOOKS table, those tables had to be joined with the ORDERS and ORDERITEMS tables in the FROM clause of the subquery to identify the books ordered by each customer. The subquery given in Figure 8–8 is then included in the CREATE TABLE...AS command to create the desired table. This command is shown in Figure 8–9.

```
CREATE TABLE secustomerorders
     AS  (SELECT customer#, state, ISBN,
         category, quantity, cost, retail
         FROM customers NATURAL JOIN orders NATURAL JOIN
              orderitems NATURAL JOIN books
         WHERE state IN ('FL', 'GA', 'AL'));
```

Figure 8-9 Command to create a table based on a subquery

The command given in Figure 8–9 is the complete SQL command to create the new table. The name of the new table is SECUSTOMERORDERS, to indicate that the table contains the order information of customers from the Southeast region. When the command is executed, Oracle9*i* will return the message "Table created" to indicate that the command was successful, as shown in Figure 8–10.

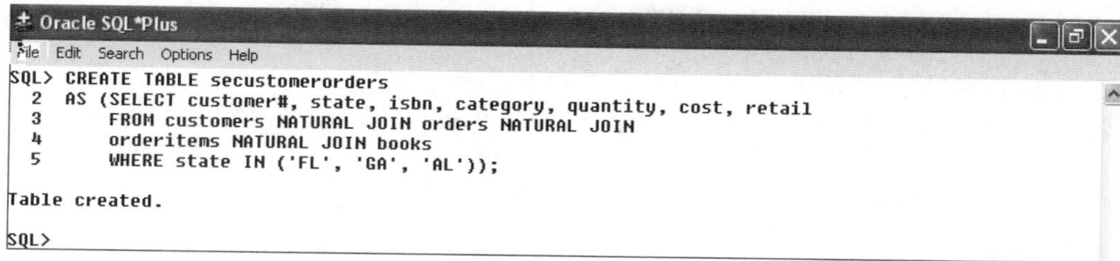

Figure 8-10 Execution of the CREATE TABLE..AS command

Although the command was executed, Oracle9*i* does not indicate how many, if any, records were actually inserted into the new table. To view the contents of the new table, simply enter **`SELECT * FROM secustomerorders;`** at the SQL prompt, as shown in Figure 8–11.

```
Oracle SQL*Plus
File Edit Search Options Help
SQL> SELECT *
  2  FROM secustomerorders;

CUSTOMER# ST ISBN       CATEGORY       QUANTITY       COST     RETAIL
--------- -- ---------- ------------ ---------- ---------- ----------
     1010 GA 9247381001 BUSINESS              1       15.4      31.95
     1010 GA 2491748320 FAMILY LIFE           1         48      89.95
     1001 FL 8843172113 COMPUTER              1       31.4      55.95
     1001 FL 1059831198 FITNESS               1      18.75      30.95
     1001 FL 3437212490 COOKING               1       12.5      19.95
     1018 GA 2147428890 LITERATURE            1      21.85      39.95
     1003 FL 9959789321 COMPUTER              1       37.9       54.5
     1010 GA 2491748320 FAMILY LIFE           1         48      89.95
     1003 FL 2491748320 FAMILY LIFE           1         48      89.95
     1015 FL 8117949391 CHILDREN              2       5.32       8.95
     1001 FL 3437212490 COOKING               1       12.5      19.95
     1001 FL 8843172113 COMPUTER              1       31.4      55.95
     1018 GA 0401140733 FAMILY LIFE           1       14.2         22

13 rows selected.

SQL>
```

Figure 8-11 Contents of the SECUSTOMERORDERS table

As shown in Figure 8–11, there are 13 records that match the criteria specified by the subquery. These records were automatically added to the new table when it was created. Whew! This will save a lot of time and eliminate data-entry errors—as long as the subquery is correct. Notice that the new table does not contain any customers from Alabama, although that state was specified in the subquery used to create the new table. Why? Simply because no customers in the CUSTOMERS table live in Alabama. Even though there are no customers in this state, the syntax of the subquery was correct and still executed.

If the command in Figure 8–10 returns an error message, or if the new table contains no records, double-check that (1) the subquery is within parentheses, and (2) all the table names are included in the FROM clause of the subquery.

Subqueries with Group Functions

In addition to the tables containing the customers for each region and their purchases, management would also like summary tables for each region. These tables should only contain the name of each customer and the total amount each customer has recently spent. A subquery to extract this information from the existing tables is shown in Figure 8–12.

```
SELECT firstname|| ' '||lastname, SUM(retail*quantity)
FROM customers NATURAL JOIN orders NATURAL JOIN
       orderitems NATURAL JOIN books
WHERE state IN ('FL', 'GA', 'AL')
GROUP BY firstname||' '||lastname;
```

Figure 8-12 Subquery to determine customer names and amount spent

The subquery in Figure 8–12 will return two columns: the name of each customer and total amount spent by each customer. Notice that the first column will be a concatenation of the Firstname and Lastname columns of the CUSTOMERS table. The concatenation of the two columns will result in a single column in the new table that includes the contents of these fields, separated by a space. This might create a problem with the column name in the new table, so the CREATE TABLE...AS command should include a new column name for this column. Also notice that the second column will be the result of a group function—it will add the total amount spent by the customer. Because the group function is used in the SELECT clause along with an individual field (the customer's name), the individual field must be listed in a GROUP BY clause.

Once the form of the subquery has been determined, it can now be embedded in the CREATE TABLE...AS command to create the new table. The new table will be called SECUSTOMERSSPENT to indicate that this table will contain the amount spent by customers in the Southeast region. There is one problem: The column names of **`firstname|| ' '||lastname`** and **`SUM(retail*quantity)`** are not appealing. In fact, Oracle9*i* will return an error message that the column names are not valid. So, you have two choices: (1) You can assign each column a column alias in the subquery, or (2) you can provide new names for the columns in a column list in the CREATE TABLE clause. Either option will have the same results. If you decide to use column aliases in the SELECT clause of the subquery, Oracle9*i* will use the aliases as the new column names in the table being created. If you provide a column list, the column names provided will be the new names for the columns. For this example, use the second option and provide a column list in the CREATE TABLE clause. Using this approach will result in the command shown in Figure 8–13.

The results of the command issued in Figure 8–13 will be a new table consisting of two columns. The first column will contain the name of each customer, and the second will contain the total amount spent by each customer—which is exactly what management wants, and all without having to re-enter any of the data already stored in the JustLee

Books database. To view the exact data now contained in the new table, simply query the new table with a SELECT command, as shown in Figure 8–14.

```
CREATE TABLE secustomersspent (name, spent)
AS (SELECT firstname||' '||lastname,
    SUM(retail*quantity)
       FROM customers NATURAL JOIN orders NATURAL JOIN
            orderitems NATURAL JOIN books
      WHERE state IN ('FL', 'GA', 'AL')
      GROUP BY firstname||' '||lastname);
```

Figure 8-13 CREATE TABLE...AS command with a column list

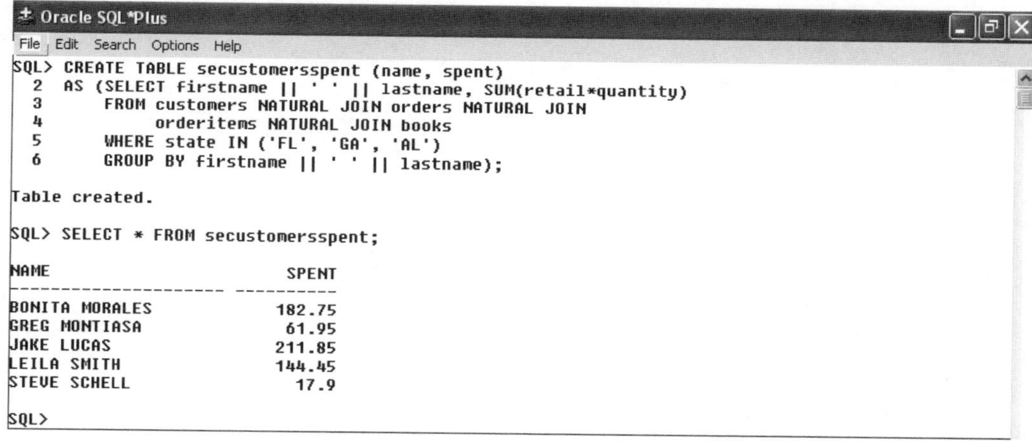

Oracle SQL*Plus
File Edit Search Options Help
SQL> CREATE TABLE secustomersspent (name, spent)
2 AS (SELECT firstname || ' ' || lastname, SUM(retail*quantity)
3 FROM customers NATURAL JOIN orders NATURAL JOIN
4 orderitems NATURAL JOIN books
5 WHERE state IN ('FL', 'GA', 'AL')
6 GROUP BY firstname || ' ' || lastname);
Table created.
SQL> SELECT * FROM secustomersspent;
NAME SPENT
BONITA MORALES 182.75
GREG MONTIASA 61.95
JAKE LUCAS 211.85
LEILA SMITH 144.45
STEVE SCHELL 17.9
SQL>

Figure 8-14 Contents of the SECUSTOMERSPENT table

MODIFYING EXISTING TABLES

There may be times when you will need to make structural changes to a table. For example, you may need to add a column, delete a column, or simply change the size of a column. Each of these changes is accomplished through the **ALTER TABLE** command. A useful feature of Oracle9*i* is that a table can be modified without having to shut down the database. Even if users are accessing a table, it can still be modified without disruption of service. The syntax for the ALTER TABLE command is shown in Figure 8–15.

```
ALTER TABLE tablename
ADD|MODIFY|DROP COLUMN| columnname [definition];
```

Figure 8-15 Syntax for the ALTER TABLE command

Whether you should use an ADD, MODIFY, or DROP COLUMN clause depends on the type of change being made. First, let's look at the ADD clause.

ALTER TABLE...ADD Command

Using an **ADD** clause with the ALTER TABLE command allows a user to add a new column to a table. The same rules that apply to creating a new column apply to defining a column during table creation. The new column must be defined by a column name and datatype (and width, if applicable). A default value can also be assigned. The difference is that the new column will be added at the end of the existing table—i.e., it will be the last column. The syntax for the ALTER TABLE command with the ADD clause is shown in Figure 8–16.

```
ALTER TABLE tablename
ADD (columnname datatype, [DEFAULT] …);
```

Figure 8-16 Syntax for the ALTER TABLE...ADD command

Suppose that after the ACCTMANAGER table was created, management requests that the table also contain each account manager's telephone number extension. The column name should be Ext. The column can consist of a maximum of four numeric digits since the column is defined as a NUMBER datatype with a precision of 4. To make this change to the ACCTMANAGER table, issue the command shown in Figure 8–17.

```
ALTER TABLE acctmanager
ADD (ext    NUMBER(4));
```

Figure 8-17 The ALTER TABLE...ADD command

Once the SQL command given in Figure 8–17 is processed, Oracle9*i* will return the message "Table altered" to indicate that the command was completed successfully. To double-check that the column was added with the correct datatype, etc., simply issue the SQL*Plus command `DESCRIBE acctmanager`, as shown in Figure 8–18.

If you need to add more than one column to the ACCTMANAGER table, list the additional column(s) in a column list, and separate each column and its datatype from the other columns with a comma, using the same format as the CREATE TABLE command.

ALTER TABLE...MODIFY Command

A **MODIFY** clause can be used with the ALTER TABLE command to change the definition of an existing column. The changes that can be made to a column include the following:

- Changing the size of a column (e.g., increase or decrease)

- Changing the datatype (e.g., VARCHAR2 to CHAR)
- Changing or adding the default value of a column

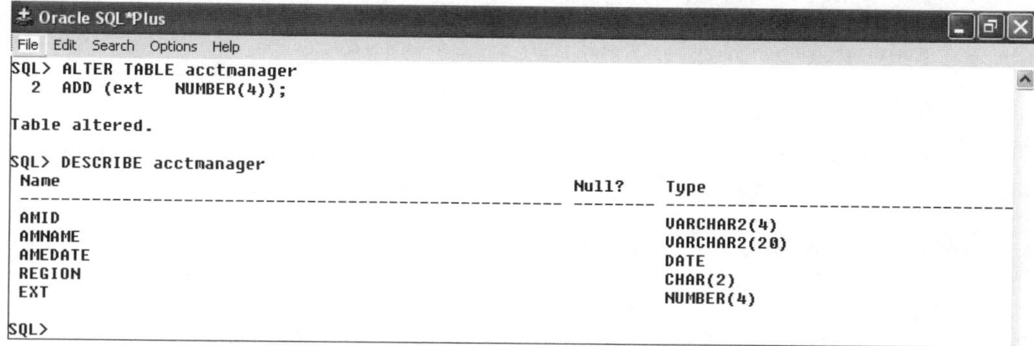

Figure 8-18 ACCTMANAGER table with new column added

The syntax for the ALTER TABLE...MODIFY command is shown in Figure 8–19.

```
ALTER TABLE tablename
MODIFY (columnname datatype [DEFAULT],…);
```

Figure 8-19 Syntax of the ALTER TABLE...MODIFY command

There are three rules you should be aware of when modifying existing columns.

1. A column must be as wide as the data fields it already contains.
2. If a NUMBER column already contains data, you can't decrease the precision or scale of the column.
3. Changing the default value of a column does not change the values of data already in the table.

Next, let's look at each of these rules.

The first rule applies when you want to decrease the size of a column that already contains data. You can only decrease the size of a column to a size that is no less than the largest width of the existing data. For example, suppose that a column has been defined as a VARCHAR2 datatype with a width of 15 characters. However, the largest entry in that particular column contains only 12 characters. Thus, you would only be able to decrease the size of the column to a width of 12 characters. If you try to decrease the size to a width less than 12, then Oracle9*i* will return an error message, and the SQL statement will fail. As shown in Figure 8–20, when a user attempts to decrease the width of the column to a size that will not accommodate the data it already contains, Oracle9*i* will return an ORA-01441 error message, and the table will not be altered.

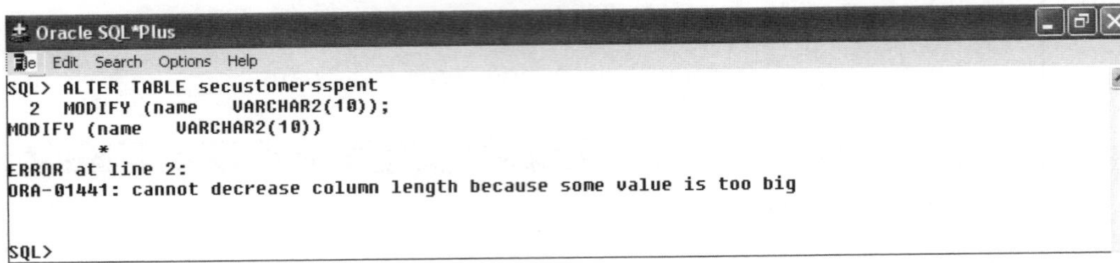

```
SQL> ALTER TABLE secustomersspent
  2  MODIFY (name   VARCHAR2(10));
MODIFY (name   VARCHAR2(10))
        *
ERROR at line 2:
ORA-01441: cannot decrease column length because some value is too big

SQL>
```

Figure 8-20 Error generated when attempting to decrease the width of a column to a size smaller than the length of the current data

On the other hand, Oracle9*i* will not allow you to decrease the precision or scale of a NUMBER column if the column contains any data. Regardless of whether the current values stored in a NUMBER column will be affected, Oracle9*i* will return an ORA-01440 error message, and the statement will fail unless the column is empty. As shown in Figure 8–21, if you attempt to change the size of the Spent column in the SECUSTOMERSSPENT table, an error message will be displayed, and the table will not be altered.

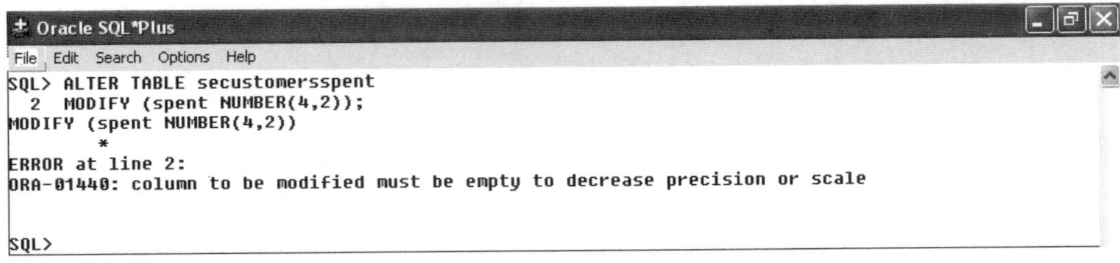

```
SQL> ALTER TABLE secustomersspent
  2  MODIFY (spent NUMBER(4,2));
MODIFY (spent NUMBER(4,2))
        *
ERROR at line 2:
ORA-01440: column to be modified must be empty to decrease precision or scale

SQL>
```

Figure 8-21 Error generated when attempting to resize a NUMBER column that already contains data

The third rule applies when you modify existing columns and decide to change the default value assigned to a column. When the default value of a column is changed, it will only change the default value assigned to *future rows* inserted into the table. The default value assigned to rows that already exist in the table will remain the same. Thus, if the default value contained in existing rows must be changed, those changes must be performed manually. (Chapter 10 will discuss how to change existing values in a row.)

Let's assume that an account manager might not have an assigned telephone extension number at the time the manager's data are entered into the ACCTMANAGER table. The supervisor has decided that if a telephone call comes in for one of those managers, the call should automatically be routed to the department's administrative assistant. The extension of the administrative assistant is 1200. As shown in Figure 8–22, a value of 1200 has been assigned as the DEFAULT for the telephone extension of the account managers. If the table already contains account managers who do not have a previously assigned telephone extension, then those rows will not be affected by the change.

However, any new manager without a phone extension who is added to the table will automatically have the value 1200 stored in the Ext column.

Figure 8-22 ALTER TABLE...MODIFY command to add a DEFAULT value to an existing column

After creating the ACCTMANAGER table, suppose that you find out that one of the account managers has a long name that will require more than the 20 spaces previously assigned to the Amname column. To accommodate the name, the Amname column must be increased to a width of 25 characters. The command to increase the width of the column is shown in Figure 8–23.


```
ALTER TABLE acctmanager
MODIFY (amname VARCHAR2(25));
```

Figure 8-23 The ALTER TABLE...MODIFY command to increase the width of the Amname column

Notice that in the MODIFY clause of the command given in Figure 8–23, the correct syntax does not explicitly state that the Amname column should be increased by five characters. Instead, the datatype and new width are stated with the width increased to the total desired width for the column. Once the command is processed, Oracle9*i* will return the message "Table altered," as shown in Figure 8–24. To make certain that change is made, you might also use the DESCRIBE command to view the new table structure.

Figure 8-24 Result of the ALTER TABLE...MODIFY command

ALTER TABLE...DROP COLUMN Command

The **DROP COLUMN** clause can be used with the ALTER TABLE command to delete an existing column from a table. The clause will delete both the column and its contents, so it should be used with extreme caution. The syntax for the ALTER TABLE...DROP COLUMN command is shown in Figure 8–25.

```
ALTER TABLE tablename
DROP COLUMN columnname;
```

Figure 8-25 Syntax for the ALTER TABLE...DROP COLUMN command

You should keep the following in mind when using the DROP COLUMN clause:

- Unlike using an ALTER TABLE command with the ADD or MODIFY clauses, a DROP COLUMN clause can only reference *one* column.
- If you drop a column from a table, the deletion is permanent. You will not be able to "undo" the damage if you accidentally delete the wrong column from a table. The only option will be to add the column back to the table and then manually re-enter all the data that it previously contained.
- You cannot delete the last remaining column in a table. If a table only contains one column, and you try to delete the column, the command will fail, and Oracle9*i* will return an error message.

Previously, you added the Ext column to store the telephone extension of each account manager. However, management has now decided that the extension is not necessary and does not want to waste the storage space that would be taken up by that column. Therefore, the Ext column needs to be deleted from the ACCTMANAGER table. The command to delete the Ext column is shown in Figure 8–26.

```
ALTER TABLE acctmanager
DROP COLUMN ext;
```

Figure 8-26 The ALTER TABLE...DROP COLUMN command

After the command in Figure 8–26 is processed, the Ext column of the ACCTMANAGER table is removed. To verify that the column no longer exists, use the DESCRIBE command to list the structure of the ACCTMANAGER table, as shown in Figure 8–27.

ALTER TABLE...SET UNUSED/DROP UNUSED COLUMNS Command

When the Oracle9*i* server drops a column from a very large table, this can slow down the processing of queries or other SQL commands from users. To avoid this problem, a **SET UNUSED** clause can be included in the ALTER TABLE command to mark the column for deletion at a later time. If a column is marked for deletion, then the column is unavailable and will not be displayed in the table structure. Since the column is unavailable, it will not appear in the results of any queries, nor can any other operation be performed on the column except the ALTER TABLE...DROP UNUSED command.

In other words, once a column is set as "unused," the column and all its contents are no longer available and cannot be recovered at a future time. It simply postpones the physical erasing of the data from the storage device until a later time—usually when the server is processing fewer queries, such as after business hours. A **DROP UNUSED** clause is used with the ALTER TABLE command to complete the deletion process for any column that has been marked as unused.

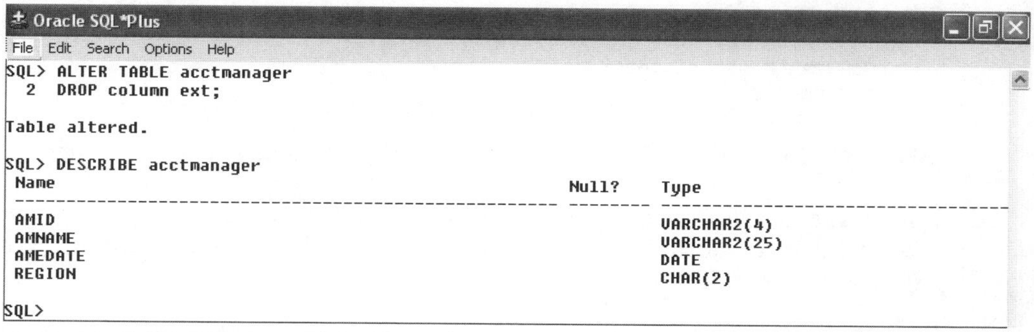

```
Oracle SQL*Plus
File Edit Search Options Help
SQL> ALTER TABLE acctmanager
  2  DROP column ext;

Table altered.

SQL> DESCRIBE acctmanager
 Name                                                  Null?    Type
 ----------------------------------------------------- -------- ------------------------------------
 AMID                                                           VARCHAR2(4)
 AMNAME                                                         VARCHAR2(25)
 AMEDATE                                                        DATE
 REGION                                                         CHAR(2)

SQL>
```

Figure 8-27 Structure of the ACCTMANAGER table after the Ext column is dropped

The syntax for the ALTER TABLE...SET UNUSED command is shown in Figure 8–28.

```
ALTER TABLE tablename
SET UNUSED (columnname);
     OR
ALTER TABLE tablename
SET UNUSED COLUMN columnname;
```

Figure 8-28 Syntax for the ALTER TABLE...SET UNUSED command

As shown in Figure 8–28, the syntax for the ALTER TABLE...SET UNUSED command has two options for the SET UNUSED clause. Regardless of the syntax used, only one column per command can be marked for deletion. The syntax to drop a column previously identified as unused is shown in Figure 8–29.

```
ALTER TABLE tablename
DROP UNUSED COLUMNS;
```

Figure 8-29 Syntax for the ALTER TABLE...DROP UNUSED COLUMNS command

When the DROP UNUSED COLUMNS clause is used, any column that has previously been set as "unused" is deleted, and any storage previously occupied by data contained in the column(s) becomes available.

Suppose that management has decided that the account managers do not need to see the actual wholesale cost of books sold by JustLee Books. To delete that column from the SECUSTOMERORDERS table previously created, you could use the DROP COLUMN option of the ALTER TABLE command. However, suppose that the table contains thousands of records and deleting a column would slow down operations for other Oracle9*i* users. In this case, you can mark the Cost column of the SECUSTOMERORDERS table as unused with the command shown in Figure 8–30.

```
ALTER TABLE secustomerorders
SET UNUSED (cost);
```

Figure 8-30 The ALTER TABLE...SET UNUSED command

To make certain that the Cost column was correctly marked for deletion, the DESCRIBE command can be used to check that the column is no longer available, as shown in Figure 8–31.

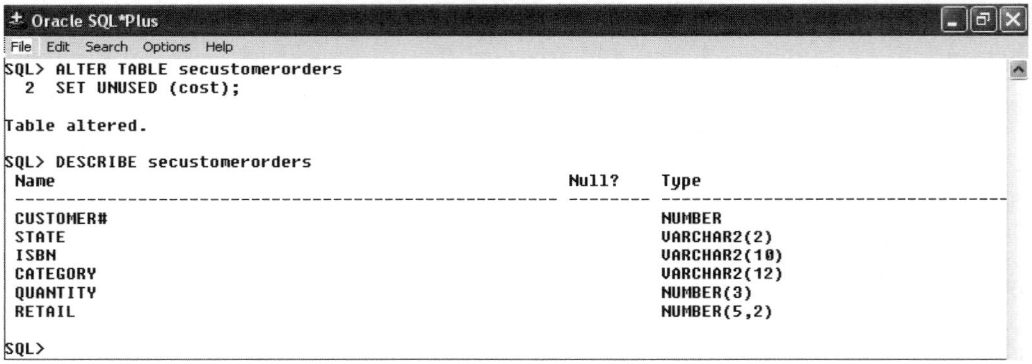

```
Oracle SQL*Plus
File Edit Search Options Help
SQL> ALTER TABLE secustomerorders
  2  SET UNUSED (cost);

Table altered.

SQL> DESCRIBE secustomerorders
 Name                                                  Null?    Type
 ----------------------------------------------------- -------- ------------------------------
 CUSTOMER#                                                      NUMBER
 STATE                                                          VARCHAR2(2)
 ISBN                                                           VARCHAR2(10)
 CATEGORY                                                       VARCHAR2(12)
 QUANTITY                                                       NUMBER(3)
 RETAIL                                                         NUMBER(5,2)

SQL>
```

Figure 8-31 SECUSTOMERORDERS table after Cost column is set as unused

Once the column has been set as unused, the storage space taken up by data contained within the Cost column can be reclaimed, using the command shown in Figure 8–32.

```
ALTER TABLE secustomerorders
DROP UNUSED COLUMNS;
```

Figure 8-32 The ALTER TABLE...DROP UNUSED COLUMNS command

Renaming a Table

Oracle9*i* will allow you to change the name of any table you own, using the **RENAME...TO** command. The syntax for the RENAME...TO command is shown in Figure 8–33.

```
RENAME oldtablename TO newtablename;
```

Figure 8-33 Syntax for the RENAME...TO command

In a previous section, a table named SECUSTOMERSSPENT was created. However, this table name is somewhat long, and the double S's could cause some users to type the name incorrectly when entering SQL commands. To avoid this problem, you can rename the table SETOTALS to reflect that the table contains the total amount spent by customers in the Southeast region. The command to make the name change is shown in Figure 8–34.

```
RENAME secustomersspent TO setotals;
```

Figure 8-34 The RENAME...TO command

Once the RENAME...TO command is executed, any queries directed to the SECUSTOMERSSPENT table will result in an error message. The table can now only be referenced as the SETOTALS table. The output is shown in Figure 8–35.

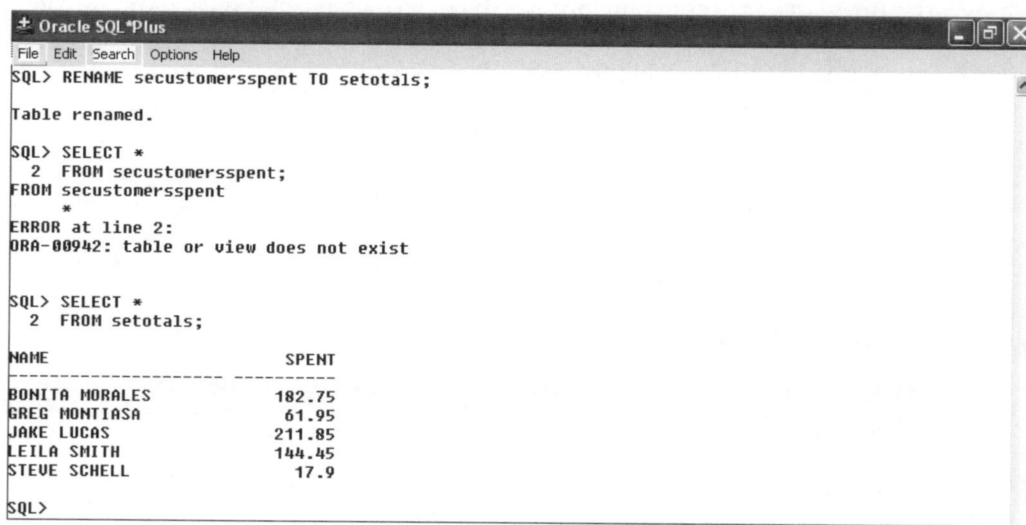

Figure 8-35 Referencing the SECUSTOMERSSPENT and SETOTAL tables after issuing the RENAME...TO command

When working in an organization, don't change the name of a table that is accessed by other users unless you first inform them of the new table name. Failure to inform users of the change will prevent them from finishing their work and will create havoc until the problem is identified. Of course, this is assuming that you didn't change the name of the table to stop someone from accessing it in the first place!

Truncating a Table

When a table is truncated, all the rows contained in the table are removed, but the table itself remains. In other words, the columns still exist even though no values are stored in them. This is basically the same as deleting all the rows in a table. However, if you simply delete all rows in a table, the storage space occupied by those rows will still be allocated to the table. To delete the rows stored within a table *and* free up the storage space that was occupied by those rows, use the **TRUNCATE TABLE** command. The syntax for the command is shown in Figure 8-36.

```
TRUNCATE TABLE tablename;
```

Figure 8-36 Syntax for the TRUNCATE TABLE command

Assume that between the time the SETOTALS table was originally created to test the CREATE TABLE...AS command and now, customers in the Southeast region have placed several new orders; thus, the data in that table are now obsolete. Because it is possible that the table structure could be reused in the future, you might only want to delete the rows currently contained in the SETOTALS table and release the storage space they occupy by using the TRUNCATE TABLE command. The command to perform the truncation is shown in Figure 8-37.

```
TRUNCATE TABLE setotals;
```

Figure 8-37 The TRUNCATE TABLE command

To verify that all rows of the SETOTALS table have been removed, perform a query to see all the rows in the table. If the table still exists but contains no rows, then Oracle9*i* will display the message "no rows selected," as shown in Figure 8-38.

```
Oracle SQL*Plus
File Edit Search Options Help
SQL> TRUNCATE TABLE setotals;

Table truncated.

SQL> SELECT *
  2  FROM setotals;

no rows selected

SQL>
```

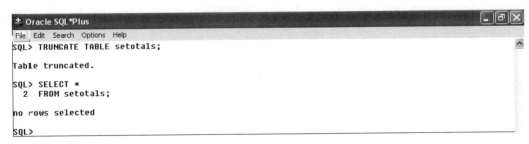

Figure 8-38 SETOTALS after performing the TRUNCATE TABLE command

Deleting a Table

A table can be removed from an Oracle9*i* database by issuing the **DROP TABLE** command. The syntax for the DROP TABLE command is shown in Figure 8-39.

```
DROP TABLE tablename;
```

Figure 8-39 Syntax for the DROP TABLE command

Always use caution when deleting, especially when it involves a table. Once a table is deleted, the table and all the data it contains are gone. In addition, any index that has been created based on that table is also dropped. (Indexes will be presented in Chapter 12.)

Suppose that after truncating the SETOTALS table, you realize that you will no longer need the table (or that numerous modifications will have to be made to the table structure so that it is not worth the trouble to make the changes). The SETOTALS table can be deleted using the DROP TABLE command, shown in Figure 8-40.

```
DROP TABLE setotals;
```

Figure 8-40 Using the DROP TABLE command to remove the SETOTALS table

Once the table has been dropped, the table name will no longer be valid, and the table cannot be accessed by any commands. To verify that the correct table was deleted, you can use the DESCRIBE command to see the structure of the SETOTALS table. If the table no longer exists, Oracle9*i* will return an error message, as shown in Figure 8-41.

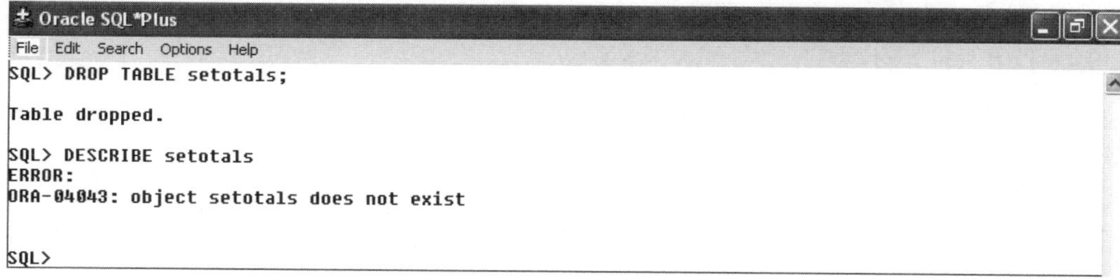

Figure 8-41 Results of dropping the SETOTALS table

Chapter Summary

- A table can be created using the CREATE TABLE command. Each column to be contained in the table must be defined in terms of the column name, datatype, and for certain datatypes, the width.
- A table can contain up to 1,000 columns.
- Each column name within a table must be unique.
- Table and column names can contain as many as 30 characters. The names must begin with a letter and cannot contain any blank spaces.
- To create a table based on the structure and data contained in existing tables, use the CREATE TABLE...AS command to use a subquery to extract the necessary data from the existing table.
- The structure of a table can be changed with the ALTER TABLE command. Columns can be added, resized, and even deleted with the ALTER TABLE command.
- When using the ALTER TABLE command with the DROP COLUMN clause, only one column can be specified for deletion.
- The SET UNUSED clause can be used to mark a column so its storage space can be freed up at a later time.
- Tables can be renamed with the RENAME...TO command.
- All the rows in a table can be deleted from a table through the TRUNCATE TABLE command.
- To completely remove both the structure of a table and all its contents, use the DROP TABLE command.

Chapter 8 Syntax Summary

The following table presents a summary of the syntax that you have learned in this chapter. You can use the table as a study guide and reference.

Syntax Guide		
Creating Tables		
Commands and Clauses	**Description**	**Example**

CREATE TABLE	Creates a new table in the database—user names columns; defaults and datatypes define/limit columns.
	To view the table, use the SQL*PLUS command DESCRIBE.
		`CREATE TABLE acctmanager`
		`(amid VARCHAR2(4),`
		`amname VARCHAR2(20),`
		`amedate DATE DEFAULT SYSDATE,`
		`region CHAR(2));`

CREATE TABLE...AS	Creates a table from existing database tables, using the AS clause and subqueries.	`CREATE TABLE secustomerorders`
		`AS (SELECT customer#, state,`
		`ISBN, category, quantity,`
		`cost, retail`
		`FROM customers NATURAL JOIN`
		`orders NATURAL JOIN`
		`order items NATURAL JOIN books`
		`WHERE state IN ('FL', 'GA',`
		`'AL'));`

Modifying Tables		
ALTER TABLE... ADD	Adds a column to a table.	`ALTER TABLE acctmanager` `ADD (ext NUMBER(4));`
ALTER TABLE... MODIFY	Changes a column size, datatype, or default value.	`ALTER TABLE acctmanager` `MODIFY (amname VARCHAR2(25));`
ALTER TABLE... DROP COLUMN	Deletes one column from a table.	`ALTER TABLE acctmanager` `DROP COLUMN ext;`
ALTER TABLE... SET UNUSED *or* SET UNUSED COLUMN	Marks a column for deletion at a later time.	`ALTER TABLE secustomerorders` `SET UNUSED (cost);`
DROP UNUSED COLUMNS	Completes the deletion of a column marked with SET UNUSED.	`ALTER TABLE secustomerorders` `DROP UNUSED COLUMNS;`
RENAME...TO	Changes a table name.	`RENAME secustomersspent TO` `setotals;`
TRUNCATE TABLE	Deletes table rows, but table name and column structure remain.	`TRUNCATE TABLE setotals;`
Deleting Tables		
DROP TABLE	Removes an entire table from the Oracle9*i* database.	`DROP TABLE setotals;`

Review Questions

To answer the following questions, refer to the tables in Appendix A.

1. Which command is used to create a table based upon data already contained in an existing table?
2. List four datatypes supported by Oracle9*i*, and provide an example of data that could be stored by each datatype.
3. What are the guidelines you should follow when naming tables and columns in Oracle9*i*?
4. What is the difference between dropping a column and setting a column as unused?
5. How many columns can be dropped in one ALTER TABLE command?
6. What happens to the existing rows of a table if the DEFAULT value of a column is changed?
7. Explain the difference between truncating a table and deleting a table.
8. If you add a new column to an existing table, where will the column appear relative to the existing columns?
9. What happens if you try to decrease the length of a VARCHAR2 column to a width that is smaller than one of the values already stored in the field?
10. What happens if you try to decrease the scale or precision of a NUMBER column to a value less than the data already stored in the field?

Multiple Choice

To answer the following questions, refer to the tables in Appendix A.

1. Which of the following is a correct statement?
   a. You can restore the data deleted with the DROP COLUMN clause, but not the data deleted with the SET UNUSED clause.
   b. You cannot create empty tables—all tables must contain at least three rows of data.
   c. A table can contain a maximum of 1,000 columns.
   d. The maximum length of a table name is 265 characters.
2. Which of the following is a valid SQL statement?
   a. `ALTER TABLE secustomersspent ADD DATE lastorder;`
   b. `ALTER TABLE secustomerorders DROP retail;`
   c. `CREATE TABLE newtable AS (SELECT * FROM customers);`
   d. `ALTER TABLE drop column *;`

3. Which of the following is not a correct statement?
   a. A table can only be modified if it does not contain any rows of data.
   b. The maximum number of characters in a table name is 30.
   c. More than one column can be added to a table at a time.
   d. The data contained in a table that has been truncated cannot be "undeleted."
4. Which of the following is not a valid SQL statement?
   a. `CREATE TABLE anothernewtable (newtableid VARCHAR2(2));`
   b. `CREATE TABLE anothernewtable (date, anotherdate) AS (SELECT orderdate, shipdate FROM orders);`
   c. `CREATE TABLE anothernewtable (firstdate, seconddate) AS (SELECT orderdate, shipdate FROM orders);`
   d. All of the above are valid statements.
5. Which of the following is true?
   a. If you truncate a table, you cannot add new data to the table.
   b. If you change the default value of an existing table, any previously stored default value will be changed to a NULL value.
   c. If you delete a column from a table, you cannot add a column to the table using the same name as the previously deleted column.
   d. If you add a column to an existing table, it will always be added as the last column of that table.
6. Which of the following will create a new table containing the order number, book title, quantity ordered, and retail price of every book that has been sold?
   a.
   ```
   CREATE TABLE newtable AS
   (SELECT order#, title, quantity, retail FROM orders);
   ```
   b.
   ```
   CREATE TABLE newtable AS
   (SELECT * FROM orders);
   ```
   c.
   ```
   CREATE TABLE newtable AS
   (SELECT order#, title, quantity, retail FROM orders
   NATURAL JOIN orderitems);
   ```
   d.
   ```
   CREATE TABLE newtable AS
   (SELECT order#, title, quantity, retail FROM orders
   NATURAL JOIN orderitems NATURAL JOIN books);
   ```

8

7. Which of the following commands will drop any columns marked as unused from the SECUSTOMERORDERS table?
   a. `DROP COLUMN FROM SECUSTOMERORDERS WHERE column_status = UNUSED;`
   b. `ALTER TABLE SECUSTOMERORDERS DROP UNUSED COLUMNS;`
   c. `ALTER TABLE SECUSTOMERORDERS DROP (unused);`
   d. `DROP UNUSED COLUMNS;`
8. Which of the following statements is correct?
   a. A table can only contain a maximum of one column that is marked as unused.
   b. You can delete a table by removing all the columns within the table.
   c. Using the SET UNUSED clause allows you to free up all the storage space used by a column.
   d. None of the above statements are correct.
9. Which of the following commands will remove all the data from a table but leave the table's structure intact?
   a. `ALTER TABLE secustomerorders DROP UNUSED COLUMNS;`
   b. `TRUNCATE TABLE secustomerorders;`
   c. `DELETE TABLE secustomerorders;`
   d. `DROP TABLE secustomerorders;`
10. Which of the following commands would change the name of a table from OLDNAME to NEWNAME?
    a. `RENAME oldname TO newname;`
    b. `RENAME table FROM oldname TO newname;`
    c. `ALTER TABLE oldname MODIFY TO newname;`
    d. `CREATE TABLE newname (SELECT * FROM oldname);`
11. The default width of a VARCHAR2 field is:
    a. 1
    b. 30
    c. 255
    d. None—there is no default width for a VARCHAR2 field.
12. Which of the following is NOT a valid statement?
    a. You can change the name of a table only if it does not contain any data.
    b. You can change the length of a column that does not contain any data.
    c. You can delete a column that does not contain any data.
    d. You can add a column to a table.

13. Which of the following can be used in a table name?
    a. _
    b. (
    c. %
    d. !
14. Which of the following is true?
    a. The database cannot be in use when an ALTER TABLE command is being executed.
    b. A column can only be added to a table as long as the table is not being queried by another user.
    c. All data contained in a table will be lost if the table is dropped.
    d. All of the above statements are true.
15. Which of the following commands is valid?
    a. `RENAME customer# TO customernumber FROM customers;`
    b. `ALTER TABLE customers RENAME customer# TO customernum;`
    c. `DELETE TABLE customers;`
    d. `ALTER TABLE customers DROP UNUSED COLUMNS;`
16. Which of the following commands will create a new table containing two columns?
    a. `CREATE TABLE newname (col1 DATE, col2 VARCHAR2);`
    b. ```
       CREATE TABLE newname AS
       (SELECT title, retail, cost FROM books);
       ```
    c. `CREATE TABLE newname (col1, col2);`
    d. ```
       CREATE TABLE newname
       (col1 DATE DEFAULT SYSDATE, col2 VARCHAR2(1));
       ```
17. Which of the following is a valid table name?
    a. 9NEWTABLE
    b. DATE9
    c. NEW"TABLE
    d. None of the above are valid table names.
18. Which of the following is a valid datatype?
    a. CHAR3
    b. VARCHAR4(3)
    c. NUMERIC
    d. NUMBER

8

19. Which of the following will create a table containing books that have a retail price of at least $30.00?
   a. `SELECT * FROM books WHERE retail >= 30;`
   b. `CREATE newtable AS`
      `(SELECT * FROM books WHERE retail >= 30);`
   c. `CREATE newtable AS`
      `(SELECT * FROM books WHERE retail > 30);`
   d. `CREATE newtable AS`
      `(SELECT * FROM books)`
      `WHERE retail >= 30;`
   e. none of the above
20. Which of the following SQL statements will change the size of the Title column in the BOOKS table from the current length of 30 characters to the needed length of 35 characters?
   a. `ALTER TABLE books CHANGE title VARCHAR(35);`
   b. `ALTER TABLE books MODIFY (title VARCHAR2(35));`
   c. `ALTER TABLE books MODIFY title (VARCHAR2(35));`
   d. `ALTER TABLE books MODIFY (title VARCHAR2(+5));`

HANDS-ON ASSIGNMENTS

To perform the following activities, refer to the tables in Appendix A.

1. Create a new table that contains the category code and description for the categories of books sold by JustLee Books. The table should be called Category. The columns should be called CatCode and CatDesc. The CatCode column should store a maximum of two characters, and the CatDesc column should store a maximum of 10 characters.
2. Create a new table that contains the following two columns: Customer# and Region. The name of the table should be CUSTOMERREGION. The Region column should be able to store character strings up to a maximum length of four. The column values should not be padded if the value has less than four characters.
3. Add a column to the CUSTOMERREGION table that will contain the date of the last order placed by each customer. The default value of the column should be the system date. The new column should be named LastODate.
4. Modify the Region column of the CUSTOMERREGION table so that it will only allow a maximum width of two characters to be stored in the column.
5. Create a new table that contains the title of each book in the BOOKS table, its ISBN, the publisher ID, and the profit generated by each book. The name of the profit column should be Profit, and the other columns should keep their original names. Name the new table ProfitGeneratedPerBook.

6. Rename the ProfitGeneratedPerBook table to ProfitPerBook.
7. Delete the PubID column from the ProfitPerBook table.
8. Mark the Title column of the ProfitPerBook table as unused. Verify that the column is no longer available.
9. Truncate the ProfitPerBook table and then verify that the ProfitPerBook table still exists.
10. Delete the ProfitPerBook table.

A CASE FOR ORACLE9*i*

To perform this activity, refer to the tables in Appendix A.

The management of JustLee Books has approved the implementation of the new commission policy for the account managers. The following changes will need to be made to the existing database:

- A new column must be added to the CUSTOMERS table to indicate the region in which each customer lives. The column should be able to store variable-length data to a maximum width of four characters. It should be named Region.
- A new table, COMMRATE, must be created to store the commission rate schedule. The following columns must exist in the table:
    1. Rate—a numeric field that can store two decimal digits (e.g., .01, .03)
    2. Minprice—a numeric field that can store the lowest retail price for a book in that price range of the commission rate
    3. Maxprice—a numeric field that can store the highest retail price for a book in that price range of the commission rate

Required: Make the necessary changes to the JustLee Books database to support the implementation of the new commission policy.

CHAPTER

9

CONSTRAINTS

Objectives

After completing this chapter, you should be able to do the following:

- Explain the purpose of constraints in a table
- Distinguish among PRIMARY KEY, FOREIGN KEY, UNIQUE, CHECK, and NOT NULL constraints and understand the appropriate use of each constraint
- Distinguish between creating constraints at the column level and at the table level
- Create PRIMARY KEY constraints for a single column and a composite primary key
- Create a FOREIGN KEY constraint
- Create a UNIQUE constraint
- Create a CHECK constraint
- Create a NOT NULL constraint, using the ALTER TABLE...MODIFY command
- Include constraints during table creation
- Use the DISABLE and ENABLE commands
- Use the DROP command

In Chapter 8, you learned how to create tables by using SQL commands. In this chapter, you will learn how to add constraints to existing tables and how to include constraints during the table creation process.

Constraints are rules used to enforce business rules, practices, and policies—and to ensure the accuracy and integrity of data. For example, suppose that a customer places an order on April 2, 2003. However, when the order is shipped, the date shipped is entered as March 31, 2003. This is impossible, and it indicates a problem with data integrity. If such errors are found in the database, management will not be able to rely on it for decision-making, or even to support day-to-day operations. Constraints can solve such problems by not allowing data to be added to tables if the data violate certain rules.

This chapter will examine constraints, listed in Figure 9–1, that can prevent errors from being entered into a database.

Constraint	Description
PRIMARY KEY	Determines which column(s) uniquely identifies each record. The primary key cannot be NULL, and the data value(s) must be unique.
FOREIGN KEY	In a one-to-many relationship, the constraint is added to the "many" table. The constraint ensures that if a value is entered into a specified column, it must already exist in the "one" table, or the record is not added.
UNIQUE	Ensures that all data values stored in a specified column are unique. The UNIQUE constraint differs from the PRIMARY KEY constraint in that it allows NULL values.
CHECK	Ensures that a specified condition is true before the data value is added to a table. For example, an order's ship date cannot be earlier than its order date.
NOT NULL	Ensures that a specified column cannot contain a NULL value. The NOT NULL constraint can *only* be created with the column-level approach to table creation.

Figure 9-1 Overview of chapter contents

Before creating constraints for JustLee Books, go to your Data Files and open the Chapter09 folder. Run the **prech9.sql** script to make certain that all necessary changes made in the previous chapter have been applied. In addition, the script will drop previously created constraints, allowing you to re-create those constraints during this chapter.

To run the script, at the SQL> prompt type **`start d:\prech9.sql`**, where **d:** represents the location (correct drive and path) of the script. If you do not have a copy of the script, ask your instructor to provide you with a copy.

Creating Constraints

Constraints can be created during table creation as part of the CREATE TABLE command, or after the table is created by using the ALTER TABLE command. When creating a constraint, you can choose one of the following options:

1. Name the constraint, using the same rules as for tables and columns.
2. Omit the constraint name, and allow Oracle9*i* to generate the name.

If the Oracle9*i* server names the constraint, it follows the format of **SYS_C*n***, where *n* is a numeric value that is assigned to make the name unique. It is always a good practice to provide a name for a constraint, so you can easily identify it in the future, if necessary.

Industry convention is to use the format *TABLENAME columnname_constrainttype* for the constraint name. Constraint types are designated by abbreviations, as shown in Figure 9–2.

Constraint	Abbreviation
PRIMARY KEY	_pk
FOREIGN KEY	_fk
UNIQUE	_uk
CHECK	_ck
NOT NULL	_nn

Figure 9-2 Constraint abbreviations

There are two ways to create a constraint: at the column level or at the table level. *Creating a constraint at the column level simply means the definition of the constraint is included as part of the column definition*, similar to assigning a default value to a column.

Creating Constraints at the Column Level

When you create constraints at the column level, the constraint being created applies to the column specified. The optional **CONSTRAINT** keyword is used *if* you want to give the constraint a specific name. The constraint type uses the following keywords to identify the type of constraint being created:

- PRIMARY KEY
- FOREIGN KEY
- UNIQUE
- CHECK
- NOT NULL

Any type of constraint can be created at the column level—unless the constraint is being defined for more than one column (e.g., a composite primary key). *If the constraint applies to more than one column, the constraint must be created at the table level.*

The general syntax for creating a constraint at the column level is shown in Figure 9–3.

```
columnname [CONSTRAINT constraintname] constrainttype,
```

Figure 9-3 Syntax for creating a column-level constraint

As you will see later in this chapter, a NOT NULL constraint can only be created at the column level.

Creating Constraints at the Table Level

The syntax for creating a constraint at the table level is shown in Figure 9–4.

```
[CONSTRAINT constraintname] constrainttype
(columnname, …),
```

Figure 9-4 Syntax for creating a table-level constraint

When a constraint is created at the table level, the constraint definition is separate from any column definitions. If the constraint is created at the same time that a table is being created, it is listed *after* all the columns have been defined. In fact, *the main difference in the syntax of a column-level constraint and a table-level constraint is that the column name(s) for the table-level constraint is provided at the end of the constraint definition in a set of parentheses, rather than at the beginning of the constraint definition*. The table-level approach can be used to create any type of constraint, except a NOT NULL constraint. A NOT NULL constraint can only be created using the column-level approach.

Although a constraint can be created at the column level or at the table level, the constraint is *always* enforced at the table level, which means the entire record cannot be added, modified, or deleted if it violates a constraint.

To simplify the examples for the different types of constraints, the following sections will demonstrate how to add constraints to an existing table. Once you have learned how to add constraints by using the ALTER TABLE command, you will learn how to include constraints at both the column level and the table level during table creation.

Using the PRIMARY KEY Constraint

A **PRIMARY KEY constraint** is used to enforce the primary key requirements for a table. Although a table can be created in Oracle9*i* without specifying a primary key, the constraint will make certain that the column(s) identified as the table's primary key is unique and *does not contain NULL values*. The syntax of the ALTER TABLE command to add a PRIMARY KEY constraint to an existing table is shown in Figure 9–5.

```
ALTER TABLE tablename
ADD [CONSTRAINT constraintname] PRIMARY KEY (columnname);
```

Figure 9-5 Syntax of the ALTER TABLE command to add a PRIMARY KEY constraint

Let's look at an example. The PROMOTION table that stores data regarding the gifts customers will receive during JustLee Books' annual Marketing promotion does not have a PRIMARY KEY constraint. By not having a primary key designated for the PROMOTION table, a new user could mistakenly enter a bookmark as a gift to those customers who purchase a book with a retail price of at least $87.00. This would not make sense because that gift is given to customers who purchase less expensive books. To designate the Gift column as the primary key for the PROMOTION table, issue the ALTER TABLE command shown in Figure 9–6.

```
ALTER TABLE promotion
ADD CONSTRAINT promotion_gift_pk PRIMARY KEY (gift);
```

Figure 9-6 Adding a PRIMARY KEY constraint

Note the following elements in Figure 9–6:

- The ADD clause instructs Oracle9*i* that a constraint will be added to the PROMOTION table, named after the ALTER TABLE keywords.
- The user has chosen the constraint name, rather than having it assigned by Oracle9*i*, and the constraint name is added as **promotion_gift_pk**.
- Because PRIMARY KEY is provided as the constraint type, Oracle9*i* will make the Gift column the primary key for the PROMOTION table.

9

If the command executes successfully, you will receive the message "Table altered," as shown in Figure 9–7. Once this command has been executed, all rows must have an entry in the Gift column, and each entry must be unique.

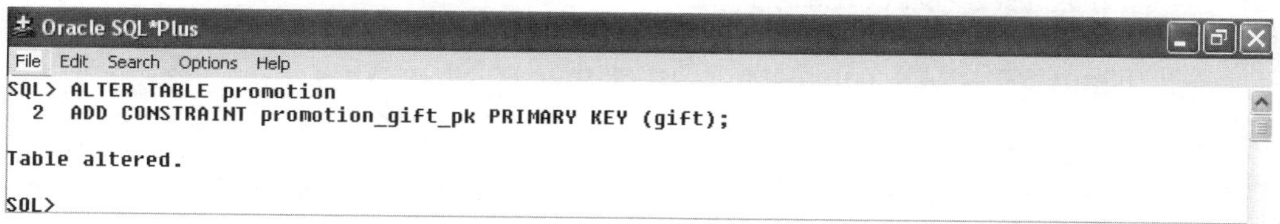

Figure 9-7 Results of adding a PRIMARY KEY constraint

Only *one* PRIMARY KEY constraint can be created for each table. If the primary key consists of more than one column (a composite primary key), it must be created at the table level. For example, the ORDERITEMS table uses two columns to uniquely identify each item on an order: Order# and Item#. To indicate that the primary key for a table will consist of more than one column, simply list the column names within parentheses after the constraint type. Commas must be used to separate the list of column names. This is shown in Figure 9–8.

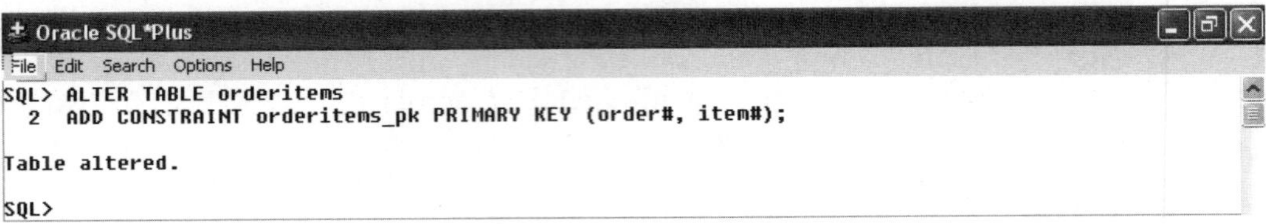

```
SQL> ALTER TABLE orderitems
  2  ADD CONSTRAINT orderitems_pk PRIMARY KEY (order#, item#);

Table altered.

SQL>
```

Figure 9-8 Adding a composite PRIMARY KEY constraint

After the constraint shown in Figure 9–8 is added to the ORDERITEMS table, all entries in the Order# and Item# columns must create a unique combination in the table, and neither column can contain a NULL value.

Because a table can only have one PRIMARY KEY constraint, the name assigned to the constraint (**`orderitems_pk`**) in Figure 9–8 does not include the names of the columns used to create the composite primary key. However, if the user wanted to include the name of the columns, the constraint name could be assigned as **`orderitems_order#item#_pk`**.

Although Oracle9*i* has already returned the message "Table altered," you can make certain that the two constraints just created actually exist by selecting the names of all constraints that you have created or own from the **USER_CONSTRAINTS** view. The SQL statement to view the names of existing constraints is shown in Figure 9–9.

```
SELECT constraint_name
FROM user_constraints;
```

Figure 9-9 SQL command to list existing constraints

After the SQL command is executed, a list of constraint names will appear. In Figure 9–10, the two constraints that were previously created, **`promotion_gift_pk`** and **`orderitems_pk`** are displayed at the end of the list. Notice, however, that there are also other constraints in the results. The output will vary, depending on the tables contained within your schema. In this case, seven constraints are displayed in the output. Since the two constraints you have created are listed in the output, you can be assured that they were successfully created. More information about this view and its contents will be provided in a later section.

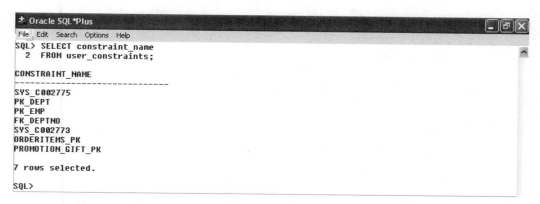

Figure 9-10 Viewing existing constraint names

Notice the two constraints that begin with **SYS_C**. Oracle9*i* named these two constraints, not the user. As previously mentioned, the constraint names assigned by the software are not very informative, and it is difficult to determine their purpose, or even which table(s) is affected by the constraints. Also notice that there are three constraints that have a name that first identifies the type of constraint (**pk** or **fk**). However, it is unclear whether the remainder of the constraint names identify a column or a table—unless you are familiar with the contents of the database. Most organizations have some type of standard naming convention that they use for constraints. Ideally, that naming convention would consist of the table name, column name, and constraint type. By using a standardized naming convention, it is much easier to identify a specific constraint for future reference.

USING THE FOREIGN KEY CONSTRAINT

Suppose that an order is placed with JustLee Books by a new customer who does not exist in the CUSTOMERS table. If the customer information was not collected, the customer's name and billing address would not be contained in the database. This would make it difficult to bill the customer for the order—not exactly what one might consider "good business practice." Or, perhaps there is a book in the BOOKS table that is published with the publisher ID of 9—a publisher ID that does not exist in the PUBLISHER table. (The publisher ID could simply be a typo, or perhaps someone neglected to add the publisher to the PUBLISHER table.)

These problems can be prevented by using a **FOREIGN KEY** constraint. To prevent someone from entering an order from a customer who does not have a record in the CUSTOMERS table, you can create a constraint that compares any entry made into the Customer# column of the ORDERS table with all customer numbers existing in the CUSTOMERS table. Thus, if a customer service representative enters a customer number not found in the CUSTOMERS table, then the corresponding entry in the ORDERS table is rejected. This would require the customer service representative first to collect and enter the customer's information into the CUSTOMERS table, and then enter the order into the ORDERS table.

The syntax to add a FOREIGN KEY constraint to a table is shown in Figure 9–11.

```
ALTER TABLE tablename
ADD [CONSTRAINT constraintname] FOREIGN KEY
    (columnname) REFERENCES referencedtablename
    (referencedcolumnname);
```

Figure 9-11 Syntax of the ALTER TABLE command to add a FOREIGN KEY constraint

The keywords FOREIGN KEY are used to identify a column that, if it contains a value, must match data contained in another table. The name of the column identified as the foreign key is contained within a set of parentheses after the FOREIGN KEY keywords. The keyword **REFERENCES** refers to **referential integrity**. Referential integrity simply means that the user is referring to something that exists in another table (e.g., the value entered into the Customer# column of the ORDERS table references a value in the Customer# column of the CUSTOMERS table). The REFERENCES keyword is used to identify the table and column that must already contain the data being entered.

To create a FOREIGN KEY constraint on the Customer# column of the ORDERS table that would ensure that any customer number entered also exists in the CUSTOMERS table before the order is accepted, use the command shown in Figure 9–12.

```
ALTER TABLE orders
ADD CONSTRAINT orders_customer#_fk FOREIGN KEY
             (customer#)REFERENCES customers (customer#);
```

Figure 9-12 Command to add a FOREIGN KEY constraint to the ORDERS table

This command instructs Oracle9*i* to add a FOREIGN KEY constraint on the Customer# column of the ORDERS table. The name chosen for the constraint is **`orders_customer#_fk`**. This constraint makes sure that an entry for the Customer# column of the ORDERS table matches a value that is stored in the Customer# column of the CUSTOMERS table. If the command is executed correctly, the message "Table altered" will be returned, as shown in Figure 9–13.

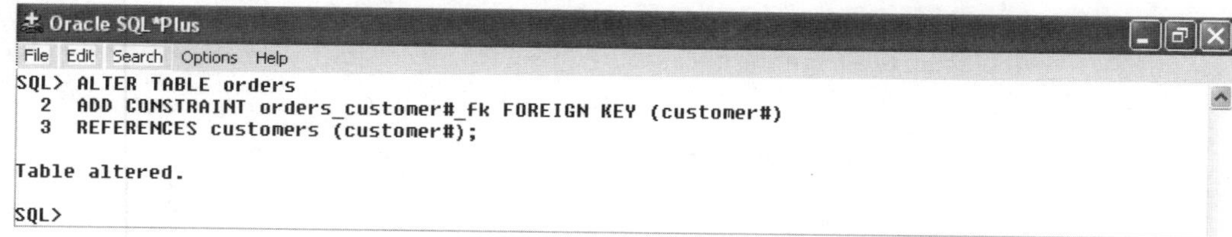

Figure 9-13 Output from adding the FOREIGN KEY constraint

The syntax for the FOREIGN KEY constraint is more complex than the PRIMARY KEY constraint because there are actually two tables involved in the constraint. The CUSTOMERS table is the referenced table (i.e., it is the "one" side of the one-to-many relationship between the CUSTOMERS and ORDERS table—each order can be placed by only one customer, but one customer can place many orders). Thus, the CUSTOMERS table is considered the parent table for the constraint; the ORDERS table is considered the **child table**.

When a FOREIGN KEY constraint exists between two tables, by default, a record cannot be deleted from the parent table if matching entries exist in the child table. In other words, *you cannot delete a customer from the CUSTOMERS table if there are orders in the ORDERS table for that customer.*

However, suppose that you do need to delete a customer from the CUSTOMERS table. Perhaps the customer has not paid for previous orders, or perhaps the customer has passed away. Your goal is to remove the customer from the database to make certain no one places an order using that customer's information. The FOREIGN KEY constraint requires that you first delete all of that customer's recent orders from the ORDERS table—the child table—and then delete the customer from the CUSTOMERS table—the parent table. If the customer has placed several orders, this process could take some time.

There is a much simpler solution, however; you can add the keywords **ON DELETE CASCADE** to the end of the command issued in Figure 9–12. If the ON DELETE CASCADE keywords are included in the constraint definition and a record is deleted from the parent table, then any corresponding records in the child table are also automatically deleted. Figure 9–14 shows a FOREIGN KEY constraint with the ON DELETE CASCADE option.

```
ALTER TABLE orders
ADD CONSTRAINT orders_customer#_fk FOREIGN KEY
               (customer#)
REFERENCES customers (customer#) ON DELETE CASCADE;
```

Figure 9-14 FOREIGN KEY constraint with the ON DELETE CASCADE option

If you attempt the command given in Figure 9–14 and receive an error message, this might be because a FOREIGN KEY constraint already exists with the same name. Enter the following command:

```
ALTER TABLE orders
DROP CONSTRAINT orders_customer#_fk;
```

Once the previous constraint has been removed from the database, you can then re-enter the command from Figure 9–14 without receiving an error message.

Thus, using the example in Figure 9–14, if a customer who has placed 20 orders is deleted from the CUSTOMERS table, then all orders placed by that customer will be deleted from the ORDERS table.

Be very cautious—if the ON DELETE CASCADE option is included in the FOREIGN KEY constraint, this could create a problem for unsuspecting users who unintentionally delete outstanding orders.

Make absolutely certain that any records that might get deleted from the child table through this option will not be needed in the future. If that possibility exists—even remotely—then do not include the ON DELETE CASCADE keywords, and force the user first to delete the entries in the child table.

If a record in a child table has a NULL value for a column that has a FOREIGN KEY constraint, the record will be accepted. For example, all the constraint ensures is that the customer number is a valid number, *not* that a customer number has been entered for an order. Basically, this means that an order could be entered into the ORDERS table without an entry in the Customer# column, and the order would still be accepted. To force the user to enter a customer number for an order, a NOT NULL constraint should also be added for the Customer# column in the ORDERS table. (You will add such a constraint in a later section.)

A FOREIGN KEY constraint cannot reference a column in a table that has not already been designated as the primary key for the referenced table.

USING THE UNIQUE CONSTRAINT

The purpose of a **UNIQUE** constraint is to ensure that two records do not have the same value stored in the same column. Although this sounds like a PRIMARY KEY constraint, there is one major difference. *A UNIQUE constraint will allow NULL values*, which are not permitted with a PRIMARY KEY constraint. The syntax to add a UNIQUE constraint to an existing table is shown in Figure 9–15.

```
ALTER TABLE tablename
ADD [CONSTRAINT constraintname] UNIQUE (columnname);
```

Figure 9-15 Syntax for adding a UNIQUE constraint to a table

As shown in Figure 9–15, the syntax to add a UNIQUE constraint is the same as the syntax for adding a PRIMARY KEY constraint, except the UNIQUE keyword is used to indicate the type of constraint being created.

For example, suppose that you want to make certain that each book in inventory has a different title. You don't want to use book titles as the primary key because users might have difficulty remembering whether a title starts with "The" or "A," for instance, but you do want each title stored in the table to be unique. To create a UNIQUE constraint on the Title column of the BOOKS table, issue the command shown in Figure 9–16.

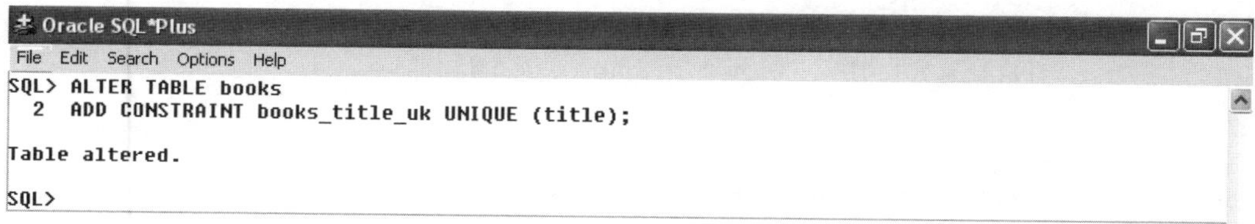

Figure 9-16 Results of issuing the command to create the UNIQUE constraint

9

Once the command is successfully issued, Oracle9*i* will not allow any entry into the Title column of the BOOKS table that will duplicate an existing entry. For example, if a second edition of a book were published with the same title as the first edition, the user would need to include the edition number in the title to make it different from the record for the previous edition (or delete the previous edition from the database entirely if it is no longer available).

USING THE CHECK CONSTRAINT

In this chapter's introduction, a scenario was presented in which an order's ship date was earlier than its order date. Data-entry errors of this type can be prevented through the use of a CHECK constraint. A **CHECK** constraint requires that a specific condition be met before a record is added to a table. With a CHECK constraint, you can, for example, make certain that a book's cost is greater than zero, its retail price is less than $200.00, and a seller's commission rate is less than 50 percent. The condition included in the constraint cannot reference certain functions, such as SYSDATE, USER, or ROWNUM, or refer to values stored in other rows; however, it can be compared to values within the *same* row. For instance, you could use the condition that the order date must be earlier than or equal to the ship date. However, you could not add a CHECK constraint that requires the ship date for an order to be the same as the current system date, because you would have to reference the SYSDATE function. The syntax for adding a CHECK constraint to an existing table is shown in Figure 9-17.

```
ALTER TABLE tablename
ADD [CONSTRAINT constraintname] CHECK (condition);
```

Figure 9-17 Syntax for adding a CHECK constraint to an existing table

The syntax to add a CHECK constraint follows the same format as the syntax to add a PRIMARY KEY or UNIQUE constraint. However, rather than simply list a column name(s) for the constraint, the condition that must be satisfied is listed after the constraint type.

To solve the problem of an incorrect ship date being entered into the table, the condition can be stated as **`(orderdate<=shipdate)`**, so the ship date entered in a record cannot be earlier than the order date. The command to add the CHECK constraint to the ORDERS table is shown in Figure 9-18.

```
ALTER TABLE orders
ADD CONSTRAINT orders_shipdate_ck
CHECK (orderdate<=shipdate);
```

Figure 9-18 Adding a CHECK constraint to the ORDERS table

If any records that are already stored in the ORDERS table violate the **`orderdate<=shipdate`** condition, Oracle9*i* will return an error message stating that the constraint has been violated, and the ALTER TABLE command will fail. This is true for all the constraint types. If you attempt to add a CHECK constraint and you receive an error message indicating that such a violation exists, the simplest solution is to issue a SELECT statement, using the same condition from the CHECK constraint as the condition for the WHERE clause (i.e., **`SELECT * FROM orders WHERE orderdate<=shipdate;`**). This will identify any records preventing the constraint from being added to the table. Once those records are identified and corrected, the ALTER TABLE command can be reissued and should be successful. Once the CHECK constraint is successfully executed, the message "Table altered" will be returned, as shown in Figure 9-19.

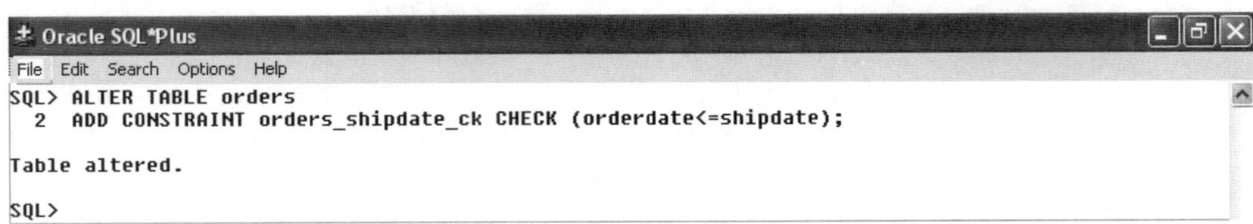

Figure 9-19 Output from adding the CHECK constraint to the ORDERS table

Using the NOT NULL Constraint

The **NOT NULL** constraint is actually a special CHECK constraint with the condition of IS NOT NULL. Basically, it prevents adding a row that contains a NULL value in the specified column. However, a NOT NULL constraint is not added to a table in the same manner as constraints previously presented in this chapter. *A NOT NULL constraint can only be added to an existing column by using the ALTER TABLE...MODIFY command.*

The syntax for adding a NOT NULL constraint is shown in Figure 9–20.

```
ALTER TABLE tablename
MODIFY (columnname [CONSTRAINT constraintname]
NOT NULL);
```

Figure 9-20 Syntax for adding a NOT NULL constraint to an existing table

The ALTER TABLE...MODIFY command is the same command used in Chapter 8 to redefine a column. However, if you are adding the NOT NULL constraint to a column, you do not have to list the column's datatype and width or any default value, if one exists. You are simply required to list the column's name and the keywords NOT NULL. For example, suppose that you want to add a NOT NULL constraint to the PubID column of the BOOKS table. The command and its successful execution, "Table altered," are shown in Figure 9–21.

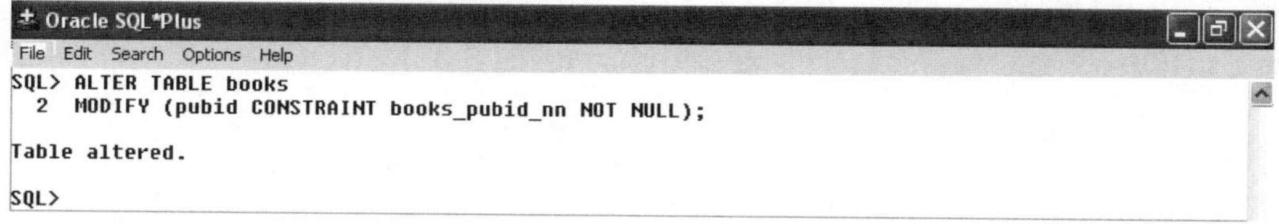

Figure 9-21 Adding a NOT NULL constraint

Because you would only expect one NOT NULL constraint for a particular column, industry rarely gives constraint names to this type of constraint (although it is still advisable if you ever need to delete the constraint in the future). If you do not want to assign a name to a NOT NULL constraint, simply omit the CONSTRAINT keyword and list the constraint type directly after the column name, as shown in Figure 9–22.

```
SQL> ALTER TABLE books
  2  MODIFY (retail NOT NULL);

Table altered.

SQL> DESCRIBE books
 Name                                      Null?    Type
 ----------------------------------------- -------- ----------------------------
 ISBN                                               VARCHAR2(10)
 TITLE                                              VARCHAR2(30)
 PUBDATE                                            DATE
 PUBID                                     NOT NULL NUMBER(2)
 COST                                               NUMBER(5,2)
 RETAIL                                    NOT NULL NUMBER(5,2)
 CATEGORY                                           VARCHAR2(12)

SQL>
```

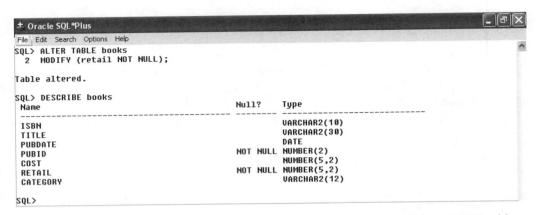

Figure 9-22 Adding a NOT NULL constraint to the Retail column of the BOOKS table

As shown in Figure 9–22, you can use the DESCRIBE command to determine whether a column can contain a NULL value. The Null? column in the results of the DESCRIBE command will contain the value NOT NULL if the column is not allowed to contain NULL values. The only problem with using this approach is that you cannot tell whether a NOT NULL constraint is being used to prevent NULL values, or if it is due to a PRIMARY KEY constraint. Methods for identifying specific information regarding the constraints for a table will be presented later in this chapter, using the USER_CONSTRAINTS view.

INCLUDING CONSTRAINTS DURING TABLE CREATION

Now that you've examined adding constraints to existing tables, let's look at adding constraints to tables during table creation. If constraints are included during a table's creation, then no data can ever be added to the table that violate those constraints—unless the user disables the constraints. If the design process for a database is thorough, all needed constraints should be identified before a table is physically created. In this case, the constraints can be included in the CREATE TABLE command, so they will not need to be added at a later time as a separate step. As previously mentioned, there are two approaches for defining constraints: at the column level or table level.

If a constraint is created at the column level as part of the CREATE TABLE command, the constraint type is simply listed after the datatype for the column. In Chapter 8, you created the ACCTMANAGER table, using the command shown in Figure 9–23.

```
CREATE TABLE acctmanager
     (amid          VARCHAR2(4),
      amname        VARCHAR2(20),
      amedate       DATE        DEFAULT SYSDATE,
      region        CHAR(2));
```

Figure 9-23 Original command to create the ACCTMANAGER table

However, this command did not include any references to a PRIMARY KEY constraint or any other type of constraint. Since the AmID column was designed to be the primary key for the ACCTMANAGER table, then a PRIMARY KEY constraint should be created to ensure that the column will never have any NULL values and that the same ID will not be assigned to more than one account manager. In addition, an account manager's name and the region to which he or she is assigned should never be left blank. With some minor modifications to the CREATE TABLE command given in Figure 9–23, these constraints can be included during the creation of the ACCTMANAGER table, as shown in Figure 9–24.

Oracle SQL*Plus
File Edit Search Options Help
SQL> CREATE TABLE acctmanager
  2 (amid VARCHAR2(4) PRIMARY KEY,
  3 amname VARCHAR2(20) NOT NULL,
  4 amedate DATE DEFAULT SYSDATE,
  5 region CHAR(2) NOT NULL);
Table created.
SQL>

Figure 9-24 Creating a table with constraints defined at the column level

The modified command in Figure 9–24 adds a PRIMARY KEY constraint to the AmID column and NOT NULL constraints to the Amname and Region columns. A NOT NULL constraint is not added to the Amedate column because a default value would automatically be assigned if the user does not enter a value for the column. In this example, the constraints are not given a name, so the Oracle server will assign a unique name for each of the three constraints. However, you should remember that this could cause a headache in the future if you ever need to delete the constraint. A name can be assigned to a constraint, using the same format as the ALTER TABLE command, by including the keyword CONSTRAINT followed by the name of the constraint.

Help

If you enter the example given in Figure 9–24 and receive an error message, indicating that a table with the same name already exists, you will need to issue the `DROP TABLE acctmanager;` command so you can re-create the table without any errors. The DROP TABLE command was included in the **prech9.sql** file referenced at the beginning of this chapter.

With the exception of the NOT NULL constraint, constraints can also be included in the CREATE TABLE command, using the table-level approach: List the constraints at the end of the command after all columns are defined, as shown in Figure 9–25.

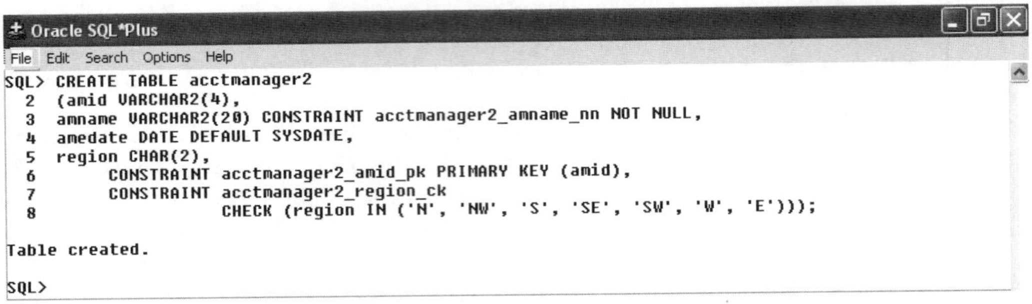

```
Oracle SQL*Plus
File Edit Search Options Help
SQL> CREATE TABLE acctmanager2
  2  (amid VARCHAR2(4),
  3  amname VARCHAR2(20) CONSTRAINT acctmanager2_amname_nn NOT NULL,
  4  amedate DATE DEFAULT SYSDATE,
  5  region CHAR(2),
  6       CONSTRAINT acctmanager2_amid_pk PRIMARY KEY (amid),
  7       CONSTRAINT acctmanager2_region_ck
  8                  CHECK (region IN ('N', 'NW', 'S', 'SE', 'SW', 'W', 'E')));

Table created.

SQL>
```

Figure 9-25 CREATE TABLE command, including constraints created with the table-level approach

In the example given in Figure 9–25, a table named ACCTMANAGER2 has been created with a structure similar to the ACCTMANAGER table. Note the following:

- The NOT NULL constraint for the Amname column has been changed to include **`acctmanager2_amname_nn`** as the constraint name.
- The PRIMARY KEY constraint is now being created using the table-level approach on line 6 of the example.
- A CHECK constraint has been added to the table to make certain that an account manager is assigned to a region using a predetermined code of N for north, NE for Northeast, etc.
- Notice the three parentheses at the end of the CREATE TABLE command. One parenthesis closes the code list used for the IN logical operator, the second parenthesis closes the condition for the CHECK constraint, and the last parenthesis closes the column list for the CREATE TABLE command.

As shown in Figure 9–25, both the column-level and table-level approaches can be used in the same command, should the need arise. However, the general practice in industry is that if a user assigns a constraint, then the constraint should be created by using the table-level approach. This is not a requirement; it is simply good practice, because a column list can become cluttered if a constraint name is provided in the middle of a list that defines all the columns. Therefore, most users define all the columns first, and then include the constraints at the end of the CREATE TABLE command to separate the column definitions from the constraints. It makes it much easier to go back and identify a problem if an error message is returned by Oracle9*i*.

Viewing Constraints

So far in this chapter, you have learned various ways to create different types of constraints. You have also used the USER_CONSTRAINT view to verify the creation of a constraint. But how, at a later time, do you determine which columns have constraints, the purpose of a constraint, etc.? The Oracle9*i* server stores information about constraints in its data dictionary. The **data dictionary** houses information about objects included in the database, even information about different users. The data dictionary will be discussed in much greater detail in a later chapter. For now, you will need to be able to access some of the information about constraints in the data dictionary. Specifically, you will need to display the constraints that have just been created for the ACCTMANAGER2 table. To view the contents of the portion of the data dictionary that references constraints, use the SELECT statement shown in Figure 9–26.

```
SELECT constraint_name, constraint_type,
       search_condition
FROM user_constraints
WHERE table_name = 'ACCTMANAGER2';
```


Figure 9-26 SELECT statement to view data about existing constraints

In the SELECT clause, note the three columns listed.

1. The first column referenced, **`constraint_name`**, lists the name of any constraint that exists in the ACCTMANAGER2 table. This is the same column you selected from the view at the beginning of the chapter.
2. The second column, **`constraint_type`**, will list a
   - P if the constraint is a PRIMARY KEY constraint,
   - C if the constraint is a CHECK or NOT NULL constraint,
   - U for a UNIQUE constraint, and
   - R for a FOREIGN KEY constraint.

   The R may seem a little strange for a FOREIGN KEY constraint; however, the purpose of the constraint is to ensure referential integrity—that you are referencing something that actually exists. Therefore, the assigned code is the letter R.
3. The third column listed in the SELECT clause, **`search_condition`**, is used to display the condition that is being used by a CHECK constraint. This column will be blank for any constraint that is not considered a CHECK constraint.

Figure 9–27 shows the output of the SELECT statement.

```
Oracle SQL*Plus
File Edit Search Options Help
SQL> SELECT constraint_name, constraint_type, search_condition
  2  FROM user_constraints
  3  WHERE table_name = 'ACCTMANAGER2';

CONSTRAINT_NAME                C SEARCH_CONDITION
------------------------------ - --------------------------------------------------------------
ACCTMANAGER2_AMNAME_NN         C "AMNAME" IS NOT NULL
ACCTMANAGER2_REGION_CK         C region IN ('N', 'NW', 'S', 'SE', 'SW', 'W', 'E')
ACCTMANAGER2_AMID_PK           P

SQL>
```

Figure 9-27 Results of the SELECT statement

In Figure 9–27, the results of the SELECT statement list three constraints for the ACCTMANAGER2 table. The first two constraints are CHECK constraints, as indicated by the **C** in the second column of the output. However, if you look at the search conditions displayed in the third column of the results, the first constraint, **`acctmanager2_amname_nn`**, is a NOT NULL constraint, since the condition specifies a NOT NULL requirement for the column. As indicated by the **P** for the constraint type of the third constraint, this is the PRIMARY KEY constraint, previously created in Figure 9–25.

If the results displayed in Figure 9–27 appear to wrap across several lines, you can change the number of characters that appear on one line by issuing the **`SET LINESIZE 150`** command.

DISABLING AND DROPPING CONSTRAINTS

Sometimes, you will want to temporarily disable or drop a constraint. In this section, you'll examine these options.

Using DISABLE/ENABLE

Whenever a constraint is created for a column(s), every time an entry is made to that column, it must be evaluated to determine whether the value is allowed in that column (i.e., it doesn't violate the constraint). If a large block of data is being added to a table, this validation process can severely slow down the Oracle server's processing speed. If you are certain that the data being added to a table adheres to the constraints, you can disable the constraints while adding that particular block of data to the table.

To **DISABLE** a constraint, you simply issue an ALTER TABLE command and change the status of the constraint to DISABLE. At a later time, you can reissue the ALTER TABLE command and change the status of the constraint back to **ENABLE**. The syntax for using the ALTER TABLE command to change the status of a constraint is shown in Figure 9–28.

```
ALTER TABLE tablename
DISABLE CONSTRAINT constraintname;

ALTER TABLE tablename
ENABLE CONSTRAINT constraintname;
```

Figure 9-28 Syntax to disable or enable an exising constraint

For example, suppose that you must set up records for account managers in the ACCT-MANAGER2 table to test some queries or other SQL operations. You know the number range that will be assigned for the IDs of managers and regions. However, you do not know the managers' names, and at this point, you do not want to make up names just for a few test runs. However, you still want the NOT NULL constraint for the Amname column to exist because it will be needed in the future. The simplest solution is to disable, or turn off, the constraint temporarily and then enable it when you are finished, as shown in Figure 9–29.

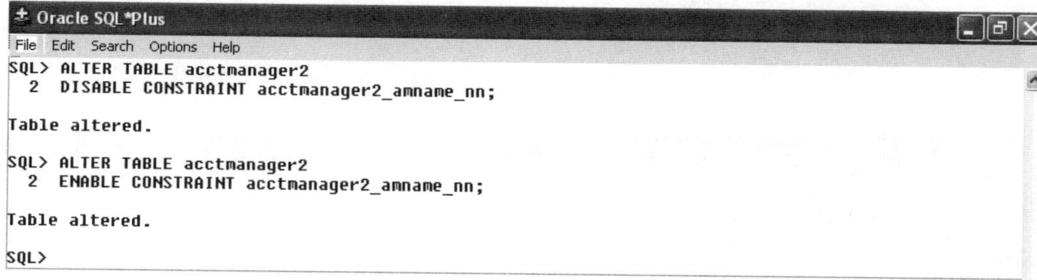

Figure 9-29 Disabling and enabling constraints

In the examples provided in Figure 9–29, the first ALTER TABLE command is used to temporarily turn off the **`acctmanager2_amname_nn constraint`**—note the keyword DISABLE. Upon successful execution of the statement, Oracle9*i* returns the message "Table altered." To instruct Oracle9*i* to begin enforcement of the constraint again, the same statement is reissued with the keyword ENABLE rather than DISABLE. After the constraint is enabled, any data added to the table will not be allowed to contain NULL values in the Amname column.

Dropping Constraints

If you create a constraint and then decide that it is no longer needed (or if you make an error when creating the constraint and need to delete your work), you can simply delete the constraint from the table with the `DROP (constraintname)` command. In addition, if you need to change or modify a constraint, your only option is to delete the constraint and then create a new one. The ALTER TABLE command is used to drop an existing constraint from a table, using the syntax shown in Figure 9–30.

```
ALTER TABLE tablename
DROP PRIMARY KEY | UNIQUE (columnname) |
CONSTRAINT constraintname;
```

Figure 9-30 Syntax for the ALTER TABLE command to delete a constraint

Note the following guidelines for the syntax shown in Figure 9–30:

- The DROP clause will vary depending on the type of constraint being deleted. If the DROP clause references the PRIMARY KEY constraint for the table, using the keywords PRIMARY KEY is sufficient because only one such clause is allowed for each table in the database.
- If a constraint is a UNIQUE constraint, then only the column name affected by the constraint is required because a column is only referenced by one UNIQUE constraint.
- Any other type of constraint must be referenced by the constraint's actual name—regardless of whether the constraint name is assigned by a user or the Oracle server.

Figure 9–31 shows the user dropping a NOT NULL constraint.

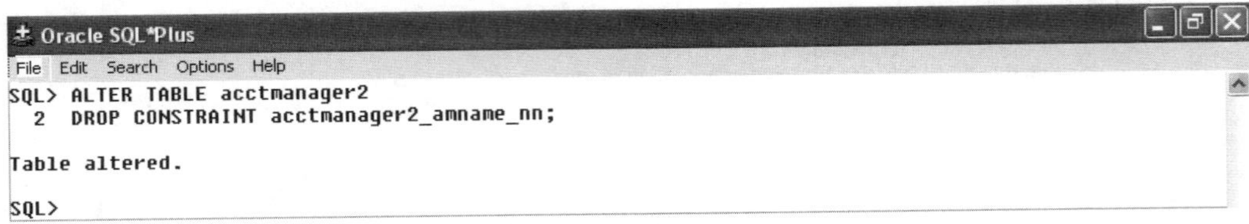

Figure 9-31 Dropping a NOT NULL constraint

In Figure 9–31, the ALTER TABLE command is issued to delete the NOT NULL constraint for the Amname column of the ACCTMANAGER2 table. If the command is executed successfully, the constraint will no longer exist, and the column will be allowed to contain NULL values.

Chapter Summary

- A constraint is a rule that is applied to data being added to a table. The constraint represents business rules, policies, and/or procedures. Data violating the constraint will not be added to the table.
- A constraint can be included during table creation as part of the CREATE TABLE command or added to an existing table, using the ALTER TABLE command.
- A constraint that is based on composite columns (i.e., consists of more than one column) must be created using the table-level approach.

- A NOT NULL constraint can only be created using the column-level approach.
- A PRIMARY KEY constraint does not allow duplicate or NULL values in the designated column.
- Only one PRIMARY KEY constraint is allowed in a table.
- A FOREIGN KEY constraint requires that the column entry match a referenced column entry in the referenced table or be NULL.
- A UNIQUE constraint is similar to a PRIMARY KEY constraint except it will allow NULL values to be stored in the specified column.
- A CHECK constraint ensures that the data meet a given condition before they are added to the table. The condition cannot reference the SYSDATE function or values stored in other rows.
- A NOT NULL constraint is a special type of CHECK constraint. It can only be added to a column using the CREATE TABLE command or the MODIFY clause of the ALTER TABLE command.
- A constraint can be disabled or enabled using the ALTER TABLE command and the DISABLE and ENABLE keywords.
- A constraint cannot be modified. To change a constraint, the constraint must first be dropped with the DROP command and then re-created.

9

Chapter 9 Syntax Summary

The following table presents a summary of the syntax that you have learned in this chapter. You can use the table as a study guide and reference.

Syntax Guide		
Constraint	**Description**	**Example**
PRIMARY KEY	Determines which column(s) uniquely identifies each record. The primary key cannot be NULL, and the data value(s) must be unique.	*Constraint created during table creation:* **CREATE TABLE newtable** **(firstcol NUMBER PRIMARY KEY,** **secondcol VARCHAR2(20));** *or* **CREATE TABLE newtable** **(firstcol NUMBER,** **secondcol VARCHAR2(20),** **PRIMARY KEY (firstcol));** *Constraint created after table creation:* **ALTER TABLE newtable** **ADD PRIMARY KEY (firstcol);**

Constraint	Description	Example

FOREIGN KEY	In a one-to-many relationship, the constraint is added to the "many" table. The constraint ensures that if a value is entered to the specified column, it must exist in the table being referred to, or the row is not added.	*Constraint created during table creation:*
		`CREATE TABLE newtable`
		`(firstcol NUMBER,`
		`secondcol VARCHAR2(20) REFERENCES`
		`anothertable(col1));`
		or
		`CREATE TABLE newtable`
		`(firstcol NUMBER,`
		`secondcol VARCHAR2(20),`
		`FOREIGN KEY (secondcol) REFERENCES`
		`anothertable(col1);`
		
		Constraint created after table creation:
		`ALTER TABLE newtable`
		`ADD FOREIGN KEY (secondcol) REFERENCES`
		`anothertable (col1);`

UNIQUE	Ensures that all data values stored in the specified column must be unique. The UNIQUE constraint differs from the PRIMARY KEY constraint in that it allows NULL values.	*Constraint created during table creation:*
		`CREATE TABLE newtable`
		`(firstcol NUMBER,`
		`secondcol VARCHAR2(20) UNIQUE);`
		or
		`CREATE TABLE newtable`
		`(firstcol NUMBER,`
		`secondcol VARCHAR2(20),`
		`UNIQUE (secondcol));`
		
		Constraint created after table creation:
		`ALTER TABLE newtable`
		`ADD UNIQUE (secondcol);`

CHECK	Ensures that a specified condition is TRUE before the data value is added to the table. For example, an order's ship date cannot be "less than" its order date.	*Constraint created during table creation:*
		`CREATE TABLE newtable`
		`(firstcol NUMBER,`
		`secondcol VARCHAR2(20),`
		`thirdcol NUMBER CHECK (BETWEEN 20`
		`AND 30));`
		or
		`CREATE TABLE newtable`
		`(firstcol NUMBER,`
		`secondcol VARCHAR2(20),`
		`thirdcol NUMBER,`
		`CHECK (thirdcol BETWEEN 20 AND 80));`
		
		Constraint created after table creation:
		`ALTER TABLE newtable`
		`ADD CHECK (thirdcol BETWEEN 20`
		`AND 80);`

Constraint	Description	Example

NOT NULL	Requires that the specified column cannot contain a NULL value. It can *only* be created with the column-level approach to table creation.	*Constraint created during table creation:*
		`CREATE TABLE newtable`
		`(firstcol NUMBER,`
		`secondcol VARCHAR2(20),`
		`thirdcol NUMBER NOT NULL);`
		
		Constraint created after table creation:
		`ALTER TABLE newtable`
		`MODIFY (thirdcol NOT NULL);`

REVIEW QUESTIONS

To answer these questions, refer to the tables in Appendix A.

1. What is the difference between a PRIMARY KEY constraint and a UNIQUE constraint?
2. All constraints are enforced at what level?
3. A table is allowed a maximum of how many PRIMARY KEY constraints?
4. Which type of constraint can be used to make certain the category for a book is included when a new book is added to inventory?
5. Which type of constraint would be required to ensure that every book has a profit margin between 15 percent and 25 percent?
6. How is adding a NOT NULL constraint to an existing table different from adding some other type of constraint?
7. When are you required to define constraints at the table level rather than at the column level?
8. To which table do you add a FOREIGN KEY constraint if you want to make certain every book ordered is actually contained in the BOOKS table?
9. What is the difference between disabling a constraint and dropping a constraint?
10. What is the simplest way to determine whether a particular column can contain NULL values?

Multiple Choice

To answer the following questions, refer to the tables in Appendix A.

1. Which of the following statements is correct?
   a. A PRIMARY KEY constraint will allow NULL values in the primary key column(s).
   b. A dropped constraint can be enabled if it is needed in the future.
   c. Every table must have at least one PRIMARY KEY constraint, or Oracle9*i* will not allow the table to be created.
   d. none of the above
2. Which of the following is not a valid constraint type?
   a. PRIMARY KEYS
   b. UNIQUE
   c. CHECK
   d. FOREIGN KEY
3. Which of the following SQL statements is invalid and will return an error message?
   a.
   ```
   ALTER TABLE books
   ADD CONSTRAINT books_pubid_uk UNIQUE (pubid);
   ```
   b.
   ```
   ALTER TABLE books
   ADD CONSTRAINT books_pubid_pk PRIMARY KEY (pubid);
   ```
   c.
   ```
   ALTER TABLE books
   ADD CONSTRAINT books_pubid_nn NOT NULL (pubid);
   ```
   d.
   ```
   ALTER TABLE books
   ADD CONSTRAINT books_pubid_fk FOREIGN KEY (pubid)
   REFERENCES publisher (pubid);
   ```
   e. all of the above
4. What is the maximum number of PRIMARY KEY constraints allowed for a table?
   a. 1
   b. 2
   c. 30
   d. 255

5. Which of the following is a valid SQL command?
   a.
   ```
   ALTER TABLE books
   ADD CONSTRAINT UNIQUE (pubid);
   ```
   b.
   ```
   ALTER TABLE books
   ADD CONSTRAINT PRIMARY KEY (pubid);
   ```
   c.
   ```
   ALTER TABLE books
   MODIFY (pubid CONSTRAINT NOT NULL);
   ```
   d.
   ```
   ALTER TABLE books
   ADD FOREIGN KEY CONSTRAINT (pubid)
   REFERENCES publisher (pubid);
   ```
   e. none of the above
6. How many NOT NULL constraints can be created at the table level, using the CREATE TABLE command?
   a. 0
   b. 1
   c. 12
   d. 30
   e. 255
7. The FOREIGN KEY constraint should be added to which table?
   a. the table representing the "one" side of a one-to-many relationship
   b. the parent table in a parent-child relationship
   c. the child table in a parent-child relationship
   d. the table that does not have a primary key
8. What is the maximum number of columns that can be defined as a primary key, using the column-level approach when creating a table?
   a. 0
   b. 1
   c. 30
   d. 255
9. Which of the following commands can be used to rename a constraint?
   a. RENAME
   b. ALTER CONSTRAINT
   c. MOVE
   d. NEW NAME
   e. none of the above

10. Which of the following is a valid SQL statement?

a.
```
CREATE TABLE table1
(col1       NUMBER PRIMARY KEY
col2        VARCHAR2(20) PRIMARY KEY,
col3        DATE        DEFAULT SYSDATE,
col4        VARCHAR2(2));
```

b.
```
CREATE TABLE table1
(col1       NUMBER PRIMARY KEY
col2        VARCHAR2(20),
col3        DATE,
col4        VARCHAR2(2) NOT NULL,
CONSTRAINT table1_col3_ck CHECK (col3 = SYSDATE));
```

c.
```
CREATE TABLE table1
(col1       NUMBER,
col2        VARCHAR2(20),
col3        DATE,
col4        VARCHAR2(2),
PRIMARY KEY (col1));
```

d.
```
CREATE TABLE table1
(col1       NUMBER,
col2        VARCHAR2(20),
col3        DATE        DEFAULT SYSDATE,
col4        VARCHAR2(2);
```

11. If a UNIQUE constraint is being created for a composite column which requires that the combination of entries in the specified columns be unique, which of the following statements is correct?

a. The constraint can only be created using the ALTER TABLE command.

b. The constraint can only be created using the table-level approach.

c. The constraint can only be created using the column-level approach.

d. The constraint can only be created using the ALTER TABLE...MODIFY command.

12. Which of the following types of constraints will only allow a record to be added to a table if it meets a specified condition?

a. PRIMARY KEY

b. UNIQUE

c. CHECK

d. NOT NULL

13. Which of the following commands can be used to enable a disabled constraint?
    a. ALTER TABLE...MODIFY
    b. ALTER TABLE...ADD
    c. ALTER TABLE...DISABLE
    d. ALTER TABLE...ENABLE
14. Which of the following keywords will allow the user to delete a record from a table even if rows in another table are referencing the record through a FOREIGN KEY constraint?
    a. CASCADE
    b. CASCADE ON DELETE
    c. DELETE ON CASCADE
    d. DROP
    e. ON DELETE CASCADE
15. Which of the following queries will display only the name of the PRIMARY KEY constraint for TABLEA?
    a. `SELECT constraint_name FROM user_constraints WHERE C = P;`
    b. `SELECT constraint_name FROM user_constraints WHERE type = P;`
    c. `SELECT constraint_name FROM user_constraints WHERE constraint_type = 'P' AND table_name = 'TABLEA';`
    d. `SELECT constraint_name FROM user_constraints WHERE constraint_type = 'p' AND table_name = 'tablea';`
    e. none of the above
16. Which of the following types of constraints cannot be created at the table level?
    a. NOT NULL
    b. PRIMARY KEY
    c. CHECK
    d. FOREIGN KEY
    e. None of the above can be created at the table level.

17. Suppose that you created a PRIMARY KEY constraint at the same time you created a table and later decide to add a name to the constraint. Which of the following commands can you use to change the name of the constraint?
    a. ALTER TABLE...MODIFY
    b. ALTER TABLE...ADD
    c. ALTER TABLE...DISABLE
    d. none of the above
18. You are creating a new table consisting of three columns: col1, col2, and col3. The column called col1 should be the primary key and cannot have any NULL values, and each entry should be unique. The column labeled col3 must not contain any NULL values either. How many total constraints will you explicitly be required to create?
    a. 1
    b. 2
    c. 3
    d. 4
19. Which of the following types of restrictions can be viewed through the DESCRIBE command?
    a. NOT NULL
    b. FOREIGN KEY
    c. UNIQUE
    d. CHECK
20. Which of the following is the valid syntax for adding a PRIMARY KEY constraint to an existing table?
    a.
    ```
    ALTER TABLE tablename
    ADD CONSTRAINT PRIMARY KEY (columnname);
    ```
    b.
    ```
    ALTER TABLE tablename
    ADD CONSTRAINT (columnname)
    PRIMARY KEY constraintname;
    ```
    c.
    ```
    ALTER TABLE tablename
    ADD [CONSTRAINT constraintname]
    PRIMARY KEY;
    ```
    d. none of the above

Hands-on Assignments

To perform these activities, refer to the tables in Appendix A.

1. You are about to create a table and remember that one of the columns has to include a NOT NULL constraint. Modify the following SQL command so the column named colC cannot contain a NULL value, and then create the table.

```
CREATE TABLE homework
(colA        NUMBER,
colB         VARCHAR2(20),
colC         DATE DEFAULT SYSDATE,
colD         VARCHAR2(2));
```

2. Change the HOMEWORK table so colA and colB together are the primary key for the table.
3. Add a column to the HOMEWORK table with the name colE. This column should only be allowed to contain numeric values between 3 and 8. Name the constraint **homework_colE_ck**.
4. Change the HOMEWORK table so colB can only contain values that are stored in the Order# column of the ORDERS table. Make certain that after you make the change, colB is still part of the composite primary key created in Step 2.
5. Issue a SELECT statement that will allow you to view the conditions for all the CHECK constraints on the HOMEWORK table.
6. Add another column to the HOMEWORK table. This column should be named colF and be able to store dates. Although the column is not part of the primary key, it should not be allowed to contain any duplicate or NULL values.
7. Disable the constraint named **homework_colE_ck**.
8. You have decided that the primary key for the HOMEWORK table should only consist of colA. Make the changes needed to make colA the only column used for the table's primary key.
9. Enable the constraint named **homework_colE_ck**.
10. Change the constraint on colB that references the ORDERS table so that if an order is deleted from the ORDERS table, any entries in the HOMEWORK table will automatically be deleted also.

A Case for Oracle9*i*

To perform this activity, refer to the tables in Appendix A.

Now that you have received training in constraints, your supervisor asks you to examine the CUSTOMERS, ORDERS, ORDERITEMS, BOOKS, PUBLISHER, and BOOKAUTHOR tables to determine whether additional constraints should be created for any of the tables. Currently, most of the tables have only PRIMARY KEY constraints. However, due to the relationships that exist among the tables, you recognize that at least a few FOREIGN KEY constraints are needed. There are also several columns that should not be allowed to contain NULL values.

You need to create a memo for your supervisor that identifies the constraints that should be added to the database and your rationale for each constraint. The memo should mention at least six constraints and provide the correct SQL command to add each constraint to the existing tables.

CHAPTER

10

DATA MANIPULATION

Objectives

After completing this chapter, you should be able to do the following:

- Add a record to an existing table
- Add a record containing a NULL value to an existing table
- Use a subquery to copy records from an existing table
- Modify the existing rows within a table
- Use substitution variables with an UPDATE command
- Issue the transaction control statements COMMIT and ROLLBACK
- Differentiate among DDL, DML, and transaction control commands
- Delete records
- Differentiate between a shared lock and an exclusive lock
- Use the SELECT...FOR UPDATE command to create a shared lock

In previous chapters, you learned how to create tables and how to add constraints to ensure the integrity of the data stored within those tables. The next step is to learn how to add, or insert, new rows into tables. This chapter will address how to add new rows and how to modify or delete existing rows. Figure 10–1 provides an overview of this chapter's contents.

Command	Description
INSERT	Adds new row(s) to a table. The user can include a subquery to copy row(s) from an existing table.
UPDATE	Adds data to, or modifies data within, an existing row
COMMIT	Permanently saves changed data in a table
ROLLBACK	Allows the user to "undo" uncommitted changes to data
DELETE	Removes row(s) from a table
LOCK TABLE	Prevents other users from making changes to a table
SELECT...FOR UPDATE	Creates a shared lock on a table to prevent another user from making changes to data in specified columns

Figure 10-1 Overview of chapter contents

Go to the Chapter10 folder in your Data Files. Before working through the examples in this chapter, run the script **prech10.sql** to ensure that all necessary tables and constraints are available. To run the script, type **`start d:\prech10.sql`** at the SQL> prompt, where **`d:`** represents the location (correct drive and path) of the script. If the script is not available, ask your instructor to provide you with a copy.

INSERTING NEW ROWS

As discussed in previous chapters, the management of JustLee Books is implementing a new commission policy for regional account managers. The ACCTMANAGER table was created to store data about account managers. Now that the table has been created and all necessary constraints for the table are in place, it is time to add data to the table. Recall that the table has the column headings shown in Figure 10–2. The data shown in Figure 10–2 is for those account managers who need to be added to the ACCTMANAGER table. (Blank spaces indicate that data have not yet been provided to the data-entry clerk.)

ID	Name	Employment Date	Region
T500	Nick Taylor	September 5, 2002	NE
L500	Mandy Lopez	October 1, 2002	
J500	Sammie Jones		NW
	Martin Giovanni	today	SW

Figure 10-2 Data for account managers

INSERT Command

You can add rows to a new or existing table with the **INSERT** command. The syntax for the INSERT command is shown in Figure 10–3.

```
INSERT INTO tablename [(columnname, …)]
VALUES (datavalue, …);
```

Figure 10-3 Syntax of the INSERT command

Note the following syntax elements in Figure 10–3:

- The keywords **INSERT INTO** are followed by the name of the table into which the rows will be entered. The table name is followed by the names of the columns that will contain the data.
- The VALUES clause identifies the data values that will be inserted into the table. The actual data are listed within parentheses after the VALUES keyword.
- If the data entered in the VALUES clause are in the same order as the columns in the table, column names can be omitted in the INSERT INTO clause. However, if you only enter data for some of the columns, or if columns are listed in a different order than they are listed in the table, the names of the columns *must* be provided in the INSERT INTO clause in the same order as they will be given in the VALUES clause. The column names must be listed within a set of parentheses after the table name in the INSERT INTO clause.
- If more than one column is listed, column names must be separated by commas.
- If more than one data value is entered, they must be separated by commas.
- As when using data values for a search condition in the WHERE clause of a SELECT statement, data inserted into a column defined for non-numeric data (i.e., the column's datatype is not NUMBER) must be enclosed in single quotation marks.

You can review the order of the columns in a table by using the `DESCRIBE` *`tablename`* command.

10

To insert the first account manager's data (Figure 10–2) into the ACCTMANAGER table, use the command shown in Figure 10–4.

```
INSERT INTO acctmanager
VALUES ('T500', 'NICK TAYLOR', '05-SEP-02', 'NE');
```

Figure 10-4 The INSERT INTO command

The INSERT INTO clause given in Figure 10–4 does not contain a list of the column names because the VALUES clause contains a valid entry for every column in the ACCTMANAGER table, and the data are given in the same order as the columns are listed in the table.

The character data in the VALUES clause are entered in all upper-case characters, so the data will be stored in the tables in the same case that is used for all the tables in the JustLee Books database. When character data are entered in a table, they will retain the case given by the user issuing the INSERT INTO command. For example, if the name of an account manager had been entered in mixed case (i.e., upper- and lower-case letters), the table would have stored the name in mixed case. This can make future record searches difficult if the data do not follow a consistent format or case. Figure 10–5 shows that the first account manager, Nick Taylor, has been added to the ACCTMANAGER table.

```
Oracle SQL*Plus
File Edit Search Options Help
SQL> INSERT INTO acctmanager
  2  VALUES ('T500', 'NICK TAYLOR', '05-SEP-02', 'NE');

1 row created.

SQL> SELECT * FROM acctmanager;

AMID AMNAME               AMEDATE   RE
---- -------------------- --------- --
T500 NICK TAYLOR          05-SEP-02 NE

SQL>
```

Figure 10-5 ACCTMANAGER table with inserted data

After the command in Figure 10–4 is issued and executed, the message "1 row created" will appear, indicating that the data have been inserted into the ACCTMANAGER table. To verify that the row was added, a SELECT statement can be used to view the table's contents.

When inserting table data, the most common error is forgetting to enclose data for non-numeric columns within single quotation marks. When this occurs, Oracle9*i* displays the following message:

```
ERROR:
ORA-01756: quoted string not properly terminated
```

You can correct the problem by simply reissuing the command with the required quotation marks.

The next record to be entered into the ACCTMANAGER table contains data about Mandy Lopez. However, as shown in Figure 10–2, she has not yet been assigned a marketing region, so that column will be left blank. There are several approaches you can take to indicate that the Region column will contain a NULL value when the row is added to the table:

1. List all the columns *except* the Region column in the INSERT INTO clause, and provide the data for the listed columns in the VALUES clause.
2. In the VALUES clause, substitute *two single quotation marks* in the position that should contain the account manager's assigned region. Oracle9*i* will interpret the quotation marks to mean that a NULL value should be stored in that column.
3. In the VALUES clause, include the keyword NULL in the position where the region should be listed. As long as the keyword NULL is not enclosed in single quotation marks, Oracle9*i* will leave the column blank. However, if the keyword is mistakenly entered as 'NULL,' the software will actually try to store the word *NULL* in the column.

Use the third approach to enter the data for Mandy Lopez in the ACCTMANAGER table. The INSERT INTO command is shown in Figure 10–6.


```
INSERT INTO acctmanager
VALUES ('L500', 'MANDY LOPEZ', '01-OCT-02', NULL);
```

Figure 10-6 Command to enter data into a table

Although the command given in Figure 10–6 is correct in terms of its syntax, Oracle9*i* will not add the record to the ACCTMANAGER table. The problem is that the table was created with a NOT NULL constraint for the Region column. This is shown in Figure 10–7. As a result, Oracle9*i* will not insert the row into the table until a value is entered for the Region column.

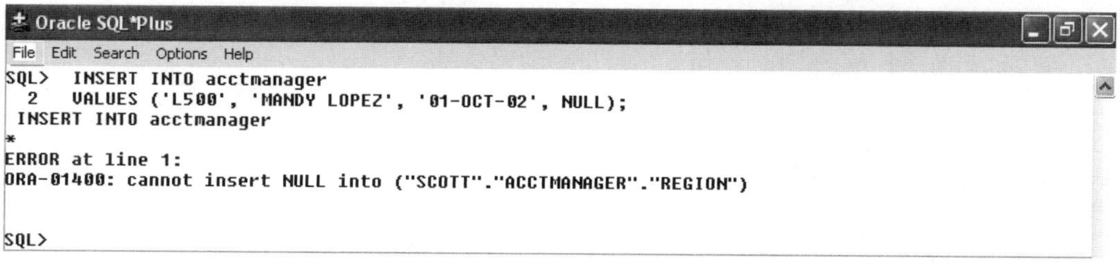

```
Oracle SQL*Plus
File Edit Search Options Help
SQL> INSERT INTO acctmanager
  2  VALUES ('L500', 'MANDY LOPEZ', '01-OCT-02', NULL);
INSERT INTO acctmanager
*
ERROR at line 1:
ORA-01400: cannot insert NULL into ("SCOTT"."ACCTMANAGER"."REGION")


SQL>
```

Figure 10-7 Adding a NULL value to a column with a NOT NULL constraint

As shown in Figure 10–7, when the command from Figure 10–6 is executed, an error message is returned, indicating that a NULL value cannot be inserted into the Region column of the ACCTMANAGER table belonging to the user named Scott, who is the individual who "owns" the table. Although the error message does not explicitly state that the command violates a NOT NULL constraint, the message displayed should, at least, lead you to examine the structure of the table. You can use the DESCRIBE command to determine whether the NULL? column indicates that nulls are not allowed.

At this point, there are two options for adding Mandy Lopez's data to the table. The first option is to disable the constraint added in the previous chapter, so NULL values are allowed in the Region table. However, constraints are added to a column to ensure that the contents of the table are valid and useful. If a NULL value is allowed, then what would be the point of having an account manager's name in the table if the name of the manager's region is not given? Another option is to e-mail or call the Personnel Department and ask whether Mandy Lopez has been assigned a region. Suppose that you do call, and you discover that an error was made: Lopez had been assigned marketing responsibilities for the Southeast (SE) region.

Once the correct region has been added to the command in Figure 10–6, Oracle9*i* will add Mandy Lopez to the ACCTMANAGER table, as shown in Figure 10–8.

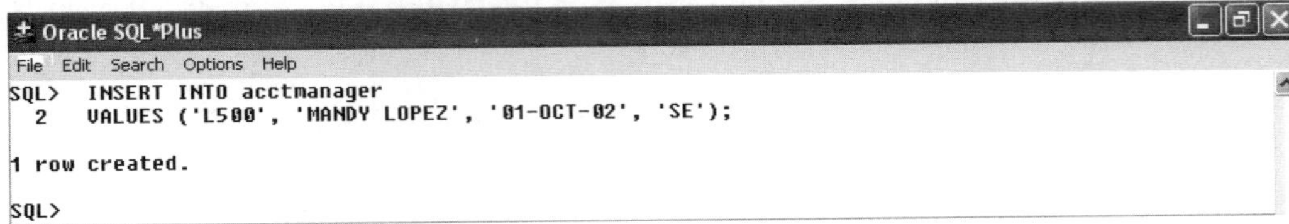

Figure 10-8 Command for insertion of Mandy Lopez's data into the ACCTMANAGER table

Next, the record for Sammie Jones must be entered into the ACCTMANAGER table. However, at the time the account managers data were provided, his actual employment date had not been determined. Luckily, however, a NOT NULL constraint does not exist for the Amedate column. To instruct the software that a value will not be entered for Sammie Jones' employment date, a column list can be provided in the INSERT INTO clause that omits the Amedate column. Using this approach, the command to enter Sammie Jones' data is shown in Figure 10–9.

```
INSERT INTO acctmanager (amid, amname, region)
VALUES ('J500', 'Sammie Jones', 'NW');
```

Figure 10-9 Command to add data to specific columns

The column list provided in the INSERT INTO clause lists all the columns of the ACCTMANAGER table except the Amedate column. Although the columns are listed in the same sequence as they appear in the actual table, this is not a requirement. What is required is that the data listed in the VALUES clause must match the *exact order* of the columns listed in the INSERT INTO clause. The order of the list is how Oracle9*i* is able to determine which data value belongs to which column. This is shown in Figure 10–10.

```
Oracle SQL*Plus
File  Edit  Search  Options  Help
SQL> INSERT INTO acctmanager (amid, amname, region)
  2  VALUES ('J500', 'Sammie Jones', 'NW');

1 row created.

SQL>  SELECT * FROM acctmanager;

AMID AMNAME                AMEDATE   RE
---- -------------------- --------- --
T500 NICK TAYLOR          05-SEP-02 NE
L500 MANDY LOPEZ          01-OCT-02 SE
J500 Sammie Jones         06-MAR-03 NW

SQL>
```

Figure 10-10 Flawed output: list of the contents of the ACCTMANAGER table

A display of the current contents of the ACCTMANAGER table shows that after the command is successfully executed, Sammie Jones is now included in the table. However, from the user's perspective, there appear to be two problems with the data. The first problem is that although no date is specified for Sammie Jones' employment date, the Amedate column has an entry. This occurs because the Amedate column had previously been assigned the default value of SYSDATE. Because no entry was made into the Amedate column, the system date of the server was assigned as the employment date. This problem must be corrected.

The actual date displayed for the system date on your computer will not be the same as the date displayed in Figure 10–10. The date stored in the Amedate column will depend upon the current date set on the computer you are using.

The second problem is that Sammie Jones' name is in mixed-case letters; however, the names of the other account managers are in upper-case letters. Although the data are technically correct, the inconsistency may present a problem in the future. For example, if someone uses the Amname column as part of a search condition and does not realize that the name for this particular account manager is stored in mixed case, then Jones' record will not be returned in query results. Therefore, this problem must be corrected before the table goes "live" and users are allowed to access the data.

In industry, a table that has moved out of the testing phase and is available as part of an online database system is commonly referred to by Oracle9*i* as a "production" table. Do not confuse the word *production*, meaning a "live" system, with the word *Production*, referring to a department within an organization. This is a common misinterpretation by novices.

The last account manager to be added to the ACCTMANAGER table is a new employee, Martin Giovanni. He has been assigned marketing activities for the Southwest (SW) region. Because he was hired today, he has not yet been assigned an ID number for the system. When entering his record into the ACCTMANAGER table, you have the following options:

- You can manually enter his employment date as today's date.
- You can use the SYSDATE keyword to instruct the software to substitute the computer's date as the employment date.
- In terms of the ACCTMANAGER table created in Chapter 9, you can choose not to make an entry into the column, because the default value for the column is SYSDATE.

To ensure that Giovanni's employment date is entered as the system date (in case someone has modified the column and removed the default date after encountering problems entering the record for Sammie Jones), you might enter the command shown in Figure 10–11.

```
INSERT INTO acctmanager (amname, amedate, region)
VALUES ('Martin Giovanni', SYSDATE, 'SW');
```

Figure 10-11 Flawed command: adding a new account manager with no ID to the ACCTMANAGER table

Because the command given in Figure 10–11 does not specify the Amid column in the INSERT INTO clause, Oracle9*i* will skip over this column and assign it a NULL value. The name of the new employee will then be entered into the Amname column. The command next assigns the computer's date to the Amedate column, and finally the value SW to the Region column.

When Oracle9*i* attempts to execute the command given in Figure 10–11, the values to be entered are compared against all existing constraints. The Amid column has a PRIMARY KEY constraint. *Because a column designated in a PRIMARY KEY constraint cannot contain NULL values, an error message is returned*, and the row is not added to the table. However, once the ID for the new employee is determined, the command can be reissued. Martin Giovanni has been assigned an ID number of D500, and it is entered into the system. The original command, revised command, and output are shown in Figure 10–12.

Since the data being entered into the table now meet all the enabled constraints, the row will be added to the table, as shown in Figure 10–13.

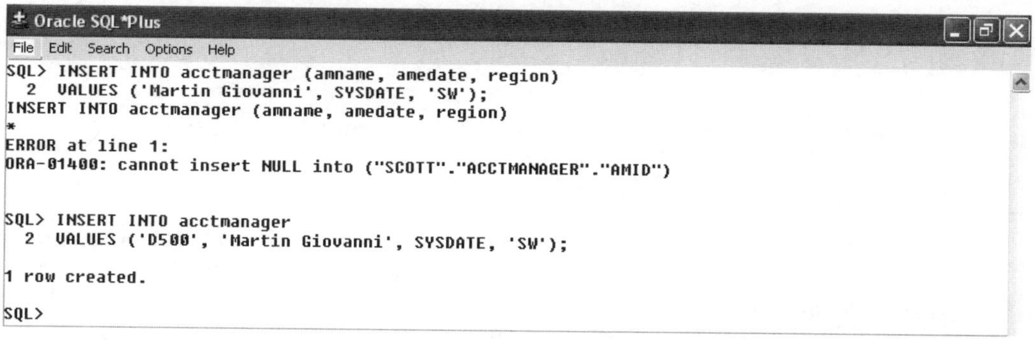

```
File  Edit  Search  Options  Help
SQL> INSERT INTO acctmanager (amname, amedate, region)
  2  VALUES ('Martin Giovanni', SYSDATE, 'SW');
INSERT INTO acctmanager (amname, amedate, region)
*
ERROR at line 1:
ORA-01400: cannot insert NULL into ("SCOTT"."ACCTMANAGER"."AMID")


SQL> INSERT INTO acctmanager
  2  VALUES ('D500', 'Martin Giovanni', SYSDATE, 'SW');

1 row created.

SQL>
```

Figure 10-12 Insertion of a new employee into the ACCTMANAGER table: original flawed command followed by correction

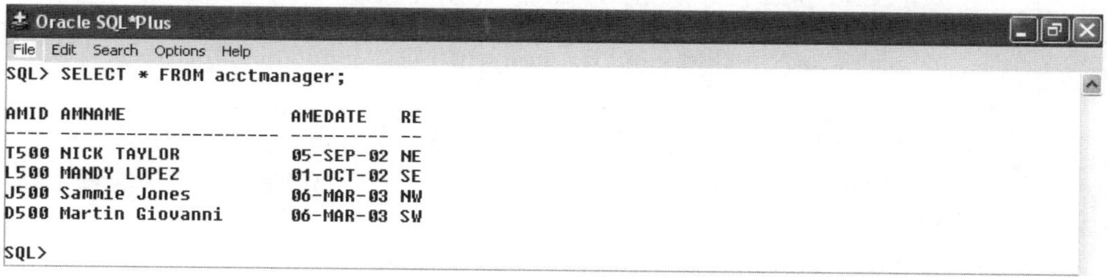

```
File  Edit  Search  Options  Help
SQL> SELECT * FROM acctmanager;

AMID AMNAME                AMEDATE   RE
---- -------------------- --------- --
T500 NICK TAYLOR           05-SEP-02 NE
L500 MANDY LOPEZ           01-OCT-02 SE
J500 Sammie Jones          06-MAR-03 NW
D500 Martin Giovanni       06-MAR-03 SW

SQL>
```

Figure 10-13 Records contained in the ACCTMANAGER table

A review of the records now stored in the ACCTMANAGER table shows that once again, the name of an account manager was entered in mixed-case rather than upper-case letters. Changing the names of two employees to all upper-case letters (and correcting the employment date for Sammie Jones) will require the user to modify the existing data, using the UPDATE command, which will be introduced later in this chapter.

Inserting Data from an Existing Table

In Chapter 8, you learned how to use the CREATE TABLE command with a subquery to create and populate a new table, based upon the structure and content of an existing table. However, what if a table already exists, and you need to add to it copies of existing records that are contained in another table? In that case, you would not be able to use the CREATE TABLE command. Because the table already exists, you would need to use the INSERT INTO command with a subquery. The syntax for combining an INSERT INTO command with a subquery is shown in Figure 10–14.

10

```
INSERT INTO tablename [(columnname, …)]
subquery;
```

Figure 10-14 Syntax for the INSERT INTO command with a subquery

Note the following elements in Figure 10–14:

- The main difference between using the INSERT INTO command with actual data values and with a subquery is that *the VALUES clause is not included when the command is used with a subquery*. The keyword VALUES indicates that the clause contains actual data values that must be inserted into the indicated table. However, there are no actual data values being entered by the user—the data are derived from the results of the subquery.
- Also, note that unlike several other commands, the INSERT INTO command does not require the subquery to be enclosed within a set of parentheses, although including parentheses will not generate an error message.

Part of JustLee Books' new commission policy is that an account manager will begin receiving a sales commission after six months of employment. Thus, the account managers who are eligible to receive a commission for orders placed after March 31, 2003, must be identified.

In Chapter 8, the ACCTMANAGER2 table was created. In this chapter, the account manager commission data must be copied into the ACCTMANAGER2 table. By copying the relevant data into the second table, SQL statements from the Payroll Department's application program that calculates commissions can be tested. This allows the Information Technology Department to have the flexibility of making changes to the original ACCTMANAGER table—without interfering with Payroll's application program testing. The command to copy the relevant rows from the ACCTMANAGER table into the ACCTMANAGER2 table is shown in Figure 10–15.

```
INSERT INTO acctmanager2
     SELECT amid, amname, amedate, region
     FROM acctmanager
     WHERE amedate <= '01-OCT-02';
```

Figure 10-15 INSERT INTO command with a subquery

Note the following elements in Figure 10–15:

- The SELECT clause of the subquery lists the columns to be copied from the ACCTMANAGER table, which is identified in the FROM clause. In this example, the INSERT INTO clause does not contain a column list because the columns returned by the subquery are in the same order as the columns in the ACCTMANAGER2 table.

- The WHERE clause provides the condition that an account manager must have been employed by JustLee Books before October 1, 2002, to receive a commission for any orders placed in April 2003.

The query in Figure 10–15 and its output are shown in Figure 10–16.

```
SQL> INSERT INTO acctmanager2
  2             SELECT amid, amname, amedate, region
  3             FROM acctmanager
  4             WHERE amedate <= '01-OCT-02';

2 rows created.

SQL> SELECT * FROM acctmanager2;

AMID AMNAME               AMEDATE   RE
---- -------------------- --------- --
T500 NICK TAYLOR          05-SEP-02 NE
L500 MANDY LOPEZ          01-OCT-02 SE

SQL>
```

Figure 10-16 Results of the INSERT INTO command with a subquery

After the command has been executed, the ACCTMANAGER2 table can be queried to determine which account managers are eligible to receive a commission for orders placed in April. As shown in the results in Figure 10–16, there are currently two eligible account managers, Nick Taylor and Mandy Lopez. Results may vary, depending on your computer's system date.

10

Modifying Existing Rows

There will be many times when record data need to be changed. For example, when customers move, their mailing addresses must be updated; when the wholesale cost of books changes, retail prices must be changed, etc. Because the INSERT INTO command can only be used to add new rows to a table, it cannot modify existing data. To alter existing table data, the UPDATE command must be used. In this section, you will also learn how to perform updates by creating interactive scripts through substitution variables.

UPDATE command

The contents of existing rows can be changed using the **UPDATE** command. The syntax for the UPDATE command is shown in Figure 10–17.

```
UPDATE tablename
SET columnname = new_datavalue
[WHERE condition];
```

Figure 10-17 Syntax for the UPDATE command

Note the following elements in Figure 10–17:

- The UPDATE clause identifies the table containing the record(s) to be changed.
- The **SET** clause is used to identify the column to be changed and the new value to be assigned to that column.
- The optional WHERE clause identifies the exact row to be changed by the UPDATE command. If the WHERE clause is omitted, then the column specified in the SET clause will be updated for *all records* contained in the table.

Next, let's fix the problems shown in the ACCTMANAGER table in Figure 10–13. First, let's correct the incorrect employment date for Sammie Jones. After a few inquires, the correct employment date is found to be January 12, 2003. Therefore, the command shown in Figure 10–18 can be issued.

```
UPDATE acctmanager
SET amedate =
     TO_DATE('JANUARY 12, 2003', 'MONTH DD, YYYY')
WHERE amid = 'J500';
```

Figure 10-18 Command to correct the Amedate column entry

Notice the SET clause in Figure 10–18. In the previous commands given in this chapter, the format of the employment date has been altered to match the default date format used by Oracle9*i*. For this example, suppose that it is absolutely imperative that the date be entered using the format MONTH DD, YYYY. (This problem usually arises if data are entered from a third-party application program.) However, if a date value is not entered in the default format DD-MON-YY, Oracle 9*i* will return an error message, and the data will not be added to the table. To allow Oracle9*i* to distinguish the correct portions of the date provided, the TO_DATE function is included when entering the date in the SET clause. Although the date is entered using the MONTH DD, YYYY format, upon execution of the command, Oracle9*i* will automatically convert the date to the default format of DD-MON-YY for internal storage.

The WHERE clause is used to identify exactly which record should be altered. In this case, the easiest way to specify that only the employment date for Sammie Jones should be changed is to include a WHERE clause with the condition that the Amid column must be equal to J500. Because the Amid column is the primary key for the ACCTMANAGER table, no two records can have the same Amid, and only the record for Sammie Jones will be affected by the update. The query and output are shown in Figure 10–19.

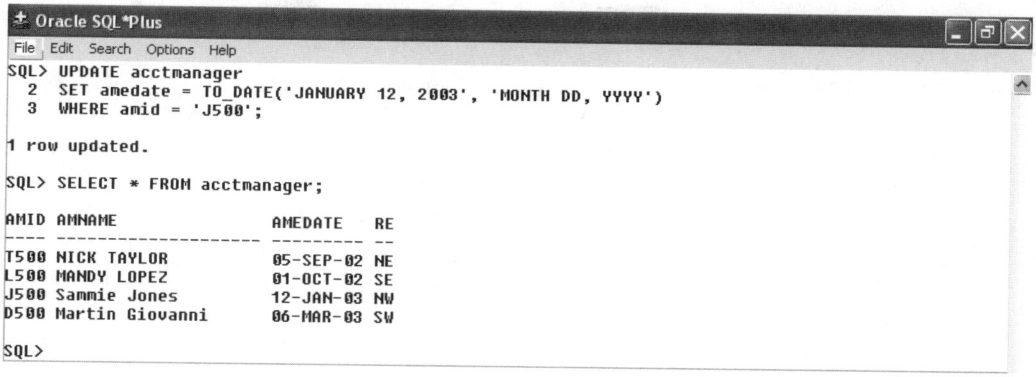

```
Oracle SQL*Plus
File Edit Search Options Help
SQL> UPDATE acctmanager
  2  SET amedate = TO_DATE('JANUARY 12, 2003', 'MONTH DD, YYYY')
  3  WHERE amid = 'J500';

1 row updated.

SQL> SELECT * FROM acctmanager;

AMID AMNAME               AMEDATE   RE
---- -------------------- --------- --
T500 NICK TAYLOR          05-SEP-02 NE
L500 MANDY LOPEZ          01-OCT-02 SE
J500 Sammie Jones         12-JAN-03 NW
D500 Martin Giovanni      06-MAR-03 SW

SQL>
```

Figure 10-19 Correct employment date entered into the ACCTMANAGER table

Next, let's convert the mixed-case letters of the account managers' names to upper-case letters. One approach is to issue a simple UPDATE command that takes the current contents of the Amname column and replaces them with the upper-case version of the same data. This can be accomplished by the command shown in Figure 10–20.


```
UPDATE acctmanager
SET amname = UPPER(amname);
```

Figure 10-20 UPDATE command to change the case of the contents of the Amname column

To specifically instruct Oracle9*i* to change the case of the value in the name column for only one account manager, the WHERE clause would be added to specify which record in particular should be modified. The revised command is shown in Figure 10–21.

```
UPDATE acctmanager
SET amname = UPPER(amname)
WHERE amid = 'J500';
```

Figure 10-21 UPDATE command to change the case of the contents of the Amname column for a specific row

When the command is executed, Oracle9*i* will first locate the record containing the value J500 in the Amid column and then take the value stored in the Amname column and convert it to all upper-case characters. To practice using the UPDATE command, try converting the contents of the Amname column for Martin Giovanni to all upper-case characters.

SUBSTITUTION VARIABLES

In some cases, it may seem like a lot of effort just to add a record to a table—and even more effort to modify an existing record. This is especially true if you need to add or modify 10 or 20 different records. For example, the CUSTOMERS table will need to contain data that identifies the marketing region for each customer. After the Region column is added to the CUSTOMERS table, using the ALTER TABLE command, every customer's record will need to be updated with the value for the new column. Depending on the strategy used, the INSERT INTO command will have to be reissued several times—at least once for every identified region. Rather than type the same command again and again for the few values that are different, it is much simpler to use a substitution variable.

A **substitution variable** in a SQL command instructs Oracle9*i* to use a substituted value in place of the variable at the time the command is actually executed. To include a substitution variable in a SQL command, simply enter an ampersand (&) followed by the name to be used for the variable in the necessary location.

First, let's look at the command to modify the records of all customers residing in Massachusetts, so that the Region column contains the value of NE, representing the Northeast region. The command is shown in Figure 10–22.

```
UPDATE customers
SET region = 'NE'
WHERE state = 'MA';
```

Figure 10-22 Command to update the Region column of the CUSTOMERS table

The Region column was added to the CUSTOMERS table when the **prech10.sql** script was run at the beginning of the chapter.

Next, to alter the command given in Figure 10–22 so it can be reused for each state in which JustLee Books has customers, the user needs to enter a substitution variable in place of the value for the State column. Furthermore, the SET clause can contain a substitution variable, so the same command could be used for every region to be entered. The new command is shown in Figure 10–23.

```
UPDATE customers
SET region = '&Region'
WHERE state = '&State';
```

Figure 10-23 UPDATE command with substitution variables

When Oracle9*i* executes the command given in Figure 10–23, the user will first be prompted to enter a value for the substitution variable named Region. The name of a substitution variable does not need to be the same as an existing column name; however, it should be an indicator of the data being requested from the user. Because the SET clause needs the value for the region to be entered by the user, the variable was named Region. After a user has entered a value for the Region column, the user is then asked for the value of the second substitution variable in the WHERE clause. The State substitution variable will be used to define exactly which rows will be updated. The output is shown in Figure 10–24.

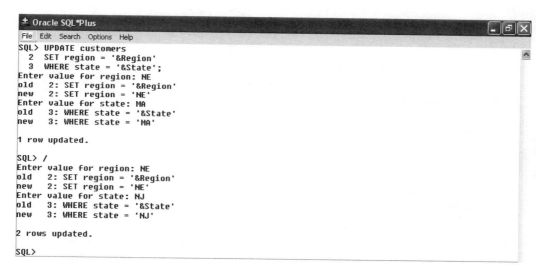

```
Oracle SQL*Plus
File  Edit  Search  Options  Help
SQL> UPDATE customers
  2  SET region = '&Region'
  3  WHERE state = '&State';
Enter value for region: NE
old   2: SET region = '&Region'
new   2: SET region = 'NE'
Enter value for state: MA
old   3: WHERE state = '&State'
new   3: WHERE state = 'MA'

1 row updated.

SQL> /
Enter value for region: NE
old   2: SET region = '&Region'
new   2: SET region = 'NE'
Enter value for state: NJ
old   3: WHERE state = '&State'
new   3: WHERE state = 'NJ'

2 rows updated.

SQL>
```

Figure 10-24 Update of the CUSTOMERS table, using substitution variables

If the display of the old and new values becomes annoying, you can issue the command SET VERIFY OFF at the SQL> prompt, and Oracle9*i* will suppress the messages. The command SET VERIFY ON can be used to instruct Oracle9*i* to start displaying the messages again.

As shown in the output displayed in Figure 10–24, when the UPDATE command is executed, Oracle9*i* asks the user to enter the data value for the Region column. Once the user responds with the requested data, Oracle9*i* replaces the variable in the command with the entered data value. After the first substitution is complete, the user will then be asked to enter a value for the second substitution variable—the state. To allow the user to verify the process that is occurring, the software displays the old and new values for the variables on the screen. After the UPDATE command has been executed, the user can re-execute the command by entering a forward slash (/) at the SQL> prompt, as shown in Figure 10–25.

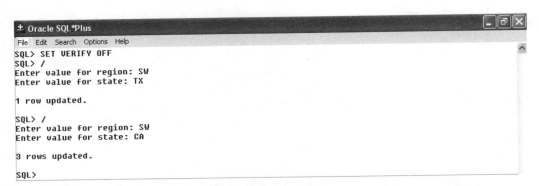

Figure 10-25 Re-execution of the UPDATE command stored in the SQL buffer

Here's how the re-execution shown in Figure 10–25 works: When the command is first executed, it is placed in the SQL buffer. The buffer always contains the last SQL command processed. The contents of the buffer can be executed by typing **RUN** or the forward slash at the SQL> prompt, and pressing the **Enter** key. Through the use of substitution variables, the statement becomes interactive, and the user can continue to update the Region column for as many states as necessary.

The buffer only contains the last SQL command entered. However, it does not store SQL *PLUS commands, such as DESCRIBE.

If a user can't complete all the customer record updates during one session, the command can be permanently stored in a script file that can be executed at a later time. To create the script file, the command in Figure 10–25 can be typed in a simple word-processing program, such as Notepad or WordPad. However, when the file is saved, it must be saved with **sql** as the file extension. Once the command is saved in a script file, it can be executed at a later time by issuing the command `START d:`*`filename`* at the SQL> prompt, where `d:` indicates the location of the file.

TRANSACTION CONTROL STATEMENTS

In Chapters 8 and 9, you issued data definition language (DDL) commands to create, alter, or drop database objects, such as the ACCTMANAGER and ACCTMANAGER2 tables. In this chapter, however, all the operations performed with the INSERT INTO and UPDATE commands have made changes to the data contained *within* tables, not to the actual structure of the table. Commands used to modify data are called **data manipulation language (DML)** commands. Changes to data made by DML commands are not permanently saved to the table when the SQL statement is executed. This allows the user the flexibility of issuing **transaction control** statements either to save the modified data or to undo the changes if they were made in error. However, you should also be aware that until the data have been permanently saved to the table, no other users will be able to view any of the changes you have made.

COMMIT and ROLLBACK Commands

When dealing with DML statements, one area of concern is that the changes are not permanently made to a table until a **COMMIT** command is either implicitly or explicitly issued. An *explicit* COMMIT occurs when the command is explicitly issued by entering **`COMMIT;`** at the SQL> prompt. The COMMIT command *implicitly* occurs when the user exits the system by issuing the **EXIT** command at the SQL> prompt. An implicit COMMIT also occurs if a DDL command, such as CREATE or ALTER TABLE, is issued. In other words, if a user adds several records to a table and then creates a new table, the records that were added before the DDL command was issued will automatically be committed. Why? The series of DML statements that were issued to add the records to the table are considered one set of **transactions**. In Oracle9*i*, a transaction is simply a series of statements that have been issued and not committed. A transaction could consist of one SQL statement or 2,000 SQL statements issued over an extended period of time. The duration of a transaction is defined by when a COMMIT implicitly or explicitly occurs.

Unless a DML operation is committed, it can be undone by issuing the **ROLLBACK** command. For example, if you have not exited Oracle9*i* since beginning to work through the examples in this chapter, typing **`ROLLBACK;`** at the SQL> prompt would reverse all the rows that you entered or altered during your work in this chapter. Thus, the ROLLBACK command will reverse all DML operations performed since the last COMMIT was performed. By contrast, *commands such as CREATE TABLE, TRUNCATE TABLE, and ALTER TABLE cannot be rolled back because they are DDL commands, and a COMMIT occurs automatically when they are executed.* Note, however, that if the system crashes, a ROLLBACK will automatically occur after Oracle9*i* restarts, and any operations not previously committed will be undone.

To ensure that all the operations performed thus far in this chapter are safe from being accidentally reversed, issue the command shown in Figure 10–26 before continuing with the remaining examples in this chapter.

```
COMMIT;
```

Figure 10-26 Command to permanently save data changes

When designing application programs for the Oracle9*i* database, developers will sometimes use the SAVEPOINT command to create a type of bookmark within a transaction. The most common use of this command is in the banking industry. For example, suppose that a customer is making both a deposit and a withdrawal through an ATM. If the customer first makes a deposit and then requests a withdrawal, but cancelled the withdrawal before the money is dispensed, is the entire customer transaction cancelled? To address this issue, a program can be designed to commit the deposit as one transaction and then begin the withdrawal process as a separate transaction. However, some designers have the application program issue the command syntax `SAVEPOINT` *`name`*`;` after the deposit is completed to identify a particular "point" within a transaction. If a subsequent portion of the transaction is cancelled, the program simply issues the command syntax `ROLLBACK TO SAVEPOINT` *`name`*;—and any SQL statements issued after the SAVEPOINT command are not permanently updated to the database. A COMMIT command would still need to be executed to update the database with any data that were added or changed by the first part of the transaction.

Deleting Rows

There will be times when rows will need to be removed from database tables. Compared to some of the other commands issued in this chapter, the **DELETE** command used to delete records from a table is incredibly simple (perhaps even too simple). The syntax for the DELETE command is shown in Figure 10–27.

```
DELETE FROM tablename
[WHERE condition];
```

Figure 10-27 Syntax for the DELETE command

The syntax for the DELETE command does not allow the user to specify any column names in the DELETE clause. This is logical because *DELETE applies to an entire row and cannot be applied to specific columns within a row*. The WHERE clause is optional and is used to identify the row(s) to be deleted from the specified table. A word of caution is appropriate here.

If you omit the WHERE clause, then all rows will be deleted from the specified table.

Suppose that you are notified that new employee Martin Giovanni might be working in the Customer Service Department and should not be added to the ACCTMANAGER table. Because Giovanni's information has already been inserted, the DELETE command will be needed to remove his row from the table. That account manager can be removed from the ACCTMANAGER table by using the command shown in Figure 10–28.

```
DELETE FROM acctmanager
WHERE amid = 'D500';
```

Figure 10-28 DELETE command to remove a row from the ACCTMANAGER table

The WHERE clause of the DELETE command is used to identify the exact record to be removed from the ACCTMANAGER table. Once the record for Martin Giovanni is deleted, the table will now contain only three records, as shown in Figure 10–29.

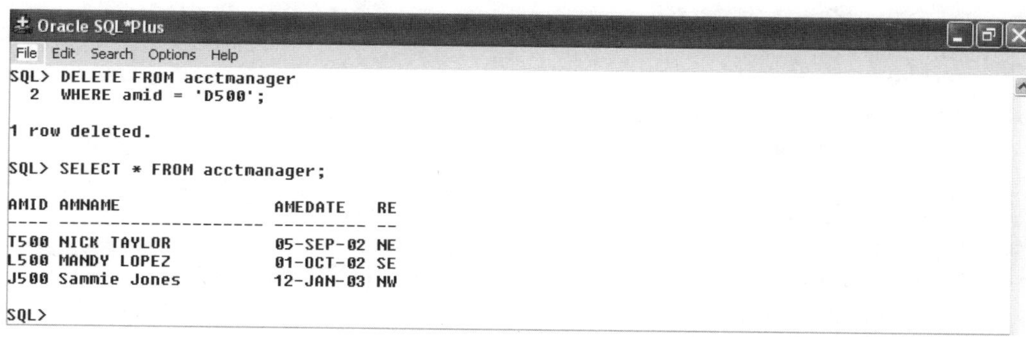

Figure 10-29 Remaining rows of the ACCTMANAGER table after deletion of Martin Giovanni's record

Immediately after the record has been removed from the table, the Human Resources Department informs you that Giovanni will be assigned as an account manager for JustLee Books and should be re-entered into the ACCTMANAGER table. Luckily, the previous deletion of his record was not permanently updated to the table, so the operation can simply be "rolled back" by issuing the command in Figure 10–30.

```
ROLLBACK;
```

Figure 10-30 ROLLBACK command to undo the previous deletion operation

Because the COMMIT command was executed immediately prior to deleting the record for Martin Giovanni, the deletion will be the only operation that is reversed, and all other changes to the table will not be affected. The reversal is shown in Figure 10–31.

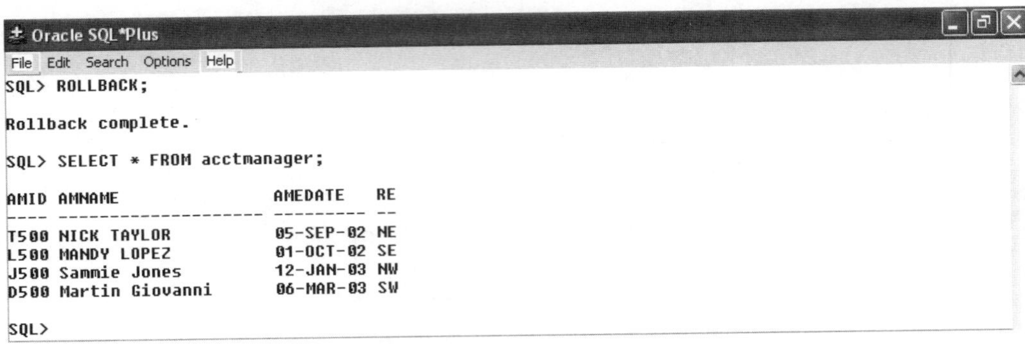
```
SQL> ROLLBACK;

Rollback complete.

SQL> SELECT * FROM acctmanager;

AMID AMNAME               AMEDATE   RE
---- -------------------- --------- --
T500 NICK TAYLOR          05-SEP-02 NE
L500 MANDY LOPEZ          01-OCT-02 SE
J500 Sammie Jones         12-JAN-03 NW
D500 Martin Giovanni      06-MAR-03 SW

SQL>
```

Figure 10-31 Results of the ROLLBACK command

Next, to see how useful the ROLLBACK command can be, issue the command in Figure 10–32 and see what happens.

```
DELETE FROM acctmanager;
```

Figure 10-32 DELETE command without the WHERE clause

Because the command in Figure 10–32 does not contain a WHERE clause, no specific record is identified for deletion. Therefore, the command will delete all the records in the ACCTMANAGER table, as shown in Figure 10–33.

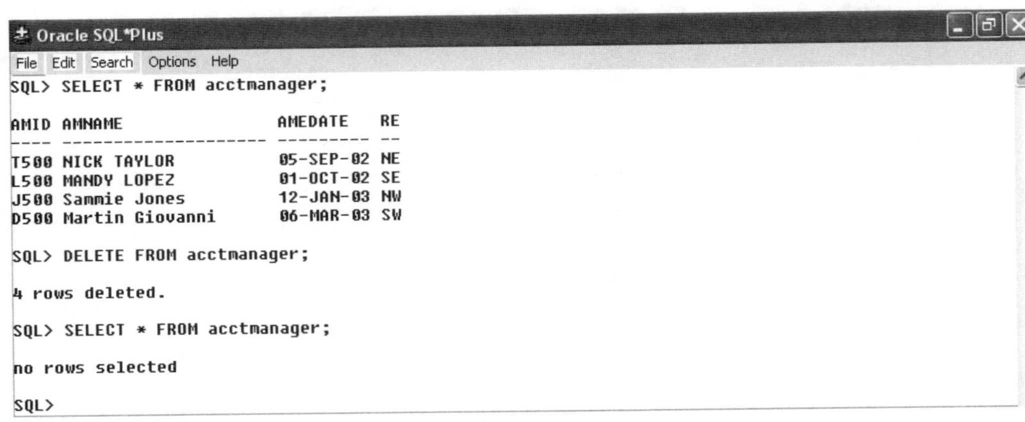
```
SQL> SELECT * FROM acctmanager;

AMID AMNAME               AMEDATE   RE
---- -------------------- --------- --
T500 NICK TAYLOR          05-SEP-02 NE
L500 MANDY LOPEZ          01-OCT-02 SE
J500 Sammie Jones         12-JAN-03 NW
D500 Martin Giovanni      06-MAR-03 SW

SQL> DELETE FROM acctmanager;

4 rows deleted.

SQL> SELECT * FROM acctmanager;

no rows selected

SQL>
```

Figure 10-33 Contents of the ACCTMANAGER table after the DELETE command

As expected, all the rows of the ACCTMANAGER table have been deleted, and the table no longer contains any data about account managers. To restore the records, the ROLLBACK command can be reissued to undo the record deletions, as shown in Figure 10–34.

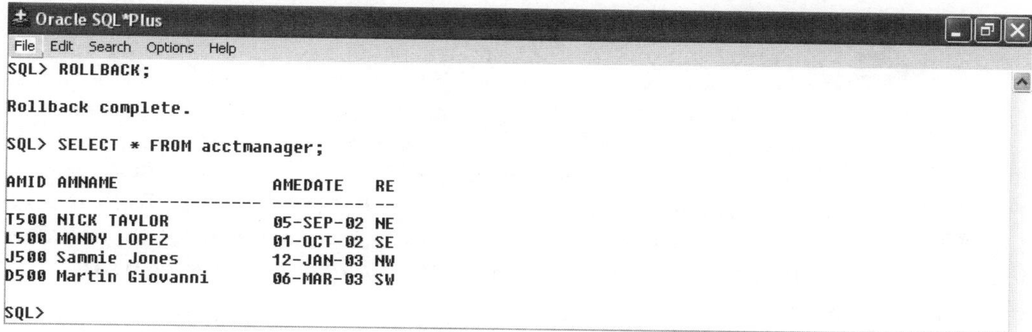

```
SQL> ROLLBACK;

Rollback complete.

SQL> SELECT * FROM acctmanager;

AMID AMNAME               AMEDATE   RE
---- -------------------- --------- --
T500 NICK TAYLOR          05-SEP-02 NE
L500 MANDY LOPEZ          01-OCT-02 SE
J500 Sammie Jones         12-JAN-03 NW
D500 Martin Giovanni      06-MAR-03 SW

SQL>
```

Figure 10-34 Contents of the ACCTMANAGER table after records are restored

TABLE LOCKS

What would happen if two users tried to change the same record at the same time? Which change would actually be saved to the table? When DML commands are issued, Oracle9*i* implicitly "locks" the row or rows being affected, so no other user can change the same row(s). This is a **table lock**. The lock will be a **shared lock** in that other users can still view the data stored in the table, but it prevents anyone from altering the structure of the table or performing other types of DDL operations.

10

LOCK TABLE command

Although rarely used outside an application program, a user can explicitly lock a table in **SHARE MODE** by issuing the **LOCK TABLE** command. The syntax for this command is shown in Figure 10–35.

```
LOCK TABLE tablename IN SHARE MODE;
```

Figure 10-35 Syntax for LOCK TABLE in SHARE MODE

When DDL operations are performed, Oracle9*i* will place an **exclusive lock** on the table, so no other user can alter the table or try to add or update the contents of the table. If an exclusive lock exists on a table, no other user can obtain an exclusive lock or a shared lock on the same table. In addition, if a user has a shared lock on a table, no other user can place an exclusive lock on the same table. If necessary, the user can instruct Oracle9*i* to lock a table in **EXCLUSIVE MODE**, using the command syntax shown in Figure 10–36.

```
LOCK TABLE tablename IN EXCLUSIVE MODE;
```

Figure 10-36 Syntax for LOCK TABLE in EXCLUSIVE MODE

Always be careful when explicitly locking a table. If one user locks a portion of a table in SHARE MODE, and another user locks a different portion of a table, and the completion of one of the commands depends on a portion of a table locked by the other user, a deadlock occurs. Usually, Oracle9*i* detects deadlocks automatically and returns an error message to one of the users. When an error message is returned, the lock is also released, so the command issued by the other user will be completed. *Locks (including exclusive locks) will automatically be released if the user issues a transaction control statement, such as ROLLBACK or COMMIT, or if the user exits the system.*

SELECT…FOR UPDATE Command

A terrifying event can occur when a user looks at the contents of a record, makes a decision based upon those contents, and then updates the record—only to find out that between the SELECT command and the UPDATE command, the contents of the record have changed. For example, suppose that you are assigned the task of increasing the retail price of certain books, and you are told to base the new retail price on a percentage of the cost of the book. As you begin to update retail prices, you realize someone has updated the cost of the books. Ugh! Now what should you do?

As previously mentioned, DML operations are not permanently stored in a table until a COMMIT command is issued. To provide a consistent view for all users accessing the table in a multi-user environment, no changes can be seen by other users until the changes have been committed. This can create major headaches when working with transaction-type tables that are constantly being changed to reflect new orders, account balances, etc.

To avoid this type of problem, the **SELECT…FOR UPDATE** command can be used to view the contents of a record when it is anticipated that the record will need to be modified. The SELECT…FOR UPDATE command places a shared lock on the record(s) to be changed and prevents any other user from acquiring a lock on the same record(s). The syntax is the same as a regular SELECT statement, except the FOR UPDATE clause is added at the end of the command, as shown in Figure 10–37.

```
SELECT columnnames,…
FROM tablename, …
[WHERE condition]
FOR UPDATE;
```

Figure 10-37 Syntax for the SELECT...FOR UPDATE command

If a user decides to update a record, a regular UPDATE command is used to perform the change. However, if the user does not change any of the data selected by the SELECT…FOR UPDATE command, a COMMIT or ROLLBACK command must still be issued, or the rows selected will remain locked, and no other users will be able to make changes to those rows. Figure 10-38 shows the COMMIT command with the SELECT...FOR UPDATE command.

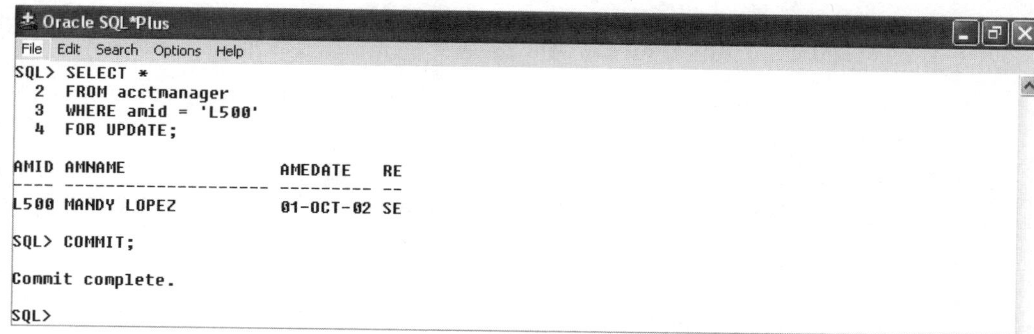

Figure 10-38 The SELECT...FOR UPDATE command

Chapter Summary

- The INSERT INTO command is used to add new rows to an existing table.
- The column list specified in the INSERT INTO clause must match the order of the data provided in the VALUES clause.
- A NULL value can be indicated in an INSERT INTO command by including the keyword NULL, omitting the column from the column list of the INSERT INTO clause, or entering two single quotation marks in the position of the NULL value.
- If rows are copied from a table and entered into an existing table through the use of a subquery in the INSERT INTO command, the VALUES clause must be omitted because it is irrelevant.
- The contents of a row can be changed with the UPDATE command.
- Substitution variables can be used to allow the same command to be executed several times with different data values.
- DML operations are not permanently stored in a table until a COMMIT command is issued or the user exits the system.
- A set of DML operations that are committed as a block is considered a transaction.
- Uncommitted DML operations can be undone by issuing the ROLLBACK command.
- The DELETE command is used to remove records from a table. If the WHERE clause is omitted, all rows in the table will be deleted.
- Table locks can be used to prevent users from mistakenly overwriting changes made by other users.

- Table locks can be in SHARE MODE or EXCLUSIVE MODE.
- EXCLUSIVE MODE is the most restrictive table lock and prevents any other user from obtaining any locks on the same table.
- A lock is released when a transaction control statement is issued, a DDL statement is executed, or the user exits the system using the EXIT command.
- SHARE MODE allows other users to obtain shared locks on other portions of the table, but it prevents any user from obtaining an exclusive lock on the table.
- The SELECT...FOR UPDATE command can be used to obtain a shared lock for a specific row or rows. The lock is not released unless a DDL command is issued or the user exits the system.

CHAPTER 10 SYNTAX SUMMARY

The following table presents a summary of the syntax that you have learned in this chapter. You can use the table as a study guide and reference.

<table>
<tr><th colspan="3">Syntax Guide</th></tr>
<tr><th>Command</th><th>Description</th><th>Example</th></tr>
<tr><td>INSERT</td><td>Adds new row(s) to a table. The user can include a subquery to copy row(s) from an existing table.</td><td><code>INSERT INTO acctmanager
VALUES ('T500', 'NICK TAYLOR',
        '05-SEP-02', 'NE');</code>
or
<code>INSERT INTO acctmanager2
SELECT amid, amname, amedate,
       region
FROM acctmanager
WHERE amedate <='01-OCT-02';</code></td></tr>
<tr><td>UPDATE</td><td>Adds data to, or modifies data within, an existing row</td><td><code>UPDATE acctmanager
SET amedate =
   TO_DATE('JANUARY 12, 2003',
           'MONTH DD, YYYY')
WHERE amid = 'J500';</code></td></tr>
<tr><td>COMMIT</td><td>Permanently saves changed data in a table</td><td><code>COMMIT;</code></td></tr>
<tr><td>ROLLBACK</td><td>Allows the user to "undo" uncommitted changes to data</td><td><code>ROLLBACK;</code></td></tr>
<tr><td>DELETE</td><td>Removes row(s) from a table</td><td><code>DELETE FROM acctmanager
WHERE amid = 'D500';</code></td></tr>
</table>

Command	Description	Example
LOCK TABLE	Prevents other users from making changes to a table	`LOCK TABLE customers IN SHARE MODE;` *or* `LOCK TABLE customers IN EXCLUSIVE MODE;`
SELECT… FOR UPDATE	Creates a shared lock on a table to prevent another user from making changes to data in specified columns	`SELECT cost FROM books WHERE category = 'COMPUTER' FOR UPDATE;`
Interactive Operator		
&	Identifies a substitution variable. Allows the user to be prompted to enter a specific value for the substitution variable.	`UPDATE customers SET region = '&Region' WHERE state = '&State';`

Review Questions

To answer the following questions, refer to the tables in Appendix A.

1. Which command is used to copy data from one table and have it added to an existing table?
2. Which command is used to change the existing data in a table?
3. When do the changes generated by DML operations become permanently stored in the database tables?
4. Explain the difference between explicit and implicit locks.
5. If you add a record to the wrong table, what is the simplest way to remove the record from the table?
6. How does Oracle9*i* identify a substitution variable in a SQL command?
7. How are NULL values included in a new record being added to a table?
8. When should the VALUES clause be omitted from the INSERT INTO command?
9. What happens if a user attempts to add data to a table, and the addition would cause the record to violate an enabled constraint?
10. How do you instruct Oracle9*i* to execute the SQL command currently stored in the SQL buffer?

Multiple Choice

To answer the following questions, refer to the tables in Appendix A.

1. Which of the following is a correct statement?
   a. A commit is implicitly issued when a user exits the system using the EXIT command.
   b. A commit is implicitly issued when a DDL command is executed.
   c. A commit is automatically issued when a DML command is executed.
   d. All of the above are correct.
   e. Both a and b are correct.
   f. Both a and c are correct.
2. Which of the following is a valid SQL statement?
   a. `SELECT * WHERE amid = 'J100' FOR UPDATE;`
   b. `INSERT INTO homework10`
      `VALUES (SELECT * FROM acctmanager);`
   c. `DELETE amid FROM acctmanager;`
   d. `rollback;`
   e. all of the above
3. Which of the following commands can be used to add rows to a table?
   a. INSERT INTO
   b. ALTER TABLE ADD
   c. UPDATE
   d. SELECT...FOR UPDATE
4. Which of the following will delete all the rows in the HOMEWORK10 table?
   a. `DELETE * FROM homework10;`
   b. `DELETE *.* FROM homework10;`
   c. `DELETE FROM homework10;`
   d. `DELETE FROM homework10 WHERE 9amid = '*';`
   e. Both c and d will delete all the rows in the HOMEWORK10 table.

5. Which of the following statements will obtain a shared lock on at least a portion of a table named HOMEWORK10?

   a. `SELECT * FROM homework10 WHERE col2 IS NULL FOR UPDATE;`

   b.
   ```
   INSERT INTO homework10 (col1, col2, col3)
   VALUES ('A', 'B', 'C');
   ```

   c.
   ```
   UPDATE homework10
   SET col3 = NULL
   WHERE col1 = 'A';
   ```

   d.
   ```
   UPDATE homework10
   SET col3 = LOWER(col3)
   WHERE col1 = 'A';
   ```

   e. all of the above

6. Assuming the HOMEWORK10 table has three columns (Col1, Col2, and Col3 in the order listed), which of the following commands will store a NULL value to the column named Col3 in the HOMEWORK10 table?

   a.
   ```
   INSERT INTO homework10
   VALUES ('A', 'B', 'C');
   ```

   b.
   ```
   INSERT INTO homework10 (col3, col1, col2)
   VALUES (NULL, 'A', 'B');
   ```

   c.
   ```
   INSERT INTO homework10
   VALUES (NULL, 'A', 'B');
   ```

   d.
   ```
   UPDATE homework10
   SET col1 = col3;
   ```

10

7. Which of the following symbols is used to designate a substitution variable?

   a. &

   b. $

   c. #

   d. _

8. Which of the following commands can be used to suppress the display of old and new values when executing a command containing substitution variables?

   a. SUPPRESS ON

   b. SET VERIFY ON

   c. SET VERIFY OFF

   d. SET DISPLAY OFF

   e. SET MESSAGE OFF

9. Which of the following commands can be used to lock the HOMEWORK10 table in EXCLUSIVE MODE?
   a. `LOCK TABLE homework10 EXCLUSIVELY;`
   b. `LOCK TABLE homework10 IN EXCLUSIVE MODE;`
   c. `LOCK TABLE homework10 TO OTHER USERS;`
   d. `LOCK homework10 IN EXCLUSIVE MODE;`
   e. both b and d
10. You issue the following command:
    ```
    INSERT INTO homework10 (col1, col2, col3) VALUES ('A',
    NULL, 'C');
    ```
    The command will fail if which of the following statements is true?
    a. Col1 has a PRIMARY KEY constraint enabled.
    b. Col2 has a UNIQUE constraint enabled.
    c. Col3 is defined as a DATE column.
    d. None of the above would cause the command to fail.
11. Which of the following will release a lock currently held by a user on the HOMEWORK10 table?
    a. A COMMIT command is issued.
    b. A DCL command is issued to end a transaction.
    c. The user exits the system.
    d. A ROLLBACK command is issued.
    e. all of the above
    f. none of the above
12. Assume you have added eight new orders to the ORDERS table. Which of the following is true?
    a. Other users can view the new orders as soon as an INSERT INTO command is executed.
    b. Other users can view the new orders as soon as a ROLLBACK command is issued.
    c. Other users can view the new orders as soon as you exit the system or execute a COMMIT command.
    d. Other users can only view the new orders if they obtain an exclusive lock on the table.

13. Which of the following commands will remove all orders placed before April 1, 2003?
    a. `DELETE FROM orders WHERE orderdate < '01-APR-03';`
    b. `DROP FROM orders WHERE orderdate < '01-APR-03';`
    c. `REMOVE FROM orders WHERE orderdate < '01-APR-03';`
    d. `DELETE FROM orders WHERE orderdate > '01-APR-03';`
14. How many rows can be added to a table per execution of the INSERT INTO...VALUES command?
    a. 1
    b. 2
    c. 3
    d. unlimited
15. You accidentally deleted all the orders in the ORDERS table. How can the error be corrected after a COMMIT command has been issued?
    a. `ROLLBACK;`
    b. `ROLLBACK COMMIT;`
    c. `REGENERATE RECORDS orders;`
    d. None of the above will restore the deleted orders.
16. Which of the following is the standard extension used for a script file?
    a. SPT
    b. SRT
    c. SCRIPT
    d. SQL
17. A ROLLBACK will not automatically occur when:
    a. a DDL command is executed.
    b. a DML command is executed.
    c. the user exits the system.
    d. the system crashes.
18. What is the maximum number of rows that can be deleted from a table at one time?
    a. 1
    b. 2
    c. 3
    d. unlimited

19. Which of the following is a correct statement?
    a. If you attempt to add a record that violates a constraint for one of a table's columns, the other valid columns for the row will be added.
    b. A subquery nested in the VALUES clause of an INSERT INTO command can only return one value without generating an Oracle9*i* error message.
    c. If you attempt to add a record that violates a NOT NULL constraint, a blank space will automatically be inserted in the appropriate column so Oracle9*i* can complete the DML operation.
    d. None of the above are correct statements.
20. What is the maximum number of records that can be modified with the UPDATE command?
    a. 1
    b. 2
    c. 3
    d. unlimited

Hands-on Assignments

To perform the following activities, refer to the tables in Appendix A.

1. Create a script that can be used to enter new orders into the ORDERS table. Name the script **SC101.sql**.
2. Add two new orders to the ORDERS table, using the **SC101.sql** script to make certain that it works properly.
3. Verify that the new orders have been properly added to the ORDERS table.
4. Create a script that can be used to update the ship date of orders in the ORDERS table. The row to be updated should be identified by the Order# column. Name the script **SC104.sql**.
5. Change the ship date of order 1012 to reflect that the order was shipped on April 9, 2003.
6. Verify that the ship date was correctly updated.
7. Use the SELECT...FOR UPDATE command to view the ship date for order 1016.
8. Use the DELETE command to remove order 1016 from the ORDERS table.
9. Undo any changes to the records in the ORDERS table.
10. Verify that the changes no longer exist in the ORDERS table.

A Case for Oracle9*i*

To perform the following activity, refer to the tables in Appendix A.

Currently, the contents of the Category column in the BOOKS table are the actual name for each category. This presents a problem if one user enters 'COMPUTER' for the Computer Category and another user enters 'COMPUTERS'. To avoid this and other problems that may occur, the database designers have decided to create a CATEGORY table that will contain a code and description for each category. The structure for the CATEGORY table should be as follows:

Column Name	Datatype	Width	Constraints
CATCODE	VARCHAR2	3	Primary Key
CATDESC	VARCHAR2	11	Not Null

The data for the CATEGORY table is as follows:

CATCODE	CATDESC
BUS	BUSINESS
CHN	CHILDREN
COK	COOKING
COM	COMPUTER
FAL	FAMILY LIFE
FIT	FITNESS
SEH	SELF HELP
LIT	LITERATURE

Required:

- Create the CATEGORY table and populate it with the given data.
- Add a column to the BOOKS table called Catcode.
- Store the correct category code in the BOOKS table, based upon each book's current category.
- Verify that the correct categories have been assigned in the BOOKS table.
- Delete the Category column from the BOOKS table.
- Add a FOREIGN KEY constraint that will require all category codes entered into the BOOKS table to already exist in the CATEGORY table.
- Commit all changes once their accuracy has been verified.

CHAPTER

11

VIEWS

Objectives

After completing this chapter, you should be able to do the following:

- Create a view, using the CREATE VIEW command or the CREATE OR REPLACE VIEW command
- Employ the FORCE and NOFORCE options
- State the purpose of the WITH CHECK OPTION constraint
- Explain the effect of the WITH READ ONLY option
- Update a record in a simple view
- Re-create a view
- Explain the implication of an expression in a view for DML operations
- Update a record in a complex view
- Identify problems associated with adding records to a complex view
- Identify the key-preserved table underlying a complex view
- Drop a view
- Explain inline views and the use of ROWNUM to perform a "TOP-N" analysis

Views show **pseudo tables**, tables that can be created to present a particular "display" of a database's contents. Views have two purposes:

- Assist users who do not have the training to issue complex SQL queries
- Restrict users' access to sensitive data

Although views are database objects, they do not actually store data. They are used to display data in the underlying base tables. You can think of a view as the result of a permanent subquery: The results are given a name that allows them to be used as the source for future queries. In fact, a view can be referenced in the FROM clause of a SELECT statement, just like any table. Views can reference one column or as many as all the columns in specified tables. Next, let's examine how views can help nontechnical users.

Data retrieval from JustLee Books' tables can require fairly complex queries. For example, suppose that an employee needs to find a customer's total balance due for an order. Just to determine that amount, the employee must simultaneously query the CUSTOMERS, ORDERS, ORDERITEMS, and BOOKS tables. The average employee probably lacks the hours of training needed to create a query that joins multiple tables, etc., to perform one seemingly simple task—to determine the total balance due for an order. To simplify the user's task, one option is to create a pseudo table, a view that contains all the information the user needs. Thus, rather than teaching the user how to use Oracle9*i*, you can merely show him or her how to perform simple queries on the pseudo table that displays customer names, order numbers, order dates, book titles, quantity of books ordered, retail prices, and the calculated extended price.

A view can be used not only to simplify complex queries, but also to restrict access to what management may consider "sensitive data." For example, the BOOKS table contains both the cost and retail price of each book in inventory. What happens if management decides that the cost of each book should not be accessible by every employee in the company? Do you delete the column from the BOOKS table? If so, then how would you calculate the profit for each book sold? Rather than providing users with access to the actual table storing all the data for the books, users can be given access to a view that displays all the data appropriate for the user, based on his or her job duties.

This chapter will present the commands and guidelines regulating DML operations for data displayed by views. Figure 11–1 provides an overview of this chapter's contents.

View Type	**Description**
Simple view	A view based upon a subquery that only references one table and does not include any group functions, expressions, or a GROUP BY clause
Complex view	A view based upon a subquery that retrieves or derives data from one or more tables—and may also contain functions or grouped data
Inline view	A subquery used in the FROM clause of a SELECT statement to create a "temporary" table that can be referenced by the SELECT and WHERE clauses of the outer statement.
Command	**Command Syntax**
Create a view	`CREATE [OR REPLACE] [FORCE|NOFORCE] VIEW` *`viewname`* `(`*`columnname, …`*`)` `AS` *`subquery`* `[WITH CHECK OPTION [CONSTRAINT` *`constraintname`*`]]` `[WITH READ ONLY];`
Drop a view	`DROP VIEW` *`viewname`*
Create an inline view	`SELECT` *`columnname, …`* `FROM (`*`subquery`*`)` `WHERE ROWNUM<=`*`N`*`;`

Figure 11-1 Overview of chapter contents

Before attempting to work through the examples provided in this chapter, open the Chapter11 folder in your Data Files. You should run the **prech11.sql** script to ensure that all necessary tables and constraints are available. To run the script, type **`start d:\prech11.sql`** at the SQL> prompt, where **d:** represents the location (correct drive and path) of the script. If the script is not available, ask your instructor to provide you with a copy.

CREATING A VIEW

A view is created with the **CREATE VIEW** command. The syntax for the CREATE VIEW command is shown in Figure 11–2.

```
CREATE [OR REPLACE] [FORCE|NOFORCE] VIEW
       viewname (columnname, …)
AS subquery
[WITH CHECK OPTION [CONSTRAINT constraintname]]
[WITH READ ONLY];
```

Figure 11-2 Syntax of the CREATE VIEW command

Let's get an overview of the syntax elements shown in Figure 11–2.

- *CREATE VIEW/CREATE OR REPLACE VIEW*: The **CREATE VIEW** keywords are used to create a view, using a name that is not being used by another database object in the current schema. *There is no way to modify or change an existing view,* so if you need to change a view, the **CREATE OR REPLACE VIEW** keywords must be used. The OR REPLACE option instructs Oracle9*i* that a view with the same name may already exist, and if it does, to replace the previous version of the view with the one that is defined in the subquery of the new command.
- *FORCE/NOFORCE*: If you attempt to create a view based upon a table(s) that does not yet exist or is currently unavailable (e.g., offline), Oracle9*i* will return an error message, and the view will not be created. However, if you include the **FORCE** keyword in the CREATE clause, Oracle9*i* will create the view in spite of the absence of any referenced tables. For example, **NOFORCE** is the default mode for the CREATE VIEW command, and that means all the tables and columns must be valid, or the view will not be created. This approach is commonly used when a new database is being developed and the data have not yet been loaded, or entered, into the database objects.
- *View name*: As previously mentioned, each view should be given a name that is not already assigned to another database object in the same schema.
- *Column names*: If you want to assign new names for the columns that are displayed by the view, the new column names can be listed after the VIEW keyword, within a set of parentheses. *The number of names listed must match the number of columns returned by the subquery.* An alternative is to use column

aliases in the subquery. If column aliases are given in the subquery, Oracle9*i* will use the aliases as the column names in the view that is created.

- *AS clause*: The subquery listed after the AS keyword must be a complete SELECT statement (both SELECT and FROM clauses are required) and can reference more than one table. The subquery can also include single-row and group functions, WHERE and GROUP BY clauses, nested subqueries, etc. However, as mentioned in Chapter 7, a subquery cannot include the ORDER BY clause. The results of the subquery will be the content of the view being created.
- *WITH CHECK OPTION*: The WITH CHECK OPTION constraint ensures that any DML operations performed on the view (e.g., adding rows, changing data) will not prevent the row from being accessed by the view because it no longer meets the condition in the WHERE clause. For example, if a view consists of only books in the Cooking Category, and the user attempts to change the category of one of the books included in the view to the Family Life Category, then the change will not be allowed if the WITH CHECK OPTION constraint was included when the view was created. Why? Because the change would mean that the book will no longer be listed in the view since it only consists of books in the Cooking Category. If the WITH CHECK OPTION constraint is omitted when the view is created, then any valid DML operation is allowed, even if the result is that the row(s) being changed would no longer be included in the view. However, if a view is being created with the sole purpose of only displaying data, the WITH READ ONLY option can be used instead to ensure that the data cannot be changed.

Next, let's look at these operations in more depth. First, you'll see how to create a simple view, then how to change a simple view, and finally how to create a complex view.

Creating a Simple View

A simple view is created from a subquery that only references one table and does not include a group function, expression (e.g., retail-cost), or GROUP BY clause. For example, when JustLee Books' service representatives assist customers with orders, they need to access the ISBN, title, and retail price of every book in JustLee's inventory. However, management does not want representatives to view the books' actual cost. The solution: Create a simple view to allow service representatives to access only the data needed to assist customers—and not access irrelevant columns, such as Publisher ID, Cost, etc.

The CREATE VIEW command to create the view needed by customer service representatives is shown in Figure 11–3.

```
CREATE VIEW inventory
AS SELECT isbn, title, retail price
   FROM books
WITH READ ONLY;
```

Figure 11-3 Command to create the INVENTORY view

As indicated in the CREATE VIEW clause, the name of the new view will be INVENTORY. Note these other elements in Figure 11–3:

1. Because another view with this name does not exist, the OR REPLACE clause is not necessary.
2. The only columns included in the view are the ISBN, Title, and Retail columns from the BOOKS table. Notice that the Retail column has been assigned the column alias of Price. Thus, whenever the Retail column is referenced through a query on the INVENTORY view, it must be called Price in the query.
3. The WITH READ ONLY clause is used so that no customer service representative can accidentally change the ISBN, title, or price of a book. Any changes made to the data contained in a simple view created without the WITH READ ONLY option will automatically update the underlying BOOKS table.

It is not uncommon for an organization to provide views to allow users to retrieve data without fear of the data being accidentally, or even intentionally, changed.

As shown in Figure 11–4, once the INVENTORY view is created, it can be referenced in the FROM clause of a SELECT statement in the same manner as any table.

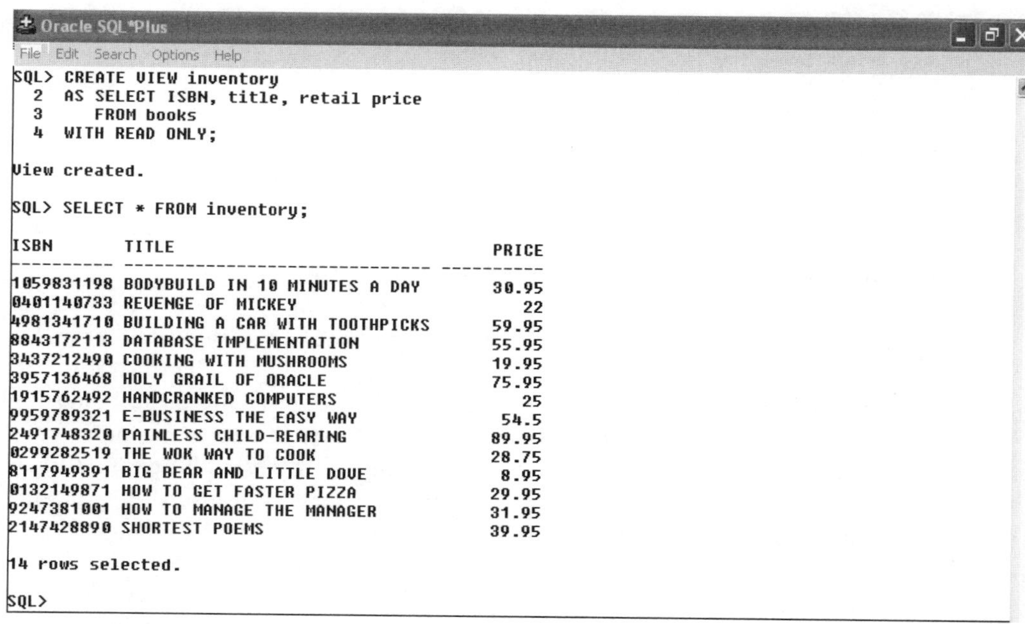

```
Oracle SQL*Plus
File Edit Search Options Help
SQL> CREATE VIEW inventory
  2  AS SELECT ISBN, title, retail price
  3     FROM books
  4  WITH READ ONLY;

View created.

SQL> SELECT * FROM inventory;

ISBN       TITLE                              PRICE
---------- ------------------------------ ----------
1059831198 BODYBUILD IN 10 MINUTES A DAY      30.95
0401140733 REVENGE OF MICKEY                     22
4981341710 BUILDING A CAR WITH TOOTHPICKS     59.95
8843172113 DATABASE IMPLEMENTATION            55.95
3437212490 COOKING WITH MUSHROOMS             19.95
3957136468 HOLY GRAIL OF ORACLE               75.95
1915762492 HANDCRANKED COMPUTERS                 25
9959789321 E-BUSINESS THE EASY WAY             54.5
2491748320 PAINLESS CHILD-REARING             89.95
0299282519 THE WOK WAY TO COOK                28.75
8117949391 BIG BEAR AND LITTLE DOVE            8.95
0132149871 HOW TO GET FASTER PIZZA            29.95
9247381001 HOW TO MANAGE THE MANAGER          31.95
2147428890 SHORTEST POEMS                     39.95

14 rows selected.

SQL>
```

Figure 11-4 Selecting all records from the INVENTORY view

If the command in Figure 11–3 returns an error message, make certain that a view with the same name does not already exist. If such a view does exist, add the keywords OR REPLACE to the CREATE VIEW clause and then execute the command again.

Although the INVENTORY view can be retrieved with a SELECT statement as if it were a regular table, the view in Figure 11–4 is created with a **WITH READ ONLY** option. This prevents any DML operations from being performed on the data. In Figure 11–5, the user is unsuccessfully attempting to update the data contained in the Price and Title columns of the INVENTORY view.

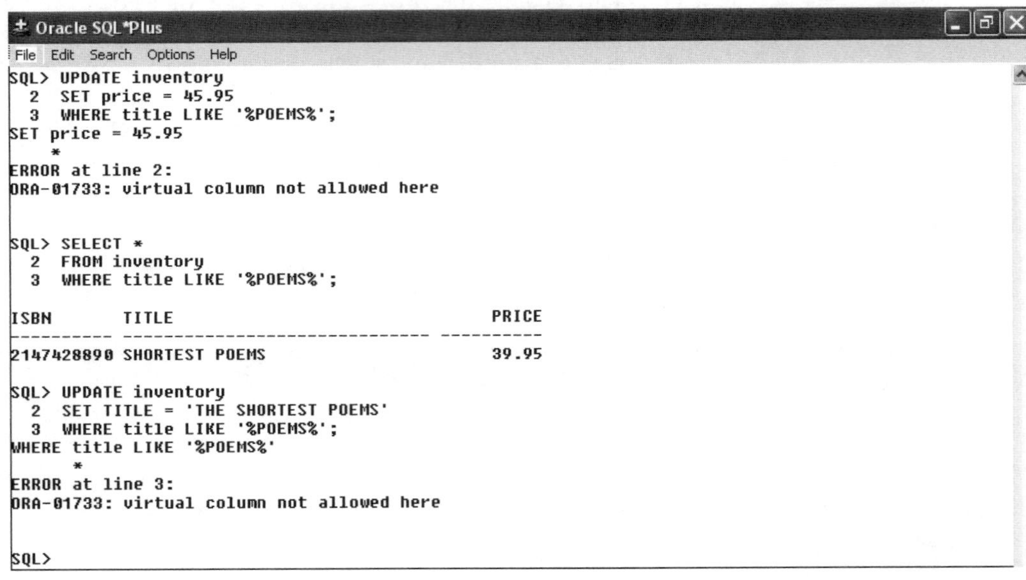

Figure 11-5 Failed updates on the INVENTORY view

Figure 11–5 shows that the user first attempts to change the retail price of the book called *Shortest Poems*. Because the user could not remember the exact title of the book, a search pattern is used in the WHERE clause to identify the book being updated. The percent signs (%) indicate that there may be characters appearing before and after the word *Poems*, but the book's title must contain the word *Poems*. However, Oracle9*i* returns the error message "virtual column not allowed here." Because the error message does not really seem to indicate the problem, the user attempts to change the title for the book to see whether the problem is due to a column alias being used for the column that is actually called Retail in the BOOKS table. However, these failed attempts at changing the data in the view are not due to any column references. Rather, it is because the view was created with the WITH READ ONLY option, and no DML operations are allowed at all. As you can see, sometimes Oracle9*i* error messages can be helpful, and sometimes they are not.

DML Operations on a Simple View

If management later decides that the INVENTORY view should be used to alter the retail prices of the books currently in inventory, it can be re-created using the CREATE OR REPLACE VIEW command—but it cannot include the WITH READ ONLY option. Since the INVENTORY view already exists in the database, the OR REPLACE keywords must be included, or Oracle9*i* will return an error message indicating that the view already exists.

In Figure 11–6, the INVENTORY view has been re-created—without a WITH READ ONLY option, so updates are allowed.

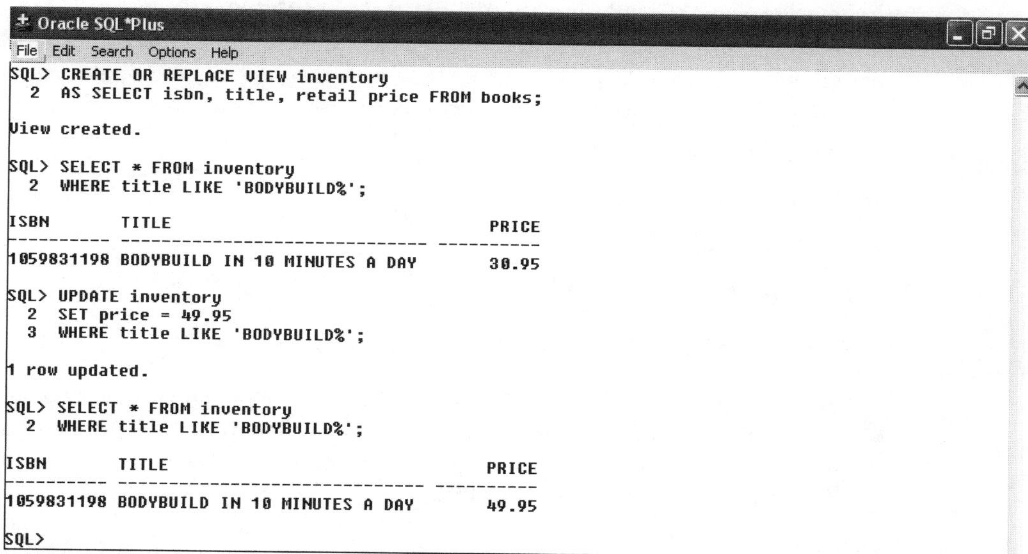

Figure 11-6 Allowing updates on the INVENTORY view

After the view shown in Figure 11–6 is re-created, anyone with access to the view can change the ISBN, title, or retail price of any book in the BOOKS table. As shown, the original retail price of *Bodybuild in 10 Minutes a Day* was $30.95. Then, an UPDATE command was issued to change the retail price of the book to $49.95. The previous version of the view would not have allowed the data to be changed.

However, as shown in Figure 11–7, not only did the view accept the change, but it also altered the price contained in the underlying BOOKS table.

```
SQL> SELECT title, retail
  2  FROM books
  3  WHERE title LIKE 'BODYBUILD%';

TITLE                              RETAIL
------------------------------ ----------
BODYBUILD IN 10 MINUTES A DAY       49.95

SQL>
```

Figure 11-7 Updated data in the BOOKS table

The basic rule for DML operations on a simple view (or a complex view) is this: As long as the view was not created with the WITH READ ONLY option, any DML operation is allowed on a simple view if it does not violate an existing constraint on the underlying base table. In essence, you can add, modify, and even delete data in an underlying base table as long as the operation is not prevented by one of the following constraints:

- PRIMARY KEY
- NOT NULL
- UNIQUE
- FOREIGN KEY
- WITH CHECK OPTION

If the SELECT statements in Figures 11-6 or 11-7 return an error message or do not select any rows, make certain the book title in the WHERE clause is enclosed in single quotation marks and includes a percent sign (%) at the end of the search pattern for BODYBUILD.

Let's look at another example. As shown in Figure 11–8, the OUTSTANDING view has been created to display all the orders contained in the ORDERS table that have not yet been shipped to the customer.

You should note that because a WITH CHECK OPTION constraint was included in the CREATE VIEW command in Figure 11–8, the OUTSTANDING view cannot be used to update the shipping date of any of the six records that have not yet been shipped because changing this would remove the record(s) from the view. Similarly, any attempt to change the ship date of an order to a non-NULL value will return an error message, and the update will fail, as shown in Figure 11–9.

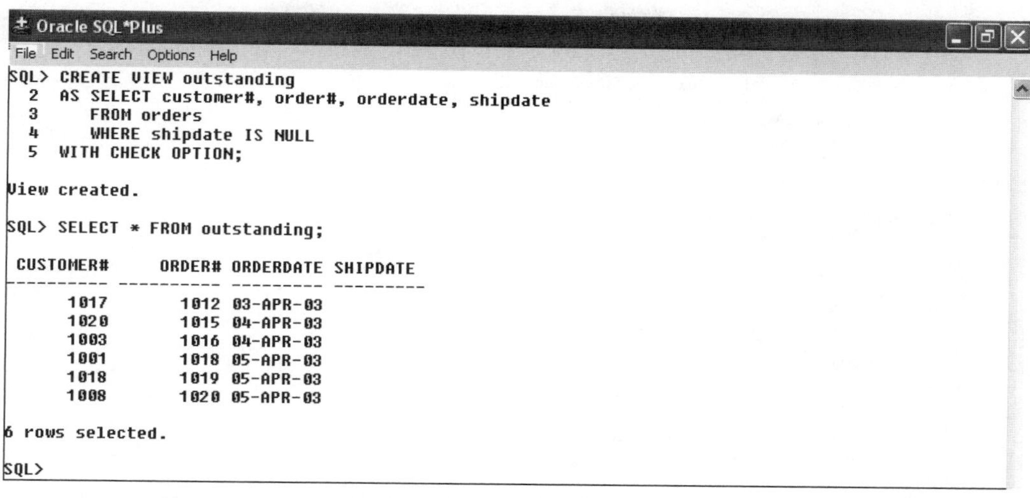

Figure 11-8 OUTSTANDING view including a WITH CHECK OPTION constraint

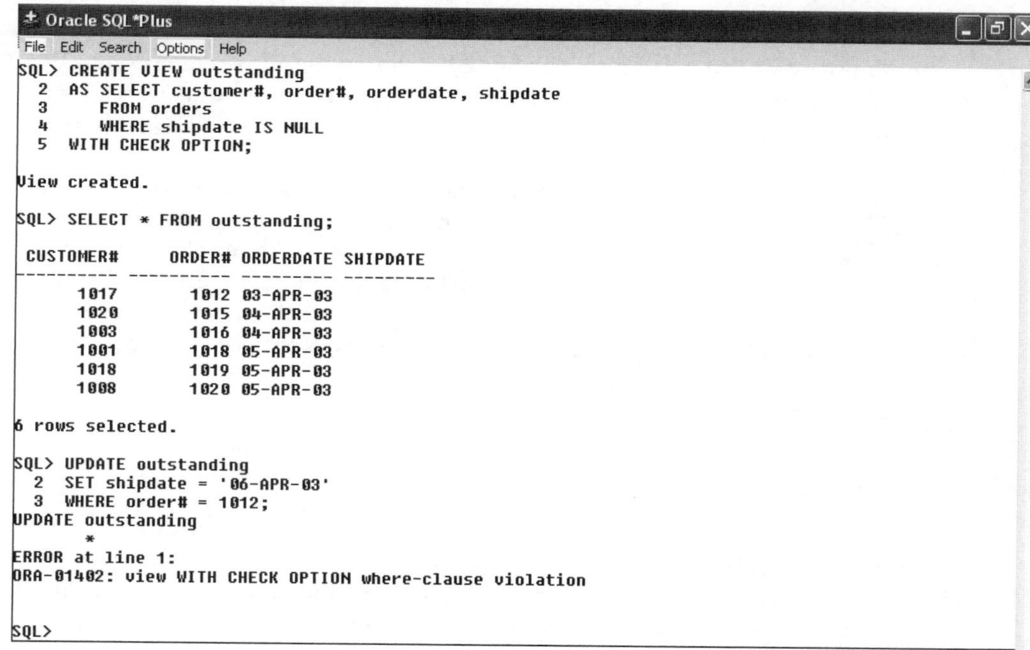

Figure 11-9 Error returned when updating the OUTSTANDING view

If the purpose of the OUTSTANDING view is to allow users to enter the ship date of an order when it is shipped, it should be re-created *without* the WITH CHECK OPTION constraint, as shown in Figure 11–10.

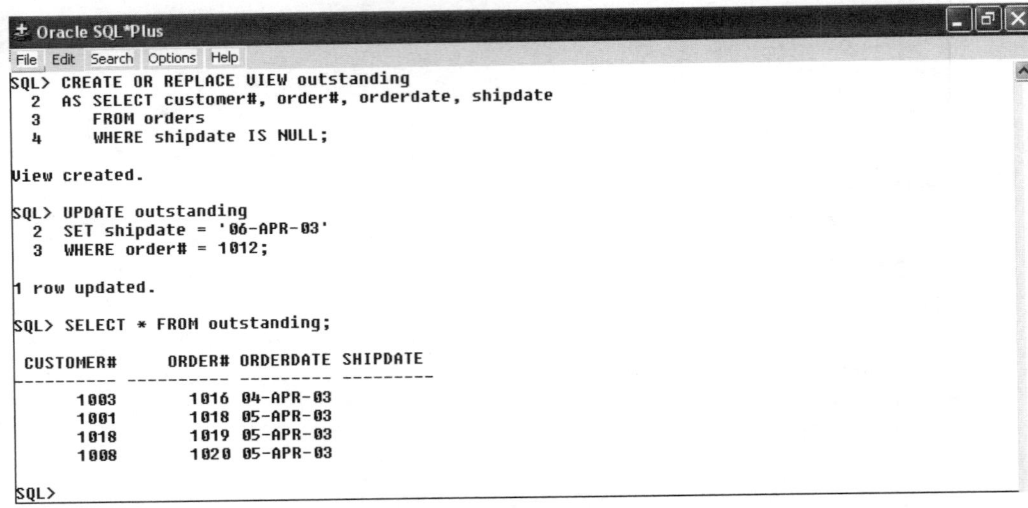

Figure 11-10 OUTSTANDING view without a WITH CHECK OPTION constraint

In Figure 11–10, the view has been re-created and a command has been issued to update the ship date of order 1012 to April 6, 2003. Since the WITH CHECK OPTION constraint no longer exists on the OUTSTANDING view, the update is allowed, although the record will no longer be included in the view after the change occurs.

You should attempt the DML operations in this chapter and experiment with variations of the examples. You can "undo" any of the DML operations by issuing the `ROLLBACK;` command without having to rebuild the views. The worst-case scenario is that you will need to rerun the **prech11.sql** script to reset the data in your tables. In a real-world situation, however, you would want to store copies of the tables in a test environment and experiment on the copies, rather than the real data.

Creating a Complex View

A complex view is created using the same CREATE VIEW command as a simple view. However, the subquery used to create a complex view retrieves or derives data from one or more tables—and may contain functions or grouped data. *The main difference between simple and complex views is that certain DML operations are not permitted with complex views.* To discuss how complex views react to various DML operations, this section will present three different views.

1. The first complex view to be presented is based on one table, but it uses an expression for one of the columns.
2. The second complex view to be presented is based on two tables, and it also uses an expression for one of the columns.

3. The third complex view to be presented is derived from four tables, and it includes a group function and a GROUP BY clause.

DML Operations on a Complex View with an Arithmetic Expression

As you will see, different factors will affect the type of DML operations that are allowed on complex views. For example, if a view contains a column that is the result of an arithmetic expression or grouped data, or if it is based upon multiple tables and it is difficult to determine exactly which table should be modified, then certain DML operations will not be allowed. Look at the complex view in Figure 11–11, which shows creating and updating a view called PRICES.

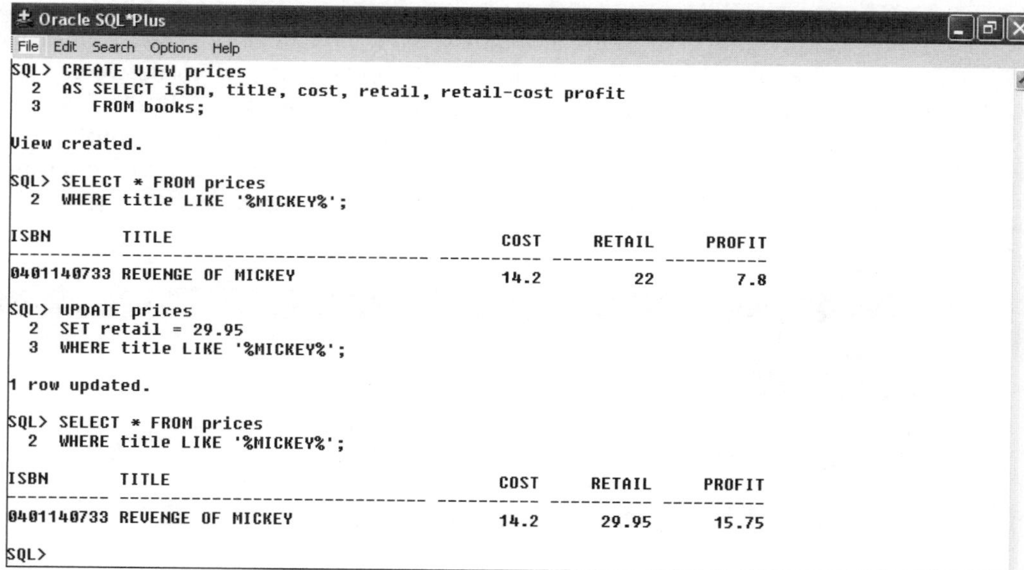

Figure 11-11 Creating and updating the PRICES view

The complex view created in Figure 11–11 almost seems like a simple view, except that it uses the expression "retail-cost" to calculate, or derive, the Profit column. Also shown is an UPDATE command using `(SET retail = 29.95)` that changes the retail price of *Revenge of Mickey* from $22.00 to $29.95. Again, the view acts like a simple view because the UPDATE command worked to make the change. But what about removing rows in the view? Look at Figure 11–12.

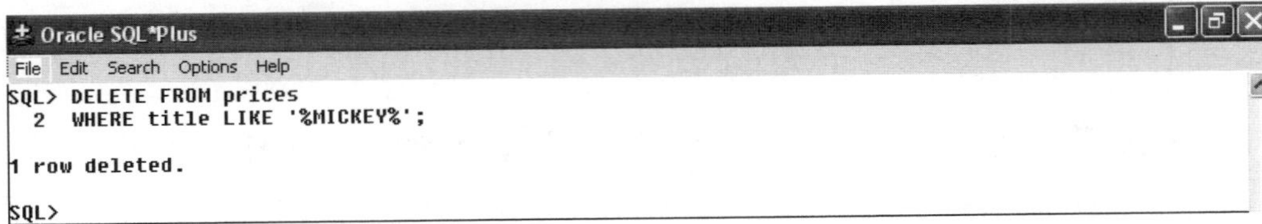

Figure 11-12 Deleting a book from the PRICES view

In Figure 11–12, the DELETE command is used to remove the book called *Revenge of Mickey* from the PRICES view, which also removes the book from the BOOKS table. So far, so good—the DELETE command works for this view. Now what about adding a record to the view?

Before attempting to add a record to the view, you need to remember that the Profit column was actually based upon the expression "retail-cost" to derive the dollar profit generated by the book. So, when you add a new book to the PRICES view, do you enter the actual profit generated, or should you let Oracle9*i* calculate the profit? Logic would support only entering values for the Cost and Retail columns, and then letting Oracle9*i* determine the profit (and if that fails, then try including the profit). Figure 11–13 shows the outcome of this logical approach.

```
Oracle SQL*Plus
File Edit Search Options Help
SQL> INSERT INTO prices
  2  VALUES (0202020202, 'A NEW BOOK', 8.95, 15.95);
INSERT INTO prices
            *
ERROR at line 1:
ORA-00947: not enough values


SQL> INSERT INTO prices
  2  VALUES (0202020202, 'A NEW BOOK', 8.95, 15.95, 7.00);
INSERT INTO prices
*
ERROR at line 1:
ORA-01733: virtual column not allowed here


SQL>
```

Figure 11-13 Failed attempts to add a record to the PRICES view

So much for logic. The first attempt to add a new record to the PRICES view returns the error message "not enough values." That seems simple enough; it appears to indicate that Oracle9*i* does want an actual value entered for the Profit column. Therefore, the second attempt in Figure 11–13 includes the amount of profit that would be generated by the new book being added to the view. However, Oracle9*i* will not accept the calculated profit entered by the user. The error message "virtual column not allowed here" is one that may appear often until you learn the rules for DML operations on complex views. In this case, the error message means that since one of the columns in the view is based upon an arithmetic expression, a value cannot be inserted into that column of the view.

The only way to add a new book to the PRICES view is to (1) add the record directly to the BOOKS table, or (2) add the data to the PRICES view by not attempting to insert a value into the Profit column. In other words, if you were to insert only the ISBN, title, cost, and retail price of the book into the PRICES view, the underlying BOOKS table would have been updated with the new book. Thus, you have now discovered one of the rules governing DML operations for complex views:

Values cannot be inserted into columns that are based on arithmetic expressions.

DML Operations on a Complex View Containing Data from Multiple Tables

To add a little more complexity to the complex view called PRICES, the view has been re-created in Figure 11–14 to include the name of the publishers from the PUBLISHER table.

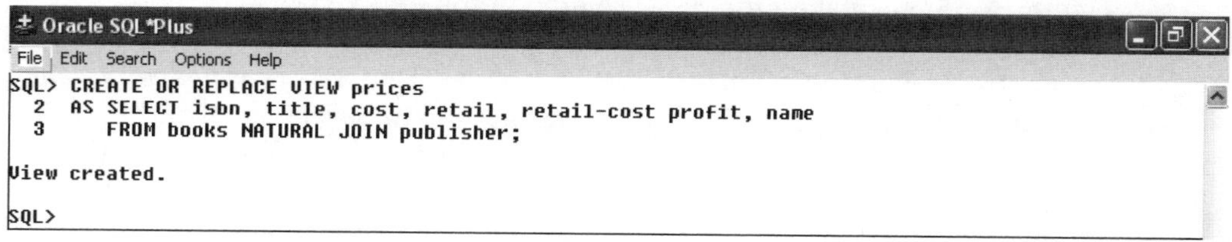

Figure 11-14 PRICES view with a table join

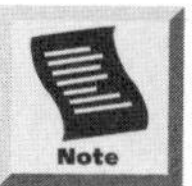

To be technically correct, a record also could not be added to the new version of the PRICES view because the Pubid column of the BOOKS table has a NOT NULL constraint. Given that the PRICES view does not contain the Pubid column, there would be no way to indicate the value to be entered in the column, which would violate the NOT NULL constraint. Therefore, even if an expression had not been included in this view, a record still could not have been added to the view.

Now let's try to perform the same type of DML operations that were performed on the previous version of the PRICES view. Because you have already discovered that records cannot be added to a view if one of the values to be inserted into a column is based on an expression, the next step is to try to update the price of one of the books, as shown in Figure 11–15.

```
Oracle SQL*Plus
File Edit Search Options Help
SQL> SELECT * FROM prices
  2  WHERE title LIKE '%BEAR%';

ISBN       TITLE                                COST     RETAIL     PROFIT NAME
---------- ------------------------------ ---------- ---------- ---------- -----------------------
8117949391 BIG BEAR AND LITTLE DOVE             5.32       8.95       3.63 REED-N-RITE

SQL> UPDATE prices
  2  SET retail = 13.95
  3  WHERE title LIKE '%BEAR%';

1 row updated.

SQL> SELECT * FROM prices
  2  WHERE title LIKE '%BEAR%';

ISBN       TITLE                                COST     RETAIL     PROFIT NAME
---------- ------------------------------ ---------- ---------- ---------- -----------------------
8117949391 BIG BEAR AND LITTLE DOVE             5.32      13.95       8.63 REED-N-RITE

SQL>
```

Figure 11-15 Updating the Retail column of the PRICE view

In Figure 11–15, the retail price of the book titled *Big Bear and Little Dove* has been changed from $8.95 to $13.95. As with the previous version of the PRICES view, the DML operation to modify a record worked (Figure 11–11). However, the change made to the view was that the name of the publisher was included from the PUBLISHER table, so what happens if the Name column is updated? Look at Figure 11–16.

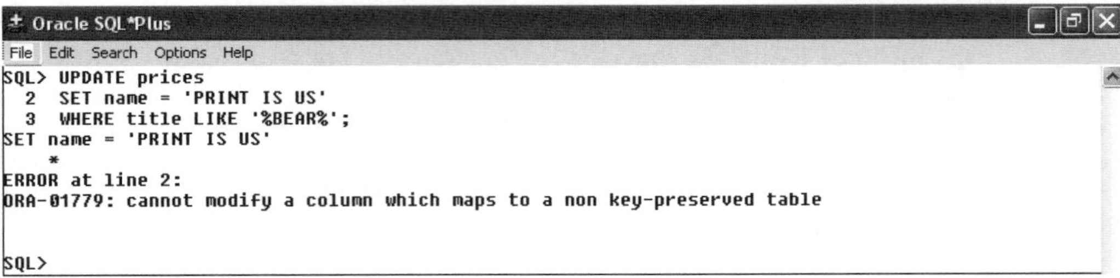

```
Oracle SQL*Plus
File Edit Search Options Help
SQL> UPDATE prices
  2  SET name = 'PRINT IS US'
  3  WHERE title LIKE '%BEAR%';
SET name = 'PRINT IS US'
    *
ERROR at line 2:
ORA-01779: cannot modify a column which maps to a non key-preserved table


SQL>
```

Figure 11-16 Failed attempt to update the publisher's name in the PRICES view

The update failed. When Oracle9*i* attempted to update the name of the publisher of *Big Bear and Little Dove*, the UPDATE command failed, and you received the error message "cannot modify a column which maps to a non key-preserved table." Perhaps taking a step back and analyzing the underlying tables will help you to understand the meaning of the error message—and what caused the error message to occur.

The PRICES view was built using columns from the BOOKS and PUBLISHER tables. When a view includes columns from more than one table, updates can only be applied to *one* table. The table that can be updated is the one that includes the primary key of an underlying table and is basically being used as the primary key for the view. The PRICES view includes the primary key for the BOOKS table, so basically any UPDATE command can be performed on the columns from the BOOKS table—if the change does not violate any constraints from that table (e.g., you cannot change the primary key if it is used as a reference for a FOREIGN KEY constraint or if it would no longer be unique). In the PRICES view, the BOOKS table is known as the **key-preserved table**. In essence, a key-preserved table is the table that contains the primary key that the view is using to uniquely identify each record being displayed by the view. By contrast, the Name column is from the PUBLISHER table. The primary key for the PUBLISHER table is the Pubid column, and it is not included in the view. However, even if the Pubid column had been included, Oracle9*i* would not have considered the column as the primary key for the PRICES view because it could have appeared more than once in the contents of the view. Therefore, Oracle9*i* will treat the data from the PUBLISHER table as coming from a **non key-preserved table** since it does not uniquely identify the records in the PRICES view.

One way to make sense of this problem is to consider that the BOOKS table actually stores the publishers' ID numbers. Thus, if Oracle9*i* were to change the name of a publisher, as instructed by the UPDATE command given in Figure 11-16, does that mean you want the Pubid column updated in the BOOKS table as well? Or would you change the name of the publisher in the PUBLISHER table? If the name is changed in the PUBLISHER table, then every book that had the same Pubid as *Big Bear and Little Dove* would now be published by the publisher Print Is Us, and that was not the intention of the command. Now you have discovered a second rule that applies to complex views:

DML operations cannot be performed on a non key-preserved table.

The rule that no DML operations can be performed on a non key-preserved table may be a little broad, considering that you have not tried to delete a record from the PRICES view. Therefore, that should be your next step before you can be certain that this second rule applies to all DML operations.

Given the results of the DELETE command in Figure 11-17, it might appear that there is a problem with the second rule. You were not allowed to update the publisher's name for the book, but you were allowed to delete the row for the book. However, what actually occurred is that the book *Big Bear and Little Dove* was deleted from the BOOKS table, which is the key-preserved table. No change occurred in the PUBLISHER table so, technically, the command did not perform a DML operation on the non key-preserved table; therefore, the rule was not violated.

Figure 11-17 Deletion of a book from the PRICES view

DML Operations on a Complex View Containing Functions or Grouped Data

Views are also considered to be complex if they contain a function or a GROUP BY clause. To determine what effect they have on DML commands, let's create a view that displays the total balance due for each order placed by a customer. Look at Figure 11–18.

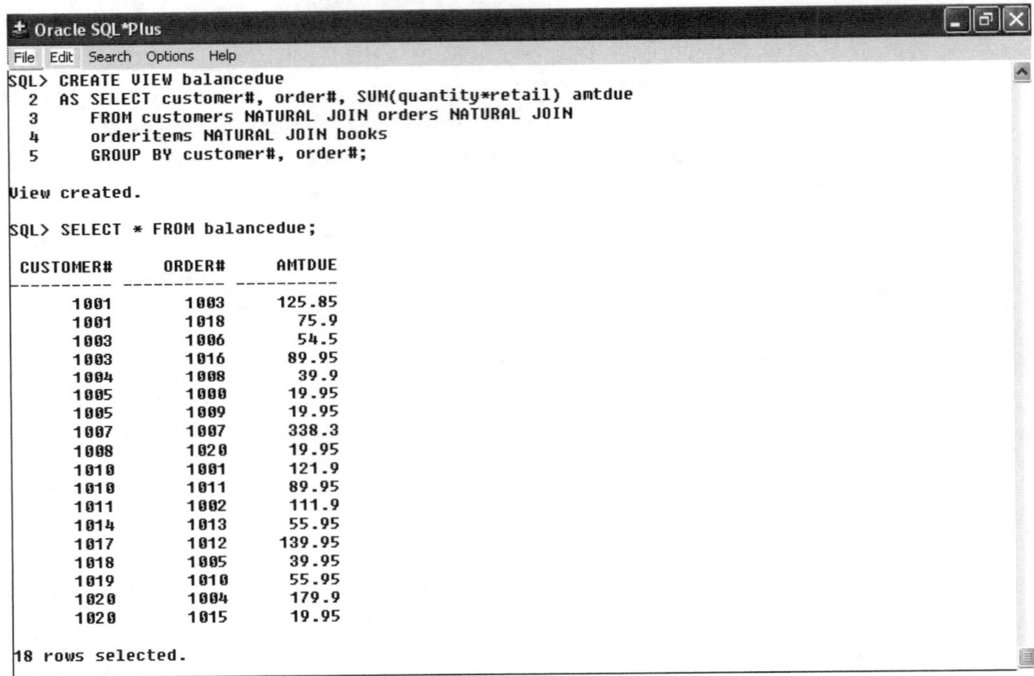

Figure 11-18 BALANCEDUE view

The BALANCEDUE view created in Figure 11–18 first groups the items on each customer's order, and then it calculates the total amount due, based on the number of books ordered and each book's retail price. Adding a record to the BALANCEDUE view is not allowed because the Amtdue column is derived from a function (it is treated the same way as an expression). Even without the SUM function, Oracle9*i* would still not allow a record to be added because of the GROUP BY clause: The data being displayed are grouped; it is not possible to add an individual record to the view. In addition, *the function and GROUP BY clause will prevent any of the data being displayed from being changed, because each record being displayed may represent more than one row in the underlying key-preserved table.* (Try a little experiment on your own and see whether you can add or modify a record, but don't be too disappointed if you receive an error message.) Thus, the question becomes, "Will Oracle9*i* allow a row to be deleted from the view?" Figure 11–19 shows an attempt.

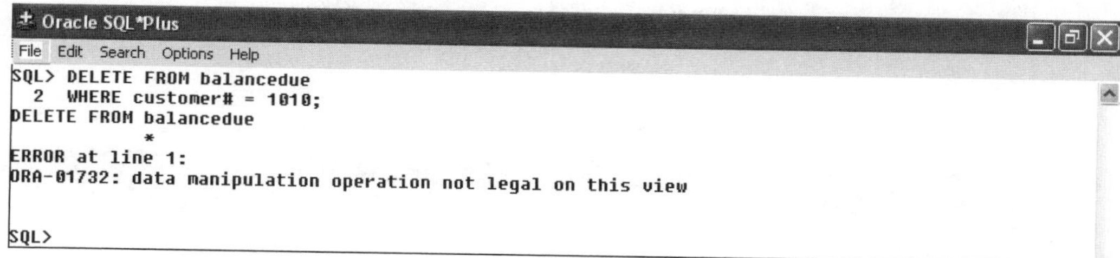

Figure 11-19 Failed DELETE command

As shown in Figure 11–19, apparently the answer is NO! The rationale behind not allowing a record to be deleted from the view is that the data are grouped, so exactly what would be deleted? Rather than allow a user to mistakenly delete what could possibly be several rows in the underlying key-preserved table, the operation is simply not allowed. Therefore, if you really want to delete a particular customer's orders, you would have to delete them manually from the ORDERS table. (Of course, if that deletion violates any existing constraints between the ORDERS and ORDERITEMS tables, then the deletion may not be possible without including an ON CASCADE DELETE option.) And now a third rule has been identified:

DML operations are not permitted if the view includes a group function or a GROUP BY clause.

DML Operations on a Complex View Containing DISTINCT or ROWNUM

Before summarizing the lessons you have learned about complex views and DML operations, you should be aware of using the DISTINCT keyword during view creation and the impact of ROWNUM on DML operations in a view. Let's look at those next.

When using the DISTINCT keyword in a subquery to create a view, remember that in a SELECT clause, the keyword instructs Oracle9*i* to suppress duplication. In other words, if more than one row of a table contains the same data, then only display the contents once in the view. In a way, if you consider each unique row as one group, it almost acts like a GROUP BY clause. Therefore, a fourth rule has emerged:

DML operations on a view that is created with the DISTINCT keyword are not permitted.

If the rationale behind not permitting DML operations doesn't seem valid, then try to create a "distinct" view of the different ISBNs in the ORDERITEMS table and attempt some DML operations.

The second point that needs to be made applies to the concept of **ROWNUM**. ROWNUM is a pseudo column that applies to every table, even though it is not displayed through a `SELECT *` command, or even the DESCRIBE command. Oracle9*i* assigns every record in a table a row number to indicate the row's position within the table.

The query issued in Figure 11–20 instructs Oracle9*i* to list the last name of each customer and the record's position in the CUSTOMERS table. However, the query also requires that the last names be presented in alphabetical order.

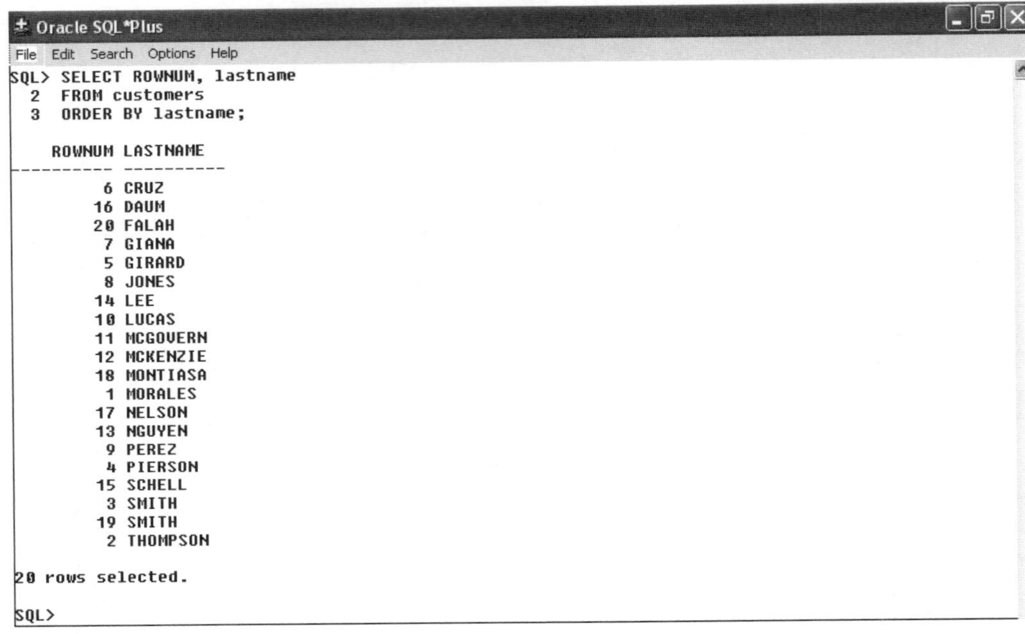

```
Oracle SQL*Plus
File Edit Search Options Help
SQL> SELECT ROWNUM, lastname
  2  FROM customers
  3  ORDER BY lastname;

    ROWNUM LASTNAME
---------- ----------
         6 CRUZ
        16 DAUM
        20 FALAH
         7 GIANA
         5 GIRARD
         8 JONES
        14 LEE
        10 LUCAS
        11 MCGOVERN
        12 MCKENZIE
        18 MONTIASA
         1 MORALES
        17 NELSON
        13 NGUYEN
         9 PEREZ
         4 PIERSON
        15 SCHELL
         3 SMITH
        19 SMITH
         2 THOMPSON

20 rows selected.

SQL>
```

Figure 11-20 Last names presented in alphabetical order

As shown in the query results in Figure 11–20, the customer with the last name of Cruz is alphabetically listed first. However, the customer is actually stored in the sixth row of the CUSTOMERS table. In fact, based on customer numbers, the customer named Morales occupies the first record in the actual CUSTOMERS table.

So, what does ROWNUM have to do with DML operations on a complex view? If the subquery of the complex view listed ROWNUM as one of the columns to be included in the view, then no DML operation will be allowed on the view. Because ROWNUM is a pseudo column that Oracle9*i* uses to assign a value to each row, Oracle9*i* will not allow any additions, deletions, or modifications on the data displayed in the view. This results in the final rule:

DML operations are not allowed on views that include the pseudo column ROWNUM.

Summary Guidelines of DML Operations on a Complex View

A summary of the guidelines regulating DML operations on complex views follows.

- DML operations that violate a constraint are not permitted.
- A value cannot be added to a column that contains an arithmetic expression.

- DML operations are not permitted on non key-preserved tables.
- DML operations are not permitted on views that include group functions, a GROUP BY clause, the ROWNUM pseudo column, or the DISTINCT keyword.

Dropping a View

A view can be dropped or deleted using the DROP VIEW command. The syntax for the command is shown in Figure 11–21.

```
DROP VIEW viewname;
```

Figure 11-21 Syntax for the DROP VIEW command

The command to drop the PRICES view, previously discussed in this chapter, is shown in Figure 11–22.

```
DROP VIEW prices;
```

Figure 11-22 Command to drop the PRICES view

Once the command is successfully executed, Oracle9*i* will return the message that the view has been dropped, as shown in Figure 11-23. However, the data that were displayed by the view are still available in the underlying tables originally used to create the view. All that has been deleted is the database object named PRICES that pointed to the data contained in the underlying tables.

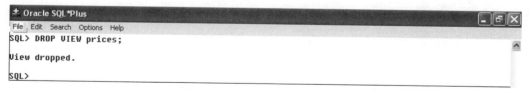

Figure 11-23 Results of the DROP VIEW command

Creating an Inline View

In Chapter 7, a subquery was used in the FROM clause of a SELECT statement to create a "temporary" table that could be referenced by the SELECT and WHERE clauses of the statement. It was considered temporary because nowhere in the database did it actually store a copy of the data returned by the subquery. That temporary table is very similar to what Oracle9*i* calls an **inline view**. The main difference between an inline view and the views that have been presented in previous sections of this chapter is that *an inline view exists only while the command is being executed*. It is not a permanent database object and cannot be referenced again by a subsequent query. It is basically used to provide a temporary data source while a command is being executed. Perhaps the most common usage for an inline view is to perform a "TOP-N" analysis.

"TOP-N" Analysis

Suppose that you want to find the five books that generate the most profit. In Chapter 6, the group function MAX was used to find the most-profitable book. However, using that function yields only the highest value in a column. How would you find the five highest values? This is where **"TOP-N" analysis** is used. Basically, the concepts of an inline view and the pseudo column ROWNUM are merged together to create a temporary list of records in a sorted order, and then the top "N," or number of records, are retrieved. An inline view must be used to perform the analysis because an ORDER BY clause must be used by the subquery to put the records in the correct order before the subquery is passed to the outer query—and ORDER BY clauses are not allowed in the CREATE VIEW command.

The syntax to perform a "TOP-N" analysis is shown in Figure 11–24.

```
SELECT columnname, …
FROM (subquery)
WHERE ROWNUM<=N;
```

Figure 11-24 Syntax for "TOP-N" analysis

To determine the five books that generate the most profit, the subquery will need to calculate the profit for each book and then sort the results in descending order by profit before passing the values to the outer query. The subquery is shown in Figure 11–25.

```
SELECT title, retail-cost profit
FROM books
ORDER BY retail-cost DESC;
```

Figure 11-25 Subquery needed to perform "TOP-N" analysis

To actually perform the analysis, the subquery must be nested in the FROM clause of a SELECT statement to create the inline view. When the sorted results are passed from the subquery, each row will be assigned a row number to identify its position. Then, a WHERE clause is added to the outer query to select only those books that have a ROWNUM that is less than or equal to the desired "N". In this case, "N" is five because you are looking for the five most-profitable books.

The command to determine the five most-profitable books is shown in Figure 11–26.

```
SELECT title, profit
FROM (SELECT title, retail-cost profit
      FROM books
      ORDER BY retail-cost DESC)
WHERE ROWNUM <=5;
```

Figure 11-26 Command to perform "TOP-N" analysis

When the command given in Figure 11–26 is executed, Oracle9*i* will display only the books that have a row number less than or equal to five. Because the data are received by the outer query in descending order based upon profit, then the top five most-profitable books will be assigned the row numbers one through five, as shown in Figure 11–27.

```
Oracle SQL*Plus
File Edit Search Options Help
SQL> SELECT title, profit
  FROM (SELECT title, retail-cost profit
        FROM books
        ORDER BY retail-cost desc)
  WHERE ROWNUM <=5;

TITLE                              PROFIT
------------------------------ ----------
PAINLESS CHILD-REARING              41.95
BODYBUILD IN 10 MINUTES A DAY        31.2
HOLY GRAIL OF ORACLE                 28.7
DATABASE IMPLEMENTATION             24.55
BUILDING A CAR WITH TOOTHPICKS      22.15

SQL>
```

Figure 11-27 Five most-profitable books

As shown in Figure 11–27, the five most-profitable books have a profit range between $22.15 and $41.95. But what if management wants to know the titles of the three least-profitable books? The simplest solution is to use a subquery to sort the data, so the book that is least profitable receives the first row number, the second least-profitable book receives the second row number, etc. This query and output are shown in Figure 11–28.

```
Oracle SQL*Plus
File Edit Search Options Help
SQL> SELECT title, profit
  FROM (SELECT title, retail-cost profit
        FROM books
        ORDER BY retail-cost)
  WHERE ROWNUM <=3;

TITLE                              PROFIT
------------------------------ ----------
HANDCRANKED COMPUTERS                 3.2
COOKING WITH MUSHROOMS               7.45
THE WOK WAY TO COOK                  9.75

SQL>
```

Figure 11-28 Three least-profitable books

The subquery in Figure 11–28 has been modified so the data will be sorted in ascending order before it is passed to the outer query. The WHERE clause has also been changed to select only the first three books that are shown in the results. In this case, the lowest profit generated is $3.20, the next lowest profit is $7.45, and the third lowest profit is $9.75. Although the query actually returns the lowest "N" values, it is still considered "TOP-N" analysis because the results consist of the top "N" row numbers.

CHAPTER SUMMARY

- A view is a pseudo table that is used to display data that exist in the underlying database tables.
- A view can be used to simplify queries or to restrict access to sensitive data.
- A view is created with the CREATE VIEW command.
- Any DML operation can be performed through a view to change table contents—as long as it does not violate any of the table's constraints.
- A view cannot be modified. To change a view, it must be dropped and then re-created, or the CREATE OR REPLACE VIEW command must be used.
- Any DML operation can be performed on a simple query if it does not violate a constraint.
- A view that contains expressions or functions, or joins multiple tables, is considered a complex view.
- A complex view can only be used to update one table. The table must be a key-preserved table.
- Oracle9*i* assigns a row number to every row in a table to indicate its position in the table. The row number can be referenced by the keyword ROWNUM.
- DML operations are not permitted on views that include group functions, a GROUP BY clause, the ROWNUM pseudo column, or the DISTINCT keyword.
- DML operations are not permitted on non key-preserved tables.
- A record cannot be added to a table that contains an expression.
- A view can be dropped with the DROP VIEW command. The data in the view are not deleted; they still exist in the original tables.
- An inline view can only be used by the current statement and can include an ORDER BY clause.
- "TOP-N" analysis uses the row number of sorted data to determine a range of top values.

Chapter 11 Syntax Summary

The following table presents a summary of the syntax that you have learned in this chapter. You can use the table as a study guide and reference.

Syntax Guide		
Element	**Command Syntax**	**Example**
Command to create a new view	`CREATE [FORCE|NOFORCE] VIEW viewname (columnname, …)] AS subquery [WITH CHECK OPTION [CONSTRAINT constraintname]] [WITH READ ONLY];`	`CREATE VIEW inventory AS SELECT isbn, title, retail price FROM books WITH READ ONLY;`
Command to replace an existing view	`CREATE OR REPLACE [FORCE|NOFORCE] VIEW viewname (columnname, …)] AS subquery [WITH CHECK OPTION [CONSTRAINT constraintname]] [WITH READ ONLY];`	`CREATE OR REPLACE VIEW inventory AS SELECT isbn, title, retail price FROM books;`
Command to drop a view	`DROP VIEW viewname;`	`DROP VIEW inventory;`
Command to create an inline view	`SELECT columnname, … FROM (subquery) WHERE ROWNUM<=N;`	`SELECT title, profit FROM (SELECT title, retail-cost profit FROM books ORDER BY retail-cost DESC) WHERE ROWNUM <=5;`

Review Questions

To answer the following questions, refer to the tables in Appendix A.

1. How is a simple view different from a complex view?
2. Under what circumstances is a DML operation not allowed on a simple view?
3. When should the FORCE keyword be used in the CREATE VIEW command?
4. What is the purpose of the WITH CHECK OPTION constraint?
5. List the guidelines regarding DML operations on complex views.
6. How can you ensure that no user will be able to change the data displayed by a view?

7. What is the difference between a key-preserved and non key-preserved table?
8. What command can be used to modify a view?
9. What is the purpose of ROWNUM in "TOP-N" analysis?
10. What happens to the data that were displayed by a view if the view is deleted?

MULTIPLE CHOICE

To answer the following questions, refer to the tables in Appendix A.

Questions 1–7 are based upon successful execution of the following statement:

```
CREATE VIEW changeaddress
AS SELECT customer#, lastname, firstname, order#, shipstreet,
          shipcity, shipstate, shipzip
   FROM customers NATURAL JOIN orders
   WHERE shipdate IS NULL
WITH CHECK OPTION;
```

1. Which of the following statements is correct?
   a. No DML operations can be performed on the CHANGEADDRESS view.
   b. The CHANGEADDRESS view is a simple view.
   c. The CHANGEADDRESS view is a complex view.
   d. The CHANGEADDRESS view is an inline view.
2. Assuming that there is only a primary key and that FOREIGN KEY constraints exist on the underlying tables, which of the following commands will return an error message?
   a. `UPDATE changeaddress SET shipstreet = '958 ELM ROAD'`
      `WHERE customer# = 1020;`
   b. `INSERT INTO changeaddress VALUES (9999, 'LAST', 'FIRST',`
      `9999, '123 HERE AVE', 'MYTOWN', 'AA', 99999);`
   c. `DELETE FROM changeaddress WHERE customer# = 1020;`
   d. all of the above
   e. only a and b
   f. only a and c
   g. none of the above
3. Which of the following is the key-preserved table for the CHANGEADDRESS view?
   a. CUSTOMERS table
   b. ORDERS table
   c. Both tables together serve as a composite key-preserved table.
   d. none of the above

4. Which of the following columns serves as the primary key for the CHANGEADDRESS view?
   a. Customer#
   b. Lastname
   c. Firstname
   d. Order#
   e. Shipstreet
5. If a record for a customer is deleted from the CHANGEADDRESS table, the customer information is then deleted from which underlying table?
   a. CUSTOMERS
   b. ORDERS
   c. CUSTOMERS and ORDERS
   d. Neither—the DELETE command cannot be used for the CHANGEADDRESS table.
6. Which of the following is correct?
   a. ROWNUM cannot be used with the view because it is not included in the results returned from the subquery.
   b. The view is a simple view because it does not include a group function or a GROUP BY clause.
   c. The data in the view cannot be presented in descending order by customer number because an ORDER BY clause is not allowed when working with views.
   d. all of the above
   e. none of the above
7. Assuming one of the orders has shipped, which of the following is true:
   a. The CHANGEADDRESS view cannot be used to update the shipping date of the order due to the WITH CHECK OPTION constraint.
   b. The CHANGEADDRESS view cannot be used to update the shipping date of the order because the column is not included in the view.
   c. The CHANGEADDRESS view cannot be used to update the shipping date of the order because the ORDERS table is not the key-preserved table.
   d. The CHANGEADDRESS view cannot be used to update the shipping date of the order because the UPDATE command cannot be used on the data in the view.

Questions 8–12 are based upon successful execution of the following command:

```
CREATE VIEW changename
AS SELECT customer#, lastname, firstname
   FROM customers
WITH CHECK OPTION;
```

Assume that the only constraint on the CUSTOMERS table is a PRIMARY KEY constraint.

8. Which of the following is a correct statement?
   a. No DML operations can be performed on the CHANGENAME view.
   b. The CHANGENAME view is a simple view.
   c. The CHANGENAME view is a complex view.
   d. The CHANGENAME view is an inline view.
9. Which of the following columns serves as the primary key for the CHANGENAME view?
   a. Customer#
   b. Lastname
   c. Firstname
   d. The view does not have or need a primary key.
10. Which of the following DML operations could never be used on the CHANGENAME view?
    a. INSERT
    b. UPDATE
    c. DELETE
    d. All of the above are valid DML operations for the CHANGENAME view.
11. The INSERT command cannot be used with the CHANGENAME view because:
    a. A key-preserved table is not included in the view.
    b. The view was created with the WITH CHECK OPTION constraint.
    c. The inserted record would not be accessible by the view.
    d. None of the above—an INSERT command can be used on the table as long as the PRIMARY KEY constraint is not violated.

12. If the CHANGENAME view needs to include the customer's zip code as a means of verifying the change (i.e., to authenticate the user), then:
    a. The CREATE OR REPLACE VIEW command can be used to re-create the view with the necessary column included in the new view.
    b. The ALTER VIEW...ADD COLUMN command can be used to add the necessary column to the existing view.
    c. The CHANGENAME view can be dropped and then the CREATE VIEW command can be used to re-create the view with the necessary column included in the new view.
    d. All of the above can be performed to include the customer's zip code in the view.
    e. Only a and b will include the customer's zip code in the view.
    f. Only a and c will include the customer's zip code in the view.
    g. None of the above will include the customer's zip code in the view.
13. Which of the following DML operations cannot be performed on a view that contains a group function?
    a. INSERT
    b. UPDATE
    c. DELETE
    d. All of the above can be performed on a view that contains a group function.
    e. None of the above can be performed on a view that contains a group function.
14. A user will not be able to perform any DML operations on which of the following?
    a. views that are created with the WITH READ ONLY option
    b. views that include the DISTINCT keyword
    c. views that include a GROUP BY clause
    d. All of the above will allow DML operations.
    e. None of the above will allow DML operations.
15. A "TOP-N" analysis is performed by determining the rows with:
    a. the highest ROWNUM values
    b. a ROWNUM value greater than or equal to N
    c. the lowest ROWNUM values
    d. a ROWNUM value less than or equal to N

16. To assign names to the columns in a view, you can:
    a. Assign aliases in the subquery, and the aliases will be used for the column names.
    b. Use the ALTER VIEW command to change the column names.
    c. Assign names for up to three columns in the CREATE VIEW clause before the subquery is listed in the AS clause.
    d. None of the above—columns cannot be assigned names for a view; they must keep their original names.
17. Which of the following is correct?
    a. The ORDER BY clause cannot be used in the subquery of a CREATE VIEW command.
    b. The ORDER BY clause cannot be used in an inline view.
    c. The DISTINCT keyword cannot be used in an inline view.
    d. The WITH READ ONLY option must be used with an inline view.
18. If you try to add a row to a complex view that includes a GROUP BY clause, you will receive which of the following error messages?
    a. virtual column not allowed here
    b. data manipulation operation not legal on this view
    c. cannot map to a column in a non key-preserved table
    d. None of the above—no error message will be returned.
19. A simple view can contain which of the following?
    a. data from one or more tables
    b. an expression
    c. a GROUP BY clause for data retrieved from one table
    d. five columns from one table
    e. all of the above
    f. none of the above
20. A complex view can contain which of the following?
    a. data from one or more tables
    b. an expression
    c. a GROUP BY clause for data retrieved from one table
    d. five columns from one table
    e. all of the above
    f. none of the above

HANDS-ON ASSIGNMENTS

To perform the following activities, refer to the tables in Appendix A.

1. Create a view that will list the name of the contact person at each publisher and the person's phone number. Do not include the publisher's ID in the view. Name the view CONTACT.
2. Change the CONTACT view so that no users can accidentally perform DML operations on the view.
3. Create a view called HOMEWORK11 that will include the columns named Col1 and Col2 from the FIRSTATTEMPT table. Make certain the view will be created even if the FIRSTATTEMPT table does not exist.
4. Attempt to view the structure of the HOMEWORK11 table.
5. Create a view that will list the ISBN and title for each book in inventory along with the name and telephone number of the individual to contact in the event the book needs to be reordered. Name the view REORDERINFO.
6. Try to change the name of one of the individuals in the REORDERINFO view to your name. Was there an error message displayed when performing this step? If so, what was the cause of the error message?
7. Select one of the books in the REORDERINFO view and try to change the ISBN of the book. Was there an error message displayed when performing this step? If so, what was the cause of the error message?
8. Delete the record in the REORDERINFO view that contains your name (if that step was not performed successfully, then delete one of the contacts already listed in the table). Was there an error message displayed when performing this step? If so, what was the cause of the error message?
9. Issue a ROLLBACK command to undo any changes made with any previous DML operations.
10. Delete the REORDERINFO view.

A Case for Oracle9*i*

To perform the following activity, refer to the tables in Appendix A.

The Marketing Department of JustLee Books is about to begin an aggressive marketing campaign to generate sales to repeat customers. Their strategy will be to look at existing customers' previous purchases, and then based on the categories from which those customers have made purchases, JustLee Books will send promotional information about other books in the same category that are highly profitable books for the company.

The Marketing Department has requested that you identify the five most frequently purchased books and the percentage of profit generated by each book. The percentage of profit can be calculated by using the formula ((retail-cost)/cost*100). The employees in the Marketing Department will use the potential profitability of the marketing campaign to determine how much money to budget for the campaign.

In a memo, provide management with a list of the five most frequently purchased books and the percentage of profit generated by each book.

CHAPTER

12

ADDITIONAL DATABASE OBJECTS

Objectives

After completing this chapter, you should be able to do the following:

- Define the purpose of a sequence and state how it can be used by an organization
- Explain why gaps may appear in the integers generated by a sequence
- Use the CREATE SEQUENCE command to create a sequence
- Identify which options cannot be changed by the ALTER SEQUENCE command
- Use NEXTVAL and CURRVAL in an INSERT command
- Explain when Oracle9*i* will automatically create an index
- Create an index, using the CREATE INDEX command
- Delete an index, using the DELETE INDEX command
- Create a Public synonym
- Delete a Public synonym
- Identify the contents of different versions of views used to access the data dictionary, based on the prefix of the view

The tables, constraints, and views created in previous chapters are all considered database objects. A **database object** is simply anything that has a name and a defined structure. Three other database objects commonly used in Oracle9*i* are sequences, indexes, and synonyms. They are the subjects of this chapter.

- A **sequence** generates sequential integers that can be used by organizations to assist with internal controls or simply to serve as a primary key for a table.
- A database **index** serves the same basic purpose as an index in a book by allowing users to quickly locate specific records.

- A **synonym** is a simple name, like a nickname, given to an object with a complex name. Some organizations' naming conventions create complex object names that are difficult for users to remember. A synonym can be created with the **CREATE SYNONYM** command to provide an alternative, simplified name to identify database objects. A synonym can be either a **private synonym**, used by an individual to reference objects owned by that person, or it can be a **public synonym** to be used by others to access an individual's database objects.

This chapter will demonstrate how to create, maintain, and delete sequences and how to create and delete indexes and synonyms. Figure 12–1 provides an overview of this chapter's contents.

Description	Command Syntax

Create a sequence to generate a series of integers	`CREATE SEQUENCE` *sequencename*
	`[INCREMENT BY` *value*`]`
	`[START WITH` *value*`]`
	`[{MAXVALUE` *value* `| NOMAXVALUE}]`
	`[{MINVALUE` *value* `| NOMINVALUE}]`
	`[{CYCLE | NOCYCLE}]`
	`[{ORDER | NOORDER}]`
	`[{CACHE` *value* `| NOCACHE}];`

Alter a sequence	`ALTER SEQUENCE` *sequencename*
	`[INCREMENT BY` *value*`]`
	`[{MAXVALUE` *value* `| NOMAXVALUE}]`
	`[{MINVALUE` *value* `| NOMINVALUE}]`
	`[{CYCLE | NOCYCLE}]`
	`[{ORDER | NOORDER}]`
	`[{CACHE` *value* `| NOCACHE}];`

Drop a sequence	`DROP SEQUENCE` *sequencename*`;`
Create an index	`CREATE INDEX` *indexname* `ON` *tablename (columnname, ...);*
Drop an index	`DROP INDEX` *indexname;*
Create a synonym	`CREATE [PUBLIC] SYNONYM` *synonymname* `FOR` *objectname;*
Drop a synonym	`DROP [PUBLIC] SYNONYM synonymname;`

Figure 12-1 Overview of chapter contents

Before attempting to work through the examples provided in this chapter, you should open the Chapter12 folder in your Data Files. Run the **prech12.sql** file to ensure that all necessary tables and constraints are available. To run the script, type **`start d:\prech12.sql`** at the SQL> prompt, where **`d:`** represents the location (correct drive and path) of the script. If the script is not available, ask your instructor to provide you with a copy.

SEQUENCES

As was mentioned, a sequence is a database object that can be used to generate a series of integers. These integers are most commonly used in organizations either (1) to generate a unique primary key for each record or (2) for internal control purposes. A brief overview of these two concepts follows.

When values generated by a sequence are used as a primary key, there is no true correlation between the number assigned to a record and the entity it represents. However, depending on the parameters used to create the sequence, database users can be assured that no two records will receive the same primary key value. This is especially important if different users are assigned the task of entering records into a database table, because the possibility exists that more than one user may attempt to assign the same primary key value to different records. For example, if several customer service representatives are entering new customers at the same time, how are customer numbers assigned? Are all the customer service representatives in the same room and asking one another, "What number did your last customer receive?" Not likely. Chances are, they are using a sequence, and thus each customer service representative can be certain that each customer's number is unique.

A sequence can also be used for internal control purposes. Every organization should have internal control procedures to avoid problems with transaction auditing, embezzlement, and accounting errors. Most organizations use sequential numbers to track checks, purchase orders, invoices, or anything else that can be used to record economic events. Through the use of sequential numbers, an auditor can determine whether items such as checks or invoices are missing, which in turn can reveal accounting problems, such as unrecorded transactions, or whether employees have obtained blank checks or invoices for their own use.

Creating a Sequence

A sequence is created with the CREATE SEQUENCE command, using the syntax shown in Figure 12–2. Optional commands are shown in square brackets. Curly brackets indicate that one of the two options shown can be used, but not both.

```
CREATE SEQUENCE sequencename
[INCREMENT BY value]
[START WITH value]
[{MAXVALUE value | NOMAXVALUE}]
[{MINVALUE value | NOMINVALUE}]
[{CYCLE | NOCYCLE}]
[{ORDER | NOORDER}]
[{CACHE value | NOCACHE}];
```

Figure 12-2 Syntax of the CREATE SEQUENCE command

Let's look at each of the syntax elements in Figure 12–2. Note that some elements, such as CACHE and NOCACHE, have two forms.

The **CREATE SEQUENCE** keywords are followed by the name used to identify the sequence. A standard naming convention for sequences is to include `_seq` at the end of the sequence name, which makes it easier to identify.

The **INCREMENT BY** clause is used to specify the intervals that should exist between two sequential values. For checks and invoices, this interval is usually one. However, for sequences that may represent credit card or bank account numbers, the interval might be much larger, so that no two account numbers are very similar (e.g., 13,519 might be a more appropriate interval). If the sequence is incremented by a positive value, then the values generated by the sequence will be in an ascending order. However, if a negative value is specified, the values generated by the sequence will be in a descending order. If the user needs the values to be in a descending order, the keywords INCREMENT BY should still be used. *If the INCREMENT BY clause is not included when the sequence is created, the software will assume that the sequence should be increased by the value of one for each integer generated.*

The **START WITH** clause is used to establish the starting value for the sequence. Oracle9*i* will begin each sequence with the value of *one* unless another value is specified in the START WITH clause. For example, if you would like all customer numbers to consist of four-digit numbers, then the START WITH clause can be assigned the value of 1,000 to avoid having the first 999 customers assigned account numbers with fewer than four digits.

If you expect (or hope) that an organization will have more than 9,000 customers, you might want to begin the sequence with 10,000, so the number of digits will remain constant despite company growth. Otherwise, you might have 9,000 customers with four-digit account numbers and 37,000 customers with five-digit account numbers.

Continuing with the syntax listed in Figure 12–2, the **MINVALUE** and **MAXVALUE** clauses are used to establish a minimum or maximum value, respectively, for the sequence. If the sequence were incremented with a positive value, the MINVALUE clause would not make sense. By the same logic, if the sequence were incremented with a negative value (e.g., the values in the sequence decrease), then a MAXVALUE clause would not be necessary, either. By default, if the minimum and maximum values are not specified, Oracle9*i* will assume **NOMINVALUE** and **NOMAXVALUE**. When the NOMINVALUE option is assumed—or assigned—then the lowest possible value for an increasing sequence is 1, and the lowest possible value for a decreasing sequence is -10^26, or -100,000,000,000,000,000,000,000,000. For the NOMAXVALUE option, 10^27, or 1,000,000,000,000,000,000,000,000,000 is the highest possible value for an ascending sequence, and -1 is the highest possible value for a descending sequence.

The **CYCLE** and **NOCYCLE** options are used to determine whether Oracle9*i* should begin reissuing values from the sequence once the minimum or maximum value has been reached. If the CYCLE option is specified and Oracle9*i* reaches the maximum value for an ascending sequence, the CYCLE option instructs Oracle9*i* to begin the cycle of numbers over again. If the sequence is being used to generate values for a primary key, this can cause problems if the sequence tries to assign a value that already exists in the table. However, some organizations tend to reuse check numbers, order numbers, etc., after an extended period of time rather than let the numbers become astronomically large. Therefore, the sequence must be allowed to reuse the same sequence of numbers. If a user does not specify a cycle option, Oracle9*i* will apply the NOCYCLE option to the sequence, because it is the default value. If the NOCYCLE option is in effect, Oracle9*i* will not generate any numbers after the minimum or maximum value has been reached, and an error message will be returned when the user requests another value from the sequence.

ORDER and **NOORDER** options are used in **application cluster environments** where multiple users may be requesting sequence values at the same time during large transactions (e.g., when printing a large quantity of checks or invoices). The ORDER option instructs Oracle9*i* to return the sequence values in the same order in which the requests were received. If the option is not specified in the CREATE SEQUENCE command, the NOORDER option is assumed by default. In cases where the sequence is being used to generate a primary key, the order of the sequence value is not a problem because each value will still be unique.

Generating sequence values can slow down the processing of requests by other users, especially if large volumes of those values are requested in a short period of time. If the **NOCACHE** option is specified when the sequence is created, then each number is generated when the request for a sequence number is received. However, if the nature of an organization's transactions requires large amounts of sequential numbers throughout a session, the **CACHE** option can be used to have Oracle9*i* pre-generate a set of values and store those values in the server's memory. Then, when a user requests a sequence value, the next available value is assigned—without Oracle9*i* having to generate the number. On the other hand, if the CACHE option is not specified when the sequence is created, Oracle9*i* will assume a default option of CACHE 20, and it will automatically store 20 sequential values in memory for users to access.

Keep this in mind when working with sequences and cached values: When a value is generated, that value has been assigned and cannot be regenerated until the sequence begins a new cycle. Therefore, if Oracle9*i* caches 20 values, then the values have been generated regardless of whether they are actually used. If the system crashes, or if the user does not use the values, then the values generated are lost. If the sequence is being used for internal control purposes, then the gaps that might appear in the sequence due to non-usage cannot be documented, and this can be a cause for concern. For example, suppose that 50 sequential numbers to be assigned as order numbers are cached. After a few orders have been received, the Oracle9*i* server crashes and has to be restarted. All unassigned numbers are now lost, and a gap will exist in the order number sequence.

Gaps may also appear if transactions are rolled back because the cached value cannot be returned for "reuse." In addition, because sequences are independent objects, the same sequence can be used by different users to generate values that are actually inserted into a number of different tables. In other words, the same sequence could be used to generate order numbers and customer account numbers. This would result in gaps in the sequence values that appear in each table. Although this would not cause concern if the values were being used for a primary key column, it would be a problem if the sequence were created for internal control purposes. If the objective of the sequence is to provide a means for internal control and make certain no check, invoices, etc., are missing, then a sequence should only be used for one table, and the values generated should never be cached.

Let's look at an example of how sequences can work. All orders placed by JustLee Books' customers are assigned a four-digit number to uniquely identify each order. However, when sales volume increases, several data-entry clerks will be entering orders into the ORDERS table. Thus, it is possible that two clerks could try to enter the same order number. Although the PRIMARY KEY constraint for the ORDERS table would prevent two orders from having the same number, it still slows down the data-entry process because one of the clerks will need to choose a different order number and then re-enter the order. To avoid this problem, a sequence can be created to generate the order numbers used in the ORDERS table. The command to create that sequence is shown in Figure 12–3.

```
CREATE SEQUENCE orders_ordernumber
INCREMENT BY 1
START WITH 1021
NOCACHE
NOCYCLE;
```

Figure 12-3 CREATE SEQUENCE command to generate a sequence for order numbers

The sequence will be named ORDERS_ORDERNUMBER to identify that the sequence was created to generate order numbers for orders in the ORDERS table. However, using this rationale is not an absolute requirement. You should note that although the name was assigned to indicate its purpose, the values can still be used in any table. The INCREMENT BY clause instructs Oracle9*i* that each number generated should be increased by the value of one. Because the last order number stored in the ORDERS table is 1020, the sequence needs to start at 1021, so there is no gap in the order numbers and no previously assigned value is duplicated. Oracle9*i* is also instructed not to cache any values in memory, which means each value will be generated only when it is requested. The final clause, NOCYCLE, instructs Oracle9*i* not to start the sequence of numbers over after the maximum value has been reached. Figure 12–4 shows the results of creating the ORDERS_ORDERNUMBER sequence. For any clause that has not been included, default values will be used.

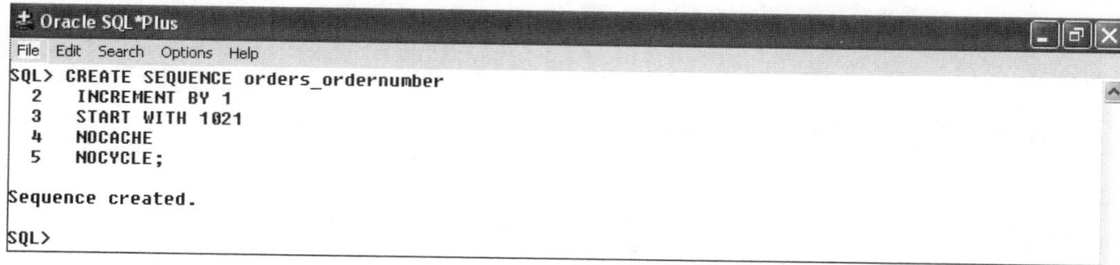

Figure 12-4 Results of creating the ORDERS_ORDERNUMBER sequence

If the CREATE SEQUENCE command is executed successfully, you should see the results displayed in Figure 12–4. After Oracle9*i* has returned the message "Sequence created," you can verify that the sequence exists by querying the **USER_OBJECTS** table in the data dictionary, using a SELECT statement, as shown in Figure 12–5.

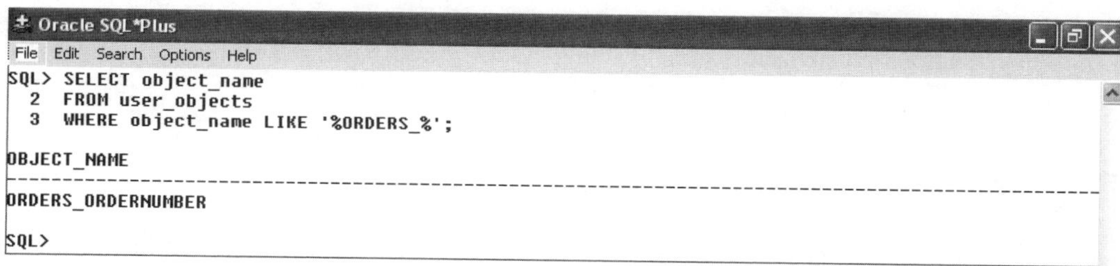

Figure 12-5 Verifying sequence creation through USER_OBJECTS

Viewing the contents of the data dictionary will be discussed later in this chapter.

To verify the individual settings for various clauses of a sequence, you can select the settings from the **USER_SEQUENCES** table of the data dictionary, as shown in Figure 12–6. This is also a quick way to identify which value is the next value to be assigned in the sequence—without accidentally generating a number.

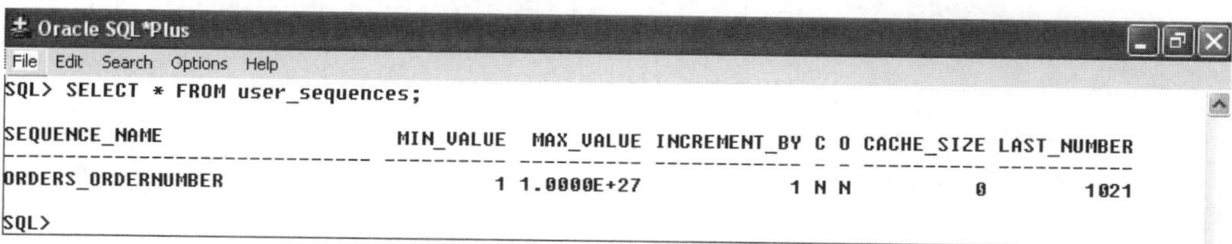

Figure 12-6 Settings of the ORDERS_ORDERNUMBER sequence

To view the next value to be assigned in a sequence created with the NOCACHE option, simply look at the column called Last_Number in the results of the SELECT query.

Using Sequence Values

Sequence values can be accessed through the two pseudocolumns NEXTVAL and CURRVAL. The pseudocolumn **NEXTVAL** (**NEXT VAL**UE) is used to actually generate the sequence value. After a value is generated, it is stored in the **CURRVAL** (**CURR**ENT **VAL**UE) pseudocolumn so it can be referenced again by a user.

Let's look at an example. After the ORDERS_ORDERNUMBER sequence is created, an order is received from customer 1010 on April 6, 2003, for one copy of *Big Bear and Little Dove*, to be shipped to 123 West Main, Atlanta, GA 30418. To process the order, the first step is to place the order information in the ORDERS table. The command to add the order to the ORDERS table is shown in Figure 12–7.

```
INSERT INTO ORDERS
VALUES (orders_ordernumber.nextval, 1010, '06-APR-03',
NULL, '123 WEST MAIN', 'ATLANTA', 'GA', 30418);
```

Figure 12-7 Inserting a record containing a sequence value

In Figure 12–7, the `orders_ordernumber.nextval` reference in the VALUES clause of the INSERT INTO command instructs Oracle9*i* to generate the next sequential value from the ORDERS_ORDERNUMBER sequence. Because the reference is listed as the first column of the VALUES clause, the value generated will be stored in the first column of the ORDERS table, which is identified in the INSERT INTO clause of the command. The NEXTVAL pseudocolumn is preceded by the name of the sequence, which identifies the sequence that should generate the value. Figure 12–8 shows an order added to the ORDERS table.

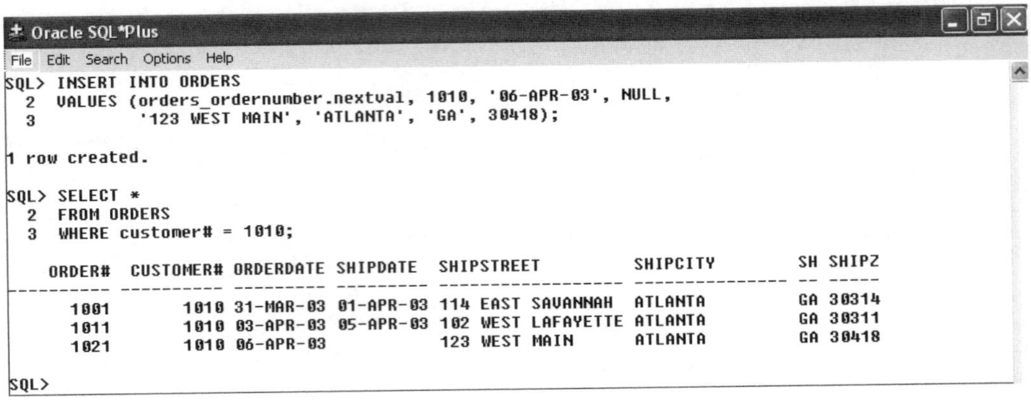

```
Oracle SQL*Plus
File Edit Search Options Help
SQL> INSERT INTO ORDERS
  2  VALUES (orders_ordernumber.nextval, 1010, '06-APR-03', NULL,
  3          '123 WEST MAIN', 'ATLANTA', 'GA', 30418);

1 row created.

SQL> SELECT *
  2  FROM ORDERS
  3  WHERE customer# = 1010;

    ORDER#  CUSTOMER# ORDERDATE SHIPDATE  SHIPSTREET          SHIPCITY        SH SHIPZ
---------- ---------- --------- --------- ------------------- --------------- -- -----
      1001       1010 31-MAR-03 01-APR-03 114 EAST SAVANNAH   ATLANTA         GA 30314
      1011       1010 03-APR-03 05-APR-03 102 WEST LAFAYETTE  ATLANTA         GA 30311
      1021       1010 06-APR-03           123 WEST MAIN       ATLANTA         GA 30418

SQL>
```

Figure 12-8 Order added to the ORDERS table

After the record has been added to the ORDERS table, the SELECT command can be used to view the order number assigned to the new order. The next step is to add the ordered item, *Big Bear and Little Dove*, to the ORDERITEMS table.

If an error message occurs when you try to insert the new order, make certain the order number does not already exist. If it does, run the **prech12.sql** script as instructed at the beginning of this chapter to reset your tables to match the contents of this chapter.

When adding a book to the ORDERITEMS table, the order number must be entered. One approach is to query the ORDERS table to determine the number assigned to the order. However, because the user did not generate another number by referencing the NEXTVAL pseudocolumn again, the assigned order number is still stored as the value in the CURRVAL pseudocolumn. Therefore, the contents of CURRVAL can be used to add the order number to the ORDERITEMS table, and the user does not have to remember the assigned order number. The INSERT INTO command shown in Figure 12–9 can be used to insert the order number, item number, ISBN, and quantity ordered into the ORDERITEMS table.

```
INSERT INTO orderitems
VALUES (orders_ordernumber.currval, 1, 8117949391, 1);
```

Figure 12-9 INSERT INTO command referencing CURRVAL

12

The command given in Figure 12–9 instructs Oracle9*i* to place the value stored in the CURRVAL pseudocolumn for the ORDERS_ORDERNUMBER sequence into the first column of the ORDERITEMS table. *Any reference to CURRVAL will not cause Oracle9i to generate a new order number.* However, if the example had referenced NEXTVAL in Figure 12–9, a new sequence number would have been generated, and the order number entered in the ORDERITEMS table would not have been the same as the order number already stored in the ORDERS table. Figure 12–10 shows CURRVAL inserted into the ORDERITEMS table.

```
Oracle SQL*Plus
File Edit Search Options Help
SQL> INSERT INTO orderitems
  2  VALUES (orders_ordernumber.currval, 1, 8117949391, 1);

1 row created.

SQL> SELECT *
  2  FROM orderitems
  3  WHERE order# = 1021;

    ORDER#      ITEM# ISBN         QUANTITY
---------- ---------- ---------- ----------
      1021          1 8117949391          1

SQL>
```

Figure 12-10 CURRVAL inserted into the ORDERITEMS table

When a user logs in to Oracle9*i*, no value has yet been generated for that session, so no value will be stored in the CURRVAL pseudocolumn and the current value will be NULL. After NEXTVAL has been used to generate a sequence value, CURRVAL will store that value until the next value is generated. *CURRVAL only contains the last value that was generated.*

Altering Sequence Definitions

The settings for a sequence can be changed by using the **ALTER SEQUENCE** command. However, any changes made will be applied only to values generated *after* the modifications are made. The only restrictions that apply to changing the sequence settings are as follows:

- The START WITH clause cannot be changed—the sequence would have to be dropped and re-created to make this change.
- The changes cannot make the existing sequence values invalid (e.g., they cannot change the defined MAXVALUE to a number less than a sequence number that has already been generated).

As shown in Figure 12–11, the ALTER SEQUENCE command follows the same syntax as the CREATE SEQUENCE command.

```
ALTER SEQUENCE sequencename
[INCREMENT BY value]
[{MAXVALUE value | NOMAXVALUE}]
[{MINVALUE value | NOMINVALUE}]
[{CYCLE | NOCYCLE}]
[{ORDER | NOORDER}]
[{CACHE value | NOCACHE}];
```

Figure 12-11 Syntax of the ALTER SEQUENCE command

Suppose that management decides that all order numbers should increase by a value of 10, rather than by 1. Figure 12–12 shows the command to change the INCREMENT BY setting for the ORDERS_ORDERNUMBER sequence.

```
ALTER SEQUENCE orders_ordernumber
INCREMENT BY 10;
```

Figure 12-12 Command to change the INCREMENT BY setting for a sequence

Because no other settings were changed in the ALTER SEQUENCE command, their values are unaffected. The new setting for the sequence can be viewed from the USER_SEQUENCES data dictionary table, as shown in Figure 12–13.

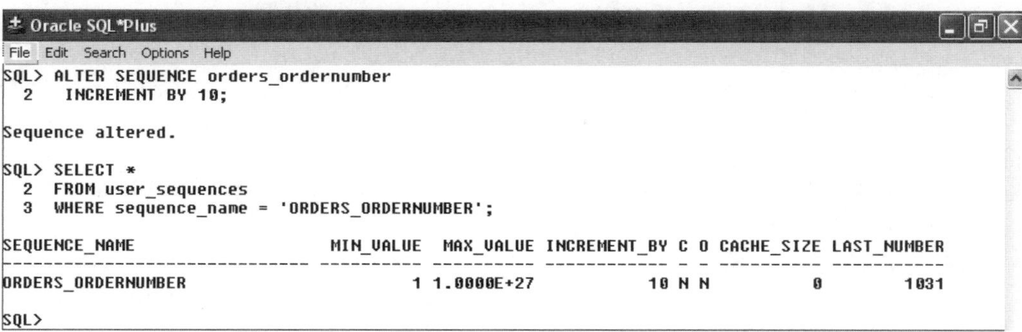

```
Oracle SQL*Plus
File Edit Search Options Help
SQL> ALTER SEQUENCE orders_ordernumber
  2   INCREMENT BY 10;

Sequence altered.

SQL> SELECT *
  2  FROM user_sequences
  3  WHERE sequence_name = 'ORDERS_ORDERNUMBER';

SEQUENCE_NAME                   MIN_VALUE  MAX_VALUE INCREMENT_BY C O CACHE_SIZE LAST_NUMBER
------------------------------ ---------- ---------- ------------ - - ---------- -----------
ORDERS_ORDERNUMBER                      1 1.0000E+27           10 N N          0        1031

SQL>
```

Figure 12-13 New settings for the ORDERS_ORDERNUMBER sequence

Dropping a Sequence

A sequence can be deleted using the **DROP SEQUENCE** command. The syntax for the command is shown in Figure 12–14.

```
DROP SEQUENCE sequencename;
```

Figure 12-14 Syntax of the DROP SEQUENCE command

When a sequence is dropped, it does not affect any values previously generated and stored in a database table. When the DROP SEQUENCE command is successfully executed, the user will receive the message "Sequence dropped," as shown in Figure 12–15.

```
Oracle SQL*Plus
File Edit Search Options Help
SQL> DROP SEQUENCE orders_ordernumber;

Sequence dropped.

SQL>
```

Figure 12-15 DROP SEQUENCE command to drop the ORDERS_ORDERNUMBER sequence

INDEXES

An Oracle9*i* **index** is a separate database object that stores a frequently referenced value and the row ID (ROWID) of the record containing that value—similar to a word and a page number in a book's index. Oracle9*i* uses an index in the same way that a reader might use a book's index. For example, if you need to find a specific item in an 800-page book, you might first check the index, and then if you find it listed in the index, go directly to that page. If you don't consult the index, finding that specific item of information will require you to skim the 800 pages until you find the information.

Similarly, Oracle9*i* uses an index to speed up the search for specific data. For example, suppose that you query Oracle9*i* for a list of all customers living in New York State. For the small tables used in the examples in this textbook, such a query would only take a second to execute and then display the results. However, what if a table contains 40 million rows? It would take Oracle9*i* longer to perform a full table scan and search each row to determine whether each customer in the database lives in New York State. An index would speed up that process.

An index can be created either (1) implicitly by Oracle9*i* or (2) explicitly by a user. Oracle9*i* will automatically create an index whenever a column (or columns) is referenced by a PRIMARY KEY or UNIQUE constraint. The purpose of the index is to allow Oracle9*i* to determine whether a value exists in a table, without having to perform a full table scan. Oracle9*i* automatically accesses the index whenever a value is inserted into or changed in a column that has been designed as a primary key column or as a column that can only contain unique values.

To speed up the retrieval of rows based on a column (or columns) that is frequently referenced in a WHERE clause, a user can also explicitly create an index. Once created, Oracle9*i* will automatically use the index whenever the indexed column(s) is being used as the search condition. Although indexes do speed up row retrieval, their use requires the following considerations:

- Because an index is a database object based on table values, it must be updated automatically by Oracle9*i* *every* time a DML operation is performed on an underlying table. Thus, if you have a table that is frequently updated, the speed for processing the update slows down because Oracle9*i* must now update both the table *and* the index. This is also true when rows are deleted from or added to a table.
- Indexes are beneficial only if a small percentage of the table is expected to be returned in query results. Because having an index requires that Oracle9*i* first examine the index to identify records that meet the criteria and then to retrieve the rows from the actual table, large result sets require Oracle9*i* to do additional work. Basically, a large result set requires Oracle9*i* to scan the entire table. This is a general rule of thumb: If more than five percent of a table's rows are expected to be returned by a query, then an index probably will *not* speed up row retrieval. Why? Because the software will need to access so many portions of the table, the time spent indexing the table will be the same as scanning the entire table. In fact, having an index might slow down data retrieval.

Depending on the type of index, the most appropriate circumstances for creating an index are as follows:

- The table is large and a particular column is frequently used in a WHERE clause.
- The column tends to contain a large number of NULL values because the index can identify rows that contain values in that column and those that do not.

- In some cases, an index might be appropriate for conditions based on a calculated value (e.g., retail-cost) or a function (e.g., UPPER) because the condition being referenced is already identified in the index in its "converted," or derived, state.

Oracle9*i* supports several types of indexes. By default, Oracle9*i* creates binary-tree (B-tree) indexes, which are appropriate for columns containing many distinct values (i.e., high cardinality). For columns containing only a few distinct values (i.e., low cardinality), a bitmap index is more appropriate. The syntax and usage of most types of indexes are covered in Oracle9*i* courses on administration and architecture.

Creating an Index

An index can be created using the **CREATE INDEX** command. The syntax for the CREATE INDEX command is shown in Figure 12–16.

```
CREATE INDEX indexname
ON tablename (columnname, …);
```

Figure 12-16 Syntax of the CREATE INDEX command

Suppose that many of the searches performed on the CUSTOMERS table are based on the last name of the customer. To speed up the retrieval process, a user may decide to create an index for the Lastname column. The command to create the index is shown in Figure 12–17.

12

```
CREATE INDEX customers_lastname_idx
ON customers(lastname);
```

Figure 12-17 Command to create an index for the Lastname column

After the index is created, any future searches based on the last name of a customer will automatically use the index CUSTOMERS_LASTNAME_IDX. (Underscores in the index name are optional and are used merely to improve readability.) The user does not need to instruct Oracle9*i* to use the index; its use will be automatic.

If the majority of the searches for a customer's account are actually based on both the customer's first and last name, then an index can be created on the combination of the two fields. If the search conditions using the Lastname and Firstname columns are combined by the AND operator in the WHERE clause, then the index, shown in Figure 12–18, will speed up row retrieval for a large table.

```
CREATE INDEX customers_name_idx
ON customers (lastname, firstname);
```

Figure 12-18 Creating an index on more than one column

After the command has been executed, Oracle9*i* will return the message "Index created," as shown in Figure 12–19.

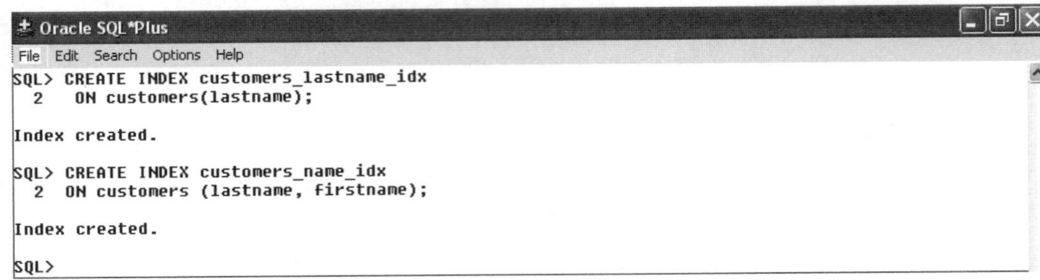
```
SQL> CREATE INDEX customers_lastname_idx
  2   ON customers(lastname);

Index created.

SQL> CREATE INDEX customers_name_idx
  2  ON customers (lastname, firstname);

Index created.

SQL>
```

Figure 12-19 Results returned from creating indexes

When Oracle9*i* references the index called CUSTOMERS_NAME_IDX to identify a specific customer, it will look for the combination of listed values in the two columns and then return the row, or rows, that contain those first and last names.

A **function-based index** can be useful if a query is based on a calculated value or a function. For JustLee Books, one commonly used search criterion is profit. Management may be looking for values that fall above or below a certain dollar profit. To speed up the retrieval of rows that meet a given condition, an index can be created that is based on the calculated profit for each book. The command for creating a function-based index is the same as the syntax in Figure 12–16. *The only difference is that the expression or function that the index is based on is provided in the ON clause rather than in just the column name.* For example, to create an index on the BOOKS table for the dollar profit returned by each book, you could use the command shown in Figure 12–20.

```
CREATE INDEX books_profit_idx
ON books(retail-cost);
```

Figure 12-20 Command to create a function-based index

Whenever the expression **`retail-cost`** is used in the WHERE clause of a SELECT statement, Oracle9*i* will automatically use the index BOOKS_PROFIT_IDX to determine which rows should be returned by the query. Figure 12–21 shows the output from a function-based index that uses the **`retail-cost`** expression.

Figure 12-21 Output from creating a function-based index

You might receive an error message, indicating that you do not have sufficient privileges, when you try to create the index in Figure 12–21. If so, your instructor (or database administrator) will need to grant you the privilege before you will be able to complete the example.

Verifying an Index

After an index has been either implicitly or explicitly created, the USER_INDEXES data dictionary view can be used to determine whether the index exists. Figure 12–22 shows the command to verify an index's existence.

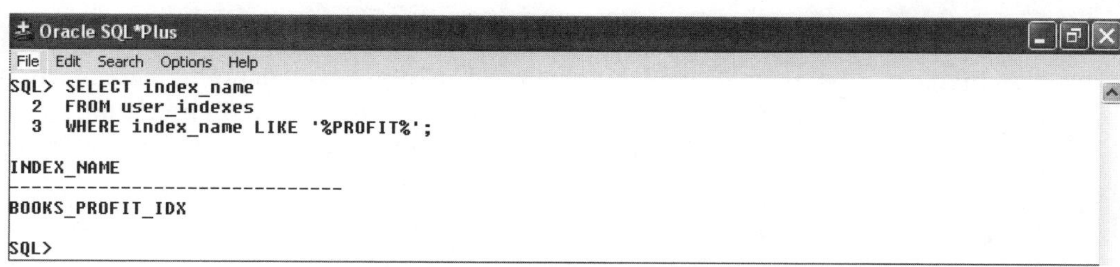

Figure 12-22 Verifying an index

Information about the data dictionary and its views is given at the end of this chapter.

If it appears that an inappropriate index is being used (e.g., queries tend to return a large number of rows or updates are slow), then an index can be deleted by using the **DROP INDEX** command. In addition, an index cannot be modified; if an existing index needs to be changed, it must be deleted and then re-created. The syntax for the DROP INDEX command is shown in Figure 12–23.

```
DROP INDEX indexname;
```

Figure 12-23 Syntax for the DROP INDEX command

As with other DROP and DDL commands, after the DROP INDEX command is executed, the statement cannot be rolled back, and the index is no longer available to Oracle9*i*. Figure 12–24 shows the command to drop the BOOKS_PROFIT_IDX index.

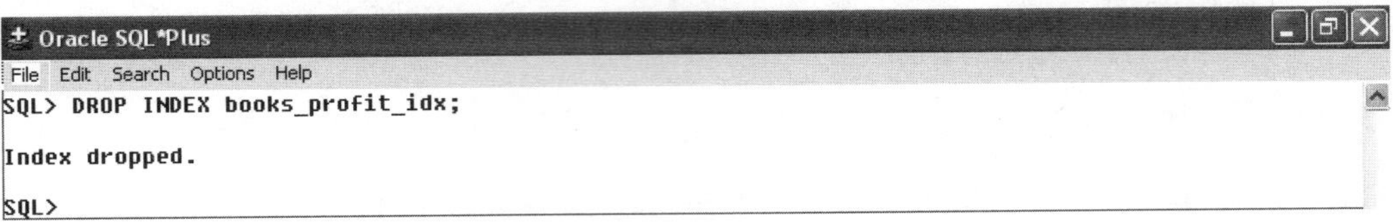

Figure 12-24 Dropping BOOKS_PROFIT_IDX

In Figure 12–24, the index called BOOKS_PROFIT_IDX, previously created in Figure 12–21, has been dropped. If the expression **`retail-cost`** is used in a subsequent query, Oracle9*i* will be required first to calculate the profit for each book and then determine the books that should be returned in the results.

SYNONYMS

When an object is created by a user, unless otherwise specified, it will belong to the schema of that user. As will be discussed in the next chapter, a schema is basically a collection of objects. By grouping objects according to the owner, multiple objects can exist in the same database that have the same object name. *This is only possible if each object belongs to a different schema.*

For example, suppose that a user named Jeff creates a table called PROFITTABLE. Unless Jeff indicates otherwise, the table is an object in the schema called Jeff. If any other user who has permission wants to access Jeff's table, the user would have to identify the table using the correct schema name in the FROM clause of the SELECT statement (e.g., Jeff.PROFITTABLE). If the name of the table is not prefixed by a schema name, then Oracle9*i* will search for the table only in the schema of the user who issued the SELECT statement. If a table with the same name does not exist in the user's schema, then Oracle9*i* will return an error message indicating that the table does not exist.

This type of scenario can cause problems if several users must frequently access a table. In the case of JustLee Books, different users (customer service representatives) enter orders into the ORDERS table. Before a user can enter an order into the table, the individual must remember who actually owns the table, and then prefix the table name with the correct schema. To simplify this process, Oracle9*i* allows synonyms to be created that serve as a substitute for an object name.

The syntax for the CREATE SYNONYM command is shown in Figure 12–25.

```
CREATE [PUBLIC] SYNONYM synonymname
FOR objectname;
```

Figure 12-25 Syntax for creating a synonym

Note the following elements in Figure 12–25:

- The synonym name given in the CREATE SYNONYM clause identifies the substitute name, or permanent alias, for the object listed in the FOR clause.
- The optional PUBLIC keyword can be used so any user in the database can use that synonym to refer to the object.
- The object listed in the FOR clause can be the name of a table, constraint, view, or any other Oracle9*i* object.

For example, if the user who owns the ORDERS table wants other users to be able to access the table without having to reference the correct schema, the synonym shown in Figure 12–26 can be created.

```
CREATE PUBLIC SYNONYM orderentry
FOR orders;
```

Figure 12-26 Command to create a public synonym

Because the PUBLIC keyword is included in the command in Figure 12–26, then any user can reference the ORDERS table by using ORDERENTRY as the table name. Because the PUBLIC keyword allows any database user to use the synonym, only someone with DBA privileges will be allowed to delete the synonym. Why? To make certain that one user doesn't delete the synonym, which could impact the work of other users.

User privileges will be discussed in Chapter 13.

Look at the SELECT statement in Figure 12–27. The statement uses a synonym.

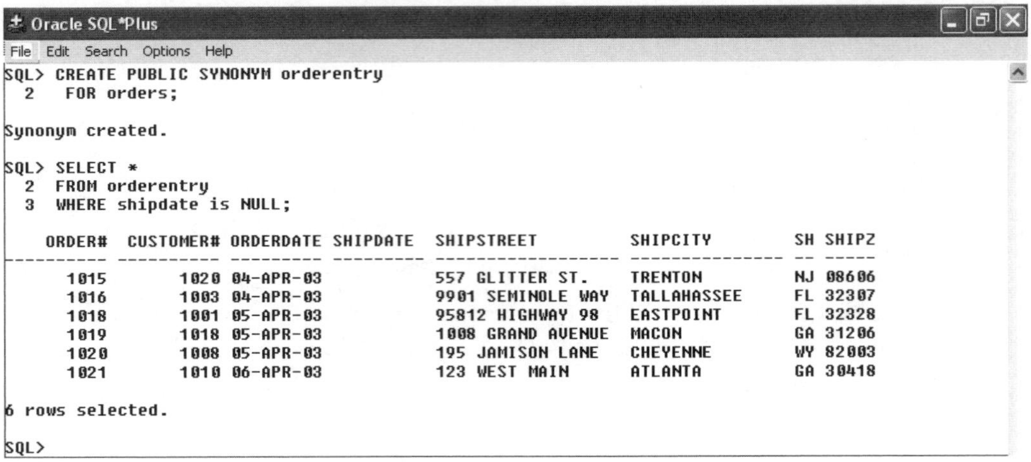

```
Oracle SQL*Plus
File Edit Search Options Help
SQL> CREATE PUBLIC SYNONYM orderentry
  2   FOR orders;

Synonym created.

SQL> SELECT *
  2  FROM orderentry
  3  WHERE shipdate is NULL;

    ORDER#  CUSTOMER# ORDERDATE SHIPDATE  SHIPSTREET         SHIPCITY        SH SHIPZ
---------- ---------- --------- --------- ------------------ --------------- -- -----
      1015       1020 04-APR-03           557 GLITTER ST.    TRENTON         NJ 08606
      1016       1003 04-APR-03           9901 SEMINOLE WAY  TALLAHASSEE     FL 32307
      1018       1001 05-APR-03           95812 HIGHWAY 98   EASTPOINT       FL 32328
      1019       1018 05-APR-03           1008 GRAND AVENUE  MACON           GA 31206
      1020       1008 05-APR-03           195 JAMISON LANE   CHEYENNE        WY 82003
      1021       1010 06-APR-03           123 WEST MAIN      ATLANTA         GA 30418

6 rows selected.

SQL>
```

Figure 12-27 SELECT statement using a synonym

As shown in Figure 12–27, after a synonym is created, the synonym can be substituted for the object name in a command. When Oracle9*i* tries to find that object in the database, the software takes this path: First, it searches for an object that has the same name as the synonym; if no object is found, then it searches for private synonyms with the synonym's name; if no private synonym is found, then it searches for public synonyms with the synonym's name; if no public synonym is found, then Oracle9*i* returns an error message, indicating that the object does not exist.

Many users like to create synonyms to avoid typing long or complex object names. Synonyms can be created for an individual user by simply omitting the PUBLIC keyword from the CREATE SYNONYM clause. *If the PUBLIC keyword is not included, then the individual who created the private synonym is the only one who can use it.* However, a private synonym can still be used to reference objects contained in someone else's schema. Therefore, if user Jane frequently references the table named PROFITTABLE in user Jeff's schema, then Jane can create a synonym for the object Jeff.PROFITTABLE, and she will not need to remember to include the schema name when she accesses the table.

Deleting a Synonym

The **DROP SYNONYM** command is used to drop both private and public synonyms. However, if the synonym being dropped is a public synonym, then you *must* include the keyword PUBLIC. The syntax for the DROP SYNONYM command is shown in Figure 12–28.

```
DROP [PUBLIC] SYNONYM synonymname;
```

Figure 12-28 Syntax of the DROP SYNONYM command

If you attempt to drop a public synonym and forget to include the PUBLIC keyword, Oracle9*i* will return an error message, indicating that a private synonym by that name does not exist. To drop the public synonym ORDERENTRY that was created in Figure 12–26, you could use the command shown in Figure 12–29.

```
DROP PUBLIC SYNONYM orderentry;
```

Figure 12-29 Command to delete the ORDERENTRY synonym

If you have DBA privileges, then the DROP SYNONYM command will execute successfully, and the synonym will no longer be available. However, if your account does not have the correct privileges, you will not be able to delete the synonym even if you are the user who created it.

After the DROP SYNONYM command has executed successfully, Oracle9*i* returns the message "Synonym dropped," as shown in Figure 12–30. After the synonym has been deleted, any user referencing ORDERENTRY in a SQL statement will receive an error message indicating that the object does not exist.

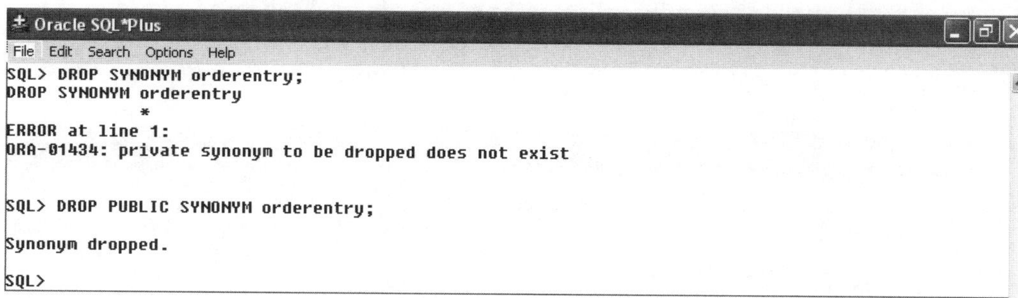

Figure 12-30 Results of dropping the ORDERENTRY synonym

THE DATA DICTIONARY

Oracle9*i* stores all information about database objects in a **data dictionary**. It is owned by the user **SYS**, a user who is automatically created when Oracle9*i* is installed. Information stored in the data dictionary includes the following:

- each object's name
- each object's type
- each object's structure
- each object's owner
- identity of users who have access to each object

12

The data dictionary cannot be queried directly (i.e., it is not a table you can view). The contents of the data dictionary are accessed through a set of predefined views that exist in Oracle9*i*. Therefore, if you need information regarding a database object, you will have to query a view that will display a portion of the data dictionary. More than 100 such views exist in Oracle9*i*. As you receive advanced Oracle9*i* DBA training, you will encounter many of these views.

Previously in this chapter, you were able to access information regarding indexes through the USER_INDEXES view. Each type of object has a separate view (e.g., INDEXES, TABLES, SEQUENCES) containing information about various objects. In addition, information regarding all objects owned by a database user can be accessed through the USER_OBJECTS view. However, you should notice the prefix that is used for the view, USER_.

A view's prefix has a special meaning. The **USER_** prefix indicates that the view contains objects that are owned by the user. Objects accessible to the user (the user has the privilege to use or query the object) have the prefix **ALL_**. For example, to see all tables to which a user can issue SELECT statements, the user can query the **ALL_TABLES** view. The **DBA_** prefix indicates all the objects within the database. For example, to see all of a database's constraints, the user can query the **DBA_CONSTRAINTS** view. Another type of view that is commonly assessed by database administrators is a **dynamic view**, prefixed with **V$**. This view is used to access statistics relating to the database's performance since the database was started. Figure 12–31 summarizes these prefixes.

Prefix	Description
USER_	Objects owned by the user
ALL_	Objects accessible by the user (also includes objects that would be listed with a view using the USER_ prefix)
DBA_	Every object contained within the database, regardless of who owns the object
V$	Dynamic views containing statistics about database performance

Figure 12-31 View prefixes

CHAPTER SUMMARY

- A sequence can be created to generate a series of integers.
- The values generated by a particular sequence can be stored in any table.
- A sequence is created with the CREATE SEQUENCE command.
- Gaps in sequences may occur if the values are stored in various tables, if numbers are cached but not used, or if a rollback occurs.
- A value is generated by using the NEXTVAL pseudocolumn.

- The CURRVAL pseudocolumn will be NULL until a value is generated by NEXTVAL.
- The USER_SEQUENCES table of the data dictionary is used to view sequence settings.
- The ALTER SEQUENCE command is used to modify an existing sequence. The only setting that cannot be modified is the START WITH option or any option that would be invalid because of previously generated values.
- The DROP SEQUENCE command deletes an existing sequence.
- An index can be created to speed up the query process.
- An index can improve query performance on large tables if fewer than five percent of the rows are retrieved.
- DML operations are always slower when indexes exist.
- Oracle9*i* automatically creates an index for PRIMARY KEY and UNIQUE constraints.
- An explicit index is created with the CREATE INDEX command.
- An index is automatically used by Oracle9*i* if a query criterion is based on a column or expression used to create the index.
- Information about an index can be retrieved from the USER_INDEXES view.
- An index can be dropped using the DROP INDEX command.
- An index cannot be modified. It must be deleted and then re-created.
- A synonym provides a type of permanent alias for a database object.
- A public synonym is available to any database user.
- A private synonym is available only to the user who created it.
- A synonym is created by using the CREATE SYNONYM command.
- A synonym is deleted by using the DROP SYNONYM command.
- Only a user with DBA privileges can drop a public synonym.
- Information about all database objects is stored in the data dictionary. The data dictionary is owned by the user SYS.
- The contents of the data dictionary are accessed through views. There are several versions of most views. The content of each version is based on the view's prefix.
- The USER_ prefix indicates that the view contains objects owned by the user.
- The ALL_ prefix indicates that the view contains objects accessible by the user.
- The DBA_ prefix indicates that the view contains all the objects in the database.
- The V$ prefix indicates that the view contains dynamic statistics about the database.

CHAPTER 12 SYNTAX SUMMARY

The following table presents a summary of the syntax that you have learned in this chapter. You can use the table as a study guide and reference.

<table>
<tr><th colspan="3">Syntax Guide</th></tr>
<tr><th>Description</th><th>Command Syntax</th><th>Example</th></tr>
<tr><td>Create a sequence to generate a series of integers</td><td>CREATE SEQUENCE sequencename
[INCREMENT BY value]
[START WITH value]
[{MAXVALUE value |
NOMAXVALUE}]
[{MINVALUE value |
NOMINVALUE}]
[{CYCLE | NOCYCLE}]
[{ORDER | NOORDER}]
[{CACHE value | NOCACHE}];</td><td>CREATE SEQUENCE
orders_ ordernumber
INCREMENT BY 1
START WITH 1021
NOCACHE
NOCYCLE;</td></tr>
<tr><td>Alter a sequence</td><td>ALTER SEQUENCE sequencename
[INCREMENT BY value]
[{MAXVALUE value |
NOMAXVALUE}]
[{MINVALUE value |
NOMINVALUE}]
[{CYCLE | NOCYCLE}]
[{ORDER | NOORDER}]
[{CACHE value | NOCACHE}];</td><td>ALTER SEQUENCE
orders_ordernumber
INCREMENT BY 10;</td></tr>
<tr><td>Drop a sequence</td><td>DROP SEQUENCE sequencename;</td><td>DROP SEQUENCE
orders_ordernumber;</td></tr>
<tr><td>Create an index</td><td>CREATE INDEX indexname
ON tablename
(columnname, ...);</td><td>CREATE INDEX
customers_lastname_idx
ON customers(lastname);</td></tr>
<tr><td>Drop an index</td><td>DROP INDEX indexname;</td><td>DROP INDEX
books_profit_idx;</td></tr>
<tr><td>Create a synonym</td><td>CREATE [PUBLIC] SYNONYM
synonymname
FOR objectname;</td><td>CREATE PUBLIC SYNONYM
orderentry
FOR orders;</td></tr>
<tr><td>Drop a synonym</td><td>DROP [PUBLIC] SYNONYM
synonymname;</td><td>DROP PUBLIC SYNONYM
orderentry;</td></tr>
</table>

<table>
<tr><th colspan="2">View Prefixes</th></tr>
<tr><th>Prefix</th><th>Description</th></tr>
<tr><td>ALL_</td><td>Displays objects accessible by the user</td></tr>
<tr><td>DBA_</td><td>Displays all objects contained in the database</td></tr>
<tr><td>USER_</td><td>Displays objects owned by the user</td></tr>
<tr><td>V$</td><td>Displays dynamic database statistics</td></tr>
</table>

Review Questions

1. How can a sequence be used in a database?
2. When is it not appropriate to use a sequence for internal control purposes?
3. How can gaps appear in a sequence?
4. How can you indicate that the values generated by a sequence should be in a descending order?
5. When is an index appropriate for a table?
6. When does Oracle9*i* automatically create an index for a table?
7. Under what circumstances should you not create an index for a table?
8. What command is used to modify an index?
9. What is the purpose of a synonym?
10. What is the search order used by Oracle9*i* to identify a referenced object?

Multiple Choice

To answer the following questions, refer to the tables in Appendix A.

1. Which of the following generates a series of integers that can be stored in a database?
   a. a number generator
   b. a view
   c. a sequence
   d. an index
   e. a synonym
2. Which syntax is appropriate for removing a public synonym?
   a. `DROP SYNONYM` *`synonymname`*`;`
   b. `DELETE PUBLIC SYNONYM` *`synonymname`*`;`
   c. `DROP PUBLIC SYNONYM` *`synonymname`*`;`
   d. `DELETE SYNONYM` *`synonymname`*`;`
3. Which of the following commands can be used to modify an index?
   a. ALTER SESSION
   b. ALTER TABLE
   c. MODIFY INDEX
   d. ALTER INDEX
   e. none of the above

4. Which of the following generates an integer in a sequence?
   a. NEXTVAL
   b. CURVAL
   c. NEXT_VALUE
   d. CURR_VALUE
   e. NEXT_VAL
   f. CUR_VAL
5. Which of the following is a valid SQL statement?
   a.
   ```
   INSERT INTO publisher
   VALUES (pubsequence.nextvalue, 'HAPPY PRINTING',
   'LAZY LARRY', NULL);
   ```
   b. `CREATE INDEX a_new_index ON (firstcolumn*.02);`
   c. `CREATE SYNONYM pub FOR publisher;`
   d. all of the above
   e. only a and c
   f. none of the above
6. Suppose that user Juan creates a table called MYTABLE that has four columns. The first column has a PRIMARY KEY constraint, the second column has a NOT NULL constraint, the third column has a CHECK constraint, and the fourth column has a FOREIGN KEY constraint. Given this information, how many indexes will Oracle9*i* automatically create when the table and constraints are created?
   a. 0
   b. 1
   c. 2
   d. 3
   e. 4
7. Given the table created in Question 6, which of the following commands can Juan use to create a synonym that will allow anyone to access the table without having to identify his schema in the table reference?
   a.
   ```
   CREATE SYNONYM thetable
   FOR juan.mytable;
   ```
   b.
   ```
   CREATE PUBLIC SYNONYM thetable
   FOR mytable;
   ```
   c.
   ```
   CREATE SYNONYM juan
   FOR mytable;
   ```
   d. none of the above

8. Which of the following statements is true?
   a. A gap can appear in a sequence created with the NOCACHE option if the system crashes before a user can commit a transaction.
   b. Any unassigned sequence values will appear in the USER_SEQUENCE data dictionary table as unassigned.
   c. Only the user who creates a sequence is allowed to delete a sequence.
   d. Only the user who created a sequence is allowed to use the value generated by the sequence.
9. When is it inappropriate to manually create an index?
   a. when queries return a large percentage of the rows in the results
   b. when the table is small
   c. when the majority of the table processes are updates
   d. all of the above
   e. only a and c
10. Which of the following prefixes is used to indicate that the view contains the objects that are accessible to the user?
   a. DBA_
   b. ALL_
   c. USER_
   d. USERS_
11. Who is the owner of the data dictionary for a database?
   a. USER
   b. SYS
   c. EVERYBODY
   d. DBA
12. Oracle9*i* automatically creates an index for which type of constraint(s)?
   a. NOT NULL
   b. PRIMARY KEY
   c. FOREIGN KEY
   d. CHECK
   e. none of the above
   f. only a and b
   g. only b and d

13. Which of the following settings cannot be modified with the ALTER SEQUENCE command?
    a. INCREMENT BY
    b. MAXVALUE
    c. START WITH
    d. MINVALUE
    e. CACHE
14. Which of the following commands can be used to instruct Oracle9*i* to use a particular index when a SELECT statement is being executed?
    a. ALTER SESSION
    b. ALTER INDEX
    c. SELECT...FOR UPDATE...USING
    d. none of the above
15. If the CACHE or NOCACHE options are not included in the CREATE SEQUENCE command, which of the following statements is correct?
    a. Oracle9*i* will automatically generate 20 integers and store those in memory.
    b. No integers will be cached by default.
    c. Only one integer will be cached at a time.
    d. The command will fail.
    e. Oracle9*i* will automatically generate 20 three-digit decimal numbers and store those in memory.
16. Which of the following is a valid command?
    a.
    ```
    SELECT * FROM books FOR UPDATE
    USING book_profit_idx WHERE (retail-cost)>10;
    ```
    b.
    ```
    CREATE INDEX book_profit_idx
    ON (retail-cost)
    WHERE (retail-cost) > 10;
    ```
    c.
    ```
    CREATE FUNCTION INDEX book_profit_idx
    ON books
    WHERE (retail-cost) > 10;
    ```
    d. both a and c
    e. none of the above

17. Which of the following can be used to determine whether or not an index exists?
    a. `DESCRIBE indexname;`
    b. the USERS_INDEXES view
    c. the indexes table
    d. the USER_INDEX view
    e. all of the above
    f. none of the above
18. Which of the following is not a valid option for the CREATE SEQUENCE command?
    a. ORDER
    b. NOCYCLE
    c. MINIMUMVAL
    d. NOCACHE
    e. All of the above are valid options.
19. Which of the following commands can be used to insert a sequence value into a record?
    a. INSERT VALUE INTO
    b. INSERT SEQUENCE.NEXTVAL INTO
    c. INSERT INTO
    d. ADD NEXTVAL TO
20. Which of the following commands will create a private synonym?
    a. CREATE PRIVATE SYNONYM
    b. CREATE NONPUBLIC SYNONYM
    c. CREATE SYNONYM
    d. CREATE PUBLIC SYNONYM

HANDS-ON ASSIGNMENTS

To perform the following activities, refer to the tables in Appendix A.

1. Create a sequence that will generate integers starting with the value 9. Each value should be three less than the previous value generated. The lowest possible value the sequence should be allowed to generate is –1, and it should not be allowed to cycle. Name the sequence MY_FIRST_SEQUENCE.
2. Issue a SELECT statement that will display NEXTVAL for MY_FIRST_SEQUENCE. Because the value is not being placed in a table, reference the DUAL table in the FROM clause of the SELECT statement.

3. Issue a SELECT statement to display the CURRVAL for MY_FIRST_SEQUENCE. Because the value is not being placed in a table, reference the DUAL table in the FROM clause of the SELECT statement.
4. Reissue the SELECT statement from Assignment 2 five more times. Which values or messages does Oracle9*i* display for each execution of the SELECT statement?
5. Change the setting of MY_FIRST_SEQUENCE so the minimum value that can be generated is –1000.
6. Create a private synonym that will allow you to reference the MY_FIRST_SEQUENCE object as NUMGEN.
7. Use a SELECT statement to view the CURRVAL of NUMGEN.
8. Create an index on the CUSTOMERS table to speed up queries that search for customers based on their state of residence. Verify that the index exists, and then delete the index.
9. Delete the NUMGEN synonym.
10. Delete MY_FIRST_SEQUENCE.

A CASE FOR ORACLE9*i*

To perform the following activity, refer to the tables in Appendix A.

Management forecasts that the volume of orders to be processed by JustLee Books will more than triple in the next few months. In fact, the Human Resources Department has already begun recruiting new data-entry clerks, account managers, and IT personnel to compensate for the anticipated growth. As the amount of work begins to increase, the database's performance could begin to suffer, and queries and DML operations may take longer to complete. In addition, some of the new users will not be familiar with some of the names that have been assigned to various views, table names, etc.

Using the training you have received, determine appropriate uses for at least three sequences, indexes, and synonyms that can make the database for JustLee Books more efficient. In a memo to management, you should identify each sequence, index, and synonym that is needed and the rationale supporting your suggestions. You should also state any drawbacks that may affect the performance of the database if the changes are implemented.

CHAPTER

13

USER CREATION AND MANAGEMENT

Objectives

After completing this chapter, you should be able to do the following:

- Explain the concept of authentication
- Create a new user account
- Grant a user the CREATE SESSION privilege
- Make a password expire
- Change the password of an existing account
- Create a role
- Grant privileges to a role
- Assign a user to a role
- Revoke privileges from a user and a role
- Drop a user

This chapter will focus on authentication of users and the security provided by Oracle9*i* software. (Operating-system and third-party authentication are covered in advanced Oracle administration courses.) Specifically, this chapter examines creating, maintaining, and dropping user accounts; granting and revoking privileges; and simplifying the administration of privileges through the use of roles.

Figure 13–1 provides an overview of the commands presented in this chapter. The commands are grouped by category, rather than by the order in which you will encounter them in this chapter.

Command Description	Command Syntax
Creating, Maintaining, and Dropping User Accounts	
Create a user	`CREATE USER` *`username`* `IDENTIFIED BY` *`password`*`;`
Connect to the Oracle server from within SQL*Plus	`CONNECT` *`username/password@connectstring`*

Change or expire a password	`ALTER USER` *`username`*
	`[IDENTIFIED BY` *`newpassword`*`]`
	`[PASSWORD EXPIRE];`

Drop a user	`DROP USER` *`username`*`;`
Granting and Revoking Privileges	

Grant object privileges to users or roles	`GRANT {`*`objectprivilege`*`|ALL} [(`*`columnname`*`),`
	`objectprivilege (columnname)``]`
	`ON` *`objectname`*
	`TO {`*`username`*`|`*`rolename`*`|PUBLIC}`
	`[WITH GRANT OPTION];`

Grant system privileges to users or roles	`GRANT` *`systemprivilege`* `[,` *`systemprivilege`*`, …]`
	`TO` *`username`*`|`*`rolename`* `[,` *`username`*`|`*`rolename`*`, …]`
	`[WITH ADMIN OPTION];`

Revoke object privileges	`REVOKE` *`objectprivilege`* `[,…`*`objectprivilege`*`]`
	`ON` *`objectname`*
	`FROM` *`username`*`|`*`rolename`*`;`

Granting and Revoking Roles	
Create a role	`CREATE ROLE` *`rolename`*`;`
Grant a role to a user	`GRANT` *`rolename`* `[,` *`rolename`*`]` `TO` *`username`* `[,` *`username`*`];`
Assign a default role to a user	`ALTER USER` *`username`* `DEFAULT ROLE` *`rolename`*`;`
Set or enable a role	`SET ROLE` *`rolename`*`;`
Add a password to a role	`ALTER ROLE` *`rolename`* `IDENTIFIED BY` *`password`*`;`
Revoke a role	`REVOKE` *`rolename`* `FROM` *`username`*`|`*`rolename`*`;`
Drop a role	`DROP ROLE` *`rolename`*`;`

Figure 13-1 Overview of chapter contents

Before attempting the examples provided in this chapter, go to the Chapter13 folder in your Data Files and access the **prech13.sql** script. To run the script, type **`start d:\prech13.sql`** at the SQL> prompt, where **d**: represents the location (correct drive and path) of the script. If the script is not available, ask your instructor to provide you with a copy. Your user account must have been assigned the DBA role in Oracle9*i* to run the script file, or you may receive an error message.

If you are using either the Oracle9*i* Enterprise or Standard Edition to complete the examples presented in this chapter, check with your instructor for alternate user names or role names to use; otherwise, you may receive error messages indicating that the names already exist in the system.

The Need for Security

Organizations once used paper copies, a "paper trail," to document everything that occurred within the organization. In the past 25 years, however, organizations have made enormous capital investments in information technology (IT) to support their business activities. As organizations have become increasingly dependent on IT, many no longer need or want a paper trail. Now, most organizations store data in an electronic format. The advantages of the electronic format are that data storage is easy and takes up little space; data retrieval is fast and accurate. The disadvantage of electronic data storage is that data can be more easily destroyed by a natural disaster—or by a few simple keystrokes.

Most organizations have disaster recovery plans and policies/procedures designed to protect their data from threats—which can take many forms. Some threats arise from natural disasters, such as floods, fires, and tornadoes. However, the biggest threat to an organization's data often comes in the form of people—computer criminals and the organization's own employees. Computer criminals, often referred to as hackers, attempt to illegally access a system and then copy, manipulate, or delete the data it contains—or a hacker might simply stop by for a quick look at the data, just to prove he or she can break into the system. By contrast, disgruntled employees—who are trusted by their employers—can purposefully create havoc with an organization's data. Disgruntled employees might see their actions as a way of "getting even" for a promotion they didn't receive, for an insult by a supervisor, or simply because they feel as if their employer doesn't value them. However, even employees with good intentions can damage an organization's data by accidentally deleting records, inserting data into the wrong table, etc.

13

Years ago, the term **hackers** described individuals who were knowledgeable about computers. However, over time, the term has generally been applied to computer criminals who gain illegal access to information systems and the data they contain. In the FBI's cyber-crime report released in 2002, organizations reported that more than $455.8 million was lost the previous year due to unauthorized network access. Approximately 74 percent of the network breaches occurred via the Internet.

Most organizations employ some type of authentication procedure as a "first line" of defense to prevent individuals from illegally accessing data. **Authentication** is the process of ensuring that the individuals trying to access the system are who they claim they are. For example, if you want to withdraw funds through an ATM or use a debit card to make purchases, usually you are required to have both an identity card and a personal identification number (PIN) to prove your identity and complete the transaction. Similarly, Oracle9*i* also requires authentication before you are allowed to "gain entry" into the database system.

The Oracle9*i* database has several methods available to prevent illegal access into the software.

- *Limit privileges to access data:* Even after someone has been authenticated and permitted into the system, the user can only perform certain operations, depending on the privileges that have been granted.
- *Use operating-system authentication:* This is the simplest level of authentication. Basically, this only requires that the operating system authenticate the user through a user name and password. If Oracle9*i* has a valid account for the user, access to Oracle9*i* is automatically provided. In other words, if an individual provides a valid user name and password and is accepted by the operating system, then the person does not have to provide a separate user name and password for the software. However, after the user is connected to the database, he or she must still have privileges in the Oracle9*i* system to perform any type of operations.
- *Support third-party security software:* Obviously, if database data are of a highly sensitive nature (e.g., a government security clearance is required, the data are critical to business operations), then an organization will require more stringent authentication procedures. Thus, Oracle9*i* is designed to support third-party software and hardware devices, such as biometrics that use fingerprints or retinal scans to establish a user's identity.

If a user is to be authenticated by the operating system, then a valid Oracle9*i* account is still required. The account is created by using the command `CREATE USER` *`username`* `IDENTIFIED EXTERNALLY;`.

Most organizations have authentication procedures that lie between the extremes of minimum and maximum security. The most common approach is to have Oracle9*i* require a valid user name and password when the user logs in to the database. However, just because a user is authenticated, that does not mean the user has the right to perform SELECT, UPDATE, CREATE TABLE, or any other operation on the database. It only means that the user has a valid account and provided the correct passwords assigned to that account. As an extra measure of security, Oracle9*i* requires that each user be granted certain privileges before he or she can actually connect to the Oracle server or perform any type of database operation. Operations such as creating user accounts and granting privileges are generally performed by the DBA.

Next, let's look at data security in the context of a newly hired employee at JustLee Books. This chapter will trace the steps for creating a "new user" database ID for that employee and then granting him privileges to access database data.

CREATING A USER

When an individual is hired at JustLee Books, the new employee's supervisor notifies the DBA of the new employee's name and requests an Oracle9*i* database account. The supervisor also tells the DBA the kind of duties the new employee will need to perform and the database objects he or she should be allowed to access.

Creating User Names and Passwords

The first step in creating a new user account is to determine the user's name and password. By default, a user name can contain up to 30 characters, including numbers, letters, and the symbols _, $, and #. A user account is created using the **CREATE USER** command with the syntax shown in Figure 13–2.

```
CREATE USER username
IDENTIFIED BY password;
```

Figure 13-2 Syntax of the CREATE USER command

The DBA for JustLee Books has the user name "scott" and the password "tiger". The following examples will require you to alternate between your existing account and the new account being created. *Whenever the "scott" account is used, you should substitute your own user name and password.* Make certain that you understand who is issuing the command in the examples provided, or you may receive error messages.

Suppose that the DBA has been notified that a new data-entry clerk, Roman Thomas, has been hired by JustLee Books and will need an Oracle9*i* user account. In most cases, the DBA will create the account by using a **coding scheme** for the user's account name, such as the user's first initial followed by his or her last name. Because the new employee's name is Roman Thomas, he will be assigned the account name of "rthomas." As shown in the syntax of the CREATE USER command in Figure 13–2, a password is entered in the **IDENTIFIED BY** clause when the account is created. Usually, a temporary password is assigned by the DBA, and the user is allowed to change the password after he or she has logged in to the database. This allows the user to create a password that is easier to remember but would still be difficult for other individuals to guess. To create the account for the new employee, use the CREATE USER command shown in Figure 13–3.

```
CREATE USER rthomas
IDENTIFIED BY little25car;
```

Figure 13-3 Command to create an account for a new employee

The CREATE USER command in Figure 13–3 will create a new user account with the user name of RTHOMAS and the password LITTLE25CAR. After Oracle9*i* executes the command, the message "User created" will be displayed.

Connecting to the Oracle Server

After the account has been created, a user can change to the new account by issuing the SQL*Plus **CONNECT** command. The CONNECT command can be used to disconnect and reconnect to the Oracle server in one step. The syntax for the CONNECT command is shown in Figure 13–4.

```
CONNECT username/password@connectstring
```

Figure 13-4 Syntax of the SQL*Plus CONNECT command

The *@connectstring* portion of the CONNECT command is not required for Oracle9*i* Personal Edition, only for the Enterprise and Standard Editions. When Oracle9*i* is installed on a network, the Oracle9*i* database and server are installed on a computer that has been designated as the server and on individual workstations for those users who have client software installed. If Oracle9*i* is installed on a standalone computer, then the Oracle9*i* database and server reside on that computer, and no connect or host string is required.

If the DBA issues the CONNECT command to verify that the account for Roman Thomas exists, Oracle9*i* will recognize that the user name and password are valid; however, as shown in Figure 13–5, it will *not* allow the user to log in to the Oracle server because no privileges have been granted to the user's account.

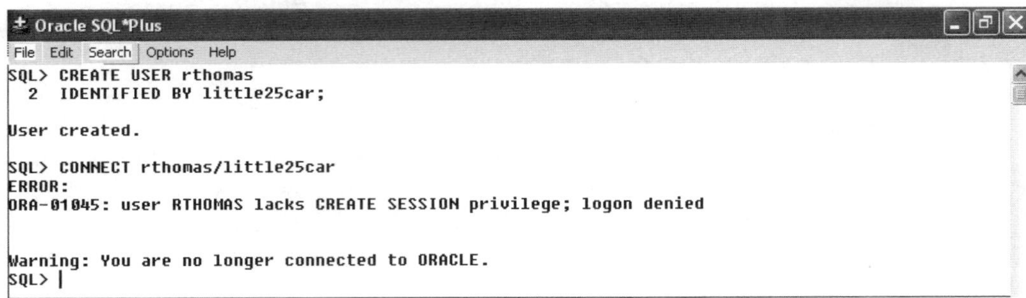

Figure 13-5 Example of a failed connection attempt due to insufficient privileges

To provide the new employee with the privilege necessary to access the Oracle9*i* database, you will first need to re-establish a connection, using the CONNECT command. Simply issue the CONNECT command with your valid user name and password, as shown in Figure 13-6.

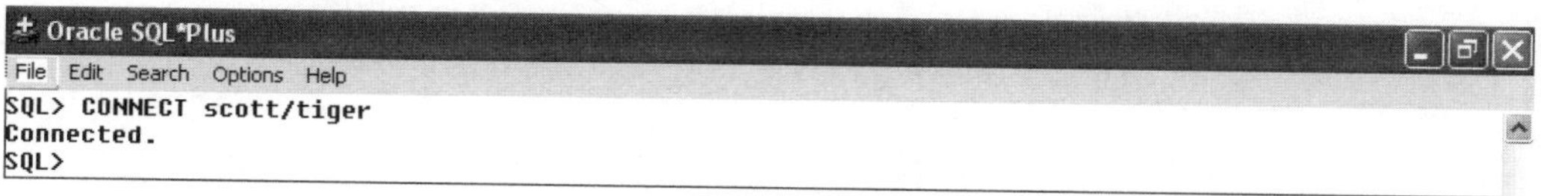

Figure 13-6 Connection re-established with the Oracle server

Granting Privileges

Privileges allow Oracle9*i* users to execute certain SQL statements. Two types of privileges exist in Oracle9*i*: system privileges and object privileges. **System privileges** allow access to the Oracle9*i* database and let users perform DDL operations such as CREATE, ALTER, and DROP on database objects (e.g., tables, views). **Object privileges** allow users to perform DML operations on the data contained within the database objects. First, let's look at object privileges.

Object Privileges

When a user creates an object, he or she automatically has all the object privileges associated with that object. However, if other users need to have access to the data contained in a database object, or the ability to manipulate the data, they must be granted the privilege to do so. The object privileges in Oracle9*i* are as follows:

1. SELECT—allows users to display data contained in a table, view, or sequence. The SELECT privilege allows a user to generate the next sequence value from a sequence by using NEXTVAL.
2. INSERT—allows users to insert data into a table or view.
3. UPDATE—allows users to modify data in a table or view.
4. DELETE—allows users to delete data in a table or view.
5. INDEX—allows users to create an index for a table.
6. ALTER—can only be used to alter the definition of a table or sequence.
7. REFERENCES—allows users to reference a table when creating a FOREIGN KEY constraint. This privilege can only be granted to a user, not to a role.

There are a total of 13 object privileges in Oracle9*i*. The remaining object privileges are: (1) **EXECUTE**, which allows users to run a stored function or procedure; (2) **READ**, which allows users to view binary files (BFILEs) in a directory; (3) **ON COMMIT REFRESH**, which allows users to refresh a view; (4) **QUERY REWRITE**, which allows users to create a view for query rewrite; (5) **UNDER**, which allows users to create a view based on an existing view; and (6) **WRITE**, which allows users to write to a directory.

Granting Object Privileges

Object privileges are assigned or granted to users, using the **GRANT** command. The syntax for the GRANT command is shown in Figure 13–7.

```
GRANT {objectprivilege|ALL} [(columnname),
       objectprivilege (columnname)]
ON objectname
TO {username|rolename|PUBLIC}
[WITH GRANT OPTION];
```

Figure 13-7 Syntax of the GRANT command for object privileges

Let's look at the individual clauses in Figure 13–7.

- The GRANT clause is used to identify the object privilege(s) being assigned. The INSERT, UPDATE, and REFERENCES privileges can also be assigned to specific columns within a table or view. If the object privilege is being assigned to a specific column, the column name should be included in the GRANT clause, within parentheses, after the privilege name. Rather than identifying individual object privileges, the ALL keyword can be substituted to indicate that all object privileges (e.g., SELECT, INSERT, UPDATE, and DELETE) are to be granted. *The curly brackets indicate that either an object privilege or the ALL keyword is used after the GRANT keyword.* Be careful when granting users all available object privileges because this will provide them with the ability to perform any DML operation on the named object.
- The **ON** clause is used to identify the object (e.g., table, view, sequence) to which the privilege(s) applies.
- The **TO** clause identifies the user or role (discussed in a later section) receiving the privilege. Multiple users or roles can receive privileges from the same GRANT command by providing the names in a list, separated by commas. If all database users should receive the privilege being granted, then the PUBLIC keyword can be used in the TO clause, rather than a list of names.
- The **WITH GRANT OPTION** gives the user the ability to grant the same object privileges to other users.

For example, to give Roman Thomas the right to select rows from and insert rows into the CUSTOMERS table, and the right to grant those privileges to other users, you can use the command shown in Figure 13–8.

```
GRANT select, insert
ON customers
TO rthomas
WITH GRANT OPTION;
```

Figure 13-8 Granting object privileges to a user

The WITH GRANT OPTION cannot be used when granting object privileges to roles; it only applies to individual users.

System Privileges

There are approximately 140 system privileges in Oracle9*i*. The ability to create, alter, and drop tables, views, and other database objects are system privileges. Any of the Oracle9*i* object privileges can be granted as system privileges if the keyword ANY is used. For example, the INSERT ANY TABLE command gives the user the ability to add rows to any table, regardless of whether he or she owns the table or has explicit permission to access that particular table. The DROP ANY TABLE command allows a user to delete or truncate any table that exists in the database.

Other system privileges apply to database access and user accounts. For example, to connect to Oracle9*i*, a user must have the **CREATE SESSION** privilege. To create new user accounts, a user must have the CREATE USER privilege. All the system privileges available in Oracle9*i* can be viewed through the data dictionary view **SYSTEM_PRIVILEGE_MAP**. Figure 13–9 displays a portion of the first screen of output of Oracle9*i*'s 140 system privileges.

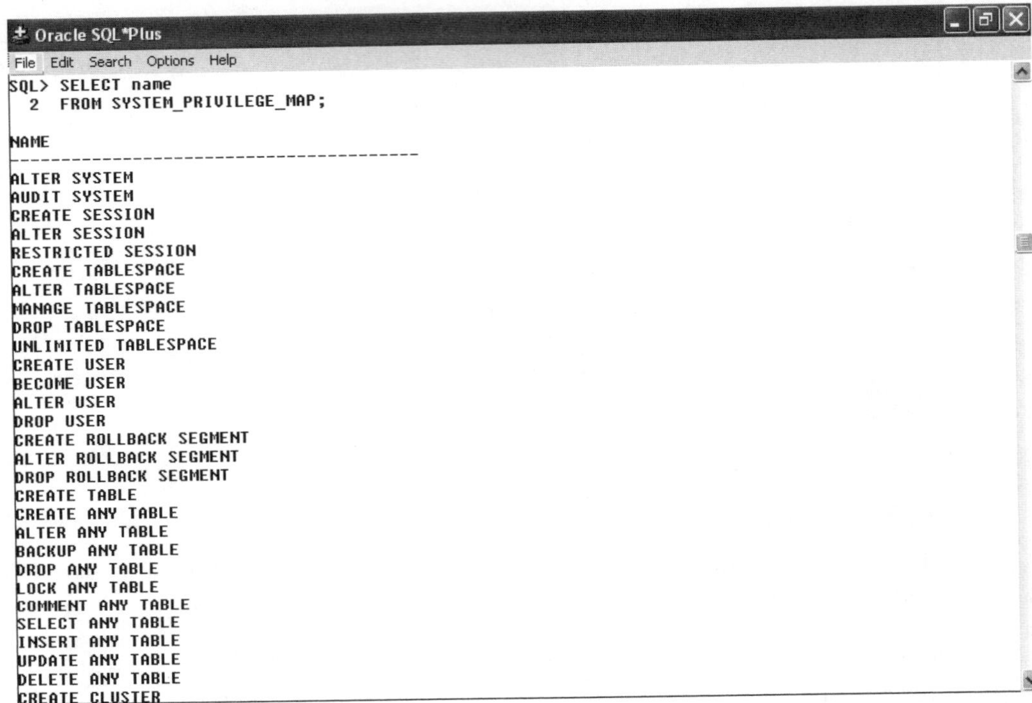

Figure 13-9 Sample of available system privileges (partial output shown)

Granting System Privileges

System privileges can also be granted using the GRANT command. Figure 13–10 shows the syntax for the GRANT command for system privileges.

```
GRANT systemprivilege [, systemprivilege, …]
TO username|rolename [,username|rolename, …]
[WITH ADMIN OPTION];
```

Figure 13-10 Syntax of the GRANT command for system privileges

Let's look at the elements of the statement in Figure 13–10.

- The system privilege being assigned is identified in the GRANT clause. If more than one system privilege is being granted, commas separate the privileges listed. Because a system privilege is not granted for a particular database object, the ON clause is not included for assigning system privileges.
- The user(s) or role(s) receiving the system privileges is identified in the TO clause.
- The **WITH ADMIN OPTION** is used to allow any user or role identified in the TO clause to grant the system privilege(s) to any other database users.

Deploying GRANT Commands

Let's look at how a DBA would deploy a series of GRANT commands to the new employee, Roman Thomas. First, the employee needs to be given the system privilege, CREATE SESSION, to connect to the Oracle9*i* database. The command is shown in Figure 13–11.

```
GRANT CREATE SESSION
TO rthomas;
```

Figure 13-11 Command to give CREATE SESSION privilege to new employee

The only privilege that has been assigned to the new user is the ability to log in to Oracle9*i*. If the user attempts to execute a SELECT statement to view the contents of the BOOKS table, an error message will be returned, as shown in Figure 13–12.

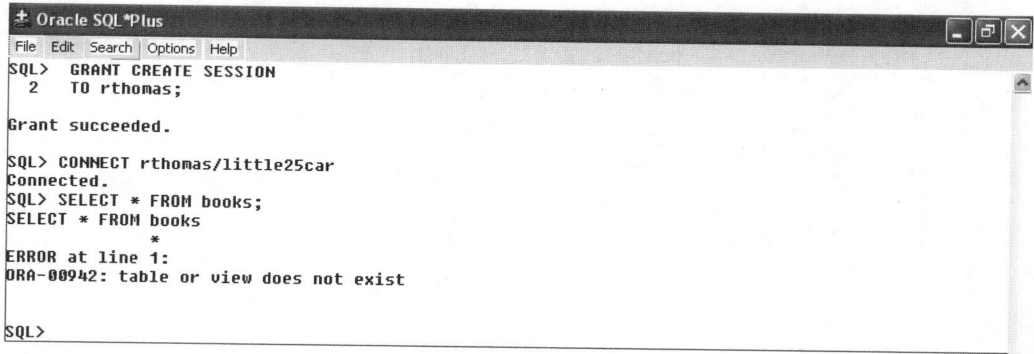

Figure 13-12 The user can log in to Orcale9*i*, but he has no object privileges.

The **prech13.sql** script executed at the beginning of the chapter created a public synonym for the BOOKS table called BOOKS. If the synonym had not existed, the table could not have been referenced without a schema prefix.

Because the new employee will need to view the contents of the BOOKS table, the GRANT command will need to be issued to provide the necessary object privilege. The command is shown in Figure 13–13.

```
GRANT SELECT
ON books
TO rthomas;
```

Figure 13-13 Command to provide the SELECT privilege

Because the command given in Figure 13–13 does not include the WITH GRANT OPTION, the user rthomas will not be allowed to assign the SELECT privilege for the BOOKS table to any other user. However, Roman Thomas can now view the contents of the BOOKS table by using a SELECT command, even though he cannot perform any DML or DDL operations. In Figure 13–14, the user rthomas has been granted the SELECT privilege and can now view the contents of the BOOKS table.

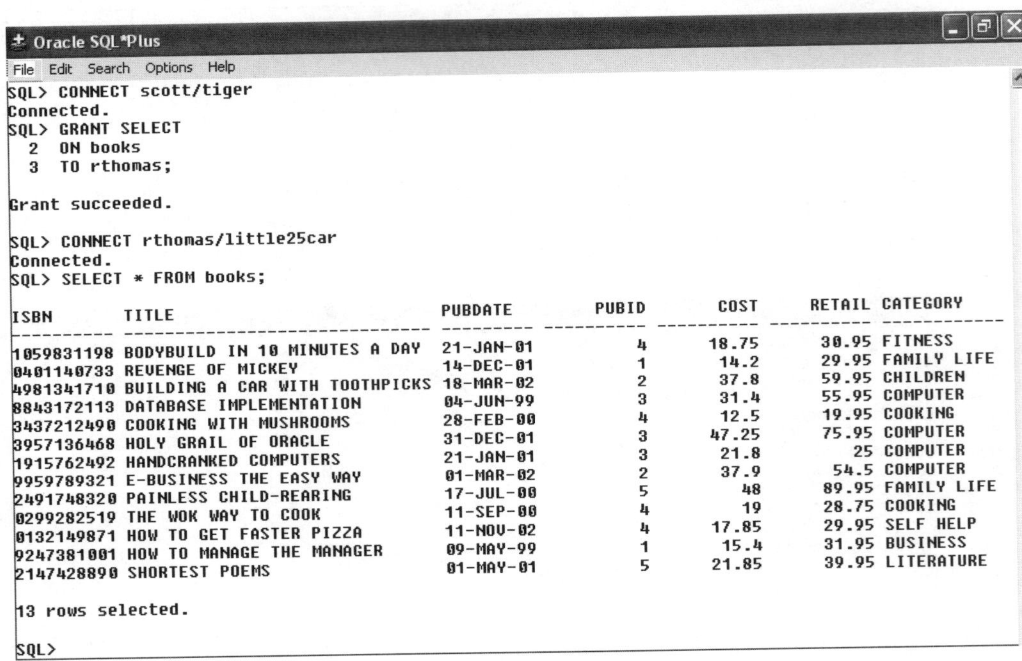

Figure 13-14 New user has received the SELECT object privilege.

Make certain that you switched back to your valid user account, or you will still be logged in to the Oracle9*i* server as "rthomas." Because "rthomas" only has the CREATE SESSION privilege, you will not be able to grant the SELECT privilege.

Changing a User Password

Now that you have verified that the new user's account has been created and he can view the BOOKS table, the user's name and password can be sent to the employee's supervisor. However, as a security precaution, it is customary to force the user to change his or her password after logging in to the database for the first time.

After an account has been created, the simplest approach for forcing a user to change a temporary password is by using the **ALTER USER** command. The DBA can set the current password as "expired," which forces the user to set up a new password as soon as the user attempts to connect to the database. The syntax for the ALTER USER command is shown in Figure 13–15.

```
ALTER USER username
[IDENTIFIED BY newpassword]
[PASSWORD EXPIRE];
```

Figure 13-15 Syntax of the ALTER USER command

If a user simply needs to change the current password of an account, the **IDENTIFIED BY** clause is used to specify the new password. However, if the DBA wants to force the user to change the current password at the time of the next login, the **PASSWORD EXPIRE** option can be used, as shown in Figure 13–16.

```
ALTER USER rthomas
PASSWORD EXPIRE;
```

Figure 13-16 ALTER USER command makes the current password expire immediately

The ALTER USER clause identifies the user whose password is to expire. The DBA is not required to actually know the user's password; the option PASSWORD EXPIRE is simply entered to instruct the software to make the user enter a new password at the time of the next login. Figure 13–17 shows the results of the PASSWORD EXPIRE option.

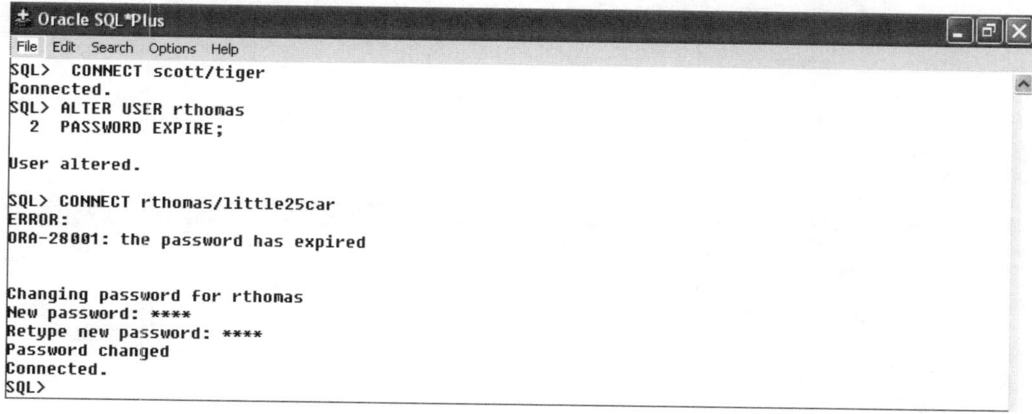

Oracle SQL*Plus
File Edit Search Options Help
SQL> CONNECT scott/tiger
Connected.
SQL> ALTER USER rthomas
  2 PASSWORD EXPIRE;
User altered.
SQL> CONNECT rthomas/little25car
ERROR:
ORA-28001: the password has expired
Changing password for rthomas
New password: ****
Retype new password: ****
Password changed
Connected.
SQL>

Figure 13-17 New password required due to the PASSWORD EXPIRE option

In Figure 13–17, the DBA has made certain that the password currently assigned to Roman Thomas has expired, requiring the user to enter a new password. To test whether the option works correctly, the DBA attempts to connect to Oracle9*i*. As expected, it requires a new entry for the password. Because asterisks are used to emulate the actual password entered, Oracle9*i* requires the user to enter the same password twice. After the same password has been entered a second time, the connection to the database is permitted.

After a user is successfully logged in to an account, the user can change the account's password, using the ALTER USER command with the IDENTIFIED BY clause. After Roman logs in to his account and decides that the new password he selected is too difficult to remember, he could issue the command shown in Figure 13–18.

```
ALTER USER rthomas
IDENTIFIED BY monstertruck42;
```

Figure 13-18 Command to change a password

The command given in Figure 13–18 can be issued by the account owner—or anyone with the ALTER USER system privilege.

When changing an account password, Oracle9*i* does not require the person issuing the command to know the current password for the account. Therefore, this privilege should be assigned with caution. Also, users should be warned that if they log in to Oracle9*i* and then leave their workstation while still connected to the database, anyone could sit down at their computer and change the password for their account. Because Oracle9*i* will allow a user multiple logins to the database for the same account at the same time, someone could be causing damage to the database while the "real" user is performing legitimate tasks. The illegal password change would go undetected until the user's next login attempt.

Once the password has been successfully changed, the message "User altered" will appear, as shown in Figure 13–19.

```
Oracle SQL*Plus
File Edit Search Options Help
SQL> ALTER USER rthomas
  2  IDENTIFIED BY monstertruck42;

User altered.

SQL>
```

Figure 13-19 ALTER USER...IDENTIFIED BY command to change a password

In addition, the user can also change his or her password after logging in to an account by simply issuing the SQL*Plus **PASSWORD** command. The user will be prompted for the old password and then the new password.

Granting Roles

In most cases, a user will need more privileges than just the CREATE SESSION system privilege and the SELECT object privilege for one table. For example, as a data-entry clerk, Roman Thomas will probably need to enter the ship date for orders, update information in the CUSTOMERS table, etc. In fact, all JustLee Books' data-entry clerks would perform those tasks. Rather than assign the same privileges again and again to users who need identical privileges, a simpler approach is to assign the privileges to a role, and then assign the role to a user.

A **role** is simply a group, or collection, of privileges. In most organizations, roles correlate to users' job duties. For example, customer service representatives might need to view all the data in each database table. However, they probably would not need to have the privilege of updating the data in the BOOKS table (ISBN, cost, retail price, etc.). By grouping individuals based on the tasks they need to perform, you can create roles that have been assigned the privileges required by each group. Thus, rather than assign each user a series of individual privileges, you can just assign a collection of privileges—a role—to the users.

Before you can assign privileges to a role, the role must be created by using the **CREATE ROLE** command. The syntax for the command is shown in Figure 13-20.

```
CREATE ROLE rolename;
```

13

Figure 13-20 Syntax of the CREATE ROLE command

After the role has been created, you can grant system and/or object privileges to the role, using the same syntax as when granting the privileges directly to the user. The *only exception* is that an object privilege cannot be granted to a role with the WITH GRANT OPTION. After all the privileges have been assigned to a role, the role can then be assigned to all relevant users, using the GRANT command. The syntax for using the GRANT command to grant a role to a user is shown in Figure 13-21.

```
GRANT rolename [, rolename]
TO username [, username];
```

Figure 13-21 Syntax to grant a role to a user

For example, suppose that each data-entry clerk should be allowed to issue the SELECT, INSERT, and UPDATE commands for any table. Rather than remember exactly which privileges are required every time a new data-entry clerk is hired, a DBA could create a role called DATAENTRY that could be assigned in lieu of the individual privileges. To create the role, connect as the user "Scott," and issue the command shown in Figure 13–22.

```
CREATE ROLE dataentry;
```

Figure 13-22 Command to create a role

After the role has been created, you can then assign the necessary privileges to the newly created DATAENTRY role, using the command shown in Figure 13–23.

```
GRANT SELECT ANY TABLE, INSERT ANY TABLE,
UPDATE ANY TABLE
TO dataentry;
```

Figure 13-23 Command to grant privileges to a role

Now that privileges have been assigned to the DATAENTRY role, the role can be assigned to any new data-entry clerk by granting the role to the employee, using the command shown in Figure 13–24.

```
GRANT dataentry
TO rthomas;
```

Figure 13-24 Command to grant a role to a user

After Roman Thomas has been assigned the DATAENTRY role, he can execute SELECT, UPDATE, and INSERT commands on any table in the database.

Users can be assigned several roles based on the different types of tasks they usually perform. For example, in this database, the user "Scott" is considered a database administrator and has been assigned the majority of the privileges available in Oracle9*i*. Many of the privileges that allow the user Scott to drop tables and views have been assigned through a role named DBA. "DBA" is a predefined role in Oracle9*i* that has all the system privileges already available upon database creation. Users other than the actual database administrator are sometimes assigned DBA privileges—and that can cause major problems if incorrect commands are issued. Usually, these users are assigned several different roles, depending on the various duties they perform. When necessary, the user can assume these different roles to perform various tasks.

A user can be assigned a default role that is automatically enabled whenever the user logs in to the database. The default role should consist of only those privileges the user will frequently need. Privileges that are rarely needed (and that could cause problems in the database if incorrectly used) should be assigned to other roles the user can assume when necessary.

After the user account has been created, an ALTER USER command can be issued to assign a default role to a user by using the syntax shown in Figure 13–25.

```
ALTER USER username DEFAULT ROLE rolename;
```

Figure 13-25 Syntax to assign a default role to a user

Although a user may have several assigned roles—and they can all be enabled at any time—only one can be set as the default role for the user.

If a user needs to assume a different role or set of privileges after connecting to the Oracle9*i* database, the user can issue a **SET ROLE** command. The syntax for the SET ROLE command is provided in Figure 13–26.

```
SET ROLE rolename;
```

Figure 13-26 Syntax of the SET ROLE command

13

Users cannot, however, set their role to any role that has not already been assigned to them. For example, a user could not issue the command **SET ROLE DBA;** unless the individual had already been assigned that role by an authorized user.

As a safety precaution, some database administrators add a password to a role. For example, suppose that employee Tom has been assigned certain privileges through the DBA role. However, he only uses the DBA role when certain types of operations need to be performed. One day, he is logged in to the Oracle9*i* database, performing day-to-day activities that only require his normal privileges. He is summoned away for a quick meeting and doesn't bother to log out of the server. Because no special privileges are currently available, this may not present a problem. However, what if a disgruntled employee knows that Tom has access to the DBA role. All that employee needs to do is sit down at Tom's computer and set Tom's role to DBA with the SET ROLE command. With those types of privileges available, the disgruntled employee could do a lot of damage in a very short period of time.

To avoid this type of problem, administrators add passwords to roles that have important and potentially dangerous privileges. To add a password to a role, simply use the **ALTER ROLE** command. The syntax for the ALTER ROLE command is provided in Figure 13–27.

```
ALTER ROLE rolename
IDENTIFIED BY password;
```

Figure 13-27 Syntax of the ALTER ROLE command

Once the ALTER ROLE command has been used to add a password to a role, any user attempting to use the role will be required to enter the password, or it will not be enabled.

A password will not be required for roles that have been assigned as a default role for a user. It is only required when the SET ROLE command is issued to enable a role.

As shown in Figure 13–28, Roman Thomas has been granted both the DATAENTRY and DBA roles. However, the DATAENTRY role is his default role. If he needs to change to the DBA role, the SET ROLE command must be issued.

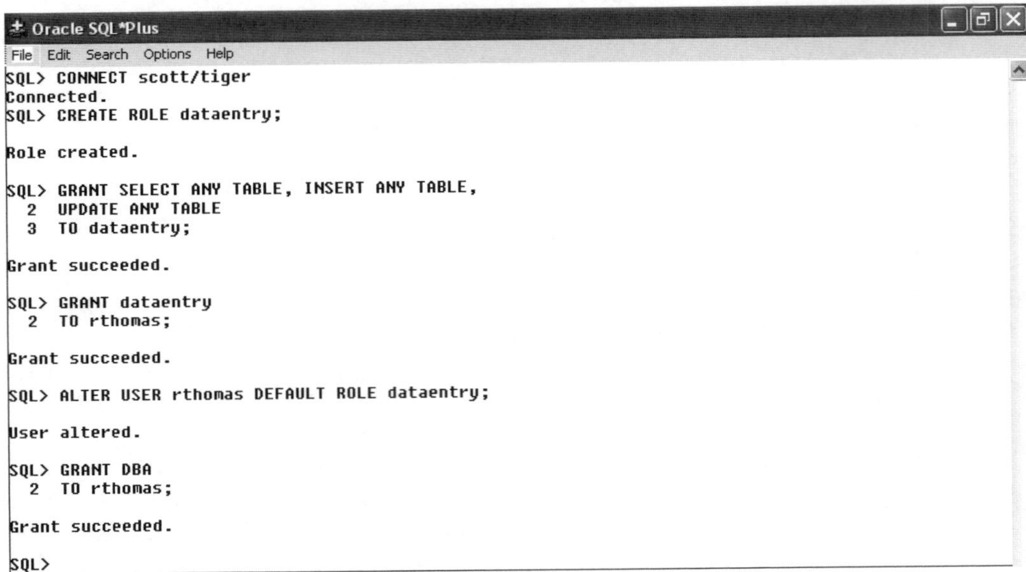

Figure 13-28 Granting and changing roles

Viewing Privileges

Various data dictionary views can be queried to determine the privileges currently assigned to a user or role. The most commonly accessed views are as follows:

- ROLE_SYS_PRIVS
- SESSION_PRIVS

The **ROLE_SYS_PRIVS** data dictionary view lists all system privileges that have been granted to roles. In Figure 13–29, a query on the ROLE_SYS_PRIVS view lists the three privileges assigned to the DATAENTRY role. The ADM column of the display indicates whether the privilege was granted with the WITH ADMIN OPTION. In this case, none of the privileges were granted with that option.

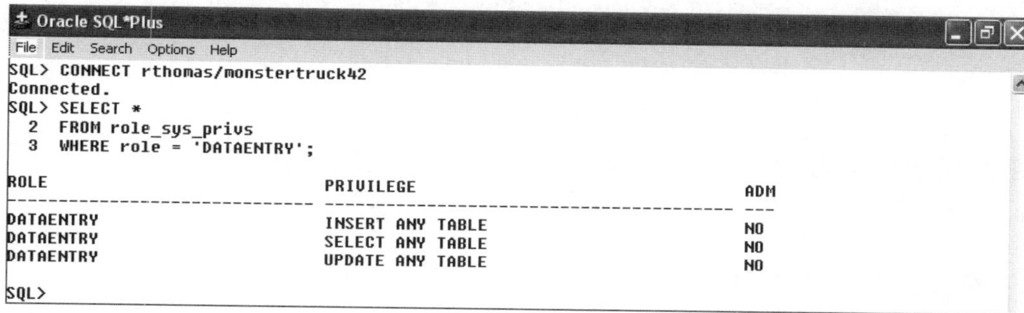
```
SQL> CONNECT rthomas/monstertruck42
Connected.
SQL> SELECT *
  2  FROM role_sys_privs
  3  WHERE role = 'DATAENTRY';

ROLE                           PRIVILEGE                                ADM
------------------------------ ---------------------------------------- ---
DATAENTRY                      INSERT ANY TABLE                         NO
DATAENTRY                      SELECT ANY TABLE                         NO
DATAENTRY                      UPDATE ANY TABLE                         NO

SQL>
```

Figure 13-29 Query of the ROLE_SYS_PRIVS table

When an individual has a habit of using the SET ROLE command throughout a session, the person may forget exactly which roles, or privileges, are currently available. The quickest way to determine exactly which privileges are currently enabled for a user is to query the **SESSION_PRIVS** view. The list in Figure 13–30 displays the privileges that are available to Roman Thomas after he logged in to Oracle9*i*. Remember that some of the privileges were granted through the DATAENTRY role, but the CREATE SESSION privilege was granted directly to the user. In addition, the **UNLIMITED TABLESPACE** privilege was granted to the user by default when the account was created.

Figure 13-30 Privileges currently available to the RTHOMAS account

Whether or not the default assignment of the UNLIMITED TABLESPACE privilege occurs will depend upon the setting of your specific Oracle9*i* installation. This privilege means that the storage space assigned to the user for database objects is not limited to a maximum size.

REVOKING AND DROPPING PRIVILEGES AND ROLES

Just as easily as privileges and roles can be given, they can also be taken away or dropped.

Revoking Privileges and Roles

Privileges granted to a user or role can be removed by using the **REVOKE** command. The syntax for the REVOKE command to remove a system privilege from a user or role is shown in Figure 13–31.

```
REVOKE systemprivilege [,…systemprivilege]
FROM username|rolename;
```

Figure 13-31 Syntax to revoke a system privilege

The REVOKE command can also be used to revoke object privileges, using the syntax shown in Figure 13–32.

```
REVOKE objectprivilege [,…objectprivilege]
ON objectname
FROM username|rolename;
```

Figure 13-32 Syntax to revoke an object privilege

In addition, the REVOKE command can be used to revoke a role from an account, using the syntax shown in Figure 13–33.

```
REVOKE rolename
FROM username|rolename;
```

Figure 13-33 Command to revoke a role from an account

For example, to revoke the DATAENTRY role from Roman Thomas, the command shown in Figure 13–34 can be issued by the DBA.

```
REVOKE dataentry
FROM rthomas;
```

Figure 13-34 Command to revoke the DATAENTRY role from a user

As shown in Figure 13–35, once the role has been revoked from the user, it will no longer be listed in the ROLE_SYS_PRIVS view and, therefore, is no longer available to the user.

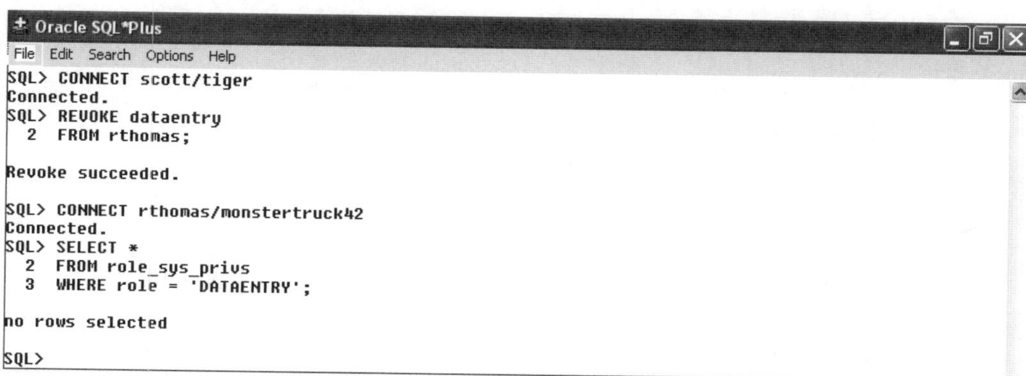

Figure 13-35 Effect of revoking the DATAENTRY role from a user

When revoking an object privilege that was originally granted using the WITH GRANT OPTION, the privilege is revoked not only from the specified user, but also from any other users to whom he or she may have subsequently granted the privilege. *Revoking a system privilege that was originally granted with the WITH ADMIN OPTION has no cascading effect on other users.*

Dropping a Role

A role can be deleted from the Oracle9*i* database through the **DROP ROLE** command. The syntax for the DROP ROLE command is shown in Figure 13-36.

```
DROP ROLE rolename;
```

Figure 13-36 Syntax of the DROP ROLE command

When a role is removed from the database, users will lose any privileges derived from that role. The only way the users will be able to use the privileges previously assigned by the role is to receive those privileges again, either directly or via another role.

For example, suppose that you decide that the DATAENTRY role should be more restrictive and specify exactly which columns can be updated in certain tables. The simplest solution is to drop the existing role and re-create it with a new DATAENTRY role. You can drop the role, using the command shown in Figure 13-37.

```
DROP ROLE dataentry;
```

Figure 13-37 Command to drop the DATAENTRY role

Now that the DATAENTRY role is not available to Roman Thomas, he no longer has access to any of the tables.

Dropping a User

Suppose that Roman's supervisor has just informed you that the correct spelling of the new employee's last name is Tomas. There is no ALTER USER option available to change the user name for an account. Instead, you will need to delete the existing account and re-create Roman's account with the correct spelling. The DROP USER command is used to remove a user account from an Oracle9*i* database. The syntax for the DROP USER command is shown in Figure 13-38.

```
DROP USER username;
```

Figure 13-38 Syntax of the DROP USER command

The command **`DROP USER rthomas;`** would drop the existing account named "rthomas".

Once the DBA has removed the user account, no one can log in to the database under the user name *rthomas*. As shown in Figure 13–39, such an attempt results in an error message, indicating that an invalid username/password was used, and the connection to the database is terminated.

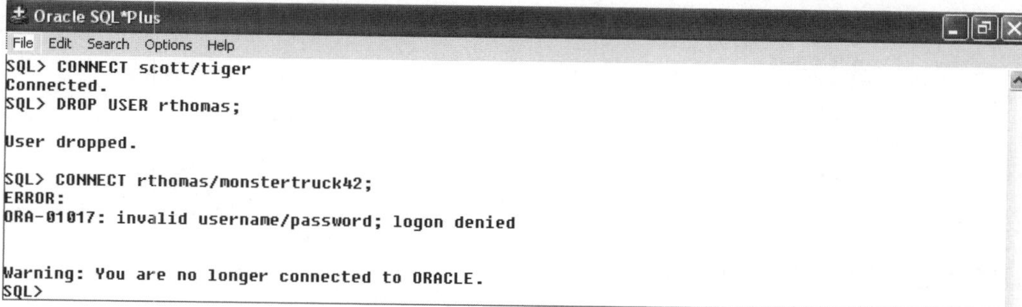
```
SQL> CONNECT scott/tiger
Connected.
SQL> DROP USER rthomas;

User dropped.

SQL> CONNECT rthomas/monstertruck42;
ERROR:
ORA-01017: invalid username/password; logon denied


Warning: You are no longer connected to ORACLE.
SQL>
```

Figure 13-39 Failed attempt to access the Oracle9*i* database

Chapter Summary

- Oracle9*i* authentication requires that a valid user account exist before a user can log in to the Oracle9*i* server.
- A new user account is created with the CREATE USER command. The IDENTIFIED BY clause contains the password for the account.
- Users can use the CONNECT command to connect to the Oracle server.
- The CREATE SESSION privilege is required before a user can access his or her account on the Oracle server.
- Object privileges allow users to manipulate data in database objects.
- System privileges are used to grant access to the database and to create, alter, and drop database objects.
- The system privileges available in Oracle9*i* can be viewed through SYSTEM_PRIVILEGE_MAP.
- Privileges are given through the GRANT command.
- The ALTER USER command, combined with the PASSWORD EXPIRE clause, can be used to force a user to change his or her password upon the next attempted login to the database.
- The ALTER USER command, combined with the IDENTIFIED BY clause, can be used to change a user's password.
- Privileges can be assigned to roles to make the administration of privileges easier.

- Roles are collections of privileges.
- The ALTER USER command, combined with the DEFAULT ROLE keywords, can be used to assign a default role to a user.
- A role can be enabled using the SET ROLE command.
- Privileges can be revoked from users and roles, using the REVOKE command.
- Roles can be revoked from users, using the REVOKE command.
- A role can be deleted, using the DROP ROLE command.
- A user account can be deleted, using the DROP USER command.

Chapter 13 Syntax Summary

The following table presents a summary of the syntax that you have learned in this chapter. You can use the table as a study guide and reference.

Syntax Guide		
Command Description	**Command Syntax**	**Example**
Creating, Maintaining, and Dropping User Accounts		
Create a user	`CREATE USER` *username* `IDENTIFIED BY` *password*`;`	`CREATE USER rthomas` `IDENTIFIED BY little25car;`
Connect to the Oracle server from SQL*Plus	`CONNECT` *username/password* *@connectstring*	`CONNECT rthomas/little25car`

Change or expire a password	`ALTER USER` *username*
	`[IDENTIFIED BY` *newpassword*`]`
	`[PASSWORD EXPIRE];`
		`ALTER USER rthomas`
		`IDENTIFIED BY`
		`monstertruck42;`

Drop a user	`DROP USER` *username*`;`	`DROP USER rthomas;`
Granting and Revoking Privileges		

Grant object privileges to users or roles	`GRANT {`*objectprivilege*`|ALL}`
	`[(`*columnname*`),`
	objectprivilege
	`(`*columnname*`)]`
	`ON` *objectname*
	`TO {`*username*`|`*rolename*`|PUBLIC}`
	`[WITH GRANT OPTION];`
		`GRANT select, insert`
		`ON customers`
		`TO rthomas`
		`WITH GRANT OPTION;`

Command Description	Command Syntax	Example
Granting and Revoking Privileges (continued)		
Grant system privileges to users or roles	GRANT *systemprivilege* [, *systemprivilege*, …] TO *username*|*rolename* [, *username*|*rolename*, …] [WITH ADMIN OPTION];	GRANT CREATE SESSION TO rthomas;
Revoke object privileges	REVOKE *objectprivilege* [,…*objectprivilege*] ON *objectname* FROM *username*|*rolename*;	REVOKE INSERT ON customers FROM rthomas;
Granting and Revoking Roles		
Create a role	CREATE ROLE *rolename*;	CREATE ROLE dataentry;
Grant a role to a user	GRANT *rolename* [, *rolename*] TO *username* [, *username*];	GRANT dataentry TO rthomas;
Assign a default role to a user	ALTER USER *username* DEFAULT ROLE *rolename*;	ALTER USER rthomas DEFAULT ROLE dataentry;
Set or enable a role	SET ROLE *rolename*;	SET ROLE DBA;
Add a password to a role	ALTER ROLE *rolename* IDENTIFIED BY *password*;	ALTER ROLE dataentry IDENTIFIED BY apassword;
Revoke a role	REVOKE *rolename* FROM *username*|*rolename*;	REVOKE dataentry FROM rthomas;
Drop a role	DROP ROLE *rolename*;	DROP ROLE dataentry;

Review Questions

To answer the following questions, refer to the tables in Appendix A.

1. What is the purpose of authentication?
2. What types of authentication methods are available in Oracle9*i*?
3. How is a user password assigned in Oracle9*i*?
4. What is a privilege?
5. If you are logged in to Oracle9*i*, how can you determine which privileges are currently available to your account?
6. What types of privileges are available in Oracle9*i*? Define each type.
7. What is the purpose of a role in Oracle9*i*?
8. How can you assign a password to a role?

9. What happens if you revoke an object privilege that was granted with the WITH GRANT OPTION? What if the privilege is revoked from a user who had already granted the same object privilege to three other users?
10. How can you remove a user account from Oracle9*i*?

Multiple Choice

To answer the following questions, refer to the tables in Appendix A.

1. Which of the following commands can be used to change a password for a user account?
   a. ALTER PASSWORD
   b. CHANGE PASSWORD
   c. MODIFY USER PASSWORD
   d. ALTER USER…PASSWORD
   e. none of the above
2. Which of the following statements assigns the role of CUSTOMERREP as the default role for Maurice Cain?
   a. `ALTER ROLE mcain DEFAULT ROLE customerrep;`
   b. `ALTER USER mcain TO customerrep;`
   c. `SET DEFAULT ROLE customerrep FOR mcain;`
   d. `ALTER USER mcain DEFAULT ROLE customerrep;`
   e. `SET ROLE customerrep FOR mcain;`
3. Which of the following statements is most accurate?
   a. Authentication procedures will prevent any data stored in the Oracle9*i* database from becoming stolen or damaged.
   b. Authentication procedures are used to limit unauthorized access to the Oracle9*i* database.
   c. Oracle9*i* authentication will not prevent anyone from accessing data in the database if the individual has a valid operating-system account.
   d. Authentication procedures restrict the type of data manipulation operations that can be executed by a user.
4. Which of the following statements will create a user account named DeptHead?
   a. `CREATE ROLE depthead IDENTIFIED BY apassword;`
   b. `CREATE USER depthead IDENTIFIED BY apassword;`
   c. `CREATE ACCOUNT depthead;`
   d. `GRANT ACCOUNT depthead;`

5. Which of the following privileges must be granted to a user's account before the user can connect to the Oracle9*i* database?
   a. CONNECT
   b. CREATE SESSION
   c. CONNECT ANY DATABASE
   d. CREATE ANY TABLE
6. Which of the following privileges allows a user to truncate tables in a database?
   a. DROP ANY TABLE
   b. TRUNCATE ANY TABLE
   c. CREATE TABLE
   d. TRUNC TABLE
7. Which of the following tables or views will display the current enabled privileges for a user?
   a. SESSION_PRIVS
   b. SYSTEM_PRIVILEGE_MAP
   c. USER_ASSIGNED_PRIVS
   d. V$ENABLED_PRIVILEGES
8. Which of the following commands will only eliminate the user ELOPEZ's ability to enter new books into the BOOKS table?
   a.
   ```
   REVOKE insert
   ON books
   FROM elopez;
   ```
   b.
   ```
   REVOKE insert
   FROM elopez;
   ```
   c.
   ```
   REVOKE INSERT INTO
   FROM elopez;
   ```
   d.
   ```
   DROP insert
   INTO books
   FROM elopez;
   ```
9. Which of the following commands is used to assign a privilege to a role?
   a. CREATE ROLE
   b. CREATE PRIVILEGE
   c. GRANT
   d. ALTER PRIVILEGE

10. Which of the following options will require a user to change his or her password at the time of the next login?
    a. CREATE USER
    b. ALTER USER
    c. IDENTIFIED BY
    d. PASSWORD EXPIRE
11. Which of the following options allows a user to grant system privileges to other users?
    a. WITH ADMIN OPTION
    b. WITH GRANT OPTION
    c. DBA
    d. ASSIGN ROLES
    e. SET ROLE
12. Which of the following is an object privilege?
    a. CREATE SESSION
    b. DROP USER
    c. INSERT ANY TABLE
    d. UPDATE
13. Which of the following privileges cannot be granted to a role, only to a user?
    a. SELECT
    b. CREATE ANY
    c. REFERENCES
    d. READ
    e. WRITE
14. Which of the following is used to grant all the object privileges for an object to a specified user?
    a. ALL
    b. PUBLIC
    c. ANY
    d. OBJECT
15. Which of the following identifies a collection of privileges?
    a. an object privilege
    b. a system privilege
    c. DEFAULT privilege
    d. a role

16. Which of the following is true?
    a. If the DBA changes the password for a user while the user is connected to the database, the connection will automatically terminate.
    b. If the DBA revokes the CREATE SESSION privilege of a user account, the user cannot connect to the database.
    c. If a user is granted the privilege to create a table and the privilege is revoked after the user creates a table, the table is automatically dropped from the system.
    d. all of the above
17. Which of the following commands can be used to eliminate the RECEPTIONIST role?
    a. `DELETE ROLE receptionist;`
    b. `DROP receptionist;`
    c. `DROP ANY ROLE;`
    d. none of the above
18. Which of the following will display a list of all system privileges available in Oracle9*i*?
    a. SESSION_PRIVS
    b. SYS_PRIVILEGE_MAP
    c. V$SYSTEM_PRIVILEGES
    d. SYSTEM_PRIVILEGE_MAP
19. Which of the following can be used to change the role that is currently enabled for a user?
    a. SET DEFAULT ROLE
    b. ALTER ROLE
    c. ALTER SESSION
    d. SET ROLE
20. Which of the following is an object privilege?
    a. DELETE ANY
    b. INSERT ANY
    c. UPDATE ANY
    d. REFERENCES

Hands-on Assignments

To perform the following activities, refer to the tables in Appendix A.

1. Create a new user account. The name of the account should be a combination of your middle initial and your last name.

2. Attempt to log in to Oracle9*i* using the newly created account.
3. Assign the privileges to the new account that would allow a user to connect to the database, create new tables, and alter an existing table.
4. Using a properly privileged account, create a role named CUSTOMERREP that would allow new rows to be inserted into the ORDERS and ORDERITEMS tables and allow rows to be deleted from those tables.
5. Assign the account created in Assignment 1 the CUSTOMERREP role.
6. Log in to Oracle9*i* using the new account created in Assignment 1. Determine the privileges currently available to the account.
7. Revoke the privilege to delete rows in the ORDERS and ORDERITEMS tables from the CUSTOMERREP role.
8. Revoke the CUSTOMERREP role from the account created in Assignment 1.
9. Delete the CUSTOMERREP role from the Oracle9*i* database.
10. Delete the user account created in Assignment 1.

A Case for Oracle9*i*

To perform the following activity, refer to the tables in Appendix A.

There are three major classifications for employees who do not work for the Information Systems Department of JustLee Books: account managers who are responsible for the marketing activities of the company (e.g., promotions based on customers' previous purchases or for specific books); data-entry clerks who enter inventory updates (e.g., add new books and publishers, change prices); and customer service representatives who are responsible for adding new customers and entering orders into the database. Each employee group has different tasks to perform; therefore, they will need different privileges for the various tables in the database. To simplify the administration of system and object privileges, a role should be created for each employee group.

Create a memo for your supervisor that contains the following information:

1. Consider the need of each group of employees to access these tables: BOOKS, CUSTOMERS, ORDERS, ORDERITEMS, AUTHOR, BOOKAUTHOR, PUBLISHER, and PROMOTION.
2. Name the privileges needed by each group of employees.
3. For each group of employees, name a role that contains the appropriate privileges for that group.
4. For each group of employees, list the exact command(s) necessary to create and assign specific privileges to their role.
5. Explain your rationale for the privileges granted to each role.

CHAPTER

14

FORMATTING READABLE OUTPUT

Objectives

After completing this chapter, you should be able to do the following:

- Add a column heading with a line break to a report
- Format the appearance of numeric data in a column
- Specify the width of a column
- Substitute a text string for a NULL value in a report
- Add a multiple-line header to a report
- Display a page number in a report
- Add a footer to a report
- Change the setting of an environment variable
- Suppress duplicate report data
- Clear changes made by the COLUMN and BREAK commands
- Perform calculations in a report

Until now, query results have been displayed as a list. Although you have manipulated columns presented in output and have used column aliases as column headings in query results, you have not yet used SQL*Plus to create an actual formatted report. Some of the formatting features available in SQL*Plus include report headers and footers, formatting models for column data, suppression of duplicate data for group reports, and calculations. Although several formatting options are available, this chapter will present only the most commonly used features. Most elaborate reports are created using various application software and report generators.

Figure 14–1 provides an overview of this chapter's contents.

Formatting Output	
Command	**Description**
START *or* @	Executes a script file
COLUMN	Defines the appearance of column headings and the format of the column data
TTITLE	Adds a header to the top of each report page
BTITLE	Adds a footer to the bottom of each report page
BREAK	Suppresses duplicated data for a specific column(s) when presented in a sorted order
COMPUTE	Performs calculations in a report based on the AVG, SUM, COUNT, MIN, or MAX statistical function
SPOOL	Redirects output to a text file
COLUMN Options	
HEADING	Adds a column heading to a specified column
FORMAT	Defines the width of columns and applies specific formats to columns containing numeric data
NULL	Indicates text to be substituted for NULL values in a specified column
SQL*Plus Environment Variables	
UNDERLINE	Specifies the symbol to be used to separate a column heading from the contents of the column
LINESIZE	Establishes the maximum number of characters that can appear on a single line of output
PAGESIZE	Establishes the maximum number of lines that can appear on one page of output

Figure 14-1 Overview of chapter contents

Before attempting to work through the examples in this chapter, you should go to the Chapter14 folder in your Data Files and run the **prech14.sql** script. Your output should be the same as shown in the examples. To run the script, type **`start d:/prech14.sql`** where ***d:*** represents the location (correct drive and path) of the script.

THE BASIC STRATEGY

The basic strategy for creating a report in SQL*Plus is to (1) enter the format of the report and (2) enter a query to retrieve the desired rows for the report. The problem, however, is that *SQL*Plus commands are executed one line at a time*. Thus, when you enter a SQL*Plus command at the SQL> prompt to format one part of the report, it is automatically executed when you press **Enter**. Unfortunately, you will probably need to format several features of the report (title, column headings, data format, etc.), not just one. To overcome this problem, the preferred approach is to store *all* the settings for the report *and* the SELECT statement needed to retrieve the records in a script file, and then execute the script file in SQL*Plus. (Although you created script files in Chapter 10, a review is provided in this chapter.)

The most common approach to creating a script file is to enter the script, or commands, in a word-processing program, such as Notepad. The key concept to remember is that *when you save the file, assign a file extension of* **.sql** *after the file name*. If you do not include the file extension, the word-processing program may assign another extension, such as **.txt**, to the file name, and you won't be able to execute the file in Oracle9*i*.

If you use a more "elaborate" word-processing program, realize that many programs will embed hidden codes in the document, and the script file will not execute properly. If you must use such a word-processing program, make certain you change the file type to either "All Files" or "ANSI file" (depending on the options available) when you save the file, or hidden codes will be included in the script.

COLUMN Command

The **COLUMN** command can be used to format both a column heading and the data being displayed within the column, depending on the option(s) used. The basic syntax for the COLUMN command is shown in Figure 14-2.

```
COLUMN [columnname|columnalias]  [option]
```

Figure 14-2 Syntax of the COLUMN command

14

Let's examine the elements of the syntax example displayed in Figure 14-2:

- The column referenced is identified immediately after the COLUMN command.
- If desired, a column can be assigned a column alias. However, if you include a column alias, it must be used to identify the column rather than the column name whenever the column is referenced.
- The option given in the COLUMN command specifies how the display will be affected.

In this section, you will learn how to use the options shown in Figure 14-3.

Option	Description
FORMAT *formatmodel*	Applies a specified format to the column data
HEADING *columnheading*	Indicates the heading to be used for a specified column
NULL *textmessage*	Identifies the text message to be displayed in the place of NULL values

Figure 14-3 Options available for the COLUMN command

FORMAT Option

The **FORMAT** option allows you to apply a format model to the data displayed within a column. For example, recall that in previous chapters, a book's retail price of $54.50 was displayed as 54.5. The FORMAT option allows you to display the insignificant zero in output. In fact, you could also include the dollar sign, so the retail price would be displayed as $54.50. Some of the formatting options available with the COLUMN command are shown in Figure 14–4.

Format Code	Description	Example	Output
9	Identifies the position of numeric data (leading zeros are suppressed)	99999	54
$	Includes a floating dollar sign in output	$9999	$54
,	Indicates where to include a comma when needed (traditionally for the thousands' and millions' position)	9,999	54 *or* 1,212
.	Indicates the number of decimal positions to be displayed	9999.99	54.50
A*n*	Identifies the width of a column in a report	A32	Assigns a width of 32 spaces to a column

Figure 14-4 Codes for column format models

Suppose that the first report you need to create is one that displays the ISBN, title, cost, and retail price for books having a retail price greater than $50. To ensure that the cost and retail price for each book are displayed with two decimal places, you can create a model to specify exactly how data in the Cost and Retail columns should appear. Look at the script shown in Figure 14–5.

```
COLUMN title FORMAT A35
COLUMN cost FORMAT $999.99
COLUMN retail FORMAT $999.99
SELECT isbn, title, cost, retail
FROM books
WHERE retail >50
ORDER BY title;
```

Figure 14-5 Script to create a report

Let's examine the elements of Figure 14–5.

- The first line formats the Title column to a width of 35 spaces. Because the BOOKS table defines the column as having a width of only 30 spaces, this will create extra blank space between the report's Title and Cost columns.
- The second and third lines apply a format model to the numeric data displayed in the report's Cost and Retail columns. The keyword FORMAT in the COLUMN command indicates that a format model is being specified.
- The format model uses a dollar sign ($) and a series of nines. The dollar sign instructs Oracle9*i* to include the symbol with each value displayed in the columns. The two nines that appear after the decimal indicate that two decimal positions should be included. This forces the inclusion of insignificant zeros, which are usually suppressed in query results.
- After the Title, Cost, and Retail columns have been defined, the actual query to retrieve the rows to be displayed in the report is provided.
- Notice that a semicolon is included at the end of the SELECT statement. Although the statement is included in a script file with a set of SQL*Plus commands, a semicolon at the end of the statement, or a slash (/) on a blank line after the statement, is required for Oracle9*i* to execute the query. *SQL*Plus commands do not require such notation and will execute automatically.*

The COLUMN command does not have a specific format model for a DATE datatype column, other than increasing the column width. To alter the format of a date (e.g., to change the DD–MON–YY format to Month DD, YYYY), include the TO_CHAR function (presented in Chapter 5) in the SELECT clause used to retrieve the data.

The script shown in Figure 14–5 can be entered into a file, using either of the following approaches:

1. Type **ed** or **edit** at the SQL> prompt in SQL*Plus to open an editor, such as Notepad. After entering the script, select **File, Save As** from the menu, and name the script. In this example, the script was named **Titlereport.sql** and was saved on the hard drive. You may need to save the script file to a

diskette in your **A:** drive or to another drive, depending on existing restrictions on your computer.

2. The second approach is to open a word-processing program outside of SQL*Plus (e.g., from the **Start** menu of your operating system), and then follow the previous procedure to create the script file.

Once the script file has been created, you can execute it in SQL*Plus by entering the SQL*Plus command shown in Figure 14–6.

```
START d:\filename
```

Figure 14-6 Command to execute a script file

In Figure 14–6, the **START** keyword is used to indicate that a script should be executed. The ***d:*** is used to specify the drive and path name where the file is stored; *`filename`* represents the actual name of the script file to be executed.

The START keyword can be substituted with the symbol commonly referred to as the "at" sign (@) to execute a script. For example:
`@d:\filename`

To execute the script file **Titlereport.sql**, enter the command shown in Figure 14–7 at the SQL> prompt, and then press **Enter**. (Don't forget to substitute the correct drive letter and path, if necessary.)

```
START c:\titlereport.sql
```

Figure 14-7 Command to execute the **Titlereport.sql** script file

Once the command has been executed, Oracle9*i* will display the report, shown in Figure 14–8.

```
Oracle SQL*Plus
File Edit Search Options Help
SQL> start c:\titlereport.sql

ISBN       TITLE                                  COST   RETAIL
---------- ------------------------------ -------- --------
4981341710 BUILDING A CAR WITH TOOTHPICKS       $37.80   $59.95
8843172113 DATABASE IMPLEMENTATION              $31.40   $55.95
9959789321 E-BUSINESS THE EASY WAY              $37.90   $54.50
3957136468 HOLY GRAIL OF ORACLE                 $47.25   $75.95
2491748320 PAINLESS CHILD-REARING               $48.00   $89.95

SQL>
```

Figure 14-8 Results of executing the **Titlereport.sql**

When viewing the report, notice the extra blank space between the contents of the Title and Cost columns. This space is created because the Title column is defined to hold a maximum of 35 characters, and the book *Building a Car with Toothpicks* is 30 characters in length. `A35`, specified in the FORMAT option of the COLUMN command for the Title column, causes the additional spacing between the Title and Cost columns. By contrast, because the ISBN column did not have a width format applied, there is a default of one space between the data displayed in the ISBN column and the Title column.

Also notice how the data are displayed in the Cost and Retail columns. Each value now contains two decimal positions and is preceded by a dollar sign. This format was established by using the format model of $999.99, shown in Figure 14–5. Because none of the values displayed has a number in the hundreds' position (all values are <=99.99), the floating dollar sign appears in front of the first digit displayed, rather than staying in a fixed location and leaving a blank space in the hundreds' position.

Suppose that after reviewing the results of the **Titlereport.sql** script file, you want to make improvements. First, you want to add additional space between the ISBN and Title columns. Second, you want to remove the dollar signs from the Cost and Retail columns. Traditionally, dollar signs are displayed only for the first amount listed in a column, and then again with calculated values, such as subtotals or totals. Having a dollar sign repeat for every currency value in a column can become distracting and make a report appear cluttered.

Using the **Titlereport.sql** script file, let's make these changes to the original script and then save them in a new file named **Titlereport1.sql**. To make these changes, use the elements shown in Figure 14–9.

```
COLUMN isbn FORMAT A15
COLUMN title FORMAT A35
COLUMN cost FORMAT 999.99
COLUMN retail FORMAT 999.99
SELECT isbn, title, cost, retail
FROM books
WHERE retail >50
ORDER BY title;
```

Figure 14-9 Revised script

The revised script in Figure 14–9 includes a COLUMN command to format the ISBN column to a width of 15 characters. In addition, the dollar signs have been removed from the format model for the Retail and Cost columns. Once the script has been saved as **Titlereport1.sql**, it can be run in SQL*Plus and will display the report shown in Figure 14–10.

14

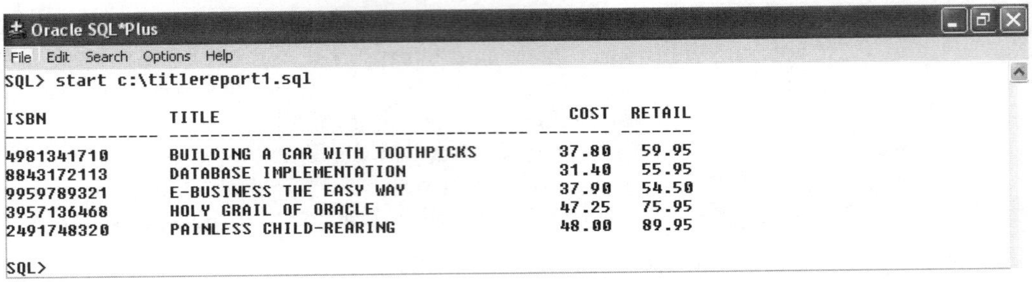

```
Oracle SQL*Plus
File Edit Search Options Help
SQL> start c:\titlereport1.sql

ISBN            TITLE                                  COST  RETAIL
--------------- ----------------------------------- ------- -------
4981341710      BUILDING A CAR WITH TOOTHPICKS        37.80   59.95
8843172113      DATABASE IMPLEMENTATION               31.40   55.95
9959789321      E-BUSINESS THE EASY WAY               37.90   54.50
3957136468      HOLY GRAIL OF ORACLE                  47.25   75.95
2491748320      PAINLESS CHILD-REARING                48.00   89.95

SQL>
```

Figure 14-10 Results of executing the **Titlereport1.sql** script file

Notice that the distance between the ISBN and Title columns has been increased. Because the ISBN column is defined in the BOOKS table as a VARCHAR2 datatype, the data are displayed left-aligned within the column. This results in excess space appearing on the right side of the ISBN column and creates a greater distance between the last digit of a book's ISBN and its corresponding title. On the other hand, columns defined as the NUMBER datatype display their contents right-aligned. Therefore, any excess space that has been assigned would appear in front of the column.

To increase the space between the Cost and Retail columns, simply increase the number of nines listed in each column's format model. For example, there are four blank spaces between the last digit in the Cost column and the first digit in the Retail column. To add an additional six spaces between the columns, simply add six nines to the format model for the Retail column. Save the revised script as **Titlereport2.sql**. The contents of the revised script file are shown in Figure 14–11.

```
COLUMN isbn FORMAT A15
COLUMN title FORMAT A35
COLUMN cost FORMAT 999.99
COLUMN retail FORMAT 999999999.99
SELECT isbn, title, cost, retail
FROM books
WHERE retail >50
ORDER BY title;
```

Figure 14-11 Contents of the **Titlereport2.sql** script file

Because no book has a retail price of more than $99.99, any excess nines in the format model for the Retail column will be represented by a blank space. In essence, the series of nines establishes the width of that column.

Once the START command is issued to execute **Titlereport2.sql**, the report shown in Figure 14–12 will be displayed.

```
SQL> start c:\titlereport2.sql

ISBN            TITLE                              COST         RETAIL
--------------- ---------------------------------- -------  -------------
4981341710      BUILDING A CAR WITH TOOTHPICKS      37.80          59.95
8843172113      DATABASE IMPLEMENTATION             31.40          55.95
9959789321      E-BUSINESS THE EASY WAY             37.90          54.50
3957136468      HOLY GRAIL OF ORACLE                47.25          75.95
2491748320      PAINLESS CHILD-REARING              48.00          89.95

SQL>
```

Figure 14-12 Results of executing the **Titlereport2.sql** script

Although the space between the Cost and Retail columns has increased, the number of dashes separating the column headings and their contents has also increased. The additional dashes in the Retail column are not balanced in appearance with the dashes in the Cost column. There is, however, a simple solution to this problem, and it will be presented in the next section.

HEADING Option

The **HEADING** option of the COLUMN command is used to specify a column heading for a particular column. Notice that in Figure 14–12, the report's column headings are simply the BOOKS table's column names. The HEADING option is similar to the use of a column alias: It provides a substitute heading for the display of the output. There are, however, two significant differences:

1. A column name assigned by the HEADING option of the COLUMN command cannot be referenced in a SELECT statement.
2. A column name assigned by the HEADING option can contain line breaks. In other words, the column name can be displayed on more than one line.

Let's work through an example. If you had assigned the column alias "Retail Price" to the Retail column of a SELECT statement, the column would have been 12 spaces wide in the output (one space for each letter and one for the space between the two words). Using the HEADING option, however, would allow you to indicate that the word "Retail" should appear on one line and the word "Price" should appear on a second line. This would give the report a more professional appearance and would not waste space. A single vertical bar (|) is used to indicate where a line break should occur in a column heading. For example, 'Retail|Price' would cause the first word, Retail, to appear on one line, and the second word, Price, to appear on the next line. (Note the use of single quotation marks to create a literal string.)

To modify the **Titlereport2.sql** file and use Retail Price as a column heading, use the script shown in Figure 14–13.

```
COLUMN isbn FORMAT A15
COLUMN title FORMAT A35
COLUMN cost FORMAT 999.99
COLUMN retail FORMAT 999999999.99 HEADING 'Retail|Price'
SELECT isbn, title, cost, retail
FROM books
WHERE retail >50
ORDER BY title;
```

Figure 14-13 Heading inserted for the Retail column

After saving the revised script in Figure 14–13 as **Titlereport3.sql**, you will receive the output shown in Figure 14–14 on execution.

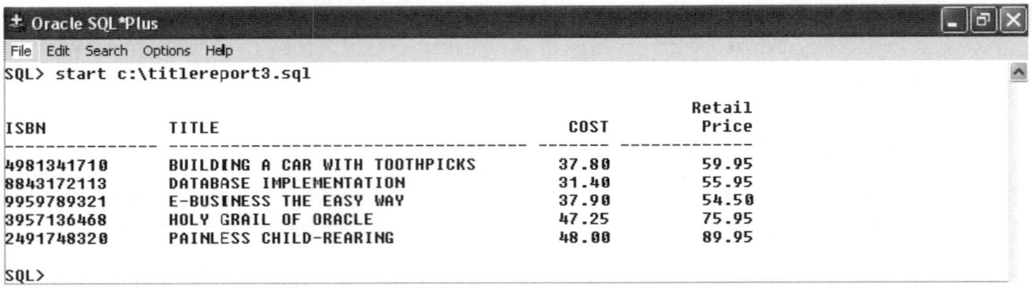

```
Oracle SQL*Plus
File Edit Search Options Help
SQL> start c:\titlereport3.sql

                                                                   Retail
ISBN            TITLE                                  COST        Price
--------------- ----------------------------------- ------- -------------
4981341710      BUILDING A CAR WITH TOOTHPICKS        37.80        59.95
8843172113      DATABASE IMPLEMENTATION               31.40        55.95
9959789321      E-BUSINESS THE EASY WAY               37.90        54.50
3957136468      HOLY GRAIL OF ORACLE                  47.25        75.95
2491748320      PAINLESS CHILD-REARING                48.00        89.95

SQL>
```

Figure 14-14 Report with two-line column headings

Of course, now the problem is that the Title and Cost columns both have upper-case column headings; however, the heading for the Retail column retains the case used in the HEADING option. To change the case of the heading of the Title and Cost columns, the simplest approach is to include a HEADING option for each of those columns and indicate the correct case.

The modified script is shown in Figure 14–15.

```
COLUMN isbn FORMAT A15
COLUMN title FORMAT A35 HEADING 'Title'
COLUMN cost FORMAT 999.99 HEADING 'Cost'
COLUMN retail FORMAT 999999999.99 HEADING 'Retail|Price'
SELECT isbn, title, cost, retail
FROM books
WHERE retail >50
ORDER BY title;
```

Figure 14-15 Using the HEADING option to indicate the case of column headings

Once the script file has been modified and saved as **Titlereport4.sql**, the script can be executed. It will display a report that has the column headings specified in the HEADING options. That output is shown in Figure 14–16.

```
SQL> start c:\titlereport4.sql

                                                                         Retail
ISBN            Title                                     Cost           Price
--------------- ---------------------------------- ------- -------------
4981341710      BUILDING A CAR WITH TOOTHPICKS            37.80          59.95
8843172113      DATABASE IMPLEMENTATION                   31.40          55.95
9959789321      E-BUSINESS THE EASY WAY                   37.90          54.50
3957136468      HOLY GRAIL OF ORACLE                      47.25          75.95
2491748320      PAINLESS CHILD-REARING                    48.00          89.95

SQL>
```

Figure 14-16 Revised report with column headings in upper- and lower-case letters

As shown in Figure 14–16, the headings for the Title, Cost, and Retail columns are now displayed with the first letter of each word capitalized and the remaining letters in lowercase. However, the problem with the dashes separating the column headings from the column data still remains.

Using dashes to separate the headings from the data is defined by the SQL*Plus environment variable **UNDERLINE**. You can use the **SET** command to change the value assigned to an environment variable. However, even if a different symbol is assigned, the symbol will still indicate the width assigned to each column in the FORMAT option.

You can, however, create the illusion that the assigned width for a column is less than it is and still leave adequate spacing between the columns. This is a two-step process. First, you need to change the UNDERLINE variable so no dashes (or any other symbol) will be displayed. After the default dashes are eliminated, you can then add the desired number of dashes to the HEADING option for each column. Follow these steps:

14

1. At the SQL> prompt, enter **`SET UNDERLINE off`** and press **Enter**. (This command basically says that no underlining should occur; that is, nothing should be used to separate the column heading from the column data in the output.)
2. To include the desired number of dashes in the column heading for each column, use the script in **Titlereport4.sql** as a starting point. Add a HEADING option for the ISBN column. The HEADING option for each of the other columns should be modified to include the dashes.
3. Save the modified file as **Titlereport5.sql**. It should have the contents shown in Figure 14–17.

```
COLUMN isbn FORMAT A15 HEADING 'ISBN|----------'
COLUMN title FORMAT A35 -
     HEADING 'Title|----------------------------'
COLUMN cost FORMAT 999.99 HEADING 'Cost|------'
COLUMN retail FORMAT 999999999.99 -
     HEADING 'Retail|Price|------'
SELECT isbn, title, cost, retail
FROM books
WHERE retail >50
ORDER BY title;
```

Figure 14-17 Modified report structure

Notice that the COLUMN command for the Title and Retail columns is now displayed on two lines. When a SQL*Plus command needs to be entered on more than one line, a dash is used at the end of the first line to indicate that the command is not complete and continues on the next line. In this case, the COLUMN command for both columns was broken just before the HEADING option and continued on the next line. The second line for each command is also indented. This is not a requirement; however, it does improve the readability of the script.

Once the modifications have been saved to the **Titlereport5.sql** file and executed, the report shown in Figure 14–18 will be displayed.

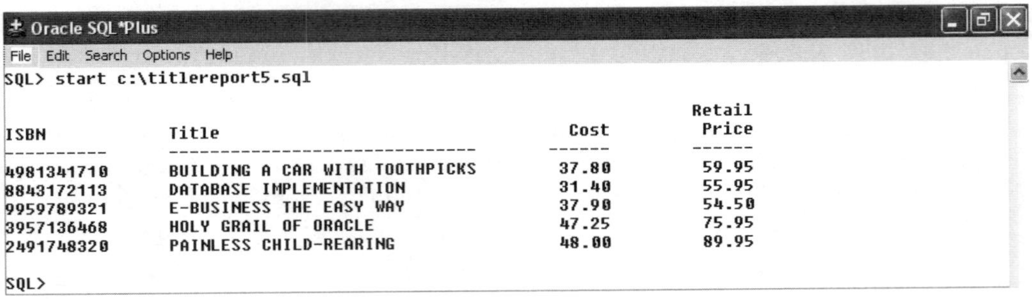

```
Oracle SQL*Plus
File Edit Search Options Help
SQL> start c:\titlereport5.sql

                                                               Retail
ISBN            Title                                 Cost      Price
----------      ------------------------------      ------     ------
4981341710      BUILDING A CAR WITH TOOTHPICKS       37.80      59.95
8843172113      DATABASE IMPLEMENTATION              31.40      55.95
9959789321      E-BUSINESS THE EASY WAY              37.90      54.50
3957136468      HOLY GRAIL OF ORACLE                 47.25      75.95
2491748320      PAINLESS CHILD-REARING               48.00      89.95

SQL>
```

Figure 14-18 Report with desired number of dashes separating column headings from data

The only part missing from the report to give it a "professional" appearance is a report heading. However, before moving on, there is one other COLUMN option that needs to be discussed.

NULL Option

Although this report does not contain NULL values, other reports might contain NULL values. In some instances, blank spaces may be appropriate. For example, consider the orders JustLee Books ships to its customers. If management wants a report to view the lag time for shipping current orders, you might want to substitute the words "Not Shipped" for any NULL values that occur in the Shipdate column. With the NULL option of the COLUMN command, this can easily be accomplished, as shown in Figure 14–19. You can create this script file as **Shippingreport.sql** to generate the new report.

```
SET UNDERLINE '='
COLUMN order# HEADING 'Order|Number'
COLUMN orderdate HEADING 'Order|Date'
COLUMN shipdate FORMAT A12 HEADING 'Shipped|Date'
       - NULL 'Not Shipped'
SELECT order#, TO_CHAR(orderdate, 'Mon DD ') orderdate,
TO_CHAR(shipdate, 'Mon DD ') shipdate
FROM orders
ORDER BY order#;
```

Figure 14-19 Script to create a shipping report

The first command in Figure 14–19 resets the symbol used to separate column headings to the equal sign. Whenever an environment variable is changed in SQL*Plus, that change will remain until the variable is reset. This includes not only the UNDERLINE variable, but also the column headings and formats that have been applied thus far. At the end of the chapter, you will be shown how to clear the settings for the columns that have been changed.

In a working environment, the usual procedure is to define, or set, the environment variables at the beginning of a script file and then clear the settings at the end of the script file. By clearing the changes at the end of the script file, you reduce the risk of affecting other reports due to changes made by a previously generated report. As an alternative, you can also clear all settings at the beginning of a script file to ensure that any modifications previously made do not affect the output of the report currently being generated. For illustrative purposes, this chapter demonstrates how to make such changes both from within a script file and at the SQL prompt.

14

Also in Figure 14–19, notice that the COLUMN command for the Shipdate column includes the NULL option. This instructs Oracle9*i* that the text message should be displayed in place of the NULL value when the report is executed. Because the text message is a literal string, it will appear in the case given, in the NULL option.

In the SELECT statement in Figure 14–19, the data in the Orderdate and Shipdate columns are retrieved by using the TO_CHAR function. The TO_CHAR function allows a format model to be applied. However, when the TO_CHAR function is used, the individual column names Orderdate and Shipdate cannot be referenced by the COLUMN command. Recall that when a function is applied to column data, the function is displayed as the column name in the output. The same thing occurs when a function is applied to a SELECT statement for a report. Therefore, a column alias must be assigned when a function is used, so the COLUMN command can reference the column through the assigned alias. In this case, the original column names are assigned as the column aliases and then used in the HEADING option of the COLUMN command to provide a column heading for each of the date columns.

If you save the script in Figure 14–19 to the file **Shippingreport.sql** and execute it from within SQL*Plus, you will receive the report shown in Figure 14–20.

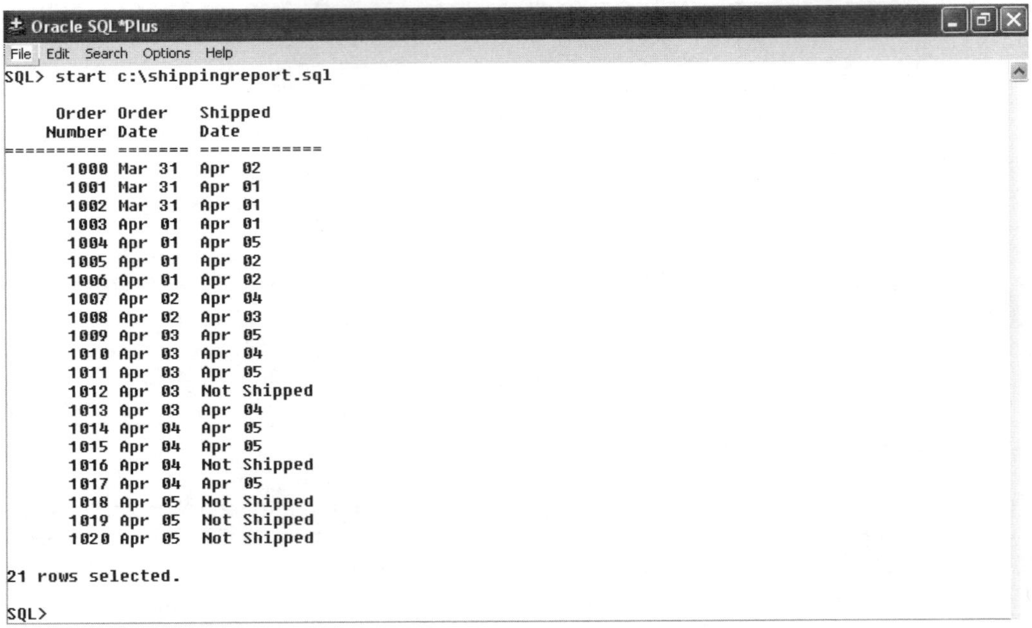

```
SQL> start c:\shippingreport.sql

     Order Order   Shipped
    Number Date    Date
========== ======= ============
      1000 Mar 31  Apr 02
      1001 Mar 31  Apr 01
      1002 Mar 31  Apr 01
      1003 Apr 01  Apr 01
      1004 Apr 01  Apr 05
      1005 Apr 01  Apr 02
      1006 Apr 01  Apr 02
      1007 Apr 02  Apr 04
      1008 Apr 02  Apr 03
      1009 Apr 03  Apr 05
      1010 Apr 03  Apr 04
      1011 Apr 03  Apr 05
      1012 Apr 03  Not Shipped
      1013 Apr 03  Apr 04
      1014 Apr 04  Apr 05
      1015 Apr 04  Apr 05
      1016 Apr 04  Not Shipped
      1017 Apr 04  Apr 05
      1018 Apr 05  Not Shipped
      1019 Apr 05  Not Shipped
      1020 Apr 05  Not Shipped

21 rows selected.

SQL>
```

Figure 14-20 Results of executing the **Shippingreport.sql** script

ENVIRONMENT VARIABLES

An **environment variable**, also called a **system variable**, is used to determine how specific elements are displayed in SQL*Plus. For example, the SQL> prompt can be changed to any sequence of characters desired by the user. In the output shown in Figure 14–20, the equal sign is used as the symbol to separate column headings from column data. What would happen if the **Titlereport5.sql** file were executed? Because this particular file did not define the symbol for the UNDERLINE variable, the equal sign would be used since it is still defined in the SQL*Plus environment by a previous report and was not cleared. As shown in Figure 14–21, the UNDERLINE environment variable would need to be turned off, so the new setting would not affect the output of the **Titlereport5.sql** file. For this reason, it is advisable to have all variables set at the beginning of each script file, so if a user alters an environment variable, the report will not have an unexpected format. Since **Titlereport5.sql** does not define the UNDERLINE environment variable, anyone running the script would have to remember to set it manually by using the SET command.

```
Oracle SQL*Plus
File  Edit  Search  Options  Help
SQL> start c:\titlereport5.sql

                                                              Retail
ISBN            Title                              Cost       Price
----------      ------------------------------     ------     ------
=============== ================================== ======= ============
4981341710      BUILDING A CAR WITH TOOTHPICKS      37.80      59.95
8843172113      DATABASE IMPLEMENTATION             31.40      55.95
9959789321      E-BUSINESS THE EASY WAY             37.90      54.50
3957136468      HOLY GRAIL OF ORACLE                47.25      75.95
2491748320      PAINLESS CHILD-REARING              48.00      89.95

SQL> set underline off
SQL> start c:\titlereport5.sql

                                                              Retail
ISBN            Title                              Cost       Price
----------      ------------------------------     ------     ------
4981341710      BUILDING A CAR WITH TOOTHPICKS      37.80      59.95
8843172113      DATABASE IMPLEMENTATION             31.40      55.95
9959789321      E-BUSINESS THE EASY WAY             37.90      54.50
3957136468      HOLY GRAIL OF ORACLE                47.25      75.95
2491748320      PAINLESS CHILD-REARING              48.00      89.95

SQL>
```

Figure 14-21 Consequences of changing the UNDERLINE environment variable

You can view a list of available environment variables by clicking the **Options** menu of the SQL*Plus menu bar and selecting **Environment**. A list of the environment variables will be provided on the left side of the subsequent dialog box.

Report Headers and Footers

Now that the columns of **Titlereport5.sql** have been formatted, headers and footers need to be applied. Technically, a header serves as the title for a report. The header and footer of a report are set with the TTITLE and BTITLE commands, respectively. The **TTITLE** command indicates the text or variables (e.g., page number) to be displayed at the top of the report. The **BTITLE** command defines what is to be printed at the bottom of the report. The syntax for both the TTITLE and BTITLE commands is shown in Figure 14–22.

```
TTITLE|BTITLE [option [text|variable]…] ON|OFF]
```

Figure 14-22 Syntax of the TTITLE and BTITLE commands

Let's look at the individual elements in Figure 14–22. You can use options to specify the formats to be applied to the data, where the data should appear, and on which line. Some of the options available with TTITLE and BTITLE are shown in Figure 14–23.

Option	Description
CENTER	Centers data to be displayed between the left and right margins of a report
FORMAT	Applies a format model to data to be displayed (uses same format model elements as the COLUMN command)
LEFT	Aligns data to be displayed to the left margin of the report
RIGHT	Aligns data to be displayed to the right margin of the report
SKIP *n*	Indicates the number of lines to skip before the display of data resumes

Figure 14-23 Options for the TTITLE and BTITLE commands

Also in Figure 14–22, `[`*`text`*`|`*`variable`*`]` means that the alignment and FORMAT options can be applied either to text entered as a literal string or to SQL*Plus variables. Some of the valid SQL*Plus variables are shown in Figure 14–24.

Variable	Description
SQL.LNO	current line number
SQL.PNO	current page number
SQL.RELEASE	current Oracle release number
SQL.USER	current user name

Figure 14-24 SQL*Plus variables

Let's look at each of these variables.

- The **SQL.LNO** variable can be included in a report to indicate a line number on a report. Many users will include this variable to determine the length of a report or how many lines can be printed on a page.
- The **SQL.PNO** variable is used to include a page number on a report.
- As a matter of documentation, some users include the software's release number in a report by using **SQL.RELEASE**. This variable is generally included if a report does not execute properly after an upgrade.
- The **SQL.USER** variable can be used to include the name of the user running a report on the report.

Next, let's add a title to the report in the **Titlereport5.sql** script file to give it a professional appearance. The title should include a report heading that indicates the report's contents. An appropriate heading for the report is "Books in Inventory," and it should be centered over the report's contents. In addition, the page number should be displayed. However, the page number should appear on a separate line and on the right side of the report.

If you do not include a format model for the title, the date and page number will appear automatically.

LINESIZE

The TTITLE command can be used to create the report heading. However, how will the center or the right side of the report be determined? The placement of aligned text will be based on the environment variable **LINESIZE**. Therefore, if LINESIZE has been set to 100, then the center of the report will be 50 characters from each margin. To ensure proper placement of the components of the heading, LINESIZE needs to be calculated to find the exact width of the report.

A review of the COLUMN settings used in Figure 14–15 shows the following spacing: ISBN column, 15 spaces; Title column, 35 spaces; Cost column, 6 spaces; Retail Price column, 12 spaces. The report contains an additional three spaces due to the extra space added between columns. Because the Cost and Retail Price columns contain numeric data, an extra space is added to each of those columns in the event a negative number needs to be displayed. Once all of these spaces have been totaled (15+35+6+12+3+1+1), a LINESIZE of 73 spaces is needed for the report's contents.

Once the necessary LINESIZE and TTITLE definitions have been added to the **Titlereport5.sql** file, the script will contain the elements shown in Figure 14–25.

```
SET LINESIZE 73
TTITLE CENTER 'Books in Inventory' SKIP -
    RIGHT 'Page: ' FORMAT 9 SQL.PNO SKIP 2
COLUMN isbn FORMAT A15 HEADING 'ISBN|----------'
COLUMN title FORMAT A35 -
    HEADING 'Title|------------------------------'
COLUMN cost FORMAT 999.99 HEADING 'Cost|------'
COLUMN retail FORMAT 999999999.99 -
    HEADING 'Retail|Price|------'
SELECT isbn, title, cost, retail
FROM books
WHERE retail >50
ORDER BY title;
```

Figure 14-25 Modified **Titlereport5.sql** file

Let's look at the individual elements in Figure 14–25:

- The first line added to the beginning of the script file defines a LINESIZE of 73 spaces to establish the right margin of the report.

14

- The second line adds a heading for the report. The first part of the TTITLE command has the words "Books in Inventory" as a centered heading. The command then advances to the next line of the report, using the **SKIP** option to begin the next portion of the heading on a separate line.
- The label "Page:" is then displayed, followed by the value stored in the SQL.PNO variable to indicate the page number. A format model is applied to the page number to indicate the expected number of digits to be displayed. If the format model is not included, several blank spaces may appear between the "Page:" label and the actual page number.
- After the page number is displayed, the report advances two more lines before beginning the column headings.

Once the modified file has been saved under the file name **Titlereport6.sql**, it can be executed, and the report shown in Figure 14–26 will be displayed.

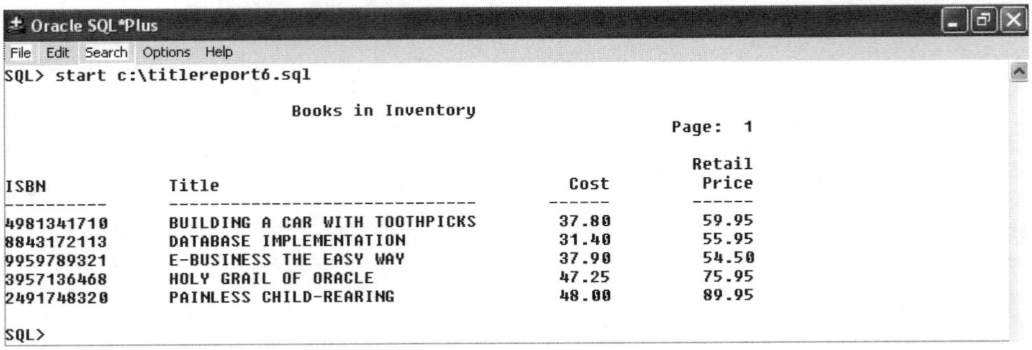

```
Oracle SQL*Plus
File Edit Search Options Help
SQL> start c:\titlereport6.sql

                        Books in Inventory
                                                        Page:  1

                                                               Retail
ISBN          Title                            Cost            Price
----------    ------------------------------   ------          ------
4981341710    BUILDING A CAR WITH TOOTHPICKS    37.80           59.95
8843172113    DATABASE IMPLEMENTATION           31.40           55.95
9959789321    E-BUSINESS THE EASY WAY           37.90           54.50
3957136468    HOLY GRAIL OF ORACLE              47.25           75.95
2491748320    PAINLESS CHILD-REARING            48.00           89.95

SQL>
```

Figure 14-26 Report with centered header for the title and right-aligned page number

Now that the report's title appears to be working, it is time to format the footer, the display at the end of the report. Technically, a report's header and footer are displayed on every page. The policy at many organizations, however, is to have a message such as "End of Report" or "End of Job" displayed at the end of a report to make certain no pages of the report are missing. However, because the Books in Inventory report is only one page in length, the message "End of Report" will be displayed as the footer, to indicate that there are no other pages to the report. In addition, the management at JustLee Books requires that the name of the user running a report also be displayed on the report for future reference. Therefore, the user name will be included as part of the footer.

PAGESIZE

Because the placement of the footer depends on the length of the page, the PAGESIZE environment variable needs to be set to indicate the length of a page. Although standard printed reports will generally have 60 lines per page (leaving a small margin at the top and bottom of the page), a standard computer monitor will display approximately 32 lines on a screen, depending on the resolution settings of the monitor. Because this report is being displayed on a computer screen, the PAGESIZE variable will be set to 20 lines. This will allow both the report header and footer to be displayed on the screen at the same time.

The footer is created using the BTITLE command. The BTITLE command has been added to the **Titlereport6.sql** file to have the words "End of Report" centered at the bottom of the page and the words "Run By:" at the right side of the page. The name of the current user is then displayed. A format model allowing five spaces has been applied to the user name, so the name will appear flush with the right margin. However, if other users might run this report, extra spaces should be allowed to accommodate longer user names. Once the changes have been made, the file should be saved with the file name **Titlereport7.sql**. The contents are shown in Figure 14–27.

```
SET LINESIZE 73
SET PAGESIZE 20
TTITLE CENTER 'Books in Inventory' SKIP -
    RIGHT 'Page: ' FORMAT 9 SQL.PNO SKIP 2
BTITLE CENTER ' End of Report ' RIGHT 'Run By: ' FORMAT A5 SQL.USER
COLUMN isbn FORMAT A15 HEADING 'ISBN|----------'
COLUMN title FORMAT A35 -
    HEADING 'Title|------------------------------'
COLUMN cost FORMAT 999.99 HEADING 'Cost|------'
COLUMN retail FORMAT 999999999.99 -
    HEADING 'Retail|Price|------'
SELECT isbn, title, cost, retail
FROM books
WHERE retail >50
ORDER BY title;
```

Figure 14-27 Contents of the **Titlereport7.sql** file

14

Once the script has been saved, it can be run in SQL*Plus and will produce the report shown in Figure 14–28.

```
Oracle SQL*Plus
File Edit Search Options Help
SQL> start c:\titlereport7.sql

                       Books in Inventory
                                                        Page:  1

                                                               Retail
ISBN         Title                              Cost        Price
----------   ------------------------------   ------       ------
4981341710   BUILDING A CAR WITH TOOTHPICKS    37.80        59.95
8843172113   DATABASE IMPLEMENTATION           31.40        55.95
9959789321   E-BUSINESS THE EASY WAY           37.90        54.50
3957136468   HOLY GRAIL OF ORACLE              47.25        75.95
2491748320   PAINLESS CHILD-REARING            48.00        89.95

                       End of Report              Run By: SCOTT

SQL>
```

Figure 14-28 Report produced by the **Titlereport7.sql** file

Suppose that after the report has been displayed, you can't remember the exact structure of the TTITLE command used to generate the report header. You can either look at the contents of the script file through the editor or word-processing program, or you can simply enter the TTITLE command at the SQL> prompt. After the **Enter** key is pressed, SQL*Plus will display the current settings for the command, as shown in Figure 14–29.

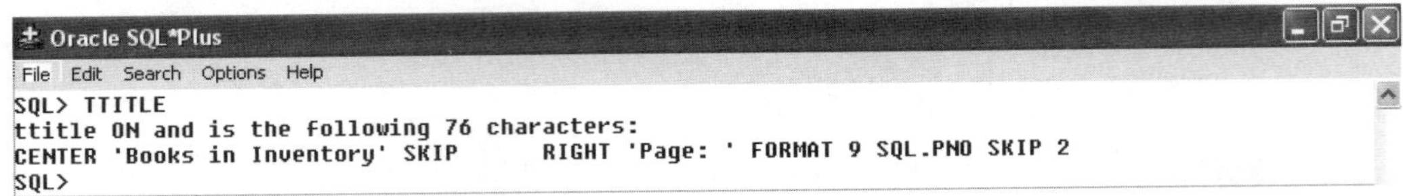
Oracle SQL*Plus
File Edit Search Options Help
SQL> TTITLE
ttitle ON and is the following 76 characters:
CENTER 'Books in Inventory' SKIP RIGHT 'Page: ' FORMAT 9 SQL.PNO SKIP 2
SQL>

Figure 14-29 Current TTITLE setting

The current settings can also be displayed for the BTITLE and COLUMN commands. If referencing a column command, make certain to list the name of the column being requested (e.g., COLUMN isbn) to display the settings for that specific column.

BREAK COMMAND

The **BREAK** command is used to suppress duplicate data in a report. This is especially useful for reports containing groups of data. For example, suppose that you are creating a list of all books in inventory, sorted by category. This would result in the Computer, Business, and other Category column names being printed multiple times. If you use the BREAK command, each Category column name will print only once. The BREAK command also includes the option of skipping lines after each group—or having each group appear on a separate page. The syntax for the BREAK command is shown in Figure 14–30.

```
BREAK ON columnname|columnalias [ON …] [skip n|page]
```

Figure 14-30 Syntax of the BREAK command

The effectiveness of the BREAK command is determined by how data are sorted in the SELECT statement. The contents of the specified column are printed once, and the printing for all subsequent rows is suppressed until the value in the column changes. For example, if a report containing a list of books sorted by category is being displayed, and the BREAK command has been applied to the Category column, then the first time the Computer Category is encountered, the word "COMPUTER" will be displayed. The category name will be suppressed until the next category is encountered. However, if the data being displayed in the report are not in a sorted order, "COMPUTER" may be displayed several times throughout the report. Therefore, as a general rule, you should always include the ORDER BY clause when using the BREAK command. Duplicate values are only suppressed when they occur in sequence. As soon as the name of the category changes, the new value will be displayed. If you want to include subgroupings, an additional ON clause can be added, followed by the name of the column to be used for the subgrouping.

Suppose that you want to produce a report that shows the amount due for orders in the ORDERS table. The orders should be sorted by customer number in the results. Rather than have customer numbers duplicated in the report, you can use the BREAK command. The script shown in Figure 14–31 can be used to create the report.

```
SET LINESIZE 32
SET PAGESIZE 27
TTITLE CENTER 'Amount Due Per Order' SKIP 2
BTITLE LEFT 'Run By: ' SQL.USER FORMAT A5
COLUMN total FORMAT 999.99
BREAK ON customer#
SELECT customer#, order#,SUM(retail*quantity) total
FROM customers NATURAL JOIN orders
  NATURAL JOIN orderitems NATURAL JOIN books
GROUP BY customer#, order#;
```

Figure 14-31 Script to create a report listing the amount due for each order

The BREAK command suppresses duplicate values in the Customer# column. Notice that an ORDER BY clause is not included in the SELECT statement used to retrieve the data for the report. Because the SELECT statement has a GROUP BY clause, the data will automatically be sorted in the order of the Customer# and Order# columns. After the script is saved, with the file name **Amountdue.sql**, the report shown in Figure 14–32 will be displayed when the file is run. Note that the current user's name is Scott.

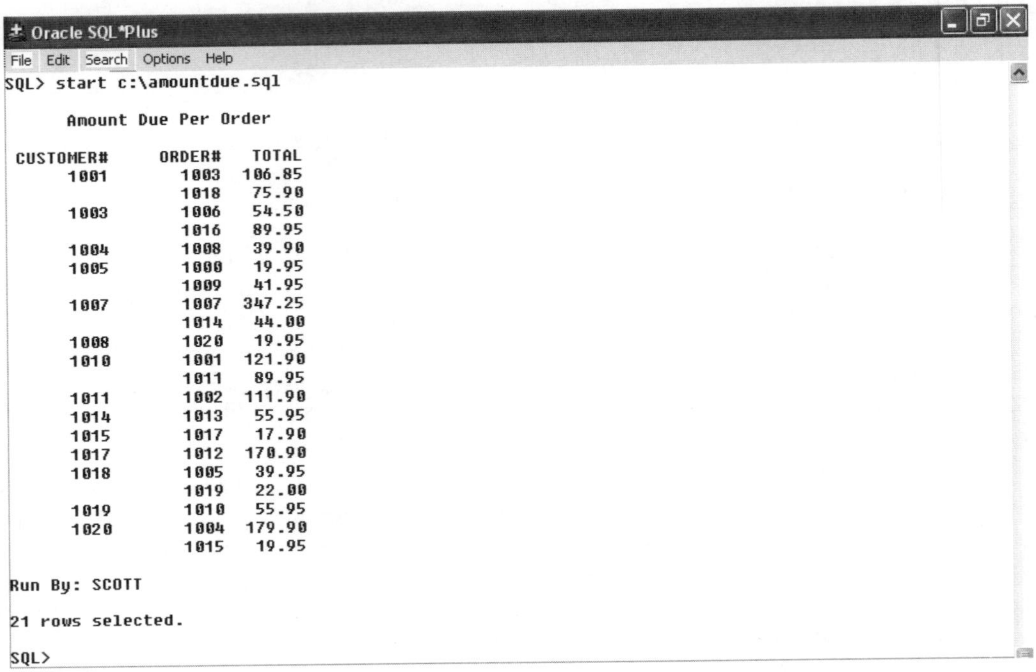

```
SQL> start c:\amountdue.sql

     Amount Due Per Order

 CUSTOMER#      ORDER#   TOTAL
      1001        1003  106.85
                  1018   75.90
      1003        1006   54.50
                  1016   89.95
      1004        1008   39.90
      1005        1000   19.95
                  1009   41.95
      1007        1007  347.25
                  1014   44.00
      1008        1020   19.95
      1010        1001  121.90
                  1011   89.95
      1011        1002  111.90
      1014        1013   55.95
      1015        1017   17.90
      1017        1012  170.90
      1018        1005   39.95
                  1019   22.00
      1019        1010   55.95
      1020        1004  179.90
                  1015   19.95

Run By: SCOTT

21 rows selected.

SQL>
```

Figure 14-32 Report with suppressed duplicate data

To make the report easier to read, the SKIP keyword could have been included with the BREAK command to have a blank line occur between each group. However, in this case, it would have caused the report to be longer than one screen of display. As a result, that option was not included.

CLEAR Command

The **CLEAR** command is used to clear the settings applied to the BREAK and COLUMN commands. The settings applied to these commands are valid even after a report has been finished. For example, if you request a list of all books stored in the BOOKS table, the Cost and Retail columns would be displayed using the format created in the **Titlereport7.sql** script file. Therefore, any time that you use the COLUMN or BREAK commands in a script, you should always clear those commands at the end of the script, so they will not adversely affect any subsequent reports. The syntax for the CLEAR command is shown in Figure 14–33.

```
CLEAR COLUMN|BREAK
```

Figure 14-33 Syntax of the CLEAR command

To instruct SQL*Plus to clear the COLUMN and BREAK settings defined in the **Amountdue.sql** script file, the CLEAR command can be added at the end of the file, as shown in Figure 14–34.

```
SET LINESIZE 32
SET PAGESIZE 27
TTITLE CENTER 'Amount Due Per Order' SKIP 2
BTITLE LEFT 'Run By: ' SQL.USER FORMAT A5
COLUMN total FORMAT 999.99
BREAK ON customer#
SELECT customer#, order#,SUM(retail*quantity) total
FROM customers NATURAL JOIN orders
  NATURAL JOIN orderitems NATURAL JOIN books
GROUP BY customer#, order#;
CLEAR BREAK
CLEAR COLUMN
```

Figure 14-34 CLEAR command added to the end of a script

Notice that the CLEAR command is issued once to clear the settings defined by the BREAK command and again to clear the settings defined by any COLUMN commands. After the CLEAR command has been added, any settings that are applied when the report is generated will automatically be cleared after the report has been displayed.

COMPUTE Command

In addition to determining the total amount due for each order, you can determine how much each customer owes by including the **COMPUTE** command. The COMPUTE command can be included with the AVG, SUM, COUNT, MAX, or MIN keywords to determine averages, totals, number of occurrences, the highest value, or the lowest value, respectively. Figure 14–35 shows the syntax for the COMPUTE command.


```
COMPUTE statisticalfunction OF columnname|REPORT ON groupname
```

Figure 14-35 Syntax of the COMPUTE command

For example, suppose that you want to determine the total amount due from each customer. The command `COMPUTE SUM OF total ON customer#` can be added to the **Amountdue.sql** script file shown in Figure 14–34 to instruct Oracle9*i* to determine the total (`SUM`) amount due (`OF total`) for each customer (`ON customer#`).

In Figure 14–36, the COMPUTE command has been included, and the LINESIZE and PAGESIZE variables have been changed to 30 and 50, respectively, to accommodate the extra data being added to the report. Because the report will now display the total amount due from each customer, the title of the report has also been changed. After these changes are made, the **Amountdue.sql** script file can be saved as **Amountdue1.sql**. It will display the output shown in Figure 14–37.

```
CLEAR COLUMN
SET LINESIZE 30
SET PAGESIZE 50
TTITLE CENTER 'Amount Due Per Customer' SKIP 2
BTITLE LEFT 'Run By: ' SQL.USER FORMAT A5
COLUMN total FORMAT 999.99
BREAK ON customer# SKIP 1
SELECT customer#, order#,SUM(retail*quantity) total
FROM customers NATURAL JOIN orders
  NATURAL JOIN orderitems NATURAL JOIN books
GROUP BY customer#, order#;
COMPUTE SUM OF total ON customer#
CLEAR BREAK
CLEAR COLUMN
```

Figure 14-36 Contents of the **Amountdue1.sql** file

The CLEAR COMPUTE command is used to clear settings declared by the COMPUTE command.

As shown in Figure 14–37, when the file is executed, the total amount due will be displayed. Partial output is shown.

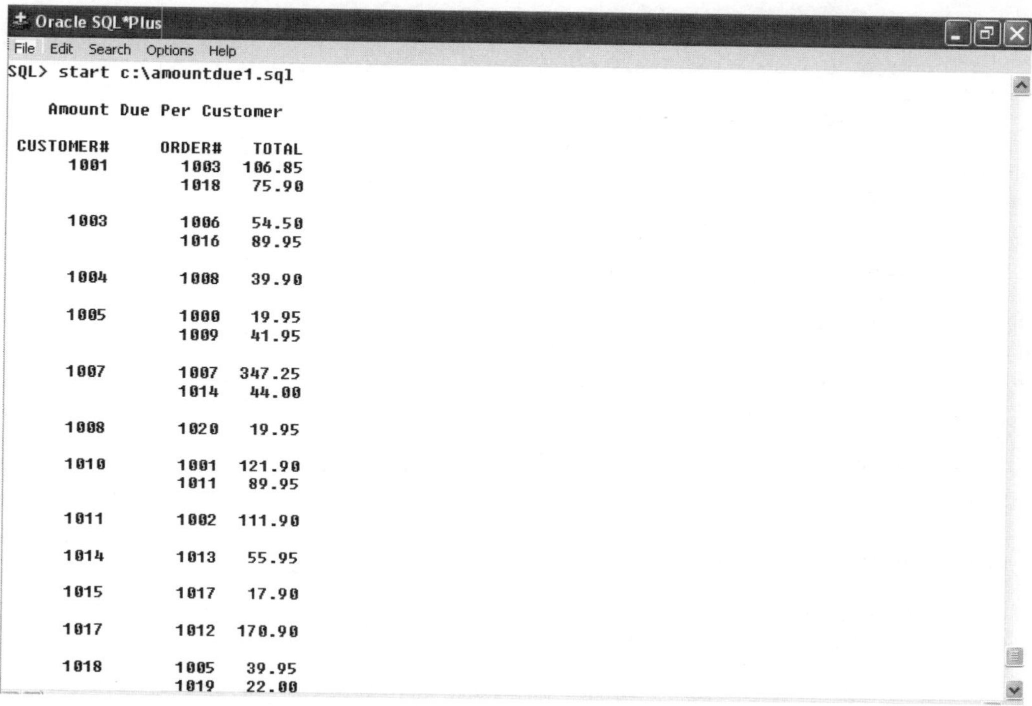

Figure 14-37 Amount due from each customer (partial output shown)

SPOOL Command

In most cases, management will want printed copies of the reports you generate. The results displayed on the computer monitor can also be redirected to a text computer file that can then be printed through a word-processing program. To save the results of a report (including the report's format) to a text file, the **SPOOL** command should be added at the beginning of the script file. The syntax for the SPOOL command is shown in Figure 14–38.

```
SPOOL d:/filename
```

Figure 14-38 Syntax of the SPOOL command

For example, if you want the results of the **Amountdue1.sql** script to be sent to a text file for printing, you would add the command `SPOOL c:\reportresults.txt` to the beginning of the **Amountdue1.sql** script, and the results will be stored in the text file **Reportresults.txt** on the hard drive. The file can then be opened with any word processor and printed. The command SPOOL OFF has been included to terminate the spooling process. Figure 14–39 shows the final version of the **Amountdue1.sql** script with the SPOOL command.

```
CLEAR COLUMN
SPOOL c:\reportresults.txt
SET LINESIZE 30
SET PAGESIZE 50
TTITLE CENTER 'Amount Due Per Customer' SKIP 2
BTITLE LEFT 'Run By: ' SQL.USER FORMAT A5
COLUMN total FORMAT 999.99
BREAK ON customer# SKIP 1
SELECT customer#, order#,SUM(retail*quantity) total
FROM customers NATURAL JOIN orders
  NATURAL JOIN orderitems NATURAL JOIN books
GROUP BY customer#, order#;
COMPUTE SUM OF total ON customer#
CLEAR BREAK
CLEAR COLUMN
```

Figure 14-39 Revised **Amountdue1.sql** script

CHAPTER SUMMARY

- Reports can be created by using a variety of formatting commands available through SQL*Plus.
- Commands to format reports are usually entered into a script file and then executed. The SELECT statement appears after the SQL*Plus formatting commands in the script.
- A script file can be run in SQL*Plus by using the START command or an "at" sign (@) before the file name.
- The COLUMN command has HEADING, FORMAT, and NULL options available to affect the appearance of columns in the report.
- The HEADING option can be used to provide a column heading. A single vertical bar (|) is used to indicate where a line break should occur in the column heading.
- The SQL*Plus environment variable UNDERLINE indicates the symbol that should be used to separate column headings from the column data.
- The FORMAT option can be used to create a format model for displaying numeric data or to specify the width of non-numeric columns.
- The NULL option indicates a text message that should be displayed if a NULL value exists in the specified column.
- The TTITLE command identifies the header, or title, that should appear at the top of the report. It can include format models, alignment options, and variables if needed.
- The BTITLE command identifies the footer, or title, that should appear at the bottom of the report. It follows the same syntax as the TTITLE command.

- Alignment used in the TTITLE and BTITLE commands will be affected by the LINESIZE and PAGESIZE variables.
- The BREAK command is used to suppress duplicate data.
- At the end of each script file that includes a COLUMN or BREAK command, the CLEAR command should be provided to clear any setting changes made by the script.
- The COMPUTE command can be used to calculate totals in a report based on groupings.
- The SPOOL command instructs Oracle9*i* to redirect output to a text file for printing through a word processor.

CHAPTER 14 SYNTAX SUMMARY

The following table presents a summary of the syntax that you have learned in this chapter. You can use the table as a study guide and reference.

Syntax Guide		
Element	**Description**	**Example**
START *or* @	Executes a script file	`START titlereport.sql` *or* `@titlereport.sql`
`TTITLE [option [`*`text`*`|`*`variable`*`]...] ON|OFF]`	Defines a report's header	`TTITLE 'Top of|Report'`
`BTITLE [option [`*`text`*`|`*`variable`*`]...] ON|OFF]`	Defines a report's footer	`BTITLE 'Bottom of|Report'`
`BREAK ON` *`columnname`*`|`*`columnalias`* `[ON ...] [skip` *`n`*`|page]`	Suppresses duplicate data in a specified column	`BREAK ON title`
`COMPUTE` *`statisticalfunction`* `OF` *`columnname`*`|REPORT ON` *`groupname`*	Performs calculations in a report based on the AVG, SUM, COUNT, MIN, and MAX statistical functions	`COMPUTE SUM OF total ON customer#`
`SPOOL` *`filename`*	Redirects the output to a text file	`SPOOL c:\reportresults.txt`
`CLEAR`	Clears settings applied with the COLUMN and BREAK commands	`CLEAR COLUMN`

14

Element	Description	Example
COLUMN Command Options		
`FORMAT` *`formatmodel`*	Applies a specified format to the column data	`COLUMN title FORMAT A30`
`HEADING` *`columnheading`*	Specifies the heading to be used for the specified column	`COLUMN title HEADING 'Book|Title'`
`NULL` *`textmessage`*	Identifies the text message to be displayed in the place of NULL values	`COLUMN shipdate NULL 'Not Shipped'`
TTITLE and BTITLE Command Options		
`CENTER`	Centers the data to be displayed between a report's left and right margins	`TTITLE CENTER 'Report Title'`
`FORMAT`	Applies a format model to the data to be displayed (uses the same format model elements as the COLUMN command)	`BTITLE 'Page:' SQL.PNO FORMAT 9`
`LEFT`	Aligns data to be displayed to a report's left margin	`TTITLE LEFT 'Internal Use Only'`
`RIGHT`	Aligns data to be displayed to a report's right margin	`TTITLE RIGHT 'Internal Use Only'`
`SKIP` *`n`*	Indicates the number of lines to skip before the display resumes	`TTITLE CENTER 'Report Title' SKIP 2 RIGHT 'Internal Use Only'`
SQL*Plus Variables		
`SQL.LNO`	current line number	`BTITLE 'End of Page on Line: ' SQL.LNO Format 99`
`SQL.PNO`	current page number	`BTITLE 'Page: ' SQL.PNO FORMAT 9`
`SQL.RELEASE`	current Oracle release number	`BTITlE 'Processed on Oracle release number:' SQL.RELEASE`
`SQL.USER`	current user name	`BTITLE 'Run By: ' SQL.USER FORMAT A8`

Element	Description	Example
	SQL*Plus Environment Variables	
`UNDERLINE x|OFF`	Specifies the symbol used to separate a column heading from the contents of the column	`SET UNDERLINE '='`
`LINESIZE n`	Establishes the maximum number of characters that can appear on a single line of output	`SET LINESIZE 35`
`PAGESIZE n`	Establishes the maximum number of lines that can appear on one page of output	`SET PAGESIZE 27`

Review Questions

1. How can you instruct SQL*Plus to place a column heading on two lines?
2. Which command and option are used to force insignificant zeros to be shown in the decimal position(s) of a numeric column?
3. What is the effect on a report's headers and footers if a report is 30 characters wide, but LINESIZE is set to 50 characters?
4. How can you display text messages in place of NULL values in a report?
5. What happens if the BREAK command is used for data that are not retrieved by the SELECT statement in any type of sorted order?
6. How is a script file run in SQL*Plus?
7. What is the purpose of the UNDERLINE variable?
8. How can a format be applied to a date column in a report?
9. How do you view the contents of a script file?
10. What is the purpose of the CLEAR command?

Multiple Choice

1. Which of the following commands can be used to clear column formats?
   a. COLUMN CLEAR
   b. CLEAR COLUMN
   c. CLEAR FORMAT
   d. all of the above
   e. None of the above—you must include a column name.

2. Which of the following commands creates a header for a report?
   a. HEADER
   b. TITLE
   c. REPORTTITLE
   d. TOPLINE
   e. none of the above
3. Which of the following format model elements cannot be applied to a column with a numeric datatype?
   a. $
   b. A*n*
   c. ,
   d. 9.99
   e. All of the above can be applied to a numeric column.
4. Which of the following commands instructs SQL*Plus to advance to the next line before resuming the display of the report?
   a. SKIP
   b. NEXT
   c. ADVANCE
   d. JUMP
   e. LINEFEED
5. Which option is used with the COLUMN command to substitute a NULL value with a text message?
   a. ON NULL
   b. FORMAT
   c. HEADING
   d. NULL
6. Which of the following commands indicates that a numeric column should be formatted to a total width of six?
   a. `FORMAT A6`
   b. `FORMAT 9999.99`
   c. `FORMAT 9999999`
   d. `FORMAT 9,999`

7. Which of the following environment variables specifies the maximum number of characters per line in SQL*Plus?
   a. PAGESIZE
   b. LINENUMBER
   c. LINESIZE
   d. CPI
8. Which of the following variables can be used to display a page number in a report?
   a. SQL.PAGE
   b. SQL.PGNO
   c. SQL.PNO
   d. SQL.Page_Number
9. Which of the following variables can be used to display the user name in a report?
   a. SQL.USER
   b. SQL.USERID
   c. SQL.NAME
   d. SQL.USERNAME
10. Which of the following can be used to display the date 31–JAN–02 in a report using the format January 31, 2002?
    a. `FORMAT ADATE`
    b. `FORMAT Month DD, YYYY`
    c. `TO_CHAR`
    d. `FORMAT AAAAAAA 99, 9999`

Questions 11–20 apply to the following script to create a report:

```
SET PAGESIZE 38
SET LINESIZE 45
TTITLE CENTER 'Listing of | Books by Publisher'
BTITLE LEFT 'Page: ' SQL.PNO FORMAT
BREAK ON "Publisher Name"
COLUMN "Publisher Name" FORMAT A30
COLUMN title
COLUMN cost HEADING 'Cost Per | Book' FORMAT 9,999.99
SELECT name, title, cost
FROM books NATURAL JOIN publisher;
CLEAR COLUMN
```

11. Which of the following commands is not valid?
    a. `BTITLE LEFT 'Page: ' SQL.PNO FORMAT`
    b. `COLUMN cost HEADING 'Cost Per | Book' FORMAT 9,999.99`
    c. `SET PAGESIZE 38`
    d. `CLEAR COLUMN`
12. When the following COLUMN command is used, which of the statements that follow it will cause an error to occur?

```
COLUMN "Publisher Name" FORMAT A30
      1 BREAK ON "Publisher Name"
      2 SELECT name, title, cost
      3 FROM books NATURAL JOIN publisher;
```

    a. line 1
    b. line 2
    c. line 3
    d. None of the above will cause an error.
13. The CLEAR COLUMN command given at the end of the script applies to which of the following?
    a. `SET PAGESIZE 38`
    b. `TTITLE CENTER 'Listing of | Books by Publisher'`
    c. `BREAK ON "Publisher Name"`
    d. `SELECT name, title, cost`
    e. none of the above
14. Which of the following clauses will cause an error to occur in a script file for a report?
    a. `SELECT name, title, cost`
    b. `SET PAGESIZE 38`
    c. `COLUMN title`
    d. `COLUMN cost HEADING 'Cost Per | Book' FORMAT 9,999.99`
15. Assuming that any errors existing in the script are corrected, how many columns will be displayed in the report?
    a. 1
    b. 2
    c. 3
    d. 4

16. What is the maximum number of characters that can appear on one line of output in the report?
    a. 38
    b. 45
    c. 30
    d. 9
17. The report will display data from how many tables?
    a. 1
    b. 2
    c. 3
    d. 4
18. Which of the following is a correct statement?
    a. No header will be displayed because there is an error in the TTITLE command.
    b. The BREAK command may not have the expected results because the SELECT statement does not include an ORDER BY clause.
    c. The exact location of the report header will be determined by the PAGESIZE variable.
    d. The exact location of the report footer will be determined by the LINESIZE variable.
19. Which of the following should be changed to correct the error in the BTITLE command?
    a. SQL.PNO
    b. A2
    c. LEFT
    d. None of the above—there is no error in the command.
20. Which of the following should be changed to correct the error in the TTITLE command?
    a. CENTER
    b. FORMAT A18 should be added to the command.
    c. The text should be enclosed in double quotation marks.
    d. None of the above—there is no error in the command.

HANDS-ON ASSIGNMENTS

To perform these activities, refer to the tables in Appendix A.

1. Create a script that will retrieve the title of each book, the name of its publisher, and the first and last name of each author of the book. Save the script in a file named **BookandAuthor1.sql**.
2. Add a title to be displayed with the query results. The title should be "Book and Author Report." Make any changes necessary to ensure that the title is centered across the top of the query results. Save the modified script in a file named **BookandAuthor2.sql**.
3. Add appropriate column headings for each column in the script file **BookandAuthor2.sql**. Save the modified script as **BookandAuthor3.sql**.
4. Modify the query in the **BookandAuthor3.sql** file so the results will be listed in ascending order by the title of the book. Save the modified script as **BookandAuthor4.sql**.
5. Change the report settings in the **BookandAuthor4.sql** file so the title of the book will not be duplicated for each author listed. Save the modified script as **BookandAuthor5.sql**.
6. Modify the **BookandAuthor5.sql** script to suppress the duplication of the publisher's name when more than one author writes a book. Save the new file as **BookandAuthor6.sql**.
7. Modify the **BookandAuthor6.sql** file and create a footer for your report that displays your user name at the bottom of each page. Save the new file as **BookandAuthor7.sql**.
8. Modify the **BookandAuthor7.sql** file to add a blank line to separate the different books listed in the report. Save the new file as **BookandAuthor8.sql**.
9. Modify the **BookandAuthor8.sql** file to clear any column formats used in the report. Save the modified file as **BookandAuthor9.sql**.
10. Modify the **BookandAuthor9.sql** file to clear any breaks used in the report. Save the modified file as **BookandAuthor10.sql**.

A CASE FOR ORACLE9*i*

To perform this activity, refer to the tables in Appendix A.

The management of JustLee Books has requested a report that shows the profit generated by each order placed between April 1, 2003, and April 5, 2003. The report should include the order number for each order, the total amount of the order, and the total profit generated by each order. A report header should be displayed that is descriptive of the report's contents, and a report footer should indicate who ran the report. Apply any other elements or formats that would make the report look professional. Save the completed report definition in a script file named **ProfitReport.sql**. Turn in the completed script to your instructor.

CHAPTER

15

INTRODUCTION TO PL/SQL

Objectives

After completing this chapter, you should be able to do the following:

- Explain the benefits of using PL/SQL blocks versus several SQL statements
- Identify the sections of a PL/SQL block and describe their contents
- Identify the mandatory and optional sections of a PL/SQL block
- Identify an anonymous block and its use
- Describe how to execute a PL/SQL block
- Explain the purpose of a variable
- Explain the difference between a constant and a variable
- Identify valid variable names
- List the valid datatypes for PL/SQL variables
- Assign a dynamic datatype for a PL/SQL variable
- Initialize a PL/SQL variable
- Use DML statements in a PL/SQL block
- Determine when it is appropriate to use an IF statement
- Identify all the clauses of an IF statement, and state when they should be used
- Create an IF statement
- Identify the purpose of a loop, and name the types of loops available in Oracle9*i*
- Create a basic loop
- Create a FOR loop
- Create a WHILE loop

In previous chapters, you issued a variety of SQL statements. When a SQL statement is issued to access an Oracle9*i* database over a network, each statement is sent to the server, processed, and then executed. After execution, the results (or a message) are returned to the user. Each SQL statement requires a minimum of two "trips" through the network—one from the user to the server, and another from the server to the user. This can generate a lot of network traffic, especially in a large organization with hundreds (or possibly thousands) of users accessing the server via an intranet.

Embedding SQL statements in a **Procedure Language SQL (PL/SQL)** program is an alternative to issuing multiple SQL statements. PL/SQL is an advanced **fourth-generation programming language (4GL)** that extends the capabilities of SQL. Although SQL is the standard database-access language, PL/SQL offers the following advantages:

- PL/SQL allows users to include error handling and control structures in addition to basic SQL statements, thus allowing more flexibility and efficiency.
- PL/SQL blocks can be stored and used by various application programs (or users) for frequently executed tasks.
- Tighter security can be maintained by granting privileges to execute stored procedures created in PL/SQL, rather than granting a user privileges directly for a table or other database object.

A PL/SQL block can be a named procedure or a function that can be referenced by a user or an application program. A **function** is a named PL/SQL block that is stored on the Oracle9*i* database server, accepts zero or more input parameters, and returns one value. A function can be used within a SQL statement, and it is generally used if the user needs to calculate one value. A **procedure** is also a named PL/SQL block, but it is used when working with several variables. A procedure can accept the parameters of input (IN), output (OUT), or both (INOUT):

- The **IN** parameter denotes that a value received by the calling application cannot be changed during the execution of the procedure.
- The **OUT** parameter denotes that a value will be calculated during the procedure's execution.
- The **INOUT** parameter denotes that a value passed to the procedure by the calling application will be changed by the procedure.

Unlike a function, *a procedure does not return a value; the OUT or INOUT parameter must be used to return the calculated value to an application program.* In addition, a procedure must be either called by a PL/SQL block or run with the EXECUTE command—it cannot be used in a SQL statement. The default parameter for a procedure is IN.

More extensive coverage of functions and procedures are found in PL/SQL application development courses. Chapters 15 and 16 of this textbook will present anonymous, or unnamed, PL/SQL blocks. An **anonymous block** is not stored by the database because there is no way to explicitly reference the PL/SQL block (e.g., it does not have a name). Instead, an anonymous block is embedded in an application program, stored in a script file, or manually entered by the user when it needs to be executed. The examples used in this chapter are designed to familiarize you with the basic concepts of PL/SQL. Once you understand the basic concepts, more complex examples will be presented in Chapter 16 to demonstrate the use of PL/SQL in a business environment.

Figure 15–1 presents an overview of this chapter's contents.

Element	Description
PL/SQL block structure	Consists of three sections: declarative, executable, and exception handling. Only the executable section is mandatory. A PL/SQL block is terminated with the END keyword, followed by a semicolon.
Declarative section	Contains the definition and initial values of all variables used within the PL/SQL block. It is identified by the keyword DECLARE.
Executable section	Contains all SQL and non-SQL statements to be executed in the PL/SQL block
Exception-handling section	Contains error handlers that are invoked when a non-syntax error occurs during the execution of the PL/SQL block
IF statement	Overrides sequential execution of statements in the executable section of a PL/SQL block, based on a specified condition or series of conditions
Basic loop	Executes a series of statements repeatedly until a specified condition is met. The loop is exited when the condition becomes TRUE.
FOR loop	Uses an implicitly declared counter to repeatedly execute a series of statements a specified number of times
WHILE loop	Executes a series of statements repeatedly until a specified condition is FALSE. Always executes at least once because the condition is evaluated at the end of the loop.

Figure 15-1 Overview of chapter contents

Before attempting to work through the examples in this chapter, you should go to the Chapter15 folder in your Data Files and run the **prech15.sql** script. Your output should be the same as that shown in the examples. To run the script, type `start d:/prech15.sql` where **d:** represents the location (correct drive and path) of the script.

BASIC STRUCTURE

15

A unit of PL/SQL code is called a **block**. As shown in Figure 15–2, each PL/SQL block can be divided into three sections: declarative, executable, and exception handling. Of the three sections, *the executable section is the only one required in every PL/SQL program*. The declarative and exception-handling sections are optional.

```
[DECLARE]      Identifies variables and constants
BEGIN          Identifies executable statements
[EXCEPTION]    Identifies error handlers
END;           Terminates the block
```

Figure 15-2 PL/SQL block structure

The **declarative section** of a PL/SQL block is identified by the **DECLARE** keyword. If any variables or constants are used within the block, they must first be identified in this section. A **variable** is used to reserve a temporary storage area in the computer's memory. When a variable is declared, the user is basically instructing the computer to set aside a portion of memory that the user can then reference by using the variable's name. This allows the user to manipulate data or perform calculations without repeatedly having to access a physical storage medium (e.g., a hard disk). A **constant** is similar to a variable except that its assigned value does not change during execution. *Variables and constants must always be declared before they are referenced anywhere in the PL/SQL block.*

The **executable section** of a PL/SQL block is identified by the **BEGIN** keyword and may consist of SQL statement(s) and/or PL/SQL statement(s). The main difference between SQL statements and PL/SQL statements is that SQL statements are used to access or manipulate data in the database tables, while PL/SQL statements focus on the data contained within the PL/SQL block.

The **exception-handling** section of a PL/SQL block is identified by the keyword **EXCEPTION** and is used to display messages or identify other actions that should be performed when an error occurs during the execution of the block. The exception-handling section is not used to resolve syntax errors that are identified when the block is compiled. Rather, it addresses errors that occur during a statement's execution, such as having no rows returned by a SELECT statement or having the software attempt to divide a number by zero.

The **END** keyword is used to close a PL/SQL block. As shown in Figure 15–2, the END keyword is followed by a semicolon. A semicolon is used to end each statement within a PL/SQL block. However, it is not included with the block identifiers DECLARE, BEGIN, or EXCEPTION.

The PL/SQL block shown in Figure 15–3 will retrieve the title and retail price of a book in the BOOKS table of JustLee Books' database. The price is then increased by 20 percent, and the new retail price of the book and its title are displayed.

```
DECLARE
      c_rateincrease  CONSTANT  NUMBER(3,2) :=1.2;
      v_title         VARCHAR2(30);
      v_retail        books.retail%TYPE;
      v_newretail     NUMBER(5,2);
BEGIN
      SELECT title, retail, retail*c_rateincrease
      INTO v_title, v_retail, v_newretail
      FROM books
      WHERE isbn = '1059831198';
      DBMS_OUTPUT.PUT_LINE
        ('The new price for ' || v_title || ' is $' || v_newretail);
END;
```

Figure 15-3 Sample PL/SQL block

Let's take a closer look at some of the elements in the declarative and executable sections of the PL/SQL block in Figure 15–3. (To simplify the example in Figure 15–3, the block does not include an exception-handling section.)

- The declarative section of the PL/SQL block includes one constant (`c_`) and three variables (`v_`).
- The executable section uses a SELECT statement to retrieve the contents of two different columns, Title and Retail, and to store those values **INTO** the variables identified in the declarative section.
- Notice that an arithmetic calculation was performed on the Retail column in the SELECT clause to determine the new price of the book. The retail price was increased by the amount stored in the **`c_ rateincrease`** variable, which is actually a value of 1.20, or 120 percent.
- After the column values have been stored in the declared variables, they can then be displayed using the **PUT_LINE** function of the **DBMS_OUTPUT** package (`DBMS_OUTPUT.PUT_LINE`).

Unlike the SELECT command you previously issued in SQL*Plus, a PL/SQL block does not display the results of a SELECT statement by default. DBMS_OUTPUT is a package that contains a set of functions that a user can reference to display the values assigned to a variable. In this case, the user would like to see the record selected by the query as well as its new price, so the PUT_LINE function of the package is used to display the results in a character string that includes the title and new retail price of the book.

If you execute the PL/SQL block displayed in Figure 15–3 and the title and new retail price are not displayed, type **`SET SERVEROUTPUT ON`** at the SQL> prompt. **SERVEROUTPUT** is an environment variable indicating that a buffer should be allocated to hold the output of the PL/SQL block. By setting the value to **ON**, the buffer is created. The PUT_LINE function of the DBMS_OUTPUT package is used to display the contents of that buffer.

Notice the structure of the declarative and executable sections in Figure 15–3. In the declarative section, the declaration of each variable ends with a semicolon. Also in the executable section, each complete statement also ends with a semicolon. Although the SELECT statement appears over several lines, Oracle9*i* evaluates everything between the SELECT keyword and the next semicolon as one statement. The end of the block is indicated by the END keyword, followed by a semicolon.

Because a PL/SQL block may contain several semicolons, how does Oracle9*i* know when the block should be executed? In SQL*Plus, the user is simply required to enter a slash (/) at the SQL> prompt or on a blank line, and the block will be executed, as shown in Figure 15–4.

```
Oracle SQL*Plus
File  Edit  Search  Options  Help
SQL> DECLARE
  2        c_rateincrease   CONSTANT  NUMBER(3,2) :=1.2;
  3        v_title          VARCHAR2(30);
  4        v_retail         books.retail%TYPE;
  5        v_newretail      NUMBER(5,2);
  6   BEGIN
  7        SELECT title, retail, retail*c_rateincrease
  8        INTO v_title, v_retail, v_newretail
  9        FROM books
 10        WHERE isbn = '1059831198';
 11        DBMS_OUTPUT.PUT_LINE
 12        ('The new price for ' || v_title || ' is $' || v_newretail);
 13   END;
 14   /
The new price for BODYBUILD IN 10 MINUTES A DAY is $37.14

PL/SQL procedure successfully completed.

SQL>
```

Figure 15-4 Execution of a PL/SQL block

As shown in Figure 15–4, when the slash (/) is entered on line 14 and the user presses the **Enter** key, the PL/SQL block is executed. The first line of output is the result of the execution of line 12 of the block. The message "PL/SQL procedure successfully completed" is always displayed by Oracle9*i* when a block is executed. If there is no statement in the block that displays output (e.g., a statement using `DBMS_OUTPUT.PUT_LINE`), then the only viewable output from the execution of a PL/SQL block is the Oracle9*i* message indicating that the block has been executed.

The remainder of this chapter will focus on the declarative and executable sections of a PL/SQL block. Chapter 16 will provide more information about the executable and exception-handling sections.

Declarative Section

The declarative section of a PL/SQL block defines the variables to be used within the block. At a minimum, every variable must (1) be given a name and (2) identify the type of data that the variable can contain. In addition, the variable can be initialized within the declarative section. When a variable is **initialized**, it is assigned a value. Unless the variable is defined as a constant, the value initially assigned can be expected to change somewhere within the PL/SQL block. The syntax for declaring a variable is shown in Figure 15–5.

```
variablename [CONSTANT] datatype [NOT NULL]
             [:= | DEFAULT value_or_expression];
```

Figure 15-5 Syntax for declaring a variable

The following sections will discuss the variable portions of the syntax given in Figure 15–5.

Variable Names

A variable name can consist of up to 30 characters, numbers, or special symbols. However, the name *must* begin with a character. In addition, the variable name should not be the same as a column name referenced within the block. If you do assign a name that is the same as a column, the Oracle server will assume that the user is referencing the column and not the variable. The prefixes in Figure 15–6 are based on standard naming conventions for various variables.

Prefix	Variable Type
c_	Constants—variables whose values do not change during the execution of the block
g_	Global variables—variables that are referenced by the host or calling environment. These variables are commonly used with application programs.
v_	Variables—used in a PL/SQL block to denote values that might change during the execution of the block

Figure 15-6 Standard variable prefixes

Constants

As previously mentioned, a constant is a variable having a value that does not change during the execution of the block. In the example given in Figure 15–3, the user could have simply entered **`retail*1.20`** in the SELECT clause, rather than using the constant **`c_rateincrease`**, without changing the results. However, what if the same 1.20 was used in a series of calculations for many books within the PL/SQL block? To apply that rate of increase to a group of books, the user would be required to go through every line of the block manually and enter the correct percentage. By declaring a constant in the declarative section and using that variable name wherever the calculation occurs, the user would only need to update the rate once, rather than multiple times throughout the entire block. This eliminates the chance of overlooking one value that must be updated and is, therefore, considered good programming practice.

15

A user can designate a constant with the optional **CONSTANT** keyword. When a variable is declared as a constant, a value must be assigned, or the Oracle server will return an error message. Even if the CONSTANT keyword is omitted, the value for the variable must still be declared, or it is simply treated as any other variable, even if the `c_` prefix is assigned. The assigned prefix is simply included for reference by the user to quickly identify that the variable is a constant.

PL/SQL Datatypes

PL/SQL variables can be broken down into four datatypes:

1. scalar
2. composite
3. reference
4. large object (LOB)

Let's examine each of these. The **scalar** datatype can be used to hold a single value and includes the same datatypes used for the columns of database tables. In addition, PL/SQL includes a Boolean datatype that can be assigned the value of TRUE, FALSE, or NULL, as well as datatypes for integers such as **BINARY_INTEGER** and **PLS_INTEGER**.

The value of a Boolean variable (TRUE, FALSE, or NULL) cannot be assigned to a column in a table with a DML statement because there is no such datatype available in Oracle9*i* SQL.

The **composite** datatype is a collection of data that can be grouped and treated as one unit. In fact, it can be used to address an entire row structure within a table, rather than having to define each individual column.

The **reference** datatype holds pointers to other program items, and the **large object (LOB)** datatype is used to hold locators that identify the location of large objects, such as maps. Only the scalar and composite datatypes will be used in this textbook. Examples using scalar datatypes will be used in this chapter, and composite datatypes will be examined briefly in Chapter 16.

Figure 15–7 provides examples of declaring variables with various datatypes.

Datatype	Example
CHAR	v_region CHAR(2);
NUMBER	v_retail NUMBER(5,2);
BOOLEAN	v_instock BOOLEAN;
DATE	v_pubdate DATE;
VARCHAR2	v_title VARCHAR2(30);
BINARY_INTEGER	v_onhand BINARY_INTEGER;
PLS_INTEGER	v_backordered PLS_INTEGER;
%TYPE	v_region customers.region%TYPE; v_newretail v_retail%TYPE;

Figure 15-7 Sample declaration of various scalar datatypes

The **%TYPE** attribute, shown in the last row of Figure 15–7, can be used to assume the same datatype as another variable declared within the same block or a column within a database table. To declare the datatype of a variable as the same datatype of another variable declared within the same block, simply prefix the %TYPE attribute with the name of the variable, as shown in Figure 15–7.

Let's look at the %TYPE attribute in more depth. In many cases, when you declare a variable for a PL/SQL block, you do so with the intention of using that variable to hold some value currently stored within a column of a table. When this occurs, you must make certain that the datatypes are the same. For example, if you declare a variable as a DATE and then try to use it to hold data stored in a NUMBER column of a table, you will receive an error message. Fortunately, the %TYPE attribute is available to copy the definition of a referenced column. For example, suppose that you cannot remember how many characters can be stored in the Title column of the BOOKS table, or whether it has been defined to store VARCHAR2 or CHAR data. One option would be to display the structure of the BOOKS table and view the definition of the Title column. Otherwise, you could define the column in the PL/SQL datatype that will hold the data as **`books.title%TYPE`**. The datatype that is currently assigned to the Title column of the BOOKS table would then be assigned to the variable being declared in the PL/SQL block. However, if the column in the database table has a NOT NULL constraint, the constraint *will not* be enforced in the PL/SQL block.

The %TYPE attribute is frequently used in industry to prevent errors that may occur in application programs or PL/SQL blocks if a user changes the definition of a referenced column. For example, suppose that a user widens a VARCHAR2 column so more characters can be stored in it. If this column is used in various PL/SQL blocks, then every variable that will hold the contents of that column must be updated. Rather than determine every possible reference to that column, most programmers will use the %TYPE attribute, so the datatype will be dynamic—based on the column referenced.

NOT NULL Constraint

To ensure that a variable will always contain a value, a variable can be assigned a NOT NULL constraint when it is declared. To assign a NOT NULL constraint to a variable, simply include the keywords NOT NULL after the datatype when the variable is declared. However, if the variable is defined as NOT NULL, it *must* be assigned a value, or initialized; otherwise, an error message will be returned when the block is executed.

Variable Initialization

Variables can be initialized, or assigned an initial value, using either the DEFAULT keyword or the PL/SQL **assignment operator** (:=). As previously mentioned, any variable with a NOT NULL constraint must be initialized when the variable is declared. Why? Because if an initial value is not assigned to a variable, then it is automatically assigned a NULL value. As with values assigned to non-numeric column datatypes, values used to initialize a variable declared with a non-numeric datatype must be enclosed in single quotation marks. Examples of initializations using the assignment operator (:=) and the DEFAULT keyword are shown in Figure 15–8.

Assignment Operator (:=)	DEFAULT keyword
`v_adate DATE NOT NULL := '04-APR-03';`	`v_adate DATE NOT NULL DEFAULT '04-APR-03';`
`c_anumber NUMBER(5) :=25;`	`c_anumber NUMBER(5) DEFAULT 25;`
`c_acharacter VARCHAR2(12) := 'Howdy';`	`c_acharacter VARCHAR2(12) DEFAULT 'Howdy';`
`v_instock BOOLEAN := TRUE;`	`v_instock BOOLEAN DEFAULT TRUE;`
`c_bnumber BOOLEAN := FALSE;`	`c_bnumber BOOLEAN := DEFAULT FALSE;`

Figure 15-8 Variable initializations

Executable Section

The executable section of a block is identified by the BEGIN keyword. As previously mentioned, it is the only mandatory section of a PL/SQL block, and it contains the SQL and PL/SQL statements to be executed when the block itself is executed. The type of SQL statements that can be executed in a PL/SQL block are almost the same as those that you can execute at the SQL> prompt. The main difference is the syntax of the SELECT statement. In each of the following sections, you will look at various statements commonly used in PL/SQL blocks.

SELECT Statement

As previously shown in Figure 15–4 and again in Figure 15–9, a SELECT statement can be included in a PL/SQL block to retrieve data from a database table.

Oracle SQL*Plus

File Edit Search Options Help

```
SQL> DECLARE
        c_rateincrease  CONSTANT  NUMBER(3,2) :=1.2;
        v_title         VARCHAR2(30);
        v_retail        books.retail%TYPE;
        v_newretail     NUMBER(5,2);
     BEGIN
        SELECT title, retail, retail*c_rateincrease
        INTO v_title, v_retail, v_newretail
        FROM books
        WHERE isbn = '1059831198';
        DBMS_OUTPUT.PUT_LINE
        ('The new price for ' || v_title || ' is $' || v_newretail);
     END;
     /
The new price for BODYBUILD IN 10 MINUTES A DAY is $37.14

PL/SQL procedure successfully completed.

SQL>
```

Figure 15-9 PL/SQL block with a SELECT statement

When a SELECT statement is used in a PL/SQL block, a slight modification must be made to the syntax of a standard SQL SELECT statement. Figure 15–10 displays the syntax of a SELECT statement in PL/SQL.

```
SELECT columnname [, columnname,…]
INTO variablename [, variablename,…]
FROM tablename
WHERE condition;
```

Figure 15-10 Syntax of a SELECT statement in PL/SQL

When a SQL statement is executed, the Oracle server allocates an area of memory that will hold not only the statement but also the results of that statement. This area of memory is called a **cursor**. There are two types of cursors: implicit and explicit. When you are working with DML statements or a SELECT statement that returns only one row of results, the Oracle server creates an **implicit cursor**. An implicit cursor is automatic and does not require any intervention on the part of the user. An **explicit cursor** must be created and managed by the user. Explicit cursors will be discussed in Chapter 16.

Because the data retrieved by a SELECT statement in a PL/SQL block must be placed in an implicit cursor, the INTO clause is used to identify the variables (i.e., areas within the implicit cursor) that will hold the retrieved data. These variables are created in the declarative section of the block. In Figure 15–10, the names of columns in the database table are listed in the SELECT clause of the statement. The INTO clause identifies the variables contained in the implicit cursor. To execute properly, the column names in the SELECT clause and the names of the variables that hold the values stored in the columns must be listed in the same order. In other words, the data are assigned to a variable, based on the sequence in which the variables are listed in the INTO clause. The FROM clause identifies the table(s) containing the referenced data.

Also notice the WHERE clause in Figure 15–10—it is not enclosed in brackets, which would indicate that it is an optional clause. Here, *the WHERE clause is not an optional clause.* Why? Because the SELECT...INTO command allows only one row to be returned from the query. When used with an implicit cursor, the SELECT...INTO statement can only return one row of results. The WHERE clause is included to restrict the results of the statement to only one row. If more than one row is returned, Oracle9*i* will display an error message, and the entire PL/SQL block will not be executed. In Chapter 16, you will learn how to use an explicit cursor to retrieve more than one row of results in a PL/SQL block.

DML Statements in PL/SQL

Although DDL statements cannot be used in a PL/SQL block, the data stored in tables can be added, updated, and deleted through DML statements within a PL/SQL block. The syntax for the INSERT, UPDATE, and DELETE commands are identical to the syntax previously shown in Chapter 10. For example, suppose that you need to add a new publisher to the PUBLISHER table in the JustLee Books' database. For illustrative purposes, you can make that addition by using a PL/SQL block.

Although technically a PL/SQL block will not explicitly execute a DDL command, it is possible to issue DDL statements in a PL/SQL block through the use of the DBMS_OUTPUT package.

As shown in Figure 15–11, a new publisher can be added to the PUBLISHER table by using the INSERT command in the executable section of a PL/SQL block. In this case, no variables are needed; therefore, the declarative section of the PL/SQL block is omitted. In addition, the COMMIT command can be included in the block to make certain that information regarding the new publisher is available to other users as soon as the INSERT command is executed.

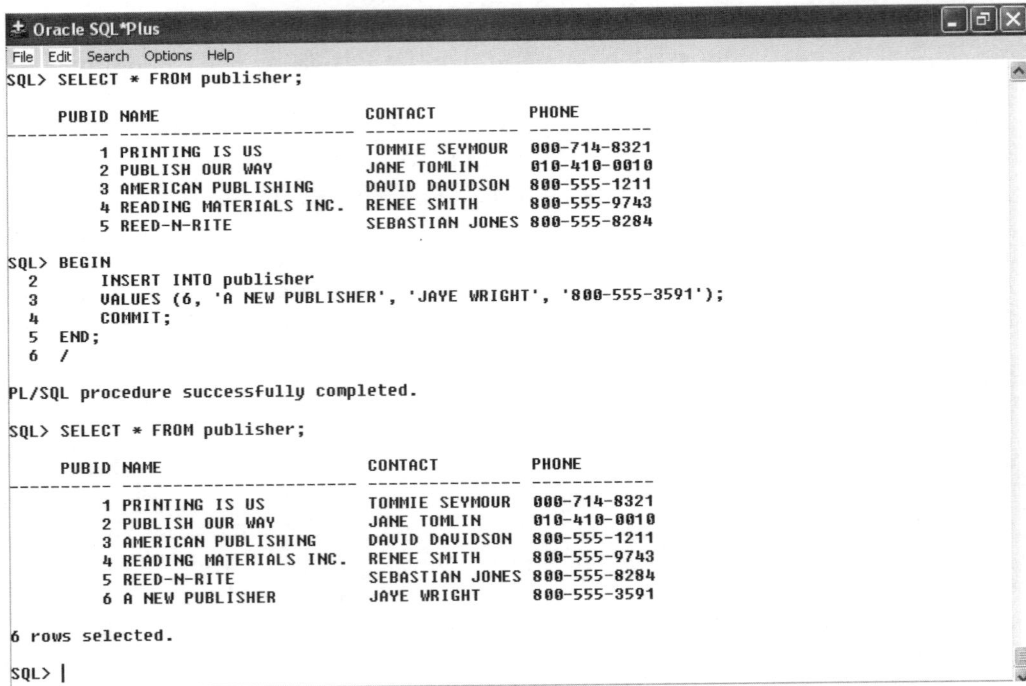


```
SQL> SELECT * FROM publisher;

    PUBID NAME                   CONTACT         PHONE
--------- ---------------------- --------------- ------------
        1 PRINTING IS US         TOMMIE SEYMOUR  000-714-8321
        2 PUBLISH OUR WAY        JANE TOMLIN     010-410-0010
        3 AMERICAN PUBLISHING    DAVID DAVIDSON  800-555-1211
        4 READING MATERIALS INC. RENEE SMITH     800-555-9743
        5 REED-N-RITE            SEBASTIAN JONES 800-555-8284

SQL> BEGIN
       INSERT INTO publisher
       VALUES (6, 'A NEW PUBLISHER', 'JAYE WRIGHT', '800-555-3591');
       COMMIT;
     END;
     /

PL/SQL procedure successfully completed.

SQL> SELECT * FROM publisher;

    PUBID NAME                   CONTACT         PHONE
--------- ---------------------- --------------- ------------
        1 PRINTING IS US         TOMMIE SEYMOUR  000-714-8321
        2 PUBLISH OUR WAY        JANE TOMLIN     010-410-0010
        3 AMERICAN PUBLISHING    DAVID DAVIDSON  800-555-1211
        4 READING MATERIALS INC. RENEE SMITH     800-555-9743
        5 REED-N-RITE            SEBASTIAN JONES 800-555-8284
        6 A NEW PUBLISHER        JAYE WRIGHT     800-555-3591

6 rows selected.

SQL> |
```

Figure 15-11 Row added with a statement in a PL/SQL block

After Oracle9*i* returns the message "PL/SQL procedure successfully completed," you can view the contents of the PUBLISHER table to verify that the row was correctly added to the table. Notice that when the INSERT INTO command is included in a PL/SQL block, no message is returned about *how many* rows were added to the PUBLISHER table. The simplest way to make certain that the command was executed correctly is to view the contents of the table.

In Chapter 16, you will learn how to validate the execution of a PL/SQL block through the use of SQL cursor attributes.

Suppose that while you were verifying that the new publisher's data were correctly added to the PUBLISHER table, you realized that the name for the publisher's contact person was misspelled. Instead of *Jaye*, it should have been spelled *Jay*. To correct this problem, you can simply create another PL/SQL block that includes an UPDATE command to update the Contact column of the PUBLISHER table. As shown in Figure 15–12, once the block has been successfully executed, you can view the contents of the PUBLISHER table with a SELECT statement to verify that the contact's name has been updated correctly.

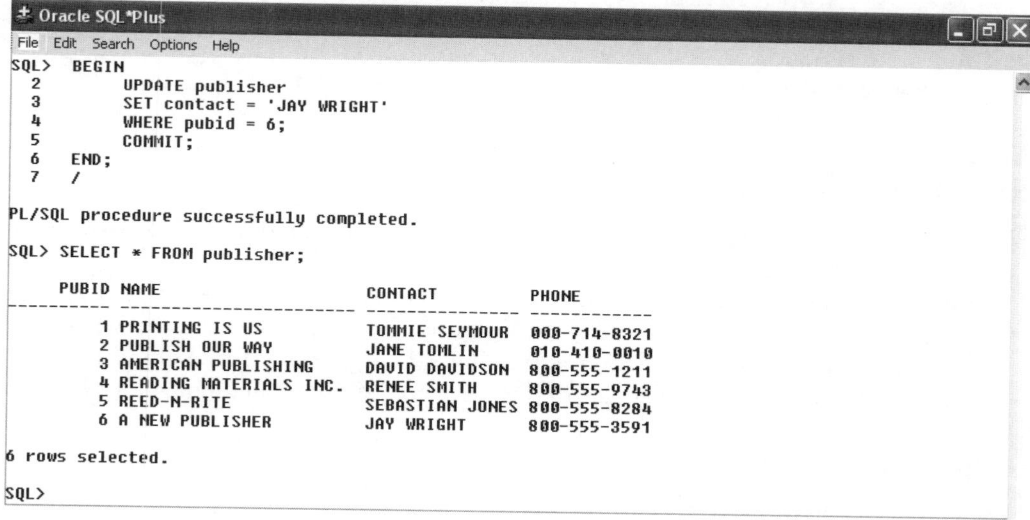

Figure 15-12 PL/SQL block to update data in a table

After you have verified that the name of the contact person for A New Publisher has been entered correctly, suppose that you are informed that the REED-N-RITE publishing company has gone out of business and should be deleted from the PUBLISHER table—of course, this is assuming that none of their books are currently in inventory. The PL/SQL block in Figure 15–13 includes a DELETE statement to remove the publisher from the PUBLISHER table. As with previous examples, once the PL/SQL block is executed, you can view the contents of the PUBLISHER table to verify that the correct publisher was removed.

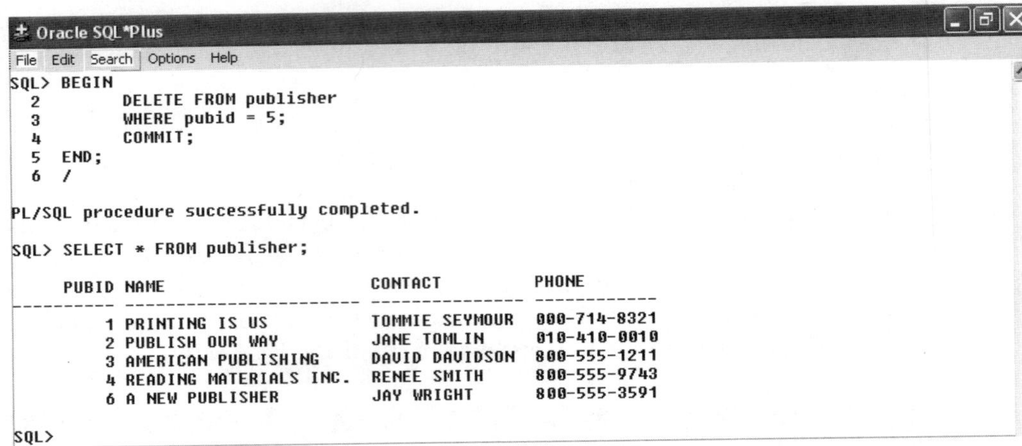

Figure 15-13 PL/SQL block to delete a row from a table

EXECUTION CONTROL

So far, the examples given in this chapter have consisted of one or two statements that are executed in the same sequence in which they are listed within the PL/SQL block. The sequential execution of statements within the executable section of a PL/SQL block can be altered by using IF statements and loops. An **IF** statement allows statements to be executed based on some condition being TRUE. This section will present various methods of controlling the execution sequence for a set of statements in a PL/SQL block.

IF Statements

If a condition is TRUE, an IF statement determines whether a statement should be executed. Figure 15–14 shows the syntax of an IF statement.

```
IF condition THEN
     statements;
[ELSIF condition THEN
     statements;]
[ELSE
     statements;]
END IF;
```

Figure 15-14 Syntax of an IF statement

The only required clause in an IF statement is the IF clause. The IF clause identifies the condition that must be TRUE for the statements listed after the **THEN** keyword to be executed. If the condition evaluates as FALSE, then Oracle9*i* goes to the first ELSIF condition, if one is provided. The **ELSIF** clause is used to identify an alternative course of action, or statement(s) that should be executed, based on a subsequent condition. If the condition listed after the ELSIF keyword is evaluated as TRUE, then the statements listed after the subsequent THEN keyword are executed. However, if the conditions provided in the IF and ELSIF clauses are FALSE, then any statements provided in the ELSE clause are executed automatically. The IF statement always ends with the **END IF** keywords.

Remember that the keyword for the ELSIF clause is one word—without a second E—and the END IF keywords are entered with a space separating the two words.

In Chapter 4, you used an inequality join to determine which gift a customer would receive, based on the retail price of the book purchased. The retail prices of books are stored in the BOOKS table, and all relevant data regarding gifts are stored in the PROMOTION table. However, suppose that the gifts change during every marketing campaign. Rather than having to delete all the rows in the PROMOTION table and then populate them with new data, a PL/SQL block can be used to determine customers' gifts, based on a book's purchase price.

The PL/SQL block in Figure 15-15 can be included in an application program to evaluate the retail price of the book being purchased, and then determine the gift that should be included with the order when it is shipped. In this particular example, however, the retail price of $29.95 has been defined in the declarative section of the block to emulate the price of the item that is being processed by the application program.

```
DECLARE
     v_gift  VARCHAR2(20);
     c_retailprice NUMBER(5, 2) := 29.95;
BEGIN
     IF c_retailprice > 56 THEN
         v_gift := 'FREE SHIPPING';
     ELSIF c_retailprice > 25 THEN
         v_gift := 'BOOKCOVER';
     ELSIF c_retailprice > 12 THEN
           v_gift := ' BOX OF BOOK LABELS';
     ELSE
           v_gift := 'BOOKMARKER';
     END IF;
     DBMS_OUTPUT.PUT_LINE ('The gift for a book costing ' ||
c_retailprice || ' is a ' || v_gift);
END;
```

Figure 15-15 IF statement in a PL/SQL block

In the PL/SQL block in Figure 15-15, the declarative section defines two variables to be used throughout the block: **`v_gift`** and **`c_retailprice`**. The **`v_gift`** variable stores the kind of gift that the customer will receive, and the **`c_retailprice`** variable stores the retail price of the book.

15

The executable section of the PL/SQL block starts with the BEGIN keyword. In this particular example, the executable section only includes an IF statement. The first clause of the IF statement, the IF clause, determines whether the retail price of the book is greater than $56.00. If the condition is TRUE, then the character string 'FREE SHIPPING' is assigned to the **`v_gift`** variable, and the ELSIF and ELSE clauses are skipped. The IF statement is exited at the END IF keyword, and the execution of the PL/SQL block ends.

However, if the condition is false (e.g., $29.95 is not greater than $56.00), then the condition in the first ELSIF clause is evaluated. If the condition is TRUE, then the statements listed immediately after the THEN keyword of the clause are executed. In this particular case, the retail price of the book is greater than $25.00, so the customer will receive a book cover with the purchase. The remaining clauses of the IF statement will be skipped, and the IF statement will be exited at the END IF keywords.

If the condition had been FALSE, the next ELSIF clause would have been evaluated. If the condition in that ELSIF clause had also been FALSE, the statement in the ELSE clause would have been executed, and the customer would have received a bookmarker. The ELSE clause does not specify a condition to be evaluated; if the clause is reached, any statements will automatically be executed. This clause can be interpreted as follows: "If everything else is FALSE, then do this." The IF statement always terminates with the END IF keywords.

As shown in Figure 15–16, after the block is executed, the message "PL/SQL procedure successfully completed" is displayed.

Oracle SQL*Plus

File Edit Search Options Help

```
SQL> DECLARE
         v_gift  VARCHAR2(20);
         c_retailprice NUMBER(5, 2) := 29.95;
     BEGIN
         IF c_retailprice > 56 THEN
                v_gift := 'FREE SHIPPING';
         ELSIF c_retailprice > 25 THEN
                v_gift := 'BOOKCOVER';
         ELSIF c_retailprice > 12 THEN
                v_gift := 'BOX OF BOOK LABELS';
         ELSE
                v_gift := 'BOOKMARKER';
     END IF;
     DBMS_OUTPUT.PUT_LINE ('The gift for a book costing ' || c_retailprice || ' is a ' || v_gift);
     END;
     /
The gift for a book costing 29.95 is a BOOKCOVER

PL/SQL procedure successfully completed.

SQL>
```

Figure 15-16 Execution of the PL/SQL block

When you are testing a PL/SQL block, especially one containing a control structure such as an IF statement, you should include a `DBMS_OUTPUT.PUT_LINE` statement that will display the contents of the variables after the END IF keywords, just to verify which clause was executed and to double-check the logic of the IF statement.

ITERATIVE CONTROL

As previously mentioned, there may be occasions when you will want statements within the executable section of a PL/SQL block to be executed repeatedly. For example, if you need to print an invoice for every order received today, you would not want to re-enter the same statements for each order. Instead, you could have the statements executed once for each order. Basically, you are creating a loop containing the statements to be executed. The loop keeps repeating until some condition is met, and then the loop is exited. There are three types of loops that you can use to repetitively execute a set of statements in the executable section of a PL/SQL block:

- basic loop
- FOR loop
- WHILE loop

Basic Loop

The **basic loop** is used to execute statements until the condition(s) stated in the EXIT clause is met. The number of times the loop will be executed may vary from one execution to the next, depending on the specified condition. The syntax for a basic loop is given in Figure 15–17.

```
LOOP
   statements;
   EXIT [WHEN condition];
END LOOP;
```

Figure 15-17 Syntax of a basic loop

As shown in Figure 15–17, the **LOOP** keyword indicates the beginning of a loop; **END LOOP** indicates the end of a loop. All statements between the LOOP and END LOOP keywords are repeatedly executed until the loop is exited. The **EXIT** keyword indicates when the loop should be exited. Notice that since the EXIT keyword is listed after the statements to be executed, any statement within the loop will automatically be executed at least once. This is referred to as a **post-test**. If you need to check the EXIT condition before any of the statements are executed (i.e., perform a **pre-test**), the EXIT clause should be listed *before* the statements, rather than after them. After the statements are executed, any condition listed in the EXIT clause is evaluated, and if the condition is TRUE, the loop ends and the remainder of the PL/SQL block is executed.

One of the simplest means of understanding how a loop works is to create a loop that prints a series of numbers. Figure 15–18 displays a PL/SQL block that contains a basic loop that will print one line of output each time the loop is executed.

```
DECLARE
v_counter NUMBER(1) := 0;
BEGIN
LOOP
    v_counter := v_counter + 1;
   DBMS_OUTPUT.PUT_LINE ('The current value of the counter
        is ' || v_counter);
   EXIT WHEN v_counter = 4;
END LOOP;
END;
```

Figure 15-18 PL/SQL block containing a basic loop

Notice that in the declarative section of the PL/SQL block, a variable has been created that will track the number of times the loop executes. The variable is initialized to zero because the loop has not yet been executed. The LOOP keyword is used in the executable section of the block to identify the beginning of the loop. The first statement in the loop increases the value of the variable by one to count the number of times the loop executes. The value contained in the variable is then displayed in output. The **EXIT WHEN** clause then checks to determine whether the desired number of loops has occurred. If the condition is TRUE, then the loop is exited; if not, the loop is repeated, beginning at the LOOP keyword. The execution of the PL/SQL block is shown in Figure 15–19.

```
Oracle SQL*Plus
File Edit Search Options Help
SQL> DECLARE
       v_counter        NUMBER(1) := 0;
     BEGIN
       LOOP
           v_counter := v_counter + 1;
           DBMS_OUTPUT.PUT_LINE ('The current value of the counter is ' || v_counter);
           EXIT WHEN v_counter = 4;
       END LOOP;
     END;
     /
The current value of the counter is 1
The current value of the counter is 2
The current value of the counter is 3
The current value of the counter is 4

PL/SQL procedure successfully completed.

SQL>
```

Figure 15-19 Execution of a basic loop

If you forget to increase the value of the counter during the execution of the loop, you will be caught in what is commonly referred to as an "infinite loop." An **infinite loop** goes on forever and usually requires you to use the operating system to stop the procedure. This condition will occur any time the condition used to end a loop cannot be achieved. Also double-check loops and make certain the condition is achievable before actually executing the loop.

Because the condition stated in the EXIT WHEN clause requires the loop to be exited when the value assigned to the counter reaches four, there are four lines of output displayed.

FOR Loop

The **FOR** loop also uses a counter to control the number of times the loop is executed. However, the counter is not a variable that must be declared in the declarative section of the PL/SQL block. The counter is implicitly declared when the LOOP is first executed. Figure 15–20 shows the syntax of a FOR loop.

```
FOR counter IN [REVERSE] lower_limit .. upper_limit LOOP
     statements;
END LOOP;
```

Figure 15-20 Syntax of a FOR loop

The FOR clause in Figure 15–20 requires the user to identify a lower and upper limit for the counter. In other words, the initial value for the counter must be specified (`lower_limit`) and the value at which the loop is terminated (`upper_limit`) must be stated. Two periods separate the lower-limit and upper-limit values. Each time the loop is executed, the counter is incremented by one. After it reaches the value defined as the upper limit for the counter, the loop is exited. The counter can also work in reverse (i.e., counting decreases rather than increases) if the REVERSE keyword is included in the clause.

To see how the FOR loop works, look at Figure 15–21.

```
BEGIN
  FOR i IN 1 .. 10 LOOP
    DBMS_OUTPUT.PUT_LINE ('The current value of the counter is ' || i);
  END LOOP;
END;
```

15

Figure 15-21 PL/SQL block with a FOR loop

Because there are no variables that need to be declared, the declarative section of the PL/SQL block has been omitted in Figure 15–21. The BEGIN keyword indicates that the block immediately starts with the executable section. In this case, the FOR clause uses the letter **i** as the counter for the loop. The **IN** keyword indicates that the loop should be continued as long as the value of the counter is in the indicated range. The initial value of the counter is one, as defined by the `lower_limit` of the range, and the loop ends when the counter reaches 10. Each time the loop is executed, the value of the counter will be displayed. After the loop has been exited (i.e., i reaches 10), the PL/SQL block is completed. As shown in Figure 15–22, the loop is executed 10 times.

```
Oracle SQL*Plus
File Edit Search Options Help
SQL> BEGIN
  2      FOR i IN 1 .. 10 LOOP
  3          DBMS_OUTPUT.PUT_LINE ('The current value of the counter is ' || i);
  4      END LOOP;
  5  END;
  6  /
The current value of the counter is 1
The current value of the counter is 2
The current value of the counter is 3
The current value of the counter is 4
The current value of the counter is 5
The current value of the counter is 6
The current value of the counter is 7
The current value of the counter is 8
The current value of the counter is 9
The current value of the counter is 10

PL/SQL procedure successfully completed.

SQL>
```

Figure 15-22 Execution of a FOR loop

WHILE Loop

The WHILE loop executes a series of statements until a condition becomes FALSE. However, unlike the previous loops, this loop is never entered if the condition is initially FALSE. The syntax of the WHILE loop is shown in Figure 15–23.

```
WHILE condition LOOP
     statements;
END LOOP;
```

Figure 15-23 Syntax of a WHILE loop

As with previous loops, the WHILE loop will execute all the statements contained within the loop until the loop is terminated. The condition provided in the WHILE clause determines when the loop will be terminated. The PL/SQL block given in Figure 15–24 uses a WHILE loop to display the value of a variable until a specific condition is FALSE.

```
DECLARE
v_counter NUMBER(2) :=0;
BEGIN
    WHILE v_counter < 15 LOOP
         DBMS_OUTPUT.PUT_LINE ('The current value of the
     counter is ' || v_counter);
     v_counter := v_counter + 1;
  END LOOP;
END;
```

Figure 15-24 PL/SQL block with a WHILE loop

The example in Figure 15–24 uses the value assigned to `v_counter` as the condition for the loop. In the WHILE clause of the WHILE loop, the condition indicates that the loop should be continued as long as `v_counter` is less than 15. Once `v_counter` reaches 15, the loop is exited. As shown in Figure 15–25, the loop is executed 14 times before it is terminated.

```
Oracle SQL*Plus
File Edit Search Options Help
SQL> DECLARE
  v_counter NUMBER(2) :=0;
  BEGIN
      WHILE v_counter < 15 LOOP
          DBMS_OUTPUT.PUT_LINE ('The current value of the counter is ' || v_counter);
           v_counter := v_counter + 1;
      END LOOP;
  END;
  /
The current value of the counter is 0
The current value of the counter is 1
The current value of the counter is 2
The current value of the counter is 3
The current value of the counter is 4
The current value of the counter is 5
The current value of the counter is 6
The current value of the counter is 7
The current value of the counter is 8
The current value of the counter is 9
The current value of the counter is 10
The current value of the counter is 11
The current value of the counter is 12
The current value of the counter is 13
The current value of the counter is 14

PL/SQL procedure successfully completed.

SQL>
```

Figure 15-25 Execution of a WHILE loop

Nested Loops

Any type of loop can be nested inside another loop. However, you must remember that *the execution of the inner loop must be completed before control is returned to the outer loop*. After control is returned to the outer loop, the outer loop is again executed as long as the loop's condition is valid, which includes the execution of the inner loop. This process will continue until the outer loop is terminated. The PL/SQL block shown in Figure 15–26 contains a nested FOR loop.

When the block in Figure 15–26 is executed, the WHILE loop is run first. However, the WHILE loop contains a nested FOR loop. This requires the FOR loop to be executed during each cycle of the WHILE loop. To make it simpler to understand, the execution of each cycle of both the inner and outer loops generates one line of output. Notice that two lines of output are first displayed from the FOR loop, and then one line of output is displayed from the WHILE loop. Once the first cycle of the WHILE loop is completed, the second cycle begins, which requires the FOR loop to be executed again. This procedure continues until the condition of the WHILE loop is evaluated as FALSE and the loop is exited.

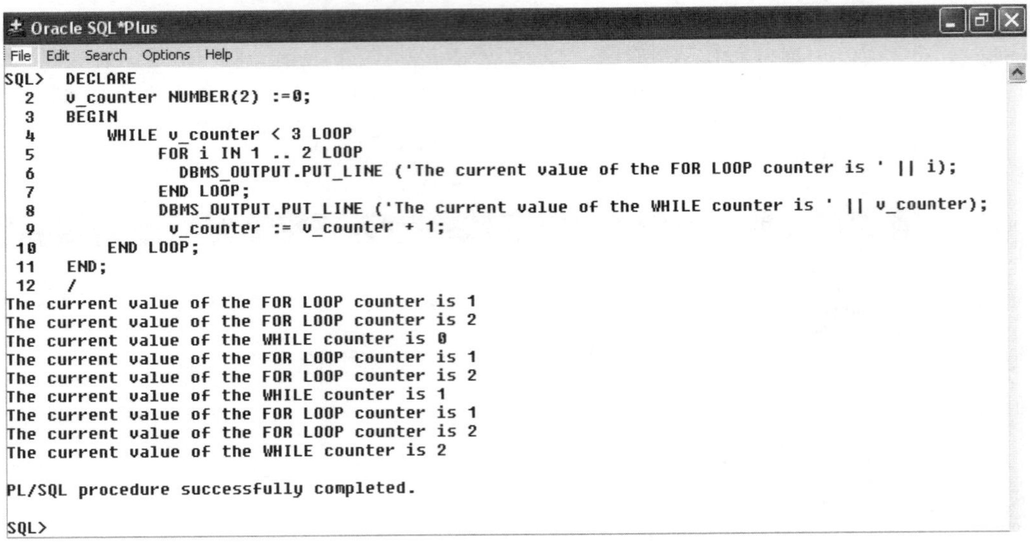

Oracle SQL*Plus

File Edit Search Options Help

```
SQL> DECLARE
     v_counter NUMBER(2) :=0;
     BEGIN
         WHILE v_counter < 3 LOOP
              FOR i IN 1 .. 2 LOOP
                DBMS_OUTPUT.PUT_LINE ('The current value of the FOR LOOP counter is ' || i);
              END LOOP;
              DBMS_OUTPUT.PUT_LINE ('The current value of the WHILE counter is ' || v_counter);
               v_counter := v_counter + 1;
         END LOOP;
     END;
     /
The current value of the FOR LOOP counter is 1
The current value of the FOR LOOP counter is 2
The current value of the WHILE counter is 0
The current value of the FOR LOOP counter is 1
The current value of the FOR LOOP counter is 2
The current value of the WHILE counter is 1
The current value of the FOR LOOP counter is 1
The current value of the FOR LOOP counter is 2
The current value of the WHILE counter is 2

PL/SQL procedure successfully completed.

SQL>
```

Figure 15-26 Execution of a nested loop

The purpose of this chapter is to familiarize you with the concept of a PL/SQL block and its structure. In Chapter 16, you will learn how to retrieve multiple rows from a table. In addition, the exception-handling section of a PL/SQL block will be discussed.

CHAPTER SUMMARY

- A PL/SQL block can combine several SQL statements and reduce network traffic.
- A PL/SQL block is divided into three sections: declarative, executable, and exception handling.
- The executable section is the only section required in a PL/SQL block.
- Every PL/SQL block must end with the END keyword, followed by a semicolon.
- A PL/SQL block is executed by entering a slash (/) on a blank line or at the SQL> prompt.
- The declarative section is identified with the DECLARE keyword.
- Variables are defined and initialized in the declarative section of a PL/SQL block.
- A variable references a holding area in the computer's memory. A constant is a variable having a value that does not change during the execution of a block.
- The declaration of each variable in the declaration section must end with a semicolon.
- The categories of variable datatypes are scalar, composite, reference, and large object (LOB).

- Valid scalar datatypes include the SQL datatypes valid for defining columns, as well as BOOLEAN, BINARY_INTEGER, and PLS_INTEGER.
- A variable can be initialized using the assignment operator (:=) or the DEFAULT keyword.
- The executable section of a PL/SQL block is identified by the BEGIN keyword.
- If a SELECT statement is included in a PL/SQL block to retrieve a row from a table, it must include the INTO clause.
- An implicit cursor is created by Oracle9*i* when a DML statement is issued or when a SELECT statement only retrieves one row from a table.
- Changes made to data by DML operations contained in a PL/SQL block cannot be viewed by other users until an implicit or explicit COMMIT occurs.
- An IF statement can be used to control the sequence for executing statements.
- The IF clause is the only required portion of the IF statement.
- The ELSIF clause is evaluated if the condition in the IF clause is FALSE.
- The ELSE clause is only executed if the previously evaluated conditions are all FALSE.
- A loop can be used to repeatedly execute a series of statements.
- The three types of loops are the basic loop, FOR loop, and WHILE loop.
- The FOR loop is the only loop that implicitly declares a counter.
- Every loop must be terminated with the END LOOP keywords.
- In a nested loop, the inner loop must complete all required cycles of execution before the execution of the next cycle of the outer loop.

Chapter 15 Syntax Summary

The following tables present a summary of the syntax that you have learned in this chapter. You can use the tables as a study guide and reference.

<table>
<tr><th colspan="3">Syntax Guide</th></tr>
<tr><th>Element</th><th>Syntax</th><th>Example</th></tr>
<tr><th colspan="3">PL/SQL Block</th></tr>
<tr><td>PL/SQL block structure</td><td><pre>[DECLARE
  used to identify variables
  and constants]
BEGIN
  executable statements
[EXCEPTION
  error handlers]
END;</pre></td><td><pre>DECLARE
  v_date DATE;
BEGIN
  v_date := SYSDATE;
END;</pre></td></tr>
<tr><td>Variable declaration</td><td><pre>variablename [CONSTANT]datatype
[NOT NULL]
[:=|DEFAULT value_or_expression];</pre></td><td><pre>c_rateincrease CONSTANT
                NUMBER(3,2) :=1.2;
v_title   VARCHAR2(30);
v_retail   books.retail%TYPE;
v_newretail   NUMBER(5,2);</pre></td></tr>
<tr><td>SELECT ...INTO</td><td><pre>SELECT columnname [, columnname,...]
INTO variablename
     [, variablename,...]

FROM tablename
WHERE condition;</pre></td><td><pre>SELECT title, retail,
  retail*c_rateincrease
INTO v_title, v_retail,
  v_newretail
FROM books
WHERE isbn = '1059831198';</pre></td></tr>
<tr><th colspan="3">Execution Control</th></tr>
<tr><td>IF statement</td><td><pre>IF condition THEN
     statements;
[ELSIF condition THEN
     statements;]
[ELSE
     statements;]
END IF;</pre></td><td><pre>DECLARE
  v_gift VARCHAR2(20);
  c_retailprice NUMBER (5, 2)
:= 29.95;
BEGIN
  IF c_retailprice > 56 THEN
     v_gift := 'FREE SHIPPING';
  ELSIF c_retailprice > 25 THEN
     v_gift := 'BOOKCOVER';
  ELSIF c_retailprice > 12 THEN
        v_gift := 'BOX OF BOOK
LABELS';
   ELSE
        v_gift := 'BOOKMARKER';
   END IF;
   DBMS_OUTPUT.PUT_LINE ('The
gift for a book costing ' ||
c_retailprice || ' is a ' ||
v_gift);
END;</pre></td></tr>
</table>

Execution Control (continued)		
Element	**Syntax**	**Example**

Basic loop	`LOOP`
	   `statements;`
	   `EXIT [WHEN` *`condition`*`];`
	`END LOOP;`
		`DECLARE`
		`v_counter  NUMBER(1) := 0;`
		`BEGIN`
		`LOOP`
		    `v_counter := v_counter + 1;`
		   `DBMS_OUTPUT.PUT_LINE ('The current value of the counter is ' || v_counter);`
		   `EXIT WHEN v_counter = 4;`
		`END LOOP;`
		`END;`

FOR loop	`FOR` *`counter`* `IN [REVERSE]`
	`lower_limit` `..` *`upper_limit`* `LOOP`
	     `statements;`
	`END LOOP;`
		`BEGIN`
		  `FOR i IN 1 .. 10 LOOP`
		    `DBMS_OUTPUT.PUT_LINE ('The`
		      `current value of the`
		      `counter is ' || i);`
		  `END LOOP;`
		`END;`

WHILE loop	`WHILE` *`condition`* `LOOP`
	     `statements;`
	`END LOOP;`
		`DECLARE`
		`v_counter NUMBER(2) :=0;`
		`BEGIN`
		    `WHILE v_counter < 15 LOOP`
		      `DBMS_OUTPUT.PUT_LINE`
		        `('The current value of the counter is ' || v_counter); v_counter := v_counter + 1;`
		  `END LOOP;`
		`END;`

Variable Prefixes		
Prefix	**Variable Type**	**Example**
`c_`	Constants—variables having values that do not change during the execution of the block	`c_date`
`g_`	Global variables—variables that are referenced by the host or calling environment. These variables are commonly used with application programs.	`g_globalvariable`
`v_`	Variables—used in a PL/SQL block to denote values that might change during the execution of the block	`v_retailprice`

Review Questions

1. List the sections of a PL/SQL block and identify what is contained in each section.
2. When are semicolons used in a PL/SQL block?
3. Which section of a PL/SQL block is mandatory? Why is it mandatory?
4. What is the relationship between a constant and a variable?
5. What keywords are used to identify each section of a PL/SQL block?
6. When are you required to assign an initial value to a variable?
7. How is a FOR loop different from a basic loop? How is a basic loop different from a WHILE loop?
8. What is the purpose of the ELSIF clause in an IF statement?
9. What is the purpose of the ELSE clause in an IF statement?
10. What is the purpose of a loop?

Multiple Choice

To answer these questions, refer to the tables in Appendix A.

1. If you are searching for a book with a retail price greater than $80, which of the following lines of code could cause incorrect results?

   ```
   1 DECLARE
   2 v_title books.title%TYPE;
   3 c_price NUMBER :=80;
   4 BEGIN
   5 SELECT title
   6 INTO v_title
   7 FROM books
   8 WHERE retail>c_price;
   9 END;
   ```

   a. line 2

   b. line 5

   c. line 6

   d. line 8

   e. None of the above—the correct results will be returned if only one book meets the stated condition.

2. Which of the following is not a valid variable declaration?

   a. `v_retail NUMBER(5,2);`

   b. `v_retail NUMBER(5,2) DEFAULT 9.2;`

   c. `v_retail NUMBER(5,2) := 9.2;`

   d. All of the above are valid.

3. Which of the following is a correct statement?
   a. Every statement in the executable section of a PL/SQL block must end with a semicolon.
   b. Every line in a PL/SQL block must end with a semicolon.
   c. Every keyword identifying a PL/SQL block must end with a semicolon.
   d. all of the above
   e. none of the above

Use the following PL/SQL block to answer questions 4–5.

```
1 DECLARE
2 v_title books.title%TYPE; v_retail books.retail%TYPE;
3 c_price NUMBER DEFAULT 1.35;
4 BEGIN
5 SELECT title, retail*c_price
6 INTO v_title, v_retail
7 FROM books;
8 END;
```

4. Which of the following lines of code contains a syntax error?
   a. line 1
   b. line 2
   c. line 3
   d. line 5
   e. line 8
5. If the syntax error identified in the previous question is corrected, the PL/SQL block will not execute unless which of the following is true?
   a. A semicolon is included at the end of the DECLARE and BEGIN keywords.
   b. A WHERE clause is included in the SELECT statement so only one row is returned in the results.
   c. The arithmetic operation in the SELECT clause is removed.
   d. There are no other errors.
6. Which of the following statements is correct?
   a. DML operations cannot be performed with a PL/SQL block.
   b. DML operations performed within a PL/SQL block are automatically committed when the block is executed.
   c. The ROLLBACK command must be issued before the execution of the PL/SQL block is completed or any changes made by a DML operation within the block are automatically committed.
   d. Any DML statement included in a PL/SQL block must end with a semicolon.

7. Which of the following is not a valid variable name?
   a. v_9price
   b. 9price
   c. price9
   d. c_9price
8. If the following variable is created in the declaration section of a PL/SQL block, what value is it assigned?

   ```
   v_anumber NUMBER(5,2);
   ```

   a. 0
   b. 5
   c. 2
   d. NULL
9. If the following variable is created in the declaration section of a PL/SQL block, what will occur?

   ```
   v_anumber NUMBER(5,2)NOT NULL;
   ```

   a. Oracle9*i* will return an error message, and the PL/SQL block will not be executed.
   b. The variable will be assigned the default value of NOT NULL.
   c. The variable will be assigned the default value of FALSE.
   d. Oracle9*i* will execute any other statements in the block that do not reference the variable.
10. Which of the following statements will cause an error if it is included in the executable section of a PL/SQL block?
    a. `v_anumber NUMBER(5);`
    b. `DELETE FROM publisher WHERE pubid = 1;`
    c. `UPDATE publisher SET contact = 'John Thomley'`
       `WHERE pubid = 1;`
    d. none of the above
11. Which statement about the following variable declaration is correct?

    ```
    v_onhand BOOLEAN :=1;
    ```

    a. It will cause an error because PL/SQL does not support the BOOLEAN datatype.
    b. It will cause an error because a BOOLEAN variable cannot be initialized.
    c. It will cause an error because the variable is being initialized with an invalid value.
    d. none of the above

12. Which of the following is the correct sequence for the sections of a PL/SQL block?
    a. BEGIN, EXCEPTION, DECLARE
    b. BEGIN, DECLARE, EXCEPTION
    c. DECLARE, EXCEPTION, BEGIN
    d. DECLARE, BEGIN, EXCEPTION
13. Which of the following can be used to dynamically assign a datatype in the declarative section of a PL/SQL block?
    a. ;=
    b. :=
    c. %TYPE
    d. %COPY
14. The statement(s) contained within the IF clause of an IF statement are only executed if:
    a. More than one row of results is returned from a SELECT statement.
    b. The condition specified in the clause is FALSE.
    c. The condition specified in the clause is NULL.
    d. The condition specified in the clause is TRUE.
15. Which of the following is used to decrement the counter of a FOR clause?
    a. a negative number
    b. the REVERSE keyword
    c. The *upper_limit* is less than the *lower_limit*.
    d. All of the above can be used.
16. Which of the following types of loops is NEVER entered if the condition is initially FALSE?
    a. basic
    b. FOR
    c. WHILE
    d. none of the above
17. An infinite loop can be created if:
    a. The value of the counter never changes in a basic or WHILE loop.
    b. The LOOP keyword is not included at the beginning of the loop.
    c. No rows are processed.
    d. The END IF keywords are not used.

18. The EXIT keyword is used in which type of loop?
    a. basic
    b. FOR
    c. WHILE
    d. It is never used in a loop.
19. To display the output produced by the use of the DBMS_OUTPUT package in a PL/SQL block, which of the following must be set?
    a. SERVEROUTPUT
    b. TERMOUT
    c. VERIFYOUT
    d. SETOUTON
20. Which of the following is used to control the sequence in which statements are executed?
    a. IF statement
    b. basic loop
    c. FOR loop
    d. WHILE loop

HANDS-ON ASSIGNMENTS

To perform these activities, refer to the tables in Appendix A. Do not cause a COMMIT event to occur unless specifically instructed.

1. Create a script file containing a PL/SQL block that will allow the user to delete a book from the BOOKS table, based on the ISBN of the book. (*Hint*: Include a substitution variable to have the user enter the ISBN for the book.) Name the script file **SC1501.sql**.
2. Execute the **SC1501.sql** script and delete the book with the ISBN of 9959789321.
3. Verify that the correct book was removed from the BOOKS table.
4. Create a script file containing a PL/SQL block that can be used to increase the retail price of every book in the BOOKS table by 10 percent. Name the script file **SC1504.sql**.
5. Execute the **SC1504.sql** script and verify that the correct changes were made to the BOOKS table.
6. Modify the **SC1504.sql** script so the user can determine the percentage of increase when the script is executed. Save the modified script as **SC1505.sql**.
7. Modify **SC1505.sql** to allow the user to change the retail price of a book based on its ISBN. Save the modified script as **SC1506.sql**.

8. Create a script containing a PL/SQL block that will display the title and price of the most expensive book in the BOOKS table. When the block is executed, it should provide the user with the title of the book and its retail price. Save the script as **SC1508.sql**.
9. Undo all the DML changes that were made to the BOOKS table.
10. Verify that the changes made to the BOOKS table no longer exist.

A Case for Oracle9*i*

To perform this activity, refer to the tables in Appendix A.

The management of JustLee Books has finally agreed on the new price increases for all books currently sold by the company. The following percentage increases will go into effect:

Category	Percentage of Increase
Computer	20
Cooking	15
Literature	10
Self Help	10

The remaining categories will all be increased by only 5 percent.

Required:

Create a PL/SQL block that will calculate and update the retail price of a book in the BOOKS table. The user should be allowed to enter the ISBN of the book that needs to be updated. The new retail price will be determined by the category of the book, based upon the stated percentage of increase. The block should determine the category of the book and then calculate the new retail price based upon the category. After the retail price of the book has been updated in the BOOKS table, the user should be notified of the title of the book and that the price has been updated. Provide your instructor with a copy of the PL/SQL block.

CHAPTER

16

CURSORS AND EXCEPTIONS

Objectives

After completing this chapter, you should be able to do the following:

- Determine when an explicit cursor is required
- Declare, open, and close an explicit cursor
- Fetch data from an explicit cursor
- Identify attributes associated with a cursor
- Use a cursor FOR loop to retrieve data from a cursor
- Declare a cursor in the subquery of a cursor FOR loop
- Evaluate Boolean conditions combined with logical operators
- Identify the purpose of the exception-handling section of a PL/SQL block
- Trap predefined exceptions in a PL/SQL block
- Trap user-defined exceptions in a PL/SQL block

In Chapter 15, you learned the correct structure of a PL/SQL block. In particular, that chapter focused on the declarative and executable sections of a PL/SQL block. In addition, you learned how to override the sequential execution of statements within the executable section of a block. In this chapter, you will learn how to use explicit cursors to retrieve more than one row for processing. In addition, you will learn how to use cursor attributes and loops that can retrieve rows from an explicit cursor. This chapter concludes with a discussion of the exception-handling section of a PL/SQL block and shows how user-defined exceptions can be raised and trapped.

Figure 16–1 presents an overview of this chapter's contents.

Element	Description
Explicit cursor	A cursor created by the user. This cursor is necessary when a SELECT statement retrieves more than one row. In most cases, the user must declare, open, and close an explicit cursor. Data are fetched from the cursor for processing.
Cursor attributes	Every cursor has four attributes: %ROWCOUNT, %FOUND, %NOT-FOUND, %ISOPEN. These attributes can be used to control looping.
Logic table	Used to determine whether a statement containing conditions joined by logical operators will be evaluated as TRUE, FALSE, or NULL
Exception handling	An exception indicates that a non-syntax error has occurred during the execution of a PL/SQL block. An error is indicated by an error number returned by the Oracle server. If the error is not trapped by the exception-handling section of the block, the error is propagated, or returned, to the calling environment.

Figure 16-1 Overview of chapter contents

Before attempting to work through the examples in this chapter, go to the Chapter16 folder in your Data Files and run the **prech16.sql** script. Your output should be the same as that shown in the examples. To run the script, type `start d:/prech16.sql` where **d:** represents the location (correct drive and path) of the script.

CURSORS

As mentioned in Chapter 15, when a PL/SQL block is used to execute a DML statement or a SELECT statement that returns only one row of results, an implicit cursor is created. You can think of a cursor as an array or as an area in memory that holds the values currently being processed. However, when a SELECT statement returns more than one row of results, the user must create an explicit cursor. The data returned by the query are then placed in the explicit cursor for processing.

Figure 16–2 shows what happens when a user does not create an explicit cursor to obtain multiple-row results.

```
Oracle SQL*Plus
File Edit Search Options Help
SQL> DECLARE
    v_title books.title%TYPE;
    v_retail books.retail%TYPE;
  BEGIN
    SELECT title, retail
    INTO v_title, v_retail
    FROM books NATURAL JOIN orderitems
    WHERE order# = 1012;
    DBMS_OUTPUT.PUT_LINE ('Book title: ' ||v_title ||' Retail price: '||  v_retail);
  END;
  /
DECLARE
*
ERROR at line 1:
ORA-01422: exact fetch returns more than requested number of rows
ORA-06512: at line 5


SQL>
```

Figure 16-2 Flawed PL/SQL block

The PL/SQL block shown in Figure 16–2 creates two variables in the declarative section to hold the title and retail price of the books included in order 1012. Because the variables were declared using the %TYPE attribute, their datatype and width will be the same as the identified Title and Retail columns from the BOOKS table. To assign values to the variables, a SELECT…INTO statement was included in the executable section of the block. However, even though the statement contained a WHERE clause to ensure that only one order was processed, *more than one book* was included in that order. As shown in Figure 16–2, if a SELECT…INTO statement returns more than one row of results, an error message will be displayed, and the block will not be executed.

When a PL/SQL block will retrieve more than one row, an explicit cursor must be used to hold the data to be processed. Unlike an implicit cursor, which is maintained automatically by Oracle9*i*, an explicit cursor must be manually declared, opened, and closed through a PL/SQL statement. The following sections will discuss the correct procedure for performing each of these steps.

Declaring an Explicit Cursor

An explicit cursor is declared in the declarative section of a PL/SQL block. The syntax for declaring a cursor is shown in Figure 16–3.

```
CURSOR cursor_name IS
       selectquery;
```

Figure 16-3 Syntax to declare an explicit cursor

The PL/SQL block shown previously in Figure 16–2 needs an explicit cursor to retrieve the title and retail price of the books in order 1012. Figure 16–4 shows the explicit cursor needed to process those rows.

```
DECLARE
    CURSOR books_cursor IS
         SELECT title, retail
         FROM books NATURAL JOIN orderitems
         WHERE order# = 1012;
```

Figure 16-4 Command to declare an explicit cursor

In Figure 16–4, an explicit cursor named BOOKS_CURSOR is declared to hold the rows. Then, the SELECT statement will retrieve the title and retail price of the books. The structure of the cursor is defined by the data retrieved with the SELECT statement. Notice that the ISBN of each book has not been included in the cursor itself, even though the column was necessary to join the BOOKS and ORDERITEMS tables to determine the correct books to select. At this point, however, the cursor has only been declared. Before any processing can occur, it must next be opened.

Opening an Explicit Cursor

When the cursor is opened, the necessary memory is allocated, the SELECT statement is executed, and then the data identified by the SELECT clause are loaded into the cursor. The syntax to open a cursor is given in Figure 16–5.

```
OPEN cursor_name;
```

Figure 16-5 Syntax to open an explicit cursor

The following portion of a PL/SQL block can be used to open the cursor created in Figure 16–4.

```
OPEN books_cursor;
```

Figure 16-6 Command to open an explicit cursor

After the cursor is opened using the statement shown in Figure 16–6, the data contained in the cursor can then be assigned to variables for processing.

Closing the Cursor

After all the data have been retrieved from a cursor, it must be explicitly closed. An explicit cursor can be closed with the **CLOSE** command in the executable section of a PL/SQL block. The syntax for the CLOSE command is shown in Figure 16–7.

```
CLOSE cursor_name;
```

Figure 16-7 Syntax to close an explicit cursor

Figure 16–8 shows the CLOSE statement needed to close the cursor that was previously opened in Figure 16–6.

```
CLOSE books_cursor;
```

Figure 16-8 Command to close an explicit cursor

Fetching Data from the Cursor

A SELECT statement cannot be used to retrieve data from an explicit cursor. Instead, you must fetch data from the cursor by using the **FETCH** command. The FETCH command retrieves rows from the cursor and assigns the values to variables. The variables must have been previously declared in the declarative section of the PL/SQL block before they can be referenced in the executable section. The syntax of the FETCH command is shown in Figure 16–9.

```
FETCH cursor_name INTO variablename [,…variablename];
```

Figure 16-9 Syntax to fetch data from an explicit cursor

The statement given in Figure 16–10 can be used to retrieve the rows stored in BOOKS_CURSOR and assign their values to the **v_title** and **v_retail** variables.

```
FETCH books_cursor INTO v_title, v_retail;
```

Figure 16-10 Command to fetch data from an explicit cursor

When you transfer data from an explicit cursor and place them in variables, you are still processing rows of data. For example, if there are four rows of data being held in the cursor when the FETCH command is executed, the first row is assigned to the variables, then the second, and so on. However, the problem is that when the second row is assigned to the variables, it overwrites the first row. In other words, the second row that is retrieved replaces the values previously assigned to the first row of variables. Figure 16–11 highlights this problem.

```
Oracle SQL*Plus
File Edit Search Options Help
SQL> DECLARE
   v_title books.title%TYPE;
   v_retail books.retail%TYPE;
   CURSOR books_cursor IS
     SELECT title, retail
     FROM books NATURAL JOIN orderitems
     WHERE order# = 1012;
  BEGIN
   OPEN books_cursor;
   FETCH books_cursor INTO v_title, v_retail;
   DBMS_OUTPUT.PUT_LINE ('Book title: ' ||v_title ||' Retail price: '||  v_retail);
   CLOSE books_cursor;
  END;
  /
Book title: BIG BEAR AND LITTLE DOVE Retail price: 8.95

PL/SQL procedure successfully completed.

SQL>
```

Figure 16-11 Flawed PL/SQL block to retrieve rows from a cursor

If you did not receive output from the DMBS_OUTPUT package reference contained in the PL/SQL block, remember to set the environmental variable SERVEROUTPUT to ON.

Let's look at the PL/SQL block in Figure 16–11 in more depth:

- The declarative section defines the variables `v_title` and `v_retail` as well as the cursor called BOOKS_CURSOR.
- The cursor is defined by the SELECT statement that retrieves the title and retail price of the books included in order 1012.
- In the executable section of the block, the previously declared cursor is opened.
- A FETCH statement is used to retrieve the rows from the cursor and assign the values to the `v_title` and `v_retail` variables.
- The `DBMS_OUTPUT` package is included to display the contents of the variables.

After the block is executed, there is one row of output displayed. This would lead the user to conclude that order 1012 was for only one book. However, as shown in Figure 16–12, execution of the SELECT statement separately from the PL/SQL block reveals that there are actually four books in that order.

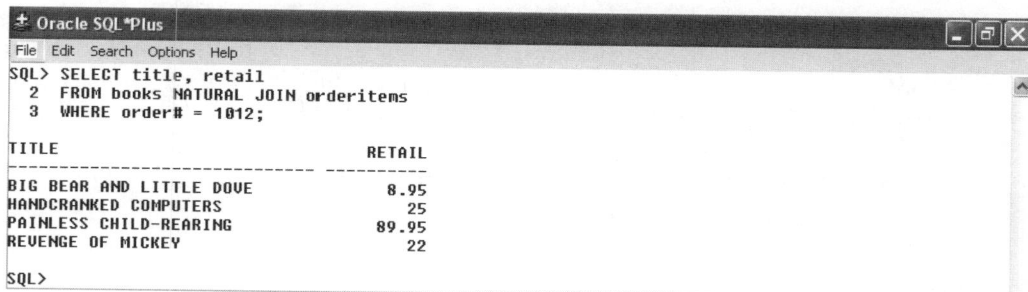
```
SQL> SELECT title, retail
  2  FROM books NATURAL JOIN orderitems
  3  WHERE order# = 1012;

TITLE                              RETAIL
------------------------------ ----------
BIG BEAR AND LITTLE DOVE             8.95
HANDCRANKED COMPUTERS                  25
PAINLESS CHILD-REARING              89.95
REVENGE OF MICKEY                      22

SQL>
```

Figure 16-12 Books in order 1012

To display the title and retail price of each book from order 1012, the `PUT_LINE` procedure of the `DBMS_OUTPUT` package needs to be executed before the next row is retrieved from the cursor. Therefore, the statements that retrieve and display the contents of the cursor should be included in a loop. However, the basic, FOR, and WHILE loops discussed in Chapter 15 all required that some type of condition be included to determine when a loop should terminate.

Every cursor (implicit and explicit) has four attributes that can be used as conditions within a PL/SQL block. The three cursor attributes most frequently used to control looping are %ROWCOUNT, %FOUND, and %NOTFOUND. The **%ROWCOUNT** attribute contains an integer value representing the number of rows processed when the block is executed. The **%FOUND** and **%NOTFOUND** attributes are used to reflect whether any rows are actually found to be processed. The %FOUND attribute will contain the Boolean value of TRUE if one or more rows are affected during the block's execution; by contrast, the %NOTFOUND attribute will be TRUE if no rows were affected. The fourth cursor attribute, **%ISOPEN**, is generally used to determine whether a cursor was closed after the execution of a PL/SQL block. A summary of these attributes is presented in Figure 16–13.

Cursor Attribute	Description
%ROWCOUNT	Identifies the number of rows processed
%FOUND	Contains the value of TRUE if one or more rows are processed—FALSE if no rows are processed
%NOTFOUND	Contains the value of TRUE if no rows are processed—FALSE if one or more rows are processed
%ISOPEN	Contains the value TRUE if a cursor is not closed after processing—FALSE if the cursor is closed. The value will always be FALSE after processing when an implicit cursor occurs because its closure is automatic.

Figure 16-13 Cursor attributes

16

If a basic loop is included in a PL/SQL block to control execution of a set of statements, a condition must also be included to terminate the loop. In theory, a customer could order any number of books, so there is no set value that can be used to determine the number of loops that should be executed. Instead, the condition can be stated as "When the cursor no longer contains any values, terminate the loop." In other words, the loop should be exited when the %NOTFOUND attribute of BOOKS_CURSOR is TRUE, or **EXIT WHEN books_cursor%NOTFOUND**.

As shown in Figure 16–14, after a basic loop is added to the PL/SQL block, there are actually four records, or books, in order 1012.

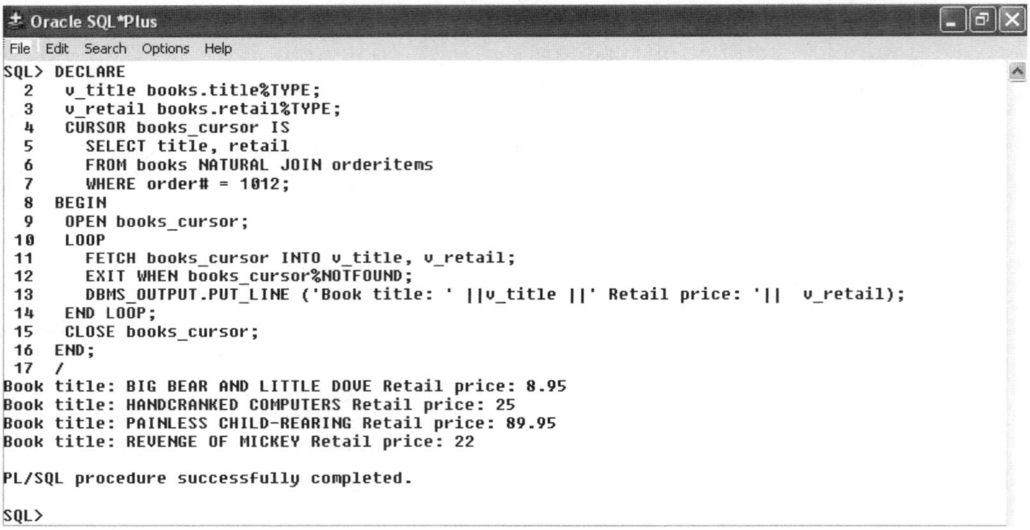

```
SQL> DECLARE
  v_title books.title%TYPE;
  v_retail books.retail%TYPE;
  CURSOR books_cursor IS
    SELECT title, retail
    FROM books NATURAL JOIN orderitems
    WHERE order# = 1012;
 BEGIN
  OPEN books_cursor;
  LOOP
    FETCH books_cursor INTO v_title, v_retail;
    EXIT WHEN books_cursor%NOTFOUND;
    DBMS_OUTPUT.PUT_LINE ('Book title: ' ||v_title ||' Retail price: '||  v_retail);
  END LOOP;
  CLOSE books_cursor;
 END;
 /
Book title: BIG BEAR AND LITTLE DOVE Retail price: 8.95
Book title: HANDCRANKED COMPUTERS Retail price: 25
Book title: PAINLESS CHILD-REARING Retail price: 89.95
Book title: REVENGE OF MICKEY Retail price: 22

PL/SQL procedure successfully completed.

SQL>
```

Figure 16-14 PL/SQL block to display the contents of order 1012

Let's look at some of the elements in Figure 16–14.

- The executable section of the PL/SQL block first opens BOOKS_CURSOR, defined in the declarative section.
- The keyword LOOP is used to indicate that the subsequent statements should be sequentially executed until the loop terminates. The first statement inside the loop fetches one row of data from the opened cursor. The second statement is used to determine whether a row is actually retrieved from the cursor.
- Notice that the name of the attribute is preceded by the name of the actual cursor being evaluated. The attribute is always prefixed with the name of the cursor in the event that a PL/SQL block contains more than one cursor.
- If a row is retrieved (e.g., %NOTFOUND is FALSE), the title and retail price of the book are displayed. The loop then continues by fetching the next row contained in the cursor.

- After all rows have been retrieved, the subsequent FETCH statement will not find any rows in the cursor, and the %NOTFOUND attribute will be TRUE.
- After the condition is evaluated as TRUE, execution advances to the `END LOOP;` statement, and the loop is terminated.
- On the next line, the `END;` statement terminates the PL/SQL block.

In most programming languages, a **primary read**, or retrieval of data, occurs before a loop is actually entered. As an extension of this concept, an initial FETCH statement is usually executed before the loop is entered to determine whether the cursor contains data. However, this step has been omitted to simplify the example.

Cursor FOR Loop

Because it is not uncommon to use loops to control the processing of rows retrieved from a cursor, Oracle9*i* includes a shortcut for processing explicit cursors in a PL/SQL block. The cursor FOR loop can be used to automatically, or implicitly, open and close a cursor, as well as fetch data from a cursor. The syntax for a cursor FOR loop is shown in Figure 16–15.

```
FOR record_name IN cursor_name LOOP
      statement;
      [statement;…]
END LOOP;
```

Figure 16-15 Syntax of the cursor FOR loop

Rather than retrieve the rows from the cursor and assign values to a variable, the rows' contents are assigned to a record. A **record** is a composite datatype that can assume the same structure as the row being retrieved. To specify that the record will have the same structure as the row being retrieved, the **%ROWTYPE** attribute can be used when the record is defined. The %ROWTYPE is similar to the %TYPE attribute presented in Chapter 15 except that the %TYPE attribute is used to define a variable that is based on a single column; by contrast, the %ROWTYPE attribute defines a record that is based on all the columns contained within a database table.

16

In Figure 16–16, the PL/SQL block previously displayed in Figure 16–14 has been modified to include a cursor FOR loop. The record R_BOOKS has been defined in the declarative section of the block, with the structure of the BOOKS table. The executable section of the block consists of the cursor FOR loop that will implicitly open the cursor named BOOKS_CURSOR and assign its contents to the R_BOOKS record. The cursor FOR loop will execute once for each row contained in the cursor named BOOKS_CURSOR. The only statement contained within the loop displays the current row being processed. Notice that the character string to be displayed by the `DBMS_OUTPUT` package references the name of the column contained within the record. Because the data values were assigned to a record rather than individual variables, the only way to display the values is to specify the name of the column containing the desired data. When the record was defined with the %ROWTYPE attribute, the column names were also assigned for each value being retrieved from the BOOKS table. To identify which column(s) should be displayed, the name of the column is prefixed with the name of the record containing the data. The column name and the record prefix are always separated by a period, as shown on line 9 in Figure 16–16. After the last row contained in BOOKS_CURSOR has been processed, the loop is terminated. After the loop has been terminated, the cursor is implicitly closed.

```
Oracle SQL*Plus
File Edit Search Options Help
SQL> DECLARE
  2   CURSOR books_cursor IS
  3     SELECT title, retail
  4     FROM books NATURAL JOIN orderitems
  5     WHERE order# = 1012;
  6   r_books books%ROWTYPE;
  7  BEGIN
  8   FOR r_books IN books_cursor LOOP
  9   DBMS_OUTPUT.PUT_LINE ('Book title: ' ||r_books.title ||' Retail price: '||  r_books.retail);
 10  END LOOP;
 11  END;
 12  /
Book title: BIG BEAR AND LITTLE DOVE Retail price: 8.95
Book title: HANDCRANKED COMPUTERS Retail price: 25
Book title: PAINLESS CHILD-REARING Retail price: 89.95
Book title: REVENGE OF MICKEY Retail price: 22

PL/SQL procedure successfully completed.

SQL>
```

Figure 16-16 PL/SQL block containing a cursor FOR loop

When a record is only being used to control a cursor FOR loop, the user is not required to define the record in the declarative section of the PL/SQL block. However, its scope will be limited to use within the loop. The record has been defined in Figure 16–16 as an example.

When using the cursor FOR loop, you also have the option of declaring a cursor, using a subquery. Rather than declare a cursor in the declarative section of a PL/SQL block, the SELECT statement can be substituted for the cursor name in the IN clause. However, because the cursor does not have a name, the cursor's attributes, such as %NOTFOUND, cannot be referenced. Figure 16–17 shows a cursor FOR loop using a subquery. Notice that the declarative section of the PL/SQL block has been omitted.

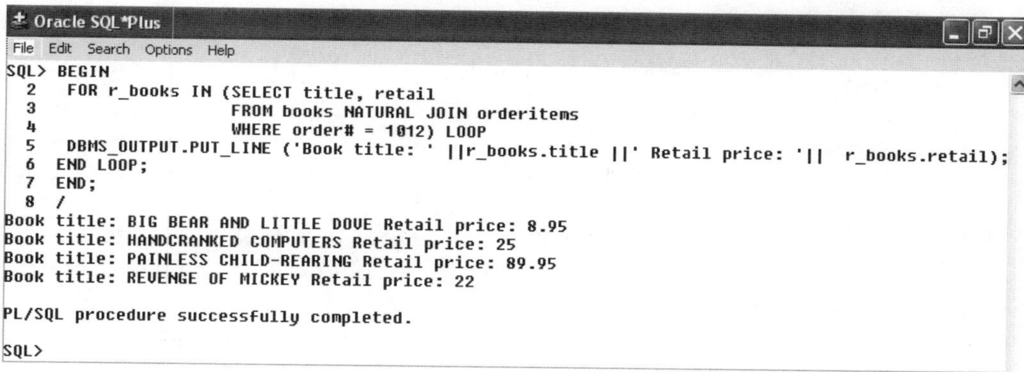

```
Oracle SQL*Plus
File Edit Search Options Help
SQL> BEGIN
 FOR r_books IN (SELECT title, retail
                 FROM books NATURAL JOIN orderitems
                 WHERE order# = 1012) LOOP
 DBMS_OUTPUT.PUT_LINE ('Book title: ' ||r_books.title ||' Retail price: '||  r_books.retail);
 END LOOP;
 END;
 /
Book title: BIG BEAR AND LITTLE DOVE Retail price: 8.95
Book title: HANDCRANKED COMPUTERS Retail price: 25
Book title: PAINLESS CHILD-REARING Retail price: 89.95
Book title: REVENGE OF MICKEY Retail price: 22

PL/SQL procedure successfully completed.

SQL>
```

Figure 16-17 Cursor FOR loop using a subquery

Because the PL/SQL block in Figure 16–17 can identify a particular order's items, with a few minor modifications, the block could also be embedded in an application program to determine not only which books were purchased, but also the total amount due for the order. In Figure 16–18, the variable **`v_ordertotal`** has been added to the block to hold the total amount due for the order. Within the cursor FOR loop, the Quantity column from the ORDERITEMS table has been included in the subquery to retrieve how many copies of each book were ordered. The expression **`v_ordertotal + r_books.retail * r_books.quantity`** has also been included in the loop to create a running total of the amount due, which is updated as each row is processed. The value calculated by the expression is then assigned to the **`v_ordertotal`** variable. Once all the rows have been processed and the loop is terminated, the `DBMS_OUTPUT` package in line 10 is used to display the value of the **`v_ordertotal`** variable. Because the statement is listed after the END LOOP keywords, it is not contained within the loop and, therefore, the statement is executed only once.

```
Oracle SQL*Plus
File Edit Search Options Help
SQL> DECLARE
  2   v_ordertotal NUMBER(5,2) := 0;
  3  BEGIN
  4   FOR r_books IN (SELECT title, retail, quantity
  5                   FROM books NATURAL JOIN orderitems
  6                   WHERE order# = 1012) LOOP
  7   DBMS_OUTPUT.PUT_LINE ('Book title: ' ||r_books.title ||' Retail price: '||  r_books.retail);
  8   v_ordertotal := v_ordertotal + r_books.retail * r_books.quantity;
  9  END LOOP;
 10   DBMS_OUTPUT.PUT_LINE ('Total Amount Due: ' || v_ordertotal );
 11  END;
 12  /
Book title: BIG BEAR AND LITTLE DOVE Retail price: 8.95
Book title: HANDCRANKED COMPUTERS Retail price: 25
Book title: PAINLESS CHILD-REARING Retail price: 89.95
Book title: REVENGE OF MICKEY Retail price: 22
Total Amount Due: 170.9

PL/SQL procedure successfully completed.

SQL>
```

Figure 16-18 PL/SQL block to display the total amount due

16

Cursor Loop Control with Logical Operators

In Figure 16–14, the %NOTFOUND attribute was used to control the number of loops executed to fetch the data from the explicit cursor. In that example, all the rows in the cursor needed to be processed, so the condition basically required rows in the cursor to be fetched until there were no more rows in the cursor. However, what if there were only a specific number of rows that needed to be processed? In the case, this most obvious cursor attribute to use would be the %ROWCOUNT attribute. However, even if the %ROWCOUNT attribute is used, there could still be a problem. For instance, you could enter a number that is greater than the number of rows contained within the cursor. In other words, not only would you need to process the specified number of rows, but also you would need to make certain that number of rows actually exists.

Previously in Chapter 11, you used "Top-N" analysis to determine the five most expensive books in the BOOKS table. You did this by creating a subquery to select books having a ROWNUM of five or fewer. Instead of creating that subquery, you could have obtained that information much more simply by imbedding a cursor in a PL/SQL block. When rows are fetched from a cursor, they are fetched in the same order as they were originally placed in the cursor. To make certain the first row in the cursor contains the book with the highest retail price, an ORDER BY clause that sorts the books in descending order, based on the Retail column, can be added to the SELECT statement that defines the cursor when it is declared. With the %ROWCOUNT attribute as the looping condition, the user can then specify the number of rows to be fetched from the cursor to view the most expensive books in the BOOKS table. To ensure that the block does not try to process more rows than there actually are, the AND logical operator can be used in the loop condition with the %FOUND attribute to make certain that looping continues only if the cursor contains additional rows. Figure 16–19 shows this block and its output.

Line 4 of the PL/SQL block in Figure 16–19 requires the user to enter how many books are to be displayed. If the user enters the numeral 3, then the three most expensive books will be displayed in the output. The first condition of the WHILE loop in line 12 controls how many rows are processed, based on the user's entry. This condition is joined to the second condition on line 13 by the AND logical operator. The second condition specifies that a row must have been found in the cursor, or the loop is to be terminated. Because the two conditions are joined with the AND logical operator, both conditions must be TRUE or the loop will not execute.

```
Oracle SQL*Plus
File Edit Search Options Help
SQL> DECLARE
  2  v_title books.title%TYPE;
  3  v_retail books.retail%TYPE;
  4  v_number NUMBER(2) := &How_Many_Books_To_Display;
  5  CURSOR books_cursor IS
  6     SELECT title, retail
  7     FROM books
  8     ORDER BY retail DESC;
  9  BEGIN
 10  OPEN books_cursor;
 11  FETCH books_cursor INTO v_title, v_retail;
 12  WHILE books_cursor%ROWCOUNT <= v_number AND
 13        books_cursor%FOUND LOOP
 14     DBMS_OUTPUT.PUT_LINE (v_title || ', ' || v_retail);
 15     FETCH books_cursor INTO v_title, v_retail;
 16  END LOOP;
 17  CLOSE books_cursor;
 18  END;
 19  /
Enter value for how_many_books_to_display: 5
old   4: v_number NUMBER(2) := &How_Many_Books_To_Display;
new   4: v_number NUMBER(2) := 5;
PAINLESS CHILD-REARING, 89.95
HOLY GRAIL OF ORACLE, 75.95
BUILDING A CAR WITH TOOTHPICKS, 59.95
DATABASE IMPLEMENTATION, 55.95
E-BUSINESS THE EASY WAY, 54.5

PL/SQL procedure successfully completed.

SQL>
```

Figure 16-19 PL/SQL block with logical operator

A FETCH statement was also included as a primary read on line 11 to make certain that a row is found before the WHILE loop was accessed. If no rows had been fetched, then the %FOUND attribute would have been evaluated as NULL, and the loop would not have been executed, regardless of how many books the user requested or how many rows were actually in the cursor. When a statement consists of two conditions combined with an AND logical operator and one of those conditions is evaluated as NULL and the other is evaluated as TRUE, then the entire statement is evaluated as NULL. Therefore, at least one row must be fetched before the WHILE loop is reached, so the value of the %FOUND attribute will be TRUE and the loop can be executed.

The logic tables in Figure 16–20 for the AND and OR logical operators list how a statement will be evaluated, based on how individual conditions are evaluated.

AND	TRUE	FALSE	NULL
TRUE	TRUE	FALSE	NULL
FALSE	FALSE	FALSE	FALSE
NULL	NULL	FALSE	NULL

OR	TRUE	FALSE	NULL
TRUE	TRUE	TRUE	TRUE
FALSE	TRUE	FALSE	NULL
NULL	TRUE	NULL	NULL

Figure 16-20 Logic tables

The upper-left cell of each table contains the logical operator AND or OR. The column below and the row to the right of the logical operator represent the value of each condition being joined by the logical operator. The intersection of a column and a row specifies how the entire statement would be evaluated, based on the values of the individual conditions. For example, in the OR logic table in Figure 16–20, a statement would be considered TRUE if one condition is TRUE and the other is NULL.

You can test the logic table by using a simple IF statement in a PL/SQL block. In Figure 16–21, the value of the first BOOLEAN variable has been initialized to NULL, but no value was assigned to the second BOOLEAN variable. Because no value was assigned to the second value, it is automatically initialized to NULL also. When the block is executed, the first three conditions in the IF statement are not evaluated as TRUE, so the ELSE clause is automatically executed. If you compare the conditions in the PL/SQL block to the logic tables given in Figure 16–20, you can determine how Oracle9*i* evaluated each condition. To gain a better understanding of the logic tables, it may also be helpful to modify the values of the variables in the PL/SQL block given in Figure 16–21 and determine how the resulting output is derived.

Oracle SQL*Plus

File Edit Search Options Help

```
SQL> DECLARE
       firstvalue BOOLEAN := NULL;
       secondvalue BOOLEAN;
     BEGIN
       IF firstvalue = TRUE AND secondvalue = TRUE THEN
          DBMS_OUTPUT.PUT_LINE('Values are TRUE');
       ELSIF firstvalue = FALSE or secondvalue = TRUE THEN
          DBMS_OUTPUT.PUT_LINE('Either the first value is FALSE or the second value is TRUE');
       ELSIF firstvalue = TRUE or secondvalue = FALSE THEN
          DBMS_OUTPUT.PUT_LINE('Either the first value is FALSE or the second value is TRUE');
       ELSE
          DBMS_OUTPUT.PUT_LINE('Value is NULL');
     END IF;
     END;
     /
Value is NULL

PL/SQL procedure successfully completed.

SQL>
```

Figure 16-21 PL/SQL block to evaluate BOOLEAN values

Exception Handling

An EXCEPTION is a signal that an error has occurred when the PL/SQL block is executed. It does not include syntax errors. The most common exceptions are **`NO_DATA_FOUND`**, **`TOO_MANY_ROWS`**, and **`ZERO_DIVIDE`**. For instance, suppose that a PL/SQL block is created to retrieve a row from the BOOKS table for a book having a retail price of more than $100. In this case, no book would be retrieved from the BOOKS table because no book has a retail price greater than $100. Therefore, when the loop is executed, there would be no row to process, and the message **NO_DATA_FOUND** would be returned.

When no rows are found to process, Oracle9*i* is said to "throw" or "raise" an exception. When an exception is raised, the exception-handing section can be used to "trap" the exception and take action. The syntax for the exception-handling section of a PL/SQL block is shown in Figure 16–22.

```
EXCEPTION
WHEN exception_type THEN
      statement;
[WHEN exception_type THEN
      statement;…]
[WHEN OTHERS THEN
      statement;…]
```

Figure 16-22 Syntax of the exception-handling section

For example, to let a user know that the faulty output is not caused by a problem with the statements in the executable section of the PL/SQL block but is a result of no rows being found, an exception-handling section can be added to display an error message. The exception-handling section begins with the keyword EXCEPTION. Each error that the user is trying to trap is included in the section. In this particular example, there is only one error you should be concerned about—no rows being returned. The WHEN OTHERS clause can be included in the last statement of the exception-handling section in the event that an unanticipated error occurs, and the error will be trapped because it is an error and did not meet the types specified previously in that section. It must be the last clause of the section because any subsequent clauses will not be reached. The statements in Figure 16–23 will display an error message if no rows meet the requirements of the SELECT statement used to create the cursor.


```
EXCEPTION
WHEN NO_DATA_FOUND THEN
DBMS_OUTPUT.PUT_LINE ('No rows were retrieved from the table');
```

Figure 16-23 Statements to display an error message when no rows are found

In Figure 16–24, the statement has been included in a PL/SQL block that is designed to retrieve a book that has a retail price of more than $100. However, as shown in the results, no rows were found.

```
SQL> DECLARE
       v_title books.title%TYPE;
       v_retail books.retail%TYPE;
     BEGIN
       SELECT title, retail
       INTO v_title, v_retail
       FROM books
       WHERE retail > 100;
     EXCEPTION
       WHEN NO_DATA_FOUND THEN
           DBMS_OUTPUT.PUT_LINE ('No rows were retrieved from the table');
     END;
     /
No rows were retrieved from the table

PL/SQL procedure successfully completed.

SQL>
```

Figure 16-24 An exception raised when no rows are retrieved from a table

If a PL/SQL block is able to trap a raised exception, then the PL/SQL block can terminate successfully. However, if the exception is not trapped because there is no statement in the exception-handling section to address that type of error, the execution of the block is aborted. Then, control is returned either to the calling environment, such as SQL*Plus or an application program, or to the outer block—if the exception is raised in a nested block. Not trapping an exception is known as **exception propagation** because the error is returned to the calling environment and can result in further errors.

User-Defined Exception Handling

Whenever an error occurs in a PL/SQL block, the Oracle server returns an error code. For example, in Figure 16–2, the error code ORA-01422 was returned to indicate that too many rows were returned from the SELECT statement. In Oracle9*i*, this particular type of error is already predefined as the **`TOO_MANY_ROWS`** exception. Predefined exception types are simple to trap because the user is simply required to specify the type in the exception-handling section of the block, and if its associated error code occurs, the exception is trapped and does not propagate outward. However, there may be instances in which certain types of exceptions occur that are not predefined. In these instances, the user must explicitly raise the exception, and then it can be trapped by the exception-handling section.

User-defined exceptions are widely used in industry to enforce business rules. For example, if the "quantity on hand" for a requested book in the INVENTORY table reaches a specified reorder point, an exception can be raised that would trigger an automatic order for the book to replenish the inventory.

When you want to trap an error message that is not predefined, three steps are required:

1. The name of the exception must be declared in the declarative section of the PL/SQL block. The datatype assigned to the exception is EXCEPTION.
2. The declared exception must be associated with the Oracle server error number using the **PRAGMA EXCEPTION_INIT** statement.
3. The declared exception must be included in the exception-handling section of the PL/SQL block.

Suppose that you are creating a PL/SQL block that will add new publishers to the PUBLISHER table. One possible problem that could occur is the assignment of an existing publisher ID to a new entry. Since a PRIMARY KEY constraint already exists for the PUBLISHER table, no two publishers can have the same ID, so an error will occur. Specifically, Oracle9*i* will return an ORA-00001 error. If this type of error occurs, you can create a message to warn a user that the ID has already been assigned to another publisher.

The first step is to choose a name for the exception. Because you are allowed to have several exception-handling statements within a PL/SQL block, the name should signify the purpose of the exception. In this case, you can use the name **ID_ALREADY_IN_USE**. The correct syntax to declare the exception is shown in Figure 16–25.

```
id_already_in_use  EXCEPTION;
```

Figure 16-25 An EXCEPTION declaration

After the exception has been declared, it must be associated with a specific Oracle9*i* error number, using the **PRAGMA EXCEPTION_INIT** statement. PL/SQL can only reference variables that have already been declared. Therefore, the exception must *ALWAYS* be declared before it is associated with an Oracle9*i* error number. The syntax for the **PRAGMA EXCEPTION_INIT** statement is given in Figure 16–26.

```
PRAGMA EXCEPTION_INIT(exception_name, errornumber);
```

16

Figure 16-26 Syntax for the PRAGMA EXCEPTION_INIT statement

In this case, the name of the exception being referenced was previously declared in Figure 16–25 as **ID_ALREADY_IN_USE** and the Oracle9*i* error number that should raise the exception is –00001. Therefore, the statement to associate the error number with the exception is **PRAGMA EXCEPTION_INIT(id_already_in_use, -00001);**.

If the ORA-00001 error occurs, the sequence of execution within the PL/SQL block should skip to the exception-handling section of the block. The Oracle server will search within the section for the associated exception name. If it is found, any statements within the corresponding clause will be executed. As shown in Figure 16–27, if a user attempts to assign an existing publisher ID, the character string specified in the exception-handling section will be displayed.

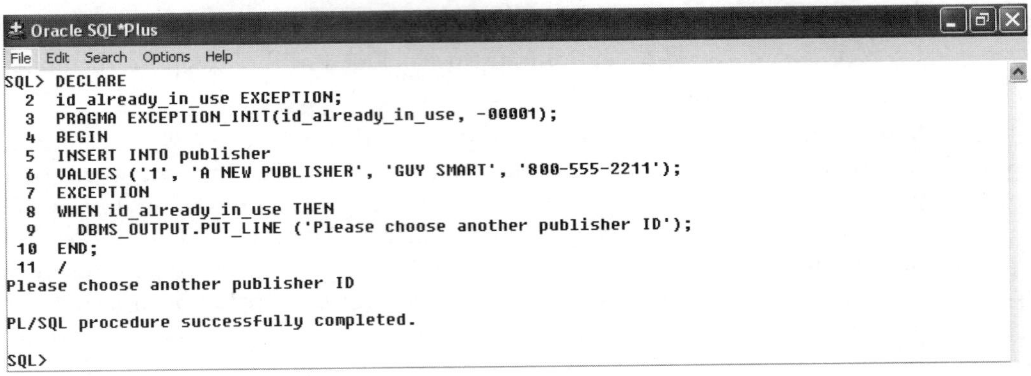

```
Oracle SQL*Plus
File Edit Search Options Help
SQL> DECLARE
  id_already_in_use EXCEPTION;
  PRAGMA EXCEPTION_INIT(id_already_in_use, -00001);
  BEGIN
  INSERT INTO publisher
  VALUES ('1', 'A NEW PUBLISHER', 'GUY SMART', '800-555-2211');
  EXCEPTION
  WHEN id_already_in_use THEN
    DBMS_OUTPUT.PUT_LINE ('Please choose another publisher ID');
  END;
  /
Please choose another publisher ID

PL/SQL procedure successfully completed.

SQL>
```

Figure 16-27 Trapping a user-defined exception

CHAPTER SUMMARY

- An explicit cursor is required when multiple rows are retrieved from a table.
- An explicit cursor must be defined and then opened before it is populated with data.
- The SELECT statement that defines the cursor is executed when the cursor is opened.
- Data are retrieved from an explicit cursor using the FETCH command.
- An explicit cursor must be closed.
- A cursor FOR loop will implicitly, or automatically, open and close an explicit cursor as well as fetch data from the cursor.
- Cursor attributes can be used to determine whether a cursor is open and how many, if any, rows were processed.
- An exception is raised if a non-syntax error occurs in the executable section of a PL/SQL block.
- The exception-handling section of a PL/SQL block is used to trap errors raised in the block.
- If an error is not trapped, execution of the block will be aborted, and the error will be propagated to any outer blocks or the calling environment.

- If an error is trapped, the block will execute successfully, and the exception will not be propagated to any outer blocks or the calling environment.
- A non-predefined exception must be declared before it can be associated with an Oracle server error number.

Chapter 16 Syntax Summary

The following tables present a summary of the syntax that you have learned in this chapter. You can use the tables as a study guide and reference.

Syntax Guide: PL/SQL Commands		
Element	**Syntax**	**Example**
CURSOR	`CURSOR cursor_name IS selectquery;`	`CURSOR books_cursor IS SELECT title, retail FROM books NATURAL JOIN orderitems WHERE order# = 1012;`
OPEN	`OPEN cursor_name;`	`OPEN books_cursor;`
CLOSE	`CLOSE cursor_name;`	`CLOSE books_cursor;`
FETCH	`FETCH cursor_name INTO variablename [,…variablename];`	`FETCH books_cursor INTO v_title, v_retail;`
PRAGMA EXCEPTION_INIT	`PRAGMA EXCEPTION_INIT (exception_name, errornumber);`	`PRAGMA EXCEPTION_INIT (id_already_in_use, -00001);`
Cursor FOR loop	`FOR record_name IN cursor_name LOOP statement; [statement;…] END LOOP;`	`FOR r_books IN books_cursor LOOP END LOOP;`
EXCEPTION Exception-handling section	`EXCEPTION WHEN exception_type THEN statement; [WHEN exception_type THEN statement;…] [WHEN OTHERS THEN statement;…]`	`EXCEPTION WHEN TOO_MANY_ROWS THEN DBMS_OUTPUT.PUT_LINE ('More than one row was returned');`

16

Cursor Attributes		
Cursor Attribute	**Description**	**Example**
%ROWCOUNT	Identifies the number of rows processed	`books_cursor%ROWCOUNT`
%FOUND	Contains the value TRUE if one or more rows were processed—FALSE if no rows were processed	`books_cursor%FOUND`
%NOTFOUND	Contains the value TRUE if no rows were processed—FALSE if one or more rows were processed	`books_cursor%NOTFOUND`
%ISOPEN	Contains the value TRUE if a cursor is not closed after processing —FALSE if the cursor is closed. The value will always be FALSE after processing for an implicit cursor because its closure is automatic.	`books_cursor%ISOPEN`

Review Questions

To answer these questions, refer to the tables in Appendix A.

1. What is the difference between an implicit and explicit cursor?
2. Why must an explicit cursor be opened before data can be fetched from the cursor?
3. What is an exception?
4. How does the cursor FOR loop differ from using a standard FOR loop to control a cursor?
5. What four attributes are commonly associated with cursors?
6. What statement can be used to associate an exception with an Oracle server error number?
7. Which cursor attribute can be used to determine the number of rows processed?
8. What is the purpose of the exception-handling section of a PL/SQL block?
9. If a statement contains one condition that is FALSE and another condition that is NULL, does the statement evaluate as being TRUE, FALSE, or NULL if the conditions are combined with the OR logical operator? With the AND operator?
10. Which cursor attribute can be used to perform Top-N analysis in a PL/SQL block?

MULTIPLE CHOICE

To answer these questions, refer to the tables in Appendix A.

1. Which of the following is a correct statement?
   a. An explicit cursor is required if a DML statement is executed in a PL/SQL block.
   b. An explicit cursor is required if a DDL statement is executed in a PL/SQL block.
   c. An explicit cursor is required if a SELECT statement retrieves only one row from a table.
   d. none of the above
2. Which of the following commands is used to create an explicit cursor?
   a. CREATE CURSOR
   b. MAKE CURSOR
   c. INSERT CURSOR
   d. none of the above
3. The keyword to denote the exception-handling section of a PL/SQL block is:
   a. EXCEPT
   b. EXCEPTION
   c. HANDLING
   d. EXCEPTION HANDLING
4. An explicit cursor is opened in which section of a PL/SQL block?
   a. declarative
   b. executable
   c. exception handling
   d. ending
5. Which of the following commands is used to open an implicit cursor?
   a. OPEN
   b. CURSER
   c. IMPLY
   d. None of the above—there is no command to open an implicit cursor.
6. Which of the following is the correct structure of the FOR clause in a cursor FOR loop?
   a. `FOR` *n* `= 10 TO 20 LOOP`
   b. `FOR n 10 TO 20 LOOP`
   c. `FOR n 10..20 LOOP`
   d. none of the above

16

7. Which of the following types of commands cannot be used with an explicit cursor?
   a. FETCH
   b. SELECT INTO
   c. OPEN
   d. CLOSE
8. Which of the following is a valid predefined Oracle9*i* exception?
   a. `NO_ROWS_RETURNED`
   b. `DIVIDE_BY_ZERO`
   c. `TOO_MANY_ROWS_RETURNED`
   d. none of the above
9. Which of the following commands is used to close an explicit cursor?
   a. CLOSED
   b. CLOSE
   c. END
   d. EXIT
10. When working with explicit cursors, which of the following commands actually executes the SELECT query and populates the cursor with the retrieved rows?
    a. OPEN
    b. FETCH
    c. CLOSE
    d. END IF
11. Which of the following can be used to copy the structure of an entire row?
    a. %ROWCOUNT
    b. %TYPE
    c. %ROWTYPE
    d. %FOUND
12. Which of the following statements is correct?
    a. DML operations can only be performed with an explicit cursor.
    b. DML operations can only be performed with an implicit cursor.
    c. An explicit cursor must be used with a query that returns more than one row of results.
    d. A cursor created in a cursor FOR loop can be explicitly referenced.

13. When testing an explicit cursor to determine if any rows were processed, the attribute should be referenced after which of the following statements is executed?
    a. OPEN
    b. FETCH
    c. CLOSED
    d. INSERT
14. When creating a non-predefined exception, the exception must be defined in which section of the PL/SQL block?
    a. DECLARE
    b. BEGIN
    c. EXCEPTION
    d. END
15. In which section of a PL/SQL block is a non-predefined exception raised?
    a. DECLARE
    b. BEGIN
    c. EXCEPTION
    d. END
16. The WHEN OTHER clause can be used in which section of a PL/SQL block?
    a. DECLARE
    b. BEGIN
    c. EXCEPTION
    d. END
17. If an exception occurs that is not trapped, which of the following will occur?
    a. The exception is propagated to the next section of the PL/SQL block.
    b. The exception is propagated to an inner block.
    c. The block is terminated and the exception will be propagated to the calling environment.
    d. The block will execute successfully and the exception will be propagated to the calling environment.
18. Which of the following is the correct sequence for processing an explicit cursor?
    a. Open the cursor, fetch rows from the cursor, declare the cursor, close the cursor.
    b. Fetch rows from the cursor, open the cursor, close the cursor, declare the cursor.
    c. Open the cursor, declare the cursor, fetch rows from the cursor, close the cursor.
    d. Declare the cursor, open the cursor, fetch rows from the cursor, close the cursor.

19. An Oracle9*i* error number is associated with an exception, using which of the following?
    a. `EXCEPTION_INIT`
    b. `PRAGMA EXCEPTION_INIT`
    c. `CURSOR`
    d. `OPEN`
20. The values contained in an explicit cursor can only be assigned after which of the following has occurred?
    a. The cursor has been declared.
    b. The cursor has been opened.
    c. The FETCH statement has been executed.
    d. The cursor has been closed.

Hands-on Assignments

To perform these activities, refer to the tables in Appendix A.

1. Create a PL/SQL block that will retrieve the contents of the BOOKS table and display the title of and profit generated by each book.
2. Create a PL/SQL block that will display the title of each book in the BOOKS table and the name of the publisher that published the book. Also include the total number of books processed in the output.
3. Create an exception-handling section that will display the message "More than one row was returned by the SELECT statement" if more than one row is retrieved by a SELECT statement. Save the code in a script file named **SC1603.sql**.
4. Test the exception-handling section stored in **SC1603.sql** by creating a PL/SQL block that will use an explicit cursor to retrieve more than one row of results. Save the block in a script file named **SC1604.sql**.
5. Create a PL/SQL block containing a WHILE loop to display the three most profitable books in the BOOKS table.
6. Create a PL/SQL block containing a WHILE loop to display the title of all the books ordered by customer 1015. Save the block in a script file named **SC1606.sql**.
7. Use a PL/SQL block containing a cursor FOR loop to display the customer ID of any customer who has recently placed an order totaling more than $100.
8. Modify the PL/SQL block contained in the script file named **SC1606.sql** to allow the user to enter the desired customer number when the block is executed. Save the new PL/SQL block in a script file named **SC1608.sql**.

9. Modify the PL/SQL block contained in the script file named **SC1608.sql** so the cursor is controlled by a cursor FOR loop rather than a WHILE loop. Save the modified file as **SC1609.sql**.
10. Execute the **SC1609.sql** script file, and enter customer 1011 to test the PL/SQL block.

A CASE FOR ORACLE9*i*

To perform this activity, refer to the tables in Appendix A.

At the end of each month, data about all shipped orders are archived into two tables: ORDERS2003 and ORDERITEMS2003. These tables were created so the ORDERS and ORDERITEMS tables did not become so large that order-processing time was slowed down. The archived data are in a separate database that is used for data warehousing. Management and analysts also use the archived data to review sales information for the entire year.

The ORDERS2003 and ORDERITEMS2003 tables were created when you executed the **prech16.sql** file at the beginning of this chapter. You can use the DESCRIBE command to view the structure of these tables.

Until now, all shipped orders that needed to be archived first had to be copied into the archive table and then purged, or deleted, from the transaction tables. Your supervisor has assigned you the task of creating a PL/SQL block that will perform the following tasks for orders that have been shipped:

- Copy customer ID number, shipping address, and order number from the ORDERS table into the ORDERS2003 table.
- Copy the order number, ISBN, and quantity ordered from the ORDERITEMS table into the ORDERITEMS2003 table.
- Copy the corresponding retail price of each ordered item from the BOOKS table into the ORDERITEMS2003 table. (The retail price of each book sold is archived with the order information because retail prices change over time, and management wants to know the retail price of the book at the time it was sold.)

After the data have been copied into the appropriate tables, the data for all shipped orders in the ORDERS and ORDERITEMS tables should be deleted. Since a FOREIGN KEY constraint exists on the ORDERITEMS table, deleting an order from the ORDERS table will automatically delete any corresponding entry in the ORDERITEMS table.

After you have created and tested the block, provide your instructor with a copy of the PL/SQL block used to perform the archiving process and the deletions from the ORDERS and ORDERITEMS tables.

TABLES FOR THE JUSTLEE BOOKS DATABASE

The tables created by the execution of the **Bookscript.SQL** file include CUSTOMERS, BOOKS, ORDERS, ORDERITEMS, AUTHOR, BOOKAUTHOR, PUBLISHER, and PROMOTION. The structure and contents for each table are provided in this appendix.

CUSTOMERS Table

Data about JustLee Books' customers are stored in the CUSTOMERS table. The structure of the CUSTOMERS table is shown in Figure A–1. The table's contents are shown in Figure A–2.

Name	Null?	Type
CUSTOMER#	NOT NULL	NUMBER(4)
LASTNAME		VARCHAR2(10)
FIRSTNAME		VARCHAR2(10)
ADDRESS		VARCHAR2(20)
CITY		VARCHAR2(12)
STATE		VARCHAR2(2)
ZIP		VARCHAR2(5)
REFERRED		NUMBER(4)

Figure A-1 Structure of the CUSTOMERS table

CUSTOMER#	LASTNAME	FIRSTNAME	ADDRESS	CITY	ST	ZIP	REFERRED
1001	MORALES	BONITA	P.O. BOX 651	EASTPOINT	FL	32328	
1002	THOMPSON	RYAN	P.O. BOX 9835	SANTA MONICA	CA	90404	
1003	SMITH	LEILA	P.O. BOX 66	TALLAHASSEE	FL	32306	
1004	PIERSON	THOMAS	69821 SOUTH AVENUE	BOISE	ID	83707	
1005	GIRARD	CINDY	P.O. BOX 851	SEATTLE	WA	98115	
1006	CRUZ	MESHIA	82 DIRT ROAD	ALBANY	NY	12211	
1007	GIANA	TAMMY	9153 MAIN STREET	AUSTIN	TX	78710	1003
1008	JONES	KENNETH	P.O. BOX 137	CHEYENNE	WY	82003	
1009	PEREZ	JORGE	P.O. BOX 8564	BURBANK	CA	91510	1003
1010	LUCAS	JAKE	114 EAST SAVANNAH	ATLANTA	GA	30314	
1011	MCGOVERN	REESE	P.O. BOX 18	CHICAGO	IL	60606	
1012	MCKENZIE	WILLIAM	P.O. BOX 971	BOSTON	MA	02110	
1013	NGUYEN	NICHOLAS	357 WHITE EAGLE AVE.	CLERMONT	FL	34711	1006
1014	LEE	JASMINE	P.O. BOX 2947	CODY	WY	82414	
1015	SCHELL	STEVE	P.O. BOX 677	MIAMI	FL	33111	
1016	DAUM	MICHELL	9851231 LONG ROAD	BURBANK	CA	91508	1010
1017	NELSON	BECCA	P.O. BOX 563	KALMAZOO	MI	49006	
1018	MONTIASA	GREG	1008 GRAND AVENUE	MACON	GA	31206	
1019	SMITH	JENNIFER	P.O. BOX 1151	MORRISTOWN	NJ	07962	1003
1020	FALAH	KENNETH	P.O. BOX 335	TRENTON	NJ	08607	

Figure A-2 Data contained in the CUSTOMERS table

BOOKS Table

Data about the books sold by JustLee Books are stored in the BOOKS table. The structure of the BOOKS table is shown in Figure A–3. The table's contents are shown in Figure A–4.

Name	Null?	Type
ISBN	NOT NULL	VARCHAR2(10)
TITLE		VARCHAR2(30)
PUBDATE		DATE
PUBID		NUMBER(2)
COST		NUMBER(5,2)
RETAIL		NUMBER(5,2)
CATEGORY		VARCHAR2(12)

Figure A-3 Structure of the BOOKS table

ISBN	TITLE	PUBDATE	PUBID	COST	RETAIL	CATEGORY
1059831198	BODYBUILD IN 10 MINUTES A DAY	21-JAN-01	4	18.75	30.95	FITNESS
0401140733	REVENGE OF MICKEY	14-DEC-01	1	14.2	22	FAMILY LIFE
4981341710	BUILDING A CAR WITH TOOTHPICKS	18-MAR-02	2	37.8	59.95	CHILDREN
8843172113	DATABASE IMPLEMENTATION	04-JUN-99	3	31.4	55.95	COMPUTER
3437212490	COOKING WITH MUSHROOMS	28-FEB-00	4	12.5	19.95	COOKING
3957136468	HOLY GRAIL OF ORACLE	31-DEC-01	3	47.25	75.95	COMPUTER
1915762492	HANDCRANKED COMPUTERS	21-JAN-01	3	21.8	25	COMPUTER
9959789321	E-BUSINESS THE EASY WAY	01-MAR-02	2	37.9	54.5	COMPUTER
2491748320	PAINLESS CHILD-REARING	17-JUL-00	5	48	89.95	FAMILY LIFE
0299282519	THE WOK WAY TO COOK	11-SEP-00	4	19	28.75	COOKING
8117949391	BIG BEAR AND LITTLE DOVE	08-NOV-01	5	5.32	8.95	CHILDREN
0132149871	HOW TO GET FASTER PIZZA	11-NOV-02	4	17.85	29.95	SELF HELP
9247381001	HOW TO MANAGE THE MANAGER	09-MAY-99	1	15.4	31.95	BUSINESS
2147428890	SHORTEST POEMS	01-MAY-01	5	21.85	39.95	LITERATURE

Figure A-4 Data contained in the BOOKS table

ORDERS Table

Data about recent orders received by JustLee Books are stored in the ORDERS table. The structure of the ORDERS table is shown in Figure A–5. The table's contents are shown in Figure A–6.

Name	Null?	Type
ORDER#	NOT NULL	NUMBER(4)
CUSTOMER#		NUMBER(4)
ORDERDATE		DATE
SHIPDATE		DATE
SHIPSTREET		VARCHAR2(18)
SHIPCITY		VARCHAR2(15)
SHIPSTATE		VARCHAR2(2)
SHIPZIP		VARCHAR2(5)

Figure A-5 Structure of the ORDERS table

ORDER#	CUSTOMER#	ORDERDATE	SHIPDATE	SHIPSTREET	SHIPCITY	SH	SHIPZ
1000	1005	31-MAR-03	02-APR-03	1201 ORANGE AVE	SEATTLE	WA	98114
1001	1010	31-MAR-03	01-APR-03	114 EAST SAVANNAH	ATLANTA	GA	30314
1002	1011	31-MAR-03	01-APR-03	58 TILA CIRCLE	CHICAGO	IL	60605
1003	1001	01-APR-03	01-APR-03	958 MAGNOLIA LANE	EASTPOINT	FL	32328
1004	1020	01-APR-03	05-APR-03	561 ROUNDABOUT WAY	TRENTON	NJ	08601
1005	1018	01-APR-03	02-APR-03	1008 GRAND AVENUE	MACON	GA	31206
1006	1003	01-APR-03	02-APR-03	558A CAPITOL HWY.	TALLAHASSEE	FL	32307
1007	1007	02-APR-03	04-APR-03	9153 MAIN STREET	AUSTIN	TX	78710
1008	1004	02-APR-03	03-APR-03	69821 SOUTH AVENUE	BOISE	ID	83707
1009	1005	03-APR-03	05-APR-03	9 LIGHTENING RD.	SEATTLE	WA	98110
1010	1019	03-APR-03	04-APR-03	384 WRONG WAY HOME	MORRISTOWN	NJ	07960
1011	1010	03-APR-03	05-APR-03	102 WEST LAFAYETTE	ATLANTA	GA	30311
1012	1017	03-APR-03		1295 WINDY AVENUE	KALMAZOO	MI	49002
1013	1014	03-APR-03	04-APR-03	7618 MOUNTAIN RD.	CODY	WY	82414
1014	1007	04-APR-03	05-APR-03	9153 MAIN STREET	AUSTIN	TX	78710
1015	1020	04-APR-03		557 GLITTER ST.	TRENTON	NJ	08606
1016	1003	04-APR-03		9901 SEMINOLE WAY	TALLAHASSEE	FL	32307
1017	1015	04-APR-03	05-APR-03	887 HOT ASPHALT ST	MIAMI	FL	33112
1018	1001	05-APR-03		95812 HIGHWAY 98	EASTPOINT	FL	32328
1019	1018	05-APR-03		1008 GRAND AVENUE	MACON	GA	31206
1020	1008	05-APR-03		195 JAMISON LANE	CHEYENNE	WY	82003

Figure A-6 Data contained in the ORDERS table

ORDERITEMS Table

Because an order can consist of more than one book, the ORDERITEMS table is used to store information about the individual books purchased on each order. It bridges the ORDERS and BOOKS tables. The structure of the ORDERITEMS table is shown in Figure A–7. The table's contents are shown in Figure A–8.

Name	Null?	Type
ORDER#	NOT NULL	NUMBER(4)
ITEM#	NOT NULL	NUMBER(2)
ISBN		VARCHAR2(10)
QUANTITY		NUMBER(3)

Figure A-7 Structure of the ORDERITEMS table

ORDER#	ITEM#	ISBN	QUANTITY
1000	1	3437212490	1
1001	1	9247381001	1
1001	2	2491748320	1
1002	1	8843172113	2
1003	1	8843172113	1
1003	2	1059831198	1
1003	3	3437212490	1
1004	1	2491748320	2
1005	1	2147428890	1
1006	1	9959789321	1
1007	1	3957136468	3
1007	2	9959789321	1
1007	3	8117949391	1
1007	4	8843172113	1
1008	1	3437212490	2
1009	1	3437212490	1
1009	2	0401140733	1
1010	1	8843172113	1
1011	1	2491748320	1
1012	1	8117949391	1
1012	2	1915762492	2
1012	3	2491748320	1
1012	4	0401140733	1
1013	1	8843172113	1
1014	1	0401140733	2
1015	1	3437212490	1
1016	1	2491748320	1
1017	1	8117949391	2
1018	1	3437212490	1
1018	2	8843172113	1
1019	1	0401140733	1
1020	1	3437212490	1

Figure A-8 Data contained in the ORDERITEMS table

AUTHOR Table

The names of book authors are stored in the AUTHOR table. The structure of the AUTHOR table is shown in Figure A–9. The table's contents are shown in Figure A–10.

Name	Null?	Type
AUTHORID	NOT NULL	VARCHAR2(4)
LNAME		VARCHAR2(10)
FNAME		VARCHAR2(10)

Figure A-9 Structure of the AUTHOR table

```
AUTH LNAME      FNAME
---- ---------- ----------
S100 SMITH      SAM
J100 JONES      JANICE
A100 AUSTIN     JAMES
M100 MARTINEZ   SHEILA
K100 KZOCHSKY   TAMARA
P100 PORTER     LISA
A105 ADAMS      JUAN
B100 BAKER      JACK
P105 PETERSON   TINA
W100 WHITE      WILLIAM
W105 WHITE      LISA
R100 ROBINSON   ROBERT
F100 FIELDS     OSCAR
W110 WILKINSON  ANTHONY
```

Figure A-10 Data contained in the AUTHOR table

BOOKAUTHOR Table

Because a book could have more than one author, and an author could have written more than one book, the BOOKAUTHOR table bridges the BOOKS and AUTHOR tables. The structure of the BOOKAUTHOR table is shown in Figure A–11. The table's contents are shown in Figure A–12.

```
Name                                                    Null?    Type
------------------------------------------------------- -------- ----------------------------------
ISBN                                                    NOT NULL VARCHAR2(10)
AUTHORID                                                NOT NULL VARCHAR2(4)
```

Figure A-11 Structure of the BOOKAUTHOR table

```
ISBN       AUTH
---------- ----
1059831198 S100
1059831198 P100
0401140733 J100
4981341710 K100
8843172113 P105
8843172113 A100
8843172113 A105
3437212490 B100
3957136468 A100
1915762492 W100
1915762492 W105
9959789321 J100
2491748320 R100
2491748320 F100
2491748320 B100
0299282519 S100
8117949391 R100
0132149871 S100
9247381001 W100
2147428890 W105
```

Figure A-12 Data contained in the BOOKAUTHOR table

PUBLISHER Table

All data about book publishers are stored in the PUBLISHER table. The structure of the PUBLISHER table is shown in Figure A–13. The table's contents are shown in Figure A–14.

```
Name                                                    Null?    Type
------------------------------------------------------- -------- ----------------------------------
PUBID                                                   NOT NULL NUMBER(2)
NAME                                                             VARCHAR2(23)
CONTACT                                                          VARCHAR2(15)
PHONE                                                            VARCHAR2(12)
```

Figure A-13 Structure of the PUBLISHER table

```
     PUBID NAME                    CONTACT         PHONE
---------- ----------------------- --------------- ------------
         1 PRINTING IS US          TOMMIE SEYMOUR  000-714-8321
         2 PUBLISH OUR WAY         JANE TOMLIN     010-410-0010
         3 AMERICAN PUBLISHING     DAVID DAVIDSON  800-555-1211
         4 READING MATERIALS INC.  RENEE SMITH     800-555-9743
         5 REED-N-RITE             SEBASTIAN JONES 800-555-8284
```

Figure A-14 Data contained in the PUBLISHER table

PROMOTION Table

The gifts that are provided to customers during JustLee Books' annual sales promotion are listed in the PROMOTION table, along with the range of retail prices required to obtain each gift. The structure of the PROMOTION table is shown in Figure A–15. The table's contents are shown in Figure A–16.

```
Name                                                    Null?    Type
------------------------------------------------------- -------- ----------------------------------
GIFT                                                             VARCHAR2(15)
MINRETAIL                                                        NUMBER(5,2)
MAXRETAIL                                                        NUMBER(5,2)
```

Figure A-15 Structure of the PROMOTION table

```
GIFT             MINRETAIL  MAXRETAIL
--------------- ---------- ----------
BOOKMARKER               0         12
BOOK LABELS          12.01         25
BOOK COVER           25.01         56
FREE SHIPPING        56.01     999.99
```

Figure A-16 Data contained in the PROMOTION table

APPENDIX

B

SQL SYNTAX GUIDE

SQL Commands		
Command	**Syntax**	**Example**
ALTER ROLE	`ALTER ROLE rolename` `IDENTIFIED BY password;`	`ALTER ROLE dataentry` `IDENTIFIED BY apassword;`

ALTER SEQUENCE	`ALTER SEQUENCE sequencename`
	`[INCREMENT BY value]`
	`[{MAXVALUE value | NOMAXVALUE}]`
	`[{MINVALUE value | NOMINVALUE}]`
	`[{CYCLE | NOCYCLE}]`
	`[{ORDER | NOORDER}]`
	`[{CACHE value | NOCACHE}];`
		`ALTER SEQUENCE`
		`orders_ordernumber`
		`INCREMENT BY 10;`

ALTER TABLE... ADD	`ALTER TABLE tablename` `ADD (columnname datatype);`	`ALTER TABLE acctmanager` `ADD (ext NUMBER(4));`
ALTER TABLE... DROP COLUMN	`ALTER TABLE tablename` `DROP COLUMN columnname;`	`ALTER TABLE acctmanager` `DROP COLUMN ext;`
ALTER TABLE... MODIFY	`ALTER TABLE tablename` `MODIFY (columnname datatype);`	`ALTER TABLE acctmanager` `MODIFY (amname VARCHAR2(25));`
ALTER TABLE... SET UNUSED *or* SET UNUSED COLUMN	`ALTER TABLE tablename` `SET UNUSED (columnname);`	`ALTER TABLE secustomerorders` `SET UNUSED (cost);`

ALTER USER	`ALTER USER username`
	`[IDENTIFIED BY newpassword]`
	`[PASSWORD EXPIRE];`
		`ALTER USER rthomas`
		`IDENTIFIED BY`
		`monstertruck42;`

ALTER USER... DEFAULT ROLE	`ALTER USER username DEFAULT ROLE` `rolename;`	`ALTER USER rthomas DEFAULT ROLE` `dataentry;`
COMMIT	`COMMIT;`	`COMMIT;`

CREATE INDEX	`CREATE INDEX indexname`
	`ON tablename (columnname, …);`
		`CREATE INDEX`
		`customers_lastname_idx`
		`ON customers(lastname);`

CREATE OR REPLACE VIEW	`CREATE OR REPLACE [FORCE|NOFORCE]`
	`VIEW viewname`
	`(columnname, …)]`
	`AS subquery`
	`[WITH CHECK OPTION [CONSTRAINT`
	`constraintname]]`
	`[WITH READ ONLY];`
		`CREATE OR REPLACE`
		`VIEW inventory`
		`AS SELECT isbn, title,`
		`retail price`
		`FROM books;`

SQL Commands (continued)		
Command	**Syntax**	**Example**
CREATE ROLE	CREATE ROLE *rolename*;	CREATE ROLE dataentry;
CREATE SEQUENCE	CREATE SEQUENCE *sequencename* [INCREMENT BY *value*] [START WITH *value*] [{MAXVALUE *value* | NOMAXVALUE}] [{MINVALUE *value* | NOMINVALUE}] [{CYCLE | NOCYCLE}] [{ORDER | NOORDER}] [{CACHE *value* | NOCACHE}];	CREATE SEQUENCE orders_ordernumber INCREMENT BY 1 START WITH 1021 NOCYCLE NOCACHE;
CREATE SYNONYM	CREATE [PUBLIC] SYNONYM *synonymname* FOR *objectname*;	CREATE PUBLIC SYNONYM orderentry FOR orders;
CREATE TABLE	CREATE TABLE *tablename* (*columnname datatype* [DEFAULT] [, *columnname* …];	CREATE TABLE acctmanager (amid VARCHAR2(4), amname VARCHAR2(20), amedate DATE DEFAULT SYSDATE, region CHAR(2));
CREATE TABLE…AS	CREATE TABLE *tablename* AS (*subquery*);	CREATE TABLE secustomerorders AS (SELECT customer#, state, ISBN, category, quantity, cost, retail FROM customers NATURAL JOIN orders NATURAL JOIN orderitems NATURAL JOIN books WHERE state IN ('FL', 'GA', 'AL'));
CREATE USER	CREATE USER *username* IDENTIFIED BY *password;*	*CREATE USER rthomas* *IDENTIFIED BY little25car;*
CREATE VIEW	CREATE [FORCE|NOFORCE] VIEW *viewname* (*columnname, …*)] AS *subquery* [WITH CHECK OPTION [CONSTRAINT *constraintname*]] [WITH READ ONLY];	CREATE VIEW inventory AS SELECT isbn, title, retail price FROM books WITH READ ONLY;
DELETE	DELETE FROM *tablename* [WHERE *expression*];	DELETE FROM acctmanager WHERE amid = 'D500';
DROP INDEX	DROP INDEX *indexname*;	DROP INDEX books_profit_idx;
DROP ROLE	DROP ROLE *rolename*;	DROP ROLE dataentry;
DROP SEQUENCE	DROP SEQUENCE *sequencename*;	DROP SEQUENCE orders_ordernumber;

SQL Commands (continued)		
Command	**Syntax**	**Example**
DROP SYNONYM	`DROP [PUBLIC] SYNONYM` *synonymname;*	`DROP PUBLIC SYNONYM orderentry;`
DROP TABLE	`DROP TABLE` *tablename;*	`DROP TABLE setotals;`
DROP UNUSED COLUMNS	`ALTER TABLE` *tablename* `DROP UNUSED COLUMNS;`	`ALTER TABLE secustomerorders` `DROP UNUSED COLUMNS;`
DROP USER	`DROP USER` *username;*	`DROP USER rthomas;`
DROP VIEW	`DROP VIEW` *viewname;*	*DROP VIEW inventory;*

GRANT...ON	`GRANT {`*objectprivilege*`|ALL}`
	`[(`*columnname*`),`
	objectprivilege
	`(`*columnname*`)]`
	`ON` *objectname*
	`TO {`*username*`|`*rolename*`|PUBLIC}`
	`[WITH GRANT OPTION];`
		`GRANT select, insert`
		`ON customers`
		`TO rthomas`
		`WITH GRANT OPTION;`

GRANT...TO (system privilege)	`GRANT` *systemprivilege* `[,`
	systemprivilege, …`]`
	`TO` *username*`|`*rolename*
	`[,`*username*`|`*rolename, ….*`]`
	`[WITH ADMIN OPTION];`
		`GRANT CREATE SESSION`
		`TO rthomas;`

GRANT...TO (role)	`GRANT` *rolename* `[,` *rolename*`]` `TO` *username* `[,`*username*`];`	`GRANT dataentry TO rthomas;`

INSERT	`INSERT INTO` *tablename*
	`[(`*columnname,…*`)]`
	`VALUES (`*datavalues, …*`)`
	`[WHERE` *expression*`];`
	or
	`INSERT INTO` *tablename;*
	`subquery;`
		`INSERT INTO acctmanager`
		`VALUES ('T500', 'NICK TAYLOR', '05-SEP-01', 'NE');`
		or
		`INSERT INTO acctmanager2`
		`SELECT amid, amname, amedate, region`
		`FROM acctmanager`
		`WHERE amedate < '01-OCT-02';`

LOCK TABLE	`LOCK TABLE` *tablename* `IN` *modetype*
	`MODE;`
		`LOCK TABLE customers IN SHARE MODE;`
		or
		`LOCK TABLE customers IN EXCLUSIVE MODE;`

RENAME...TO	`RENAME` *oldtablename* `TO` *newtablename;*	`RENAME secustomersspent TO setotals;`
REVOKE...FROM	`REVOKE` *rolename* `FROM` *username*`|`*rolename;*	`REVOKE dataentry` `FROM rthomas;`

REVOKE...ON	`REVOKE` *objectprivilege*
	`[,…`*objectprivilege*`]`
	`ON` *objectname*
	`FROM` *username*`|`*rolename;*
		`REVOKE INSERT`
		`ON customers`
		`FROM rthomas;`

SQL Commands (continued)		
Command	**Syntax**	**Example**
ROLLBACK	`ROLLBACK;`	`ROLLBACK;`
SELECT	SELECT *columnname* [, …] FROM *tablename* [WHERE *condition*] [ORDER BY *columnname*] [GROUP BY *columnname* [, *columnname*, …]] [HAVING *groupfunction* *comparisonoperator value*];	SELECT category, AVG(cost) FROM books WHERE retail >25 GROUP BY category HAVING AVG(cost)>21;
SELECT…FOR UPDATE	SELECT *statement* FOR UPDATE;	SELECT cost FROM books WHERE category = 'COMPUTER' FOR UPDATE;
SET ROLE	SET ROLE *rolename*;	SET ROLE DBA;
TRUNCATE TABLE	TRUNCATE TABLE *tablename*;	TRUNCATE TABLE setotals;
UPDATE	UPDATE *tablename* SET *columnname* = *datavalue* [WHERE *expression*];	UPDATE acctmanager SET amedate = TO_DATE('JANUARY 12, 2002','MONTH DD, YYYY') WHERE amid = 'J500';

<table>
<tr><th colspan="3">SQL Keywords and Symbols</th></tr>
<tr><th>Keyword/Symbol</th><th>Description</th><th>Example</th></tr>
<tr><td>ll concatenation)</td><td>Combines the display of contents from multiple columns into a single column</td><td>SELECT city || state
FROM customers;</td></tr>
<tr><td>*</td><td>Returns all data in a table when used in a SELECT clause</td><td>SELECT *
FROM books;</td></tr>
<tr><td>%</td><td>A “wildcard” character used with the LIKE operator to perform pattern searches. It represents zero or more characters.</td><td>WHERE lastname LIKE 'T%'</td></tr>
<tr><td>_</td><td>A “wildcard” character used with the LIKE operator to perform pattern searches. It represents exactly one character in the specified position.</td><td>WHERE lastname LIKE 'MONT_ASA'</td></tr>
<tr><td>* multiplication
/ division
+ addition
- subtraction</td><td>Solves arithmetic operations (Oracle9i first solves * and /, then solves + and -.)</td><td>SELECT title, retail-cost profit
FROM books;</td></tr>
<tr><td>,</td><td>Separates column names in a list when retrieving multiple columns from a table</td><td>SELECT title, pubdate
FROM books;</td></tr>
<tr><td>' '
(string literal)</td><td>Indicates the exact set of characters, including spaces, to be displayed</td><td>SELECT city || ' ' ||state
FROM customers;</td></tr>
</table>

SQL Keywords and Symbols (continued)		
Keyword/Symbol	**Description**	**Example**
" "	Preserves spaces, symbols, or case in a column heading alias	`SELECT title AS "Book Name"` `FROM books;`

AS	Indicates a column alias to change the heading of a column in output—optional	`SELECT title AS titles, pubdate`
		`FROM books;`
		or
		`SELECT title titles, pubdate`
		`FROM books;`

CHR(10)	Inserts a line break	`SELECT customer# ||CHR(10)`
		`||city || ' ' ||state`
		`FROM customers;`

CROSS JOIN	Matches each record in one table with each record in another table. Also known as a *Cartesian product* or *Cartesian join* Syntax: `SELECT` *`columnname`*`[,…]` `FROM` *`tablename1`* `CROSS JOIN` *`tablename2;`*	`SELECT title, name` `FROM books CROSS JOIN publisher;`
DISTINCT	Eliminates duplicate lists	`SELECT DISTINCT state` `FROM customers;`

JOIN…ON	The JOIN keyword is used in the FROM clause. The ON clause identifies the column to be used to join the tables.
	
	Syntax:
	`SELECT` *`columnname`* `[,…]`
	`FROM` *`tablename1`* `JOIN` *`tablename2`*
	`ON` *`tablename1.columnname`*
	`<comparison operator>`
	`tablename2.columnname``;`
		`SELECT customers.customer#,`
		`order#`
		`FROM customers  JOIN orders`
		`ON customers.customer# =`
		`orders.customer#;`

JOIN…USING	The JOIN keyword is used in the FROM clause and, combined with the USING clause, it identifies the common column to be used to join the tables. It is normally used if the tables have more than one commonly named column, and only one is being used for the join.
	
	Syntax:
	`SELECT` *`columnname`* `[,…]`
	`FROM` *`tablename1`* `JOIN` *`tablename2`*
	`USING (`*`columnname`*`);`
		`SELECT customer#, order#`
		`FROM customers  JOIN orders`
		`USING (customer#);`

SQL Keywords and Symbols (continued)		
Keyword/Symbol	**Description**	**Example**

NATURAL JOIN	These keywords are used in the FROM clause to join tables containing a common column with the same name and definition.
	
	Syntax:
	`SELECT columnname[,…]`
	`FROM tablename1 NATURAL JOIN tablename2;`
		`SELECT customer#, order#`
		`FROM customers NATURAL JOIN orders;`

OUTER [RIGHT|LEFT|FULL] JOIN	This indicates that at least one of the tables does not have a matching row in the other table.
	
	Syntax:
	`SELECT columnname [,…]`
	`FROM tablename1 [RIGHT|LEFT|FULL] OUTER JOIN tablename2`
	`ON tablename1.columnname = tablename2.columnname;`
		`SELECT customer#, order#`
		`FROM customers RIGHT JOIN orders ON customers.customer# = orders.customer#;`

UNIQUE	Eliminates duplicate lists	`SELECT UNIQUE state` `FROM customers;`

SQL Single-Row Functions		
Function	**Description**	**Syntax**
ADD_MONTHS	Adds months to a date to signal a target date in the future	`ADD_MONTHS(d, m)` *d* = date (beginning) for the calculation *m* = months—the number of months to add to the date
CONCAT	Concatenates two data items	`CONCAT(c1, c2)` *c1* = first data item to be concatenated *c2* = second data item to be included in the concatenation
DECODE	Takes a given value and compares it to values in a list. If a match is found, then the specified result is returned. If no match is found, then a default result is returned. If no default result is defined, a NULL is returned as the result.	`DECODE(V, L1, R1, L2, R2,…, D)` *V* = value sought *L1* = the first value in the list *R1* = result to be returned if *L1* and *V* match *D* = default result to return if no match is found
INITCAP	Converts words to mixed case, with initial capital letters	`INITCAP(c)` *c* = character string or field to be converted to mixed case
LENGTH	Returns the numbers of characters in a string	`LENGTH(c)` *c* = character string to be analyzed
LOWER	Converts characters to lower-case letters	`LOWER(c)` *c* = character string or field to be converted to lower case

SQL Single-Row Functions (continued)		
Function	**Description**	**Syntax**
LPAD/RPAD	Pad, or fills in, the area to the **L**eft (or **R**ight) of a character string, using a specific character—or even a blank space	`LPAD(c, l, s)` *c* = character string to be padded *l* = length of character string after being padded *s* = symbol or character to be used as padding
MONTHS_BETWEEN	Determines the number of months between two dates	`MONTHS_BETWEEN(d1, d2)` *d1* and *d2* = dates in question *d2* is subtracted from *d1*
NEXT_DAY	Determines the next day—a specific day of the week after a given date	`NEXT_DAY(d, DAY)` *d* = date (starting) *DAY* = the day of the week to be identified
NVL	Solves problems arising from arithmetic operations having fields that may contain NULL values. When a NULL value is used in a calculation, the result is a NULL value. The NVL function is used to substitute a value for the existing NULL.	`NVL(x, y)` *y* = the value to be substituted if *x* is NULL
NVL2	Provides options based on whether a NULL value exists	`NVL2(x, y, z)` *y* = what should be substituted if *x* is not NULL *z* = what should be substituted if *x* is NULL
REPLACE	Used to perform a search-and-replace operation on displayed results	`REPLACE(c, s, r)` *c* = the data or column to be searched *s* = the string of characters to be found *r* = the string of characters to be substituted for *s*
ROUND	Rounds numeric fields	`ROUND(n, p)` *n* = numeric data, or a field, to be rounded *p* = position of the digit to which the data should be rounded
RTRIM/LTRIM	Trims, or removes, a specific string of characters from the **R**ight (or **L**eft) of a set of data	`LTRIM(c, s)` *c* = characters to be affected *s* = string to be removed from the left of the data
SOUNDEX	Converts alphabetic characters to their phonetic representation, using an alphanumeric algorithm	`SOUNDEX(c)` *c* = characters to be phonetically represented
SUBSTR	Returns a substring, or portion of a string, in output	SUBSTR(*c, p, l*) *c* = character string *p* = position (beginning) for the extraction *l* = length of output string
TO_CHAR	Converts dates and numbers to a formatted character string	`TO_CHAR(n, 'f')` *n* = number or date to be formatted *f* = format model to be used

SQL Single-Row Functions (continued)		
Function	**Description**	**Syntax**
TO_DATE	Converts a date in a specified format to the default date format	`TO_DATE(d,f)` *d* = date entered by the user *f* = format of the entered date
TRUNC	Truncates, or cuts, numbers to a specific position	`TRUNC(n, p)` *n* = numeric data, or a field, to be truncated *p* = position of the digit to which the data should be truncated
UPPER	Converts characters to upper-case letters	`UPPER(c)` *c* = character string or field to be converted to upper case

<table>
<tr><th colspan="3">SQL Group (Multiple-Row) Functions</th></tr>
<tr><th>Function</th><th>Description</th><th>Syntax</th></tr>
<tr><td>AVG</td><td>Returns the average value of the selected numeric field. Ignores NULL values.</td><td><code>AVG([DISTINCT|ALL] n)</code></td></tr>
<tr><td>COUNT</td><td>Returns the number of rows that contain a value in the identified column. Rows containing NULL values in the column will not be included in the results. To count all rows, including those with NULL values, use an * rather than a column name.</td><td><code>COUNT(*|[|DISTINCT|ALL] c)</code></td></tr>
<tr><td>MAX</td><td>Returns the highest (maximum) value from the selected field. Ignores NULL values.</td><td><code>MAX([DISTINCT|ALL] c)</code></td></tr>
<tr><td>MIN</td><td>Returns the lowest (minimum) value from the selected field. Ignores NULL values.</td><td><code>MIN([DISTINCT|ALL] c)</code></td></tr>
<tr><td>STDDEV</td><td>Returns the standard deviation of the selected numeric field. Ignores NULL values.</td><td><code>STDDEV([DISTINCT|ALL] n)</code></td></tr>
<tr><td>SUM</td><td>Returns the sum, or total value, of the selected numeric field. Ignores NULL values.</td><td><code>SUM([DISTINCT|ALL] n)</code></td></tr>
<tr><td>VARIANCE</td><td>Returns the variance of the numeric field selected. Ignores NULL values.</td><td><code>VARIANCE([DISTINCT|ALL] n)</code></td></tr>
</table>

SQL Comparison Operators		
Operator	**Description**	**Example**
=	Equality operator—requires an exact match of the record data and the search value	`WHERE cost = 55.95`
>	"Greater than" operator—requires a record to be greater than the search value	`WHERE cost > 55.95`
<	"Less than" operator—requires a record to be less than the search value	`WHERE cost < 55.95`

SQL Comparison Operators (continued)		
Operator	**Description**	**Example**

<>, !=, *or* ^=	"Not equal to" operator—requires a record not to match the search value	`WHERE cost <> 55.95`
		or
		`WHERE cost != 55.95`
		or
		`WHERE cost ^= 55.95`

<=	"Less than or equal to" operator—requires a record to be less than or an exact match with the search value	`WHERE cost <= 55.95`
>=	"Greater than or equal to" operator—requires record to be greater than or an exact match with the search value	`WHERE cost >= 55.95`
[NOT] BETWEEN x AND y	Searches for records in a specified range of values	`WHERE cost BETWEEN 40 AND 65`
[NOT] IN(x,y,...)	Searches for records that match one of the items in the list	`WHERE cost IN(22, 55.95, 13.50)`
[NOT] LIKE	Searches for records that match a search pattern—used with wildcard characters	`WHERE lastname LIKE '_A%'`
IS [NOT] NULL	Searches for records with a NULL value in the indicated column	`WHERE referred IS NULL`

>ALL	More than the highest value returned by the subquery	`WHERE cost >ALL`
		`(SELECT MAX(cost)`
		`FROM books`
		`GROUP BY category);`

<ALL	Less than the lowest value returned by the subquery	`WHERE cost <ALL`
		`(SELECT MAX(cost)`
		`FROM books`
		`GROUP BY category);`

<ANY	Less than the highest value returned by the subquery	`WHERE cost <ANY`
		`(SELECT MAX(cost)`
		`FROM books`
		`GROUP BY category);`

>ANY	More than the lowest value returned by the subquery	`WHERE cost >ANY`
		`(SELECT MAX(cost)`
		`FROM books`
		`GROUP BY category);`

=ANY	Equal to any value returned by the subquery (same as IN)	`WHERE cost >ALL`
		`(SELECT MAX(cost)`
		`FROM books`
		`GROUP BY category);`

[NOT] EXISTS	Row must match a value in the subquery	`WHERE isbn EXISTS`
		`(SELECT DISTINCT isbn`
		`FROM orderitems);`

SQL Logical Operators

Operator	Description	Example
AND	Combines two conditions together—a record must match both conditions	`WHERE cost > 20` `AND retail < 50`
OR	Requires a record to only match one of the search conditions	`WHERE cost > 20` `OR retail < 50`

Other SQL Operators

Operator	Description	Example

INTERSECT	Lists only the results returned by both queries	`SELECT customer# FROM customers`
		`INTERSECT`
		`SELECT customer# FROM orders;`

MINUS	Lists only the results returned by the first query and not returned by the second query	`SELECT customer# FROM customers`
		`MINUS`
		`SELECT customer# FROM orders;`

Outer Join Operator (+)	Used to indicate the table containing the deficient rows. The operator is placed next to the table that should have NULL rows added to create a match.	`SELECT lastname, firstname,`
		`order#`
		`FROM customers c, orders o`
		`WHERE c.customer# =`
		`o.customer#(+);`

UNION	Used to combine the distinct results returned by multiple SELECT statements	`SELECT customer# FROM customers`
		`UNION`
		`SELECT customer# FROM orders;`

UNION ALL	Used to combine all the results returned by multiple SELECT statements	`SELECT customer# FROM customers`
		`UNION ALL`
		`SELECT customer# FROM orders;`

&	Identifies a substitution variable. Allows the user to be prompted to enter a specific value for the substitution variable.	`UPDATE customers`
		`SET region = '&Region'`
		`WHERE state = '&State';`

SQL Constraints

Constraint	During Table Creation	After Table Creation

CHECK
 	
Ensures that a specified condition is true before the data value is added to the table. For example, an order's ship date cannot be "less than" its order date.
	`CREATE TABLE newtable`
	`(firstcol NUMBER,`
	`secondcol VARCHAR2(20),`
	`thirdcol NUMBER CHECK BETWEEN`
	`20 AND 30);`
	or
	`CREATE TABLE newtable`
	`(firstcol NUMBER,`
	`secondcol VARCHAR2(20),`
	`thirdcol NUMBER,`
	`CHECK (thirdcol BETWEEN 20`
	`AND 80));`
		`ALTER TABLE newtable`
		`ADD CHECK (thirdcol BETWEEN`
		`20 AND 80);`

SQL Constraints (continued)		
Constraint	**During Table Creation**	**After Table Creation**

FOREIGN KEY
 	
In a one-to-many relationship, the constraint is added to the "many" table. The constraint ensures that if a value is entered to the specified column, it must exist in the table being referred to, or the row is not added.
	`CREATE TABLE newtable`
	`(firstcol NUMBER,`
	`secondcol VARCHAR2(20) REFERENCES`
	`anothertable(col1));`
	or
	`CREATE TABLE newtable`
	`(firstcol NUMBER,`
	`secondcol VARCHAR2(20),`
	`FOREIGN KEY (secondcol)`
	`REFERENCES anothertable(col1);`
		`ALTER TABLE newtable`
		`ADD FOREIGN KEY (secondcol)`
		`REFERENCES anothertable (col1);`

NOT NULL
 	
Requires that the specified column cannot contain a NULL value. It can only be created with the column-level approach to table creation.
	`CREATE TABLE newtable`
	`(firstcol NUMBER,`
	`secondcol VARCHAR2(20),`
	`thirdcol NUMBER NOT NULL);`
		`ALTER TABLE newtable`
		`MODIFY (thirdcol NOT NULL);`

PRIMARY KEY
 	
Determines which column(s) uniquely identifies each record. The primary key cannot be NULL, and the data value(s) must be unique.
	`CREATE TABLE` *`newtable`*
	`(firstcol NUMBER PRIMARY KEY,`
	`secondcol VARCHAR2(20));`
	or
	`CREATE TABLE newtable`
	`(firstcol NUMBER,`
	`secondcol VARCHAR2(20),`
	`PRIMARY KEY (firstcol));`
		`ALTER TABLE newtable`
		`ADD PRIMARY KEY (firstcol);`

UNIQUE
 	
Ensures that all data values stored in the specified column must be unique. The UNIQUE constraint differs from the PRIMARY KEY constraint in that it allows NULL values.
	`CREATE TABLE newtable`
	`(firstcol NUMBER,`
	`secondcol VARCHAR2(20) UNIQUE);`
	or
	`CREATE TABLE newtable`
	`(firstcol NUMBER,`
	`secondcol VARCHAR2(20),`
	`UNIQUE (secondcol));`
		`ALTER TABLE newtable`
		`ADD UNIQUE (secondcol);`

SQL*Plus Commands

Command	Description	Example

/	Executes a SQL statement—must be on a separate line	`SELECT zip`
		`FROM customers`
		`/`

;	View contents in the buffer	`SQL>;`
CONN *or* CONNECT	Used to connect to the Oracle9*i* database from within SQL*Plus `CONNECT` *`username/password@connectstring`*	`CONNECT rthomas/little25car`

DESC
or
DESCRIBE
	Displays the structure of table	`DESCRIBE books`
		or
		`DESC books`

L
or
LIST
	Displays the contents of the buffer	`SQL> LIST`
		or
		`SQL>L`

R
or
RUN
	Executes a SQL statement stored in the buffer	`SQL>run (Enter)`
		or
		`SQL>r (Enter)`

SQL*Plus Editing Commands

Command	Description	Example
A *or* APPEND	Adds the entered text to the end of the active line	`A WHERE retail < 42.89`

C *old**new*\\
or
CHANGE *old**new*\\
	Finds an indicated string of characters and replaces it with a new string of characters	`C \42.89\52.35\`
		or
		`CHANGE \42.89\52.35\`

DEL *n*
or
DELETE *n*
	Deletes the indicated line number from the SQL statement	`DEL 3`
		or
		`DELETE 3`

I
or
INPUT
	Allows new rows to be entered after the current active line	`I OR cost > 15`
		or
		`INPUT OR cost > 15`

n	Sets the current active line in the SQL*Plus buffer	`2`

SQL*Plus Environment Variables

Variable	Description	Example
`UNDERLINE` *`x`*`|OFF`	Specifies the symbol used to separate a column heading from the contents of the column	`SET UNDERLINE '='`
LINESIZE *n*	Establishes the maximum number of characters that can appear on a single line of output	`SET LINESIZE 35`
PAGESIZE *n*	Establishes the maximum number of lines that can appear on one page of output	`SET PAGESIZE 27`

SQL*Plus Report Commands		
Command/Syntax	**Description**	**Example**
START *or* @	Executes a script file	`START titlereport.sql` *or* `@titlereport.sql`
`TTITLE [option [text|variable] …] ON|OFF]`	Defines a report's header	`TTITLE 'Top of|Report'`
`BTITLE [option [text|variable] …] ON|OFF]`	Defines a report's footer	`BTITLE 'Bottom of|Report'`
`BREAK ON columnname| columnalias [ON …] [skip n|page]`	Suppresses duplicate data in a specified column	`BREAK ON title`
`COMPUTE statistical function OF columnname| REPORT ON groupname`	Performs calculations in a report based on the AVG, SUM, COUNT, MIN, and MAX statistical functions	`COMPUTE SUM OF total ON customer#`
`SPOOL filename`	Redirects the output to a text file	`SPOOL c:\reportresults.txt`
`CLEAR`	Clears settings applied with the COLUMN and BREAK commands	`CLEAR COLUMN`
COLUMN Command Options		
`FORMAT formatmodel`	Applies a specified format to the column data	`COLUMN title FORMAT A30`
`HEADING columnheading`	Specifies the heading to be used for the specified column	`COLUMN title HEADING 'Book|Title'`
`NULL textmessage`	Identifies the text message to be displayed lin the place of NULL values	`COLUMN shipdate NULL 'Not Shipped'`
TTITLE and BTITLE Command Options		
CENTER	Centers the display between a report's left and right margins	`TTITLE CENTER 'Report Title'`
FORMAT	Applies a format model to the data to be displayed. It uses the same format model elements as the COLUMN command.	`BTITLE 'Page: ' SQL.PNO FORMAT 9`
LEFT	Aligns data to be displayed to a report's left margin	`TTITLE LEFT 'Internal Use Only'`
RIGHT	Aligns data to be displayed to a report's right margin	`TTITLE RIGHT 'Internal Use Only'`
SKIP *n*	Indicates the number of lines to skip before the display resumes	`TTITLE CENTER 'Report Title' SKIP 2 RIGHT 'Internal Use Only'`

<table>
<tr><th colspan="3">PL/SQL</th></tr>
<tr><th>Element</th><th>Syntax</th><th>Example</th></tr>
<tr><td>PL/SQL block structure</td><td>[DECLARE]
identifies variables and constants
BEGIN
identifies executable statements
[EXCEPTION]
identifies error handlers
END;
terminates the block</td><td>DECLARE
v_date DATE;
BEGIN
V_date := SYSDATE;
END;</td></tr>
<tr><td>Variable declaration</td><td>variablename [CONSTANT]datatype
[NOT NULL]
[:=|DEFAULT value_or_expression];</td><td>c_rateincrease CONSTANT
NUMBER(3,2) :=1.2;
v_title VARCHAR2(30);
v_retail books.retail%TYPE;
v_newretail NUMBER(5,2);</td></tr>
<tr><td>SELECT...INTO</td><td>SELECT columnname
[, columnname,...]
INTO variablename
[, variablename,...]
FROM tablename
WHERE condition;</td><td>SELECT title, retail,
retail*c_rateincrease
INTO v_title, v_retail,
v_newretail
FROM books
WHERE isbn = '1059831198';</td></tr>
<tr><td>IF statement</td><td>IF condition THEN
statements;
[ELSIF condition THEN
statements;]
[ELSE
statements;]
END IF;</td><td>DECLARE
v_gift VARCHAR2(20);
c_retailprice NUMBER(5, 2)
:= 29.95;
BEGIN
IF c_retailprice >= 56 THEN
v_gift := 'FREE
SHIPPING';
ELSIF c_retailprice >= 25
THEN
v_gift := 'BOOKCOVER';
ELSIF c_retailprice >= 12
THEN
v_gift := ' BOX OF
BOOK LABELS';
ELSE
v_gift :=
'BOOKMARKER';
END IF;
DBMS_OUTPUT.PUT_LINE ('The
gift for a book costing ' ||
c_retailprice || ' is a ' ||
v_gift);
END;</td></tr>
</table>

PL/SQL (continued)		
Element	**Syntax**	**Example**

Basic loop	LOOP
	statements;
	EXIT [WHEN *condition*];
	END LOOP;
		DECLARE
		v_counter NUMBER(1) := 0;
		BEGIN
		LOOP
		v_counter := v_counter + 1;
		DBMS_OUTPUT.PUT_LINE ('The current value of the counter is ' || v_counter);
		EXIT WHEN v_counter = 4;
		END LOOP;
		END;

FOR loop	FOR *counter* IN [REVERSE]
	lower_limit .. *upper_limit* LOOP
	statements;
	END LOOP;
		BEGIN
		FOR i IN 1 .. 10 LOOP
		DBMS_OUTPUT.PUT_LINE ('The current value of the counter is ' || i);
		END LOOP;
		END;

WHILE loop	WHILE *condition* LOOP
	statements;
	END LOOP;
		DECLARE
		v_counter NUMBER(2) :=0;
		BEGIN
		WHILE v_counter < 15 LOOP
		DBMS_OUTPUT.PUT_LINE ('The current value of the counter is ' || v_counter);
		v_counter := v_counter + 1;
		END LOOP;
		END;

CURSOR	CURSOR *cursor_name* IS *selectquery;*	CURSOR books_cursor IS SELECT title, retail FROM books NATURAL JOIN orderitems WHERE order# = 1012;
OPEN	OPEN *cursor_name;*	OPEN books_cursor;
CLOSE	[CLOSE *cursor_name;*	CLOSE books_cursor;
FETCH	FETCH *cursor_name* INTO *variablename* [,...*variablename*];	FETCH books_cursor INTO v_title, v_retail;
PRAGMA EXCEPTION_INIT	PRAGMA EXCEPTION_INIT (*exception_name*, *errornumber*);	PRAGMA EXCEPTION_INIT (id_already_in_use, -00001);
Cursor FOR loop	FOR *record_name* IN *cursor_name* LOOP *statement;* [*statement;*...] END LOOP;	FOR r_books IN books_cursor LOOP END LOOP;

Cursor Attributes		
Cursor Attribute	**Description**	**Example**
%ROWCOUNT	Identifies the number of rows processed	`books_cursor%ROWCOUNT`
%FOUND	Contains the value of TRUE if one or more rows are processed—FALSE if no rows are processed	`books_cursor%FOUND`
%NOTFOUND	Contains the value of TRUE if no rows are processed—FALSE if one or more rows are processed	`books_cursor%NOTFOUND`
%ISOPEN	Contains the value TRUE if a cursor is not closed after processing—FALSE if the cursor is closed. The value will always be FALSE after processing for an implicit cursor occurs because its closure is automatic.	`books_cursor%ISOPEN`

APPENDIX

C

ORACLE9*i* PRACTICE EXAMS (A–E)

Practice Exam A

The questions in this practice exam test your knowledge of the concepts presented in Chapters 2-4 of this textbook. Solutions to the exam questions are available from your instructor. When necessary, reference the table structures that follow. Full column names are listed after the tables.

EMP Table	
Column Name	**Definition**
Empno	NUMBER(4), PK
Ename	VARCHAR2(10)
Job	VARCHAR2(9)
Mgr	NUMBER(4)
Hiredate	DATE
Sal	NUMBER(7,2)
Comm	NUMBER(7,2)
Deptno	NUMBER(2), FK

DEPT Table	
Column Name	**Definition**
Deptno	NUMBER(2)
Dname	VARCHAR2(14)
Loc	VARCHAR2(13)

- Empno — Employee
- Ename — Employee Name
- Job — Job Title
- Mgr — Manager of Employee
- Hiredate — Hire Date
- Sal — Salary (monthly)
- Comm — Commission (for sales)
- Deptno — Department Number
- Dname — Department Name
- Loc — Location of Operation

1. Which two of the following queries can be used to determine the name of the department in which employee Blake works?

   a. `SELECT dname FROM dept WHERE ename = ('BLAKE');`

   b.
   ```
   SELECT d.dname FROM dept d NATURAL JOIN emp e
   WHERE e.ename = 'BLAKE';
   ```

   c.
   ```
   SELECT dname FROM dept d, emp e
   WHERE e.ename = ('BLAKE')
   AND d.deptno = e.deptno;
   ```

   d.
   ```
   SELECT dname FROM dept JOIN emp USING (dname)
   WHERE ename = 'BLAKE';
   ```

   e.
   ```
   SELECT dname FROM dept JOIN emp
   ON dept.deptno = emp.deptno
   WHERE ename = 'BLAKE';
   ```

2. Which of the following queries will display the annual salary for each employee in the EMP table if the Sal column contains each employee's monthly salary?

   a. `SELECT sal * 12 'Annual Salary' FROM emp;`

   b. `SELECT salary*12 annual FROM emp;`

   c. `SELECT annual sal*12 FROM emp;`

   d. `SELECT sal*12 FROM emp;`

3. Which of the following queries will display all data stored in the EMP table?

   a. `SELECT * FROM emp;`

   b. `SELECT % FROM emp;`

   c. `SELECT ^ FROM emp;`

   d. `SELECT _ FROM emp;`

4. Which statement reflects what will occur when the following query is executed?

   ```
   SELECT ename
   FROM emp e, emp m
   WHERE e.mgr = m.empno;
   ```

   a. The query will result in a self-join that will display the name of each employee's manager.

   b. An ambiguity error will be displayed and the statement will not be executed.

   c. The query will execute a full outer join and the names of employees who have not been assigned a manager will be displayed.

   d. The query will execute a right outer join and the names of employees who are not considered managers will be displayed.

5. Which of the following keywords can be used to create a non-equality join? Choose all that apply.

   a. NATURAL JOIN

   b. JOIN...USING

   c. OUTER JOIN

   d. JOIN...ON

   e. None—a non-equality join cannot be created using any of the JOIN keywords.

6. Which of the following queries will return only the department numbers contained in the DEPT table that are not also listed in the EMP table?

   a.
   ```
   SELECT deptno FROM dept NATURAL JOIN emp
   WHERE deptno NOT IN emp;
   ```

   b.
   ```
   SELECT deptno FROM dept MINUS deptno FROM emp;
   ```

   c.
   ```
   SELECT deptno FROM dept MINUS
   SELECT deptno FROM emp;
   ```

   d.
   ```
   SELECT deptno FROM dept JOIN emp
   ON dept.deptno <> emp.deptno;
   ```

   e.
   ```
   SELECT deptno FROM emp MINUS
   SELECT deptno FROM dept;
   ```

7. Which of the following queries will display the employee number of the employee named King?

   a.
   ```
   SELECT empno FROM emp WHERE ename = '%KING';
   ```

   b.
   ```
   SELECT empno FROM emp WHERE ename = '_ING';
   ```

   c.
   ```
   SELECT empno FROM emp WHERE ename LIKE KING;
   ```

   d.
   ```
   SELECT empno FROM emp WHERE ename = KING;
   ```

   e. none of the above

8. Which of the following queries will display all employees in the Sales Department who were hired in 1981?

   a.
   ```
   SELECT * FROM emp
   WHERE dname = 'SALES' AND hiredate LIKE '%81';
   ```

   b.
   ```
   SELECT * FROM emp NATURAL JOIN dept
   WHERE dname = 'SALES' AND hiredate LIKE '%81';
   ```

   c.
   ```
   SELECT * FROM emp
   WHERE dname = 'SALES' OR hiredate LIKE '%81';
   ```

   d.
   ```
   SELECT * FROM emp NATURAL JOIN dept
   WHERE dname = 'SALES' AND hiredate LIKE '%1981';
   ```

9. Which of the following queries will display the name and job title of each employee stored in the EMP table? Choose all that apply.
   a. `SELECT ename, job AS "Job Title" FROM emp;`
   b. `SELECT ename, job "Job Title" FROM emp;`
   c. `SELECT ename, job FROM emp;`
   d. `SELECT ename, job 'JOB TITLE' FROM emp;`
   e. `SELECT ename, job 'Job Title' FROM emp;`

10. Which sentence most accurately describes the results of the following SELECT statement?

```
SELECT DISTINCT job, ename FROM emp;
```

   a. Each row returned in the results will be unique.
   b. Each job title will only be displayed once in the results.
   c. Each job title will be displayed once, along with the names of each employee assigned to that job.
   d. The results will be sorted in order of employee names.

11. Which of the following clauses is used to project certain columns from a table?
   a. SELECT
   b. FROM
   c. WHERE
   d. ORDER BY

12. Which of the following queries will display the names of all employees who earn an annual salary of at least $10,000?
   a. `SELECT ename FROM emp WHERE sal*12 > 10,000;`
   b. `SELECT ename FROM emp WHERE sal*12 > '10,000';`
   c. `SELECT ename FROM emp WHERE sal*12 => 10000;`
   d. `SELECT ename FROM emp WHERE sal *12 >= 10000.00;`
   e. The correct statement is not given.

13. Which of the following queries will display each employee's number in a sorted order, by employee name? Choose all that apply.
   a. `SELECT empno, ename FROM emp ORDER BY empno;`
   b. `SELECT empno, ename FROM emp ORDER BY ename;`
   c. `SELECT empno, ename FROM emp ORDER BY 1;`
   d. `SELECT empno, ename FROM emp ORDER BY 2;`
   e. `SELECT empno, ename ORDER BY ename;`

14. Which of the following queries will display the name of each employee who earns a salary of at least $1,200 a month but less than $2,000 a month?

    a. `SELECT ename FROM emp`
       `WHERE sal BETWEEN (1200, 2000);`

    b. `SELECT ename FROM emp`
       `WHERE sal BETWEEN 1200 AND 2000;`

    c. `SELECT ename FROM emp`
       `WHERE sal >=1200 AND <2000;`

    d. `SELECT ename FROM emp`
       `WHERE sal >=1200 AND sal<2000;`

    e. `SELECT ename FROM emp`
       `WHERE sal >1200 AND sal<2000;`

15. Which of the following clauses is used to restrict the rows returned by a query?

    a. SELECT

    b. FROM

    c. WHERE

    d. ORDER BY

16. Which of the following operators is used to perform pattern searches?

    a. IN

    b. BETWEEN

    c. IS NULL

    d. LIKE

17. Which of the following queries will not include any employees in Department 30 in its results? Choose all that apply.

    a. `SELECT * FROM emp WHERE deptno !=30;`

    b. `SELECT * FROM emp WHERE deptno <>30;`

    c. `SELECT * FROM emp WHERE deptno ^30;`

    d. `SELECT * FROM emp WHERE deptno =30;`

18. Which of the following queries will display all employees who do not earn a commission?

    a. `SELECT ename FROM emp WHERE comm = NULL;`

    b. `SELECT ename FROM emp WHERE comm IS NULL;`

    c. `SELECT ename FROM emp WHERE comm LIKE NULL;`

    d. `SELECT ename FROM emp WHERE comm LIKE 'NULL';`

19. Which of the following queries will return the names of all employees who work in the Sales Department or Accounting Department and earn at least $2,000 a month? Choose all that apply.

    a.
    ```
    SELECT ename FROM emp NATURAL JOIN dept
    WHERE dname IN ('SALES', 'ACCOUNTING')
    AND sal >= 2000;
    ```
    b.
    ```
    SELECT ename FROM emp JOIN dept
    ON emp.deptno = dept.deptno
    WHERE sal >= 2000 AND dname = 'SALES'
    OR dname = 'ACCOUNTING';
    ```
    c.
    ```
    SELECT ename FROM emp JOIN dept USING (deptno)
    WHERE sal >= 2000 AND (dname = 'SALES' OR dname =
    'ACCOUNTING');
    ```
    d.
    ```
    SELECT ename FROM emp NATURAL JOIN dept
    WHERE sal >= 2000 AND (dname = 'SALES' OR dname =
    'ACCOUNTING');
    ```

20. Which of the following clauses is used to present a query's results in sorted order?

    a. SELECT

    b. FROM

    c. WHERE

    d. ORDER BY

Practice Exam B

The questions in this practice exam test your knowledge of the concepts presented in Chapters 5-7 of this textbook. Solutions to the exam questions are available from your instructor. When necessary, reference the table structures that follow. Full column names are listed after the tables.

EMP Table	
Column Name	**Definition**
Empno	NUMBER(4), PK
Ename	VARCHAR2(10)
Job	VARCHAR2(9)
Mgr	NUMBER(4)
Hiredate	DATE
Sal	NUMBER(7,2)
Comm	NUMBER(7,2)
Deptno	NUMBER(2), FK

DEPT Table	
Column Name	**Definition**
Deptno	NUMBER(2)
Dname	VARCHAR2(14)
Loc	VARCHAR2(13)

- Empno — Employee
- Ename — Employee Name
- Job — Job Title
- Mgr — Manager of Employee
- Hiredate — Hire Date
- Sal — Salary (monthly)
- Comm — Commission (for sales)
- Deptno — Department Number
- Dname — Department Name
- Loc — Location of Operation

1. Which of the following queries will display the name of each department in lower-case letters?
   a. `SELECT LOW(dname) FROM dept;`
   b. `SELECT LOWER(dname) FROM dept;`
   c. `SELECT LOWERCASE(dname) FROM dept;`
   d. `SELECT NOTUPPER(dname) FROM dept;`

2. Which of the following queries will display the gross salary of each employee if the Sal column contains the salary of each employee and the Comm column contains the commission earned by the sales representatives? Choose all that apply.
   a. `SELECT ename, sal + NVL(comm, 0) AS "Gross Salary" FROM emp;`
   b. `SELECT ename, NVL2(sal+comm, sal, comm) "Gross" FROM emp;`
   c. `SELECT ename, NVL(comm,0) + sal FROM emp;`
   d. `SELECT ename, NVL(sal+comm, sal) FROM emp;`
3. Which of the following queries will return the total salary earned by all individuals working in Department 10? Choose all that apply.
   a. `SELECT SUM(sal) FROM emp WHERE deptno = 10;`
   b. `SELECT TOTAL(sal) FROM emp WHERE deptno = 10;`
   c. `SELECT SUM(sal) FROM emp WHERE deptno = 10 GROUP BY deptno;`
   d. `SELECT SUM(sal) FROM emp HAVING deptno = 10;`
   e. `SELECT SUM(sal) FROM emp HAVING deptno = 10 GROUP BY deptno;`
4. Which of the following queries will display the name of all employees who work in the same department as an employee named King?
   a. `SELECT ename FROM emp WHERE ename = 'KING';`
   b. `SELECT ename FROM emp WHERE deptno = (SELECT deptno FROM emp WHERE ename = "KING";`
   c. `SELECT ename FROM emp WHERE ename = (SELECT deptno FROM emp WHERE ename = 'KING');`
   d. `SELECT ename FROM emp WHERE deptno = (SELECT deptno FROM emp WHERE ename = 'KING');`

5. Which of the following queries will display the name of the department whose employees earn an average monthly salary of at least $1,500?

   a.
   ```
   SELECT dname, AVERAGE(sal)
   FROM dept NATURAL JOIN emp
   WHERE AVERAGE(sal) > 1500;
   ```
   b.
   ```
   SELECT dname, AVERAGE(sal)
   FROM dept NATURAL JOIN emp
   HAVING AVERAGE(sal) > 1500;
   ```
   c.
   ```
   SELECT dname, AVG(sal)
   FROM dept NATURAL JOIN emp
   WHERE AVG(sal) > 1500;
   ```
   d.
   ```
   SELECT dname, AVG(sal)
   FROM dept NATURAL JOIN emp
   GROUP BY dname
   HAVING AVG(sal) > 1500;
   ```

6. Which of the following queries will display the fourth number in each employee's employee number?

   a. `SELECT ename, SUBSTR(empno, 4, 1) FROM emp;`

   b. `SELECT ename, LENGTH(empno, 4) FROM emp;`

   c. `SELECT ename, TRUNC(empno, 4) FROM emp;`

   d. `SELECT ename, SOUNDEX(empno, 4, 1) FROM emp;`

7. Which of the following queries will return the total monthly salary of all employees in the company?

   a. `SELECT SUM(sal) FROM emp GROUP BY deptno;`

   b. `SELECT SUM(sal) FROM emp;`

   c. `SELECT SUM(DISTINCT sal) FROM emp;`

   d. `SELECT TOTAL(sal) FROM emp WHERE sal IS NOT NULL;`

8. Which of the following queries will return the names of only those employees who are clerks?

   a.
   ```
   SELECT UPPER(ename) FROM emp
   WHERE LOWER(job) = 'CLERK';
   ```
   b.
   ```
   SELECT LOWER(ename) FROM emp
   WHERE LOWER(job) = 'CLERK';
   ```
   c.
   ```
   SELECT UPPER(ename) FROM emp
   WHERE LOWER(job) = 'clerk';
   ```
   d.
   ```
   SELECT LOWER(ename) FROM emp
   WHERE UPPER(job) = 'clerk';
   ```

9. Which of the following statements is accurate?
   a. Group functions are used to calculate multiple values per row, but single-row functions are used to calculate only one value per row.
   b. A query containing a group function must also contain a GROUP BY clause.
   c. Group functions return one value per group of rows processed, but single-row functions return one value for each row processed.
   d. A HAVING clause cannot be used in a query that contains a single-row function.
10. Which of the following describes a scenario that would call for the use of a subquery?
    a. You need to know all the employees who have a salary higher than employee Blake's salary.
    b. You need to know the name of all the clerks who earn more than $1,000 per month.
    c. You need a list of all employees who work in Department 30.
    d. You need to find the average salary for all the employees in each department.
11. Which of the following queries will calculate the difference between today's date and the date on which an employee was hired?
    a. `SELECT ename, MONTH_BETWEEN(SYSDATE, hiredate)`
       `FROM emp;`
    b. `SELECT ename, SYSDATE-hiredate`
       `FROM emp;`
    c. `SELECT ename, DIFF(SYSDATE, hiredate)`
       `FROM emp;`
    d. `SELECT ename, TO_DATE(SYSDATE, hiredate)`
       `FROM emp;`
12. Which of the clauses that come after the following query will cause the query to return an error message?

```
SELECT ename
FROM emp
WHERE sal >
      (SELECT AVG(sal)
       FROM emp
       GROUP BY deptno);
```

    a. `SELECT ename`
    b. `WHERE sal >`
    c. `SELECT AVG(sal)`
    d. `GROUP BY deptno`

13. Which of the following queries will return the number of employees who have the same job title?
    a. `SELECT COUNT(*), job FROM emp GROUP BY job;`
    b. `SELECT COUNT(job) FROM emp;`
    c. `SELECT COUNT(DISTINCT job) FROM emp;`
    d. `SELECT SUM(job) FROM emp;`

14. Which of the following queries will display the lowest salary earned by an employee?
    a. `SELECT MIN(ename) FROM emp;`
    b. `SELECT LOW(ename) FROM emp;`
    c. `SELECT LOWER(sal) FROM emp;`
    d. `SELECT MIN(sal) FROM emp;`
    e. `SELECT MIN(sal) FROM emp GROUP BY job;`

15. Which of the following queries will display the names of all employees who are in the same department as an employee named Smith, but who earn a higher salary than Smith?

    a.
    ```
    SELECT ename FROM emp
    WHERE deptno = 'SMITH' AND sal > 'SMITH';
    ```
    b.
    ```
    SELECT ename FROM emp
    WHERE (deptno, sal) >
        (SELECT deptno, sal FROM emp
         WHERE ename = 'SMITH');
    ```
    c.
    ```
    SELECT ename FROM emp WHERE deptno =
    (SELECT deptno FROM emp WHERE ename = 'SMITH')
    AND sal > (SELECT sal FROM emp WHERE ename = 'SMITH');
    ```
    d.
    ```
    SELECT ename FROM emp
    WHERE (deptno, sal) >ANY
        (SELECT deptno, sal FROM emp
         WHERE ename = 'SMITH');
    ```

16. Which of the following is a valid multiple-row operator?
    a. ANY
    b. OR
    c. =
    d. >

17. Which of the following queries will display the name of all employees who work in the city of Boston? Choose all that apply.

    a. `SELECT ename FROM emp NATURAL JOIN dept`
       `WHERE loc = 'BOSTON';`

    b. `SELECT ename FROM emp WHERE loc = 'BOSTON';`

    c. `SELECT ename FROM dept WHERE loc = 'BOSTON';`

    d. `SELECT ename FROM emp WHERE deptno =`
       `(SELECT deptno FROM dept WHERE loc = 'BOSTON');`

    e. `SELECT ename FROM emp WHERE deptno = 'BOSTON';`

18. Which of the following operators is equivalent to using the IN operator in a multiple-row subquery?

    a. =ANY

    b. =ALL

    c. >ANY

    d. <ANY

19. Assuming the Comm column can contain NULL values, which of the following queries will display how many employees in the company earn a commission? Choose all that apply.

    a. `SELECT COUNT(comm) FROM emp;`

    b. `SELECT COUNT(comm) FROM emp`
       `WHERE comm IS NULL;`

    c. `SELECT COUNT(*) FROM emp`
       `WHERE comm IS NOT NULL;`

    d. `SELECT COUNT(*) FROM emp`
       `WHERE comm IS NULL;`

20. Which of the following is a single-row function?

    a. AVERAGE

    b. VARIANCE

    c. SUM

    d. ADD_MONTHS

Practice Exam C

The questions in this practice exam test your knowledge of the concepts presented in Chapters 8-10 of this textbook. Solutions to the exam questions are available from your instructor. When necessary, reference the table structures that follow. Full column names are listed after the tables.

EMP Table	
Column Name	**Definition**
Empno	NUMBER(4), PK
Ename	VARCHAR2(10)
Job	VARCHAR2(9)
Mgr	NUMBER(4)
Hiredate	DATE
Sal	NUMBER(7,2)
Comm	NUMBER(7,2)
Deptno	NUMBER(2), FK

DEPT Table	
Column Name	**Definition**
Deptno	NUMBER(2)
Dname	VARCHAR2(14)
Loc	VARCHAR2(13)

- Empno — Employee
- Ename — Employee Name
- Job — Job Title
- Mgr — Manager of Employee
- Hiredate — Hire Date
- Sal — Salary (monthly)
- Comm — Commission (for sales)
- Deptno — Department Number
- Dname — Department Name
- Loc — Location of Operation

1. Which of the following SQL statements will create a new table containing data for only the employees from Department 30?

   a. `CREATE TABLE ee30`
      `AS (SELECT * FROM emp WHERE deptno = 30);`

   b. `CREATE TABLE ee30,`
      `AS (SELECT * FROM emp WHERE deptno = 30);`

   c. `CREATE TABLE  (SELECT * FROM emp WHERE deptno = 30);`

   d. `CREATE TABLE 30department`
      `AS (SELECT * FROM emp WHERE deptno = 30);`

2. Which of the following SQL statements will remove all rows from the DEPT table and release the storage space occupied by the rows?

   a. `DROP TABLE dept;`

   b. `DELETE FROM dept;`

   c. `TRUNCATE TABLE dept;`

   d. `DELETE *.* FROM dept;`

3. Which of the following SQL statements will add a numeric column named SSN to the EMP table?

   a. `ALTER TABLE emp MODIFY (add SSN NUMBER(9));`

   b. `ALTER TABLE emp ADD (SSN NUMBER(9);`

   c. `ALTER TABLE emp MODIFY (SSN NUMBER(9));`

   d. `ALTER TABLE emp ADD (SSN NUMBER(9));`

4. Which of the following SQL statements will change the name of the DEPT table to DEPARTMENT?

   a. `ALTER TABLE dept RENAME AS department;`

   b. `RENAME TO department FROM dept;`

   c. `RENAME dept TO department;`

   d. `RENAME dept AS department;`

5. Which of the following statements is correct? Choose all that apply.

   a. A column that has been marked as unused cannot be reclaimed or unmarked at a later time.

   b. When a column is dropped, the contents of the column can be restored by using the ROLLBACK command.

   c. When a column is dropped, the contents of the column cannot be restored by using the ROLLBACK command.

   d. A column that has been marked as unused can be reclaimed or unmarked at a later time.

6. Which of the following SQL statements will add a new department to the DEPT table?

   a.
   ```
   UPDATE dept
   SET deptno = 65, dname = 'HR', loc = 'SEATTLE';
   ```
   b. `INSERT VALUES (65, HR, SEATTLE) INTO dept;`
   c. `INSERT INTO dept VALUES (65, HR, SEATTLE);`
   d. None of these SQL statements will add a new department to the table.

7. Which of the following statements is correct? Choose all that apply.

   a. To ensure that an employee is assigned to a department that already exists in the DEPT table, a FOREIGN KEY constraint must exist on the DEPT table.
   b. To ensure that an employee is assigned to a department that already exists in the DEPT table, a FOREIGN KEY constraint must exist on the EMP table.
   c. To ensure that an employee is assigned to a department that already exists in the DEPT table, a NOT NULL constraint must exist on the DEPT table.
   d. To ensure that an employee is assigned to a department that already exists in the DEPT table, a UNIQUE constraint must exist on the EMP table.

8. Assuming that the PRIMARY KEY constraint for the EMP table is called EMP_EMPNO_PK, which SQL statement will remove the constraint?

   a. `DROP CONSTRAINT emp_empno_pk;`
   b. `ALTER TABLE emp DROP emp_empno_pk;`
   c. `ALTER TABLE emp DROP CONSTRAINT emp_empno_pk;`
   d. `ALTER TABLE emp DROP PRIMARY KEY;`

9. Which of the following SQL statements will add a NOT NULL constraint to the Sal column of the EMP table?

   a. `ALTER TABLE emp ADD NOT NULL(sal);`
   b. `ALTER TABLE emp MODIFY (sal NOT NULL);`
   c. `ALTER TABLE emp MODIFY NOT NULL(sal);`
   d. `ALTER TABLE emp ADD (sal NOT NULL);`

10. Which of the following statements is correct? Choose all that apply.
    a. A NOT NULL constraint can only be created using the column-level approach.
    b. A constraint that is composed of multiple columns must be created using the column-level approach.
    c. If a PRIMARY KEY constraint consists of more than one column, the constraint can be added to each column individually, using the column-level approach.
    d. A PRIMARY KEY constraint that consists of more than one column must be created using the table-level approach.
    e. To change the condition used by a CHECK constraint, the change must be made using the MODIFY clause of the ALTER TABLE command.
11. Which of the following letters is used to denote a NOT NULL constraint type in the USER_CONSTRAINTS view?
    a. FK
    b. NN
    c. R
    d. C
    e. U
12. Which of the following letters is used to denote a FOREIGN KEY constraint type in the USER_CONSTRAINTS view?
    a. FK
    b. NN
    c. R
    d. C
    e. U
13. Which of the following SQL*Plus commands is used to view the structure of a table?
    a. DESCRIBE
    b. LIST
    c. VIEW
    d. DISPLAY
    e. STRUCTURE

14. Which of the following SQL statements will add a new employee, Gary Lito, to the EMP table?
    a. `INSERT INTO emp VALUES (1462, 'GARY LITO');`
    b. `INSERT INTO emp (empno, ename) VALUES (1462, 'GARY LITO', NULL, NULL, NULL, NULL, NULL, NULL);`
    c. `INSERT INTO emp (empno, ename) VALUES (1462, 'GARY LITO');`
    d. `UPDATE emp SET empno = 1462 WHERE ename = 'GARY LITO';`
15. Which of the following symbols is used to indicate a substitution variable?
    a. _
    b. &
    c. %
    d. *
16. Which of the following is not a valid table name?
    a. #DeptEE
    b. EE#
    c. Dept_EE
    d. Dept30
17. Which of the following SQL statements will remove all data within the DEPT table as well as permanently delete the entire structure of the DEPT table?
    a. `DROP TABLE dept;`
    b. `DELETE TABLE dept;`
    c. `TRUNCATE TABLE dept;`
    d. `DELETE *.* FROM dept; [END CODE]`
18. If you do not specify a name for a constraint when it is created, Oracle9*i* will automatically use which naming convention to internally assign a name to the constraint?
    a. *n*_pk
    b. SYSC_*n*
    c. SYS_C*n*
    d. C_SYS*n*

19. Which of the following types of constraints is used to ensure referential integrity?
    a. NOT NULL
    b. PRIMARY KEY
    c. FOREIGN KEY
    d. CHECK
    e. UNIQUE
20. Executing which of the following commands will release any table locks previously held by the user? Choose all that apply.
    a. `COMMIT;`
    b. `EXIT`
    c. `ALTER TABLE emp ADD UNIQUE(ename);`
    d. `UPDATE emp SET sal = 3000 WHERE ename = 'SMITH';`

Practice Exam D

The questions in this practice exam test your knowledge of the concepts presented in Chapters 11-13 of this textbook. Solutions to the exam questions are available from your instructor. When necessary, reference the table structures that follow. Full column names are listed after the tables.

EMP Table	
Column Name	**Definition**
Empno	NUMBER(4), PK
Ename	VARCHAR2(10)
Job	VARCHAR2(9)
Mgr	NUMBER(4)
Hiredate	DATE
Sal	NUMBER(7,2)
Comm	NUMBER(7,2)
Deptno	NUMBER(2), FK

DEPT Table	
Column Name	**Definition**
Deptno	NUMBER(2)
Dname	VARCHAR2(14)
Loc	VARCHAR2(13)

- Empno — Employee
- Ename — Employee Name
- Job — Job Title
- Mgr — Manager of Employee
- Hiredate — Hire Date
- Sal — Salary (monthly)
- Comm — Commission (for sales)
- Deptno — Department Number
- Dname — Department Name
- Loc — Location of Operation

1. Which of the following SQL statements will create a new user with ACCTSUPER as the user name and SUPERPWORD as the password?
   a. `CREATE USER acctsuper PASSWORD superpword;`
   b. `CREATE USER acctsuper PASS superpword;`
   c. `CREATE USER acctsuper IDENTIFIED BY superpword;`
   d. `CREATE acctsuper WITH PASSWORD superpword;`

2. Which of the following SQL statements will generate the next value in the EMP_EMPNO sequence?
   a. `SELECT emp_empno.nextvalue FROM dual;`
   b. `SELECT emp_empno.currentvalue FROM dual;`
   c. `SELECT emp_empno.nextval FROM dual;`
   d. `SELECT emp_empno.currentval FROM dual;`
3. Which of the following SQL statements will modify the existing view EMP_SAL_VU so the data it displays cannot be updated by a user?
   a.
   ```
   CREATE OR REPLACE VIEW emp_sal_vu
   AS SELECT  empno, ename, sal, comm FROM emp
   WITH READ ONLY;
   ```
   b. `REPLACE VIEW emp_sal_vu WITH READ ONLY;`
   c. `ALTER VIEW emp_sal_vu READ ONLY;`
   d.
   ```
   CREATE OR REPLACE emp_sal_vu
   AS SELECT  empno, ename, sal, comm FROM emp
   WITH CHECK OPTION;
   ```
4. Oracle9*i* will automatically create an index when which of the following occurs? Choose all that apply.
   a. A sequence is created.
   b. A PRIMARY KEY constraint is created.
   c. The CREATE INDEX command is successfully executed.
   d. A PUBLIC synonym is created.
5. Which of the following can be included in a simple view?
   a. grouped data
   b. joined tables
   c. SUM function
   d. column alias
6. Which of the following SQL commands grants a system privilege to the user named ACCTSUPER? Choose all that apply.
   a. `GRANT INSERT ON emp TO acctsuper;`
   b. `GRANT CREATE TABLE TO acctsuper;`
   c. `GRANT SELECT ON emp TO acctsuper;`
   d. `GRANT UPDATE ANY TABLE TO acctsuper;`
   e. `GRANT CREATE SESSION TO acctsuper;`

7. Which of the following SQL commands will create a view that will prevent a user from performing any operation that will make a row currently being displayed by the view inaccessible to the view in the future?
   a.
   ```
   CREATE VIEW eejobs30
   AS SELECT empno, ename, job
         FROM emp WHERE deptno = 30
   WITH CHECK OPTION;
   ```
   b.
   ```
   CREATE OR REPLACE VIEW eejobs30
   AS SELECT empno, ename, job
         FROM emp WHERE deptno = 30
   WITH READ ONLY;
   ```
   c.
   ```
   CREATE VIEW eejobs30
   AS SELECT empno, ename, job
         FROM emp WHERE deptno = 30
   WITH READ ONLY;
   ```
   d.
   ```
   CREATE OR REPLACE VIEW eejobs30
   AS SELECT empno, ename, job
         FROM emp WHERE deptno = 30;
   ```
8. Which of the following statements about simple and complex views is not valid?
   a. DML operations are not permitted on non-key-preserved tables.
   b. A row cannot be added to a table through a view if it violates an underlying constraint.
   c. NULL values cannot be added to a table through a view.
   d. DML operations are allowed on simple views that contain the pseudocolumn ROWNUM.
9. Which of the following SQL statements will require a user to create a new password the next time the user accesses his or her account?
   a. `CREATE USER acctsuper IDENTIFIED BY NULL;`
   b. `ALTER USER acctsuper PASSWORD EXPIRE;`
   c. `ALTER USER acctsuper EXPIRE PASSWORD;`
   d. `CREATE USER acctsuper IDENTIFIED BY PASSWORD EXPIRE;`
10. Which of the following SQL statements creates an inline view when executed?
   a.
   ```
   CREATE FORCE VIEW inline_grosspay
   AS SELECT empno, ename, sal + NVL(comm,0)
         FROM emp;
   ```
   b.
   ```
   CREATE VIEW inline_grosspay AS
   inline SELECT empno, ename, sal + NVL(comm,0)
         FROM emp;
   ```
   c.
   ```
   CREATE VIEW inline_grosspay
   AS SELECT empno, ename, sal + NVL(comm,0)
         FROM emp;
   ```
   d.
   ```
   SELECT empno, ename, dname
   FROM (SELECT * FROM emp NATURAL JOIN dept);
   ```

11. Which of the following SQL statements will delete the PUBLIC synonym name EMPLOYEE?
    a. `DELETE SYNONYM employee;`
    b. `DROP SYNONYM employee;`
    c. `DROP PUBLIC SYNONYM employee;`
    d. `DELETE PUBLIC SYNONYM employee;`
    e. `DROP PUBLIC employee;`
12. Which of the following statements about indexes is correct?
    a. Row retrievals are always slower when an index is used.
    b. DML operations are always faster when an index exists on the primary key of a table.
    c. An index always slows down DML operations.
    d. A function-based index will automatically result in slower query executions.
13. Which of the following terms applies to a collection or group of privileges?
    a. schema
    b. role
    c. data dictionary
    d. permission
    e. group account
14. Which of the following commands will grant the CONNECT role to two users, Smith and Blake?
    a. `GRANT CONNECT ON database TO SMITH BLAKE;`
    b. `GRANT CONNECT TO SMITH BLAKE;`
    c. `GRANT CONNECT TO 'SMITH', 'BLAKE';`
    d. `GRANT CONNECT TO SMITH, BLAKE;`
15. Which of the following commands will create a view even if the underlying table(s) does not exist?
    a.
    ```
    CREATE FORCE VIEW inline_grosspay
    AS SELECT empno, ename, sal + NVL(comm,0)
          FROM emp;
    ```
    b.
    ```
    CREATE VIEW inline_grosspay AS
    inline SELECT empno, ename, sal + NVL(comm,0)
          FROM emp;
    ```
    c.
    ```
    CREATE VIEW inline_grosspay
    AS SELECT empno, ename, sal + NVL(comm,0)
          FROM emp;
    ```
    d.
    ```
    SELECT empno, ename, dname
    FROM (SELECT * FROM emp NATURAL JOIN dept);
    ```

16. Which of the following terms is used to describe a collection of objects that belong to a particular user?
    a. user account
    b. role
    c. data dictionary
    d. schema
17. Which of the following commands will allow a user to change his or her password? Choose all that apply.
    a. ALTER USER...IDENTIFIED BY
    b. PASSWORD
    c. CREATE USER...IDENTIFIED BY
    d. MODIFY USER
    e. ALTER USER...PASSWORD
18. Which of the following will prevent the user SMITH from viewing the data stored in the EMP table but will still allow him access to the DEPT table?
    a. `REVOKE select ON emp FROM smith;`
    b. `LOCK TABLE emp FROM smith;`
    c. `REVOKE SELECT ANY TABLE FROM smith;`
    d. `ALTER USER smith RESTRICTED ACCESS on emp;`
19. Which of the following commands can be used to modify a view?
    a. ALTER VIEW
    b. MODIFY VIEW
    c. ALTER VIEW...MODIFY
    d. ALTER TABLE...MODIFY VIEW
    e. Views cannot be modified.
20. Which of the following clauses cannot be used with the ALTER SEQUENCE command?
    a. INCREMENT BY
    b. START WITH
    c. CACHE
    d. NOMAXVALUE

Practice Exam E

The questions in this practice exam test your knowledge of the concepts presented in Chapters 14–16 of this textbook. Solutions to the exam questions are available from your instructor.

1. Which of the following indicates numeric data in a column format model?
   a. #
   b. 9
   c. _
   d. &
   e. %
2. Which of the following sections is required in a PL/SQL block?
   a. declarative
   b. executable
   c. exception handling
   d. ending
3. Which of the following is *not* an accepted parameter for a procedure?
   a. INPUT
   b. OUT
   c. INOUT
   d. IN
4. If a variable is declared with a NUMBER datatype in a PL/SQL block but is not initialized when it is created, what value will be assigned to the variable?
   a. 0
   b. 1
   c. NULL
   d. 999
5. Which of the following types of loops does not require a user to issue the OPEN command to open an explicit cursor?
   a. basic loop
   b. WHILE loop
   c. IF loop
   d. cursor FOR loop

6. Which of the following variables can be used to include the user's account name on a report?
   a. SQL.PNO
   b. SQL.LINENO
   c. SQL.USERID
   d. SQL.USER
7. Which of the following types of loops uses an implicit counter to control the execution of the loop?
   a. basic loop
   b. WHILE loop
   c. FOR loop
   d. cursor FOR loop
8. Which of the following commands will suppress duplicate data within a column of a report?
   a. SUPPRESS
   b. AVOIDDUP
   c. BREAK
   d. GROUP BY
9. Which package can be used to display the contents of a variable in a PL/SQL block?
   a. PUT_LINE
   b. DBMS_OUTPUT
   c. SERVEROUTPUT
   d. DISPLAY_OUT
10. Which of the following attributes can be used to identify the number of rows processed by a cursor?
    a. %ROWCOUNT
    b. %ISOPEN
    c. %FOUND
    d. %NOTFOUND

11. Which of the following attributes can be used to assign the structure of a row to a record?
    a. %FOUND
    b. %TYPE
    c. %STRUCTURE
    d. %ROWTYPE
12. Which of the following commands identifies the title that should appear at the top of a report?
    a. BTITLE
    b. TITLE
    c. TTITLE
    d. HEADING
13. Which of the following COLUMN command options is used to identify the text message in place of a NULL value?
    a. TEXT
    b. NULL
    c. MESSAGE
    d. NLMSG
14. Which of the following attributes can be used to control the execution of a loop? Choose all that apply.
    a. %ROWCOUNT
    b. %ISOPEN
    c. %FOUND
    d. %NOTFOUND
15. Which of the following can be used to initialize a variable in a PL/SQL block? Choose all that apply.
    a. DEFAULT
    b. =
    c. :=
    d. <>

16. Which of the following commands is used to reset any formatting applied by the COLUMN command in a report?
    a. CLS
    b. CLEAR
    c. COLUMN
    d. STOP
17. Assuming the condition provided in an IF clause is FALSE, which of the following would be evaluated next, provided that it exists in the IF statement?
    a. ELSEIF
    b. ENDIF
    c. ELSIF
    d. END
18. Which of the following commands is used to retrieve data from an explicit cursor?
    a. GET
    b. OPEN
    c. FETCH
    d. RETRIEVE
19. Which of the following clauses *must* be used in a SELECT statement that retrieves data from an implicit cursor?
    a. GROUP BY
    b. ORDER BY
    c. HAVING
    d. INTO
20. Which of the following symbols can be used to execute a script file in Oracle9*i*?
    a. %
    b. $
    c. @
    d. !

ORACLE RESOURCES

The following list of resources may assist students receiving Oracle9*i* database training or considering careers that will require interaction with various Oracle products.

Oracle Academic Initiative (OAI)

The Oracle Academic Initiative (OAI) is a result of an effort by Oracle Corporation to provide curricula and other resources to the higher-education community. Individuals enrolled at an institution participating in the OAI receive benefits such as discount vouchers for certification exams and a free subscription to *Oracle Magazine*. Visit the Oracle Web site at *http://oai.oracle.com* for more information.

Oracle Certification Program (OCP)

The Oracle Certification Program (OCP) provides certification for both database administrators and application developers. Certification is based on successful completion of a series of exams. Current information about the OCP can be found at the Web site for this textbook or at *http://www.oracle.com/education/certification*.

Oracle Technology Network (OTN)

The Oracle Technology Network (OTN) provides several services to registered members. For example, members can download trial versions of various Oracle software products and access discussion groups for help with technical issues. In addition, the site also has an extensive documentation area, including reference manuals for SQL, PL/SQL, installation procedures, etc. Some of the documentation is available in PDF format and is available for download. Membership is free. This Web site can be accessed at *http://otn.oracle.com*.

International Oracle Users Group (IOUG)

The International Oracle Users Group (IOUG) is composed of more than 100 local and regional user groups who meet on a regular basis to share information about Oracle products. Members of IOUG can access the repository of knowledge accumulated from individuals who work with various Oracle products on a daily basis. In addition, members can receive publications, discounts, and special offers from various vendors, and they can have access to discussion forums. Contact information for user groups, conference information, etc., is provided on the IOUG Web site at *http://www.ioug.org*.

Glossary

aggregate functions — *See* **group functions**.

American National Standards Institute (ANSI) — One of two industry-accepted committees that sets standards for SQL.

anonymous block — An unnamed PL/SQL block that is embedded in an application program, stored in a script file, or manually entered by the user when it needs to be executed.

application cluster environment — A high-volume work environment in which multiple users simultaneously request data from a database.

argument — Values listed within parentheses in a function.

assignment operator — The symbol (:=) in PL/SQL. Variables can be assigned an initial value, using either the DEFAULT keyword or the PL/SQL assignment operator (:=).

authentication — The process of validating the identity of computer users.

basic loop — Used in PL/SQL to execute statements until the condition(s) stated in the EXIT clause is met.

block — A unit of PL/SQL code that is divided into three sections: declarative, executable, and exception handling.

bridging table — A table created to eliminate a many-to-many relationship between two tables.

Cartesian join — Links table data so each record in the first table is matched with each individual record in the second table. Also called a *Cartesian product* or *cross join*.

Cartesian product — *See* **Cartesian join**.

case conversion functions — Allow a user to temporarily alter the case of data stored in a column or character string.

character — The basic unit of data. It can be a letter, number, or special symbol.

character field — A field composed of non-numeric data. This field will not display a heading longer than the width of the data stored in the field.

character functions — Used to change the case of characters or manipulate characters.

child table — A table having data that reference data within a parent table. Considered the "many" side of a one-to-many relationship.

clause — Each section of a statement that begins with a keyword (SELECT clause, FROM clause, WHERE clause, etc.).

coding scheme — When a database administrator creates a user account, the user's identity is set by using a code; the "scheme" of the code often consists of the user's first initial followed by last name. Used widely in industry for part numbers, customer numbers, etc.

column — In a relational database, fields are commonly represented as columns and may be referred to as "columns."

column alias — A name substituted for a column name. A column alias is created in a query and displayed in the results.

column qualifier — Indicates the table containing a referenced column.

common column — A column that exists in two or more tables and contains equivalent data.

common field — A column that exists in two tables and is used to "join" two tables.

comparison operator — A search condition that indicates how data should relate to a given search value (equal to, greater than, less than, etc.). Common comparison operators include >, <, >=, and <= .

composite — A datatype that represents a collection of data that can be grouped and treated as one unit in PL/SQL.

composite primary key — A combination of columns that uniquely identifies a record in a database table.

concatenation — The combining of the contents of two or more columns or character strings. Two vertical bars, or pipes (| |), instruct Oracle9*i* to concatenate the output of a query.

condition — A portion of a SQL statement that identifies what must exist, or a requirement that must be met. When a query is executed, any record meeting the given condition will be returned in query results.

constant — An assigned data value that does not change during execution of a PL/SQL block.

constraints — Rules that ensure the accuracy and integrity of data. Constraints prevent data that violate these rules from being added to tables. Constraints include PRIMARY KEY, FOREIGN KEY, UNIQUE, CHECK, and NOT NULL.

correlated subquery — A subquery that references a column in the outer query. The outer query executes the subquery once for every row in the outer query.

cross join — *See* **Cartesian join**.

cursor — An area of memory in an Oracle server that holds an executed statement and its results. Cursors can be either *implicit* (created automatically when processing a DML statement or a SELECT statement that retrieves a single row) or *explicit* (created and managed by the user).

data definition language (DDL) — Commands, basically SQL commands, that create or modify database tables or other objects.

data dictionary — Where Oracle9*i* stores all information about database objects. Stored information includes an object's name, type, structure, owner, and the identity of users who have access to the object.

data manipulation language (DML) — Commands used to modify data. Changes to data made by DML commands are not accessible to other users until the data are committed.

data redundancy — Refers to having the same data in different places within a database, which wastes space and complicates updates and changes.

database — A collection of interrelated files.

database management system (DBMS) — A generic term that applies to a software product that allows users to interact with a database to create and maintain the structure of the database, and then to enter, manipulate, and retrieve the data it stores.

database object — A defined, self-contained structure in Oracle9*i*. Database objects include tables, sequences, indexes, and synonyms.

datatype — Identifies the type of data Oracle9*i* will be expected to store in a column.

declarative section — The portion of a PL/SQL block, identified by the DECLARE keyword, that identifies any variables or constants used within the block.

dynamic view — Used to access statistics relating to a database's performance. *See* **view**.

editor — A word-processing program. The default editor for most Windows operating systems is Notepad.

entity — Any person, place, or thing with characteristics or attributes that will be included in a database. In the E-R Model, an entity is usually represented as a square or rectangle.

Entity-Relationship (E-R) Model — A diagram that identifies the entities and data relationships in a database. The model is a logical representation of the physical system to be built.

environment variable — Used to determine how specific elements are displayed in SQL*Plus. Also called a *system variable*.

equality join — Links table data in two (or more) tables having equivalent data stored in a common column. Also called an *equijoin* or *simple join*.

equality operator — A search condition that evaluates data for exact, or equal, values. The equality operator symbol is the equal sign (=).

equijoins — *See* **equality join**.

exception-handling section — A portion of a PL/SQL block, identified by the keyword EXCEPTION, used to display messages or identify other actions that should be performed when an error occurs during execution of the block.

exception propagation — In PL/SQL, a condition that arises when an exception is not trapped and an error is returned to the calling environment, where it can result in further errors.

exclusive lock — When DDL operations are performed, Oracle9*i* places this lock on a table so no other user can alter the table or place a lock on it. *See* **table lock**.

executable section — A portion of a PL/SQL block, identified by the BEGIN keyword, that may consist of a SQL statement(s) and/or PL/SQL statement(s).

explicit cursor — *See* **cursor**.

field — One attribute or characteristic of a database entity.

file — A group of records about the same type of entity.

first-normal form (1NF) — The first step in the normalization process in which repeating groups of data are removed from database records.

foreign key — When a common column exists in two tables, it will usually be a primary key in one table and will be called a foreign key in the second table.

fourth-generation programming language (4GL) — A programming language (such as PL/SQL) that extends the capabilities of SQL.

function — A named PL/SQL block, or predefined block of code, that accepts zero or more input parameters and returns one value.

function-based index — Can be used when a query is based on a calculated value or a function. *See* **index**.

group functions — Process groups of rows, returning only one result per group of rows processed. Also called *multiple-row functions* and *aggregate functions*.

hackers — A slang term generally applied to computer criminals who gain illegal access to information systems.

implicit cursor — *See* **cursor**.

index — A separate database object that stores frequently referenced values so they can be quickly located. An index can either be created implicitly by Oracle9*i* or explicitly by a user.

infinite loop — In PL/SQL, a loop that does not end because the condition to end it is never met.

initialize — When a PL/SQL variable is *initialized*, it is assigned a value.

inline view — A temporary view of underlying database tables that exists only while a command is being executed. It is not a permanent database object and cannot be referenced again by a subsequent query.

inner join — Joins that display data if there was a corresponding record in each table queried. *Equality joins*, *non-equality joins*, and *self-joins* are all classified as inner joins.

International Standards Organization (ISO) — One of two industry-accepted committees that sets standards for SQL.

Julian date — Represents the number of days that have passed between a specified date and January 1, 4712, B.C.

key-preserved table — A table that contains the primary key that a view uses to uniquely identify each record displayed by the view.

keywords — Words used in a SQL query that have a predefined meaning to Oracle9*i*. Common keywords include SELECT, FROM, and WHERE.

large object (LOB) — A datatype that holds locators that identify the location of large objects, such as maps, in PL/SQL.

logical operators — Used to combine two or more search conditions. The logical operators include AND and OR. The NOT operator reverses the meaning of search conditions.

manipulation functions — Allow the user to control data (e.g., determine the length of a string, extract portions of a string) to yield a desired query output.

multiple-column subquery — A nested query that returns more than one column of results to the outer query. It can be listed in the FROM, WHERE, or HAVING clause.

multiple-row functions — *See* **group functions**.

multiple-row subquery — A nested query that returns more than one row of results to the parent query. It is most commonly used in a WHERE and HAVING clause and requires a multiple-row operator.

non-equality join — Links data in two tables that do not have equivalent rows of data.

non key-preserved table — Does not uniquely identify the records in a view.

normal distribution — When a large number of data values are obtained for statistical analysis, they tend to cluster around some "average" value This dispersion of values is called normal distribution.

normalization — A multistep process that allows designers to take the raw data about an entity and evolve the data into a form that will reduce a database's data redundancy.

NULL value — Means no value has been stored in that particular field. Indicates the absence of data, not a blank space.

numeric column — A column composed of only numeric data. In output, the column will display the entire column heading, regardless of the width of the field. (Also known as a *numeric field*.)

numeric field — *See* **numeric column**.

object privileges — Allow users to perform DML or retrieval operations on the data contained within database objects.

object relational database management system (ORDBMS) — Oracle9*i* is an ORDBMS because it can reference individual data elements and objects composed of individual data elements.

optional keyword — In a SQL query, the keyword AS is optional when creating a column alias.

outer join — Links data in tables that do not have equivalent rows. An outer join can be created in either the WHERE clause with an outer join operator (+) or by using the OUTER JOIN keywords.

outer join operator — The plus (+) symbol enclosed in parentheses, used in an outer join operation.

outer query — The main query in a SQL statement; a subquery passes its results back to the parent query, also known as the outer query. An outer query incorporates the value obtained from a subquery into its processing to determine the final output.

parent query — *See* **outer query**.

partial dependency — When the fields contained within a record are dependent on only one portion of the primary key.

post-test — In PL/SQL, refers to checking the terminating condition of a loop after the execution of statements within a loop.

pre-test — In PL/SQL, refers to checking the EXIT condition of a loop before any of the statements are executed.

primary key — A field that serves to uniquely identify a record in a database table.

primary read — A retrieval of data that occurs before a loop is entered.

primary sort — When only one column is specified in the ORDER BY clause, data are ordered, or sorted, based on the data organization within the specified column.

private synonym — An alias used by an individual to reference objects owned by that individual. *See* **synonym**.

privileges — Allow database access to users. Oracle9*i* has *system privileges* and *object privileges*.

procedure — A named PL/SQL block that is used when working with several variables. A procedure can accept the parameters of input (IN), output (OUT), or both (INOUT).

Procedure Language SQL (PL/SQL) — A program using embedding SQL statements as an alternative to issuing multiple SQL statements.

projection — Choosing specific column(s) in a SELECT statement.

pseudo tables — Created to present a particular "view" of a database's contents. Does not actually store data, but is referenced like a table in SQL statements.

public synonym — An alias that can be used by others to access an individual's database objects. *See* **synonym**.

query — A question posed to the database.

record — A collection of fields describing the attributes of one database element. In PL/SQL, a *record* is a composite datatype that can assume the same structure as the row being retrieved.

reference — A datatype that holds pointers to other program items in PL/SQL.

referential integrity — When a user refers to something that exists in another table, the REFERENCES keyword is used to identify the table and column that must already contain the data being entered.

relational database management system (RDBMS) — A software program used to create a relational database. It has functions that allow users to enter, manipulate, and retrieve data.

role — A group, or collection, of privileges. In most organizations, roles correlate to users' job duties.

row — A group of column values for a specific occurrence of an entity. In a database, records are commonly represented as rows.

scalar — A datatype used to hold a single value in PL/SQL.

schema — A collection of database objects owned by one user. By grouping objects according to the owner, multiple objects that have the same object name can exist in the same database.

secondary sort — When two or more columns are specified in the ORDER BY clause, data in the second column (or additional columns) provide an alternative field on which to order data if an exact match occurs between two or more rows in the first, or primary, sort.

second-normal form (2NF) — The second step in the normalization process in which partial dependencies are removed from database records.

self-join — Links data within a table to other data within the same table. A self-join can be created with a WHERE clause or by using the JOIN keyword with the ON clause.

sequence — A database object that generates sequential integers that can be used for an organization's internal controls. A sequence can also serve as a primary key for a table.

set operators — Combine the results of two (or more) SELECT statements. Valid set operators in Oracle9*i* are UNION, UNION ALL, INTERSECT, and MINUS.

shared lock — A table lock that lets other users access portions of a table but not alter the structure of the table. *See* **table lock**.

simple join — *See* **equality join**.

single-row functions — Return one row of results for each record processed.

single-row subquery — A nested subquery that can return to the outer query only one row of results and consist of only one column. The output of a single-row subquery is a single value.

single value — The output of a single-row subquery.

SQL buffer — A portion of the computer's memory that holds the SQL statement being executed. The buffer only holds one SQL statement at a time.

SQL*Plus — A tool enabling users to interact with a database. Through SQL*Plus, users can enter SQL commands, set or alter environmental variables, display the structure of tables, and execute interactive scripts.

standard deviation — A calculation used to determine how closely individual values are to the mean, or average, of a group of numbers.

statistical group functions — Perform basic statistical calculations for data analysis. Oracle9*i*'s functions include standard deviation and variance.

string literal — Alphanumeric data, enclosed within single quotation marks, that instructs the software to interpret "literally" what has been entered and to show it in the resulting display.

structured query language (SQL) — The industry standard for interacting with a relational database. It is a data sublanguage, and unlike a programming language, it processes sets of data as groups and can navigate data stored within various tables.

substitution variable — Instructs Oracle9*i* to use a substituted value in place of a specific variable at the time a command is executed. Used to make SQL statements or PL/SQL blocks interactive.

substring — A portion of a string of data.

synonym — An alternative name given to a database object with a complex name. Synonyms can be either private or public.

syntax — The basic structure, pattern, or rules for a SQL statement. For a SQL statement to execute properly, the correct syntax must be used.

system privileges — Allow access to the Oracle9*i* database and let users perform DDL operations such as CREATE, ALTER, and DROP database objects. An object privilege combined with the keyword ANY is also considered a system privilege.

system variable — *See* **environment variable**.

Systems Development Life Cycle (SDLC) — A series of steps for the design and development of a system.

table — Common term for a database table.

table alias — A temporary name for a table, given in the FROM clause. Table aliases are used to reduce memory requirements or the number of keystrokes needed when specifying a table throughout a SQL statement.

table lock — When DML commands are issued, Oracle9*i* implicitly "locks" the row or rows being affected, so no other user can change the same data.

third-normal form (3NF) — The third step in the normalization process in which transitive dependencies are removed from database records.

"TOP-N" analysis — When an inline view and a pseudocolumn ROWNUM are merged together to create a temporary list of records in a sorted order, then the top "n," or number of records, are retrieved.

transaction — A series of DML statements is considered to be one transaction. In Oracle9*i*, a transaction is simply a series of statements that have been issued and not committed. The duration of a transaction is defined by when a COMMIT implicitly or explicitly occurs.

transaction control — Data control statements that either save modified data or undo uncommitted changes made in error.

uncorrelated subquery — A subquery that follows this method of processing: The subquery is executed, then the results of the subquery are passed to the outer query, and finally the outer query is executed.

unnormalized — Refers to database records that contain repeating groups of data (multiple entries for a single column).

variable — A data value in a PL/SQL block that changes during the execution of a PL/SQL block.

view — Displays data in the underlying base tables. Views are used to provide a shortcut for users not having SQL training or to restrict users' access to sensitive data. Views are database objects, but they do not store data.

wildcard characters — Symbols used to represent one or more alphanumeric characters. The wildcard characters in Oracle9*i* are the percent sign (%) and the underscore symbol (_). The percent sign is used to represent any number of characters; the underscore symbol represents one character.

Index

M

N

P

S